Coulson and Richardson's
Chemical Engineering

Coulson and Richardson's Chemical Engineering

Volume 3A: Chemical and Biochemical Reactors and Reaction Engineering

Fourth Edition

Edited by

R. Ravi
Indian Institute of Technology Madras, Chennai, India

R. Vinu
Indian Institute of Technology Madras, Chennai, India

S.N. Gummadi
Indian Institute of Technology Madras, Chennai, India

Butterworth-Heinemann
An imprint of Elsevier

Butterworth-Heinemann is an imprint of Elsevier
The Boulevard, Langford Lane, Kidlington, Oxford OX5 1GB, United Kingdom
50 Hampshire Street, 5th Floor, Cambridge, MA 02139, United States

Library of Congress Cataloging-in-Publication Data
A catalog record for this book is available from the Library of Congress

British Library Cataloguing-in-Publication Data
A catalogue record for this book is available from the British Library

ISBN: 978-0-08-101096-9

For information on all Butterworth-Heinemann publications visit our website at
https://www.elsevier.com/books-and-journals

Working together
to grow libraries in
developing countries

www.elsevier.com • www.bookaid.org

Publisher: John Fedor
Acquisition Editor: Anita A. Koch
Editorial Project Manager: Ashlie M. Jackman
Production Project Manager: Mohanapriyan Rajendran
Cover Designer: Victoria Pearson

Typeset by TNQ Books and Journals

Contents

Chapter 6: Biochemical Reaction Engineering 451

Sathyanarayana N. Gummadi

List of Contributors

Sathyanarayana N. Gummadi Indian Institute of Technology Madras, Chennai, India
Ramamurthy Ravi Indian Institute of Technology Madras, Chennai, India
Ravikrishnan Vinu Indian Institute of Technology Madras, Chennai, India

Preface

Volume 3 has now been split into two separate volumes 3A and 3B with 3A containing chapters on chemical and biochemical reaction engineering and 3B consisting of process dynamics and control.

In volume 3A, there are now a total of six chapters with two on biochemical reaction engineering (Chapters 5 and 6) compared to a single chapter in the third edition. Overall, the basic structure of the third edition has been retained. At the outset, we express our gratitude to the contributors of the third edition—Professors J. C. Lee, W. J. Thomas, R. Lovitt, and M. Jones—for providing an enduring framework for us to build on. We highlight here the few changes we have made in the presentation of some of the material. These changes have been motivated by our own personal experience in teaching reaction engineering and also by the evolution of the presentation of the subject in various textbooks over the past two decades. Each chapter in the present edition starts with a set of Learning Outcomes, which we hope would enable the reader to assess the benefits that would accrue from reading the chapter.

In Chapter 1, greater emphasis is placed on deriving the material and energy balance equations from fundamental continuum principles. This, we hope, would enable the reader to more clearly see the assumptions inherent in the design equations of the ideal reactors and also be better equipped to handle more complicated reactors, if the need arises. Further, the section on obtaining kinetics from batch reactor data has been expanded.

In Chapter 2, on nonideal reactors, introduction of the two principal distribution functions is decoupled from the experimental methods used to determine them. The convolution theorem forms the centerpiece of the developments leading to the determination of the distribution functions from step and impulse responses. A detailed exposition of the zero parameter models is undertaken to illustrate the interplay of "macromixing," "micromixing," and kinetics in determining conversions in nonideal reactors.

Chapter 3 now includes more information related to the derivation of kinetic rate expressions for heterogeneous catalytic reactions where the importance of elementary steps is highlighted. The mechanism of catalyst poisoning is discussed. The section on gas—solid noncatalytic reactions is expanded to include the effect of different controlling resistances on the overall conversion of the solid. In Chapter 4, a rigorous framework is provided for the derivation of rate equations for gas—liquid reactions so that the reader is able to appreciate the importance of various nondimensional numbers.

Chapters 5 and 6 deal with biochemical engineering. It is important for the reader to first understand biomolecules (carbohydrates, proteins, nucleic acids, fats, and lipids), and hence this is covered before introducing the principles of microbiology. Chapter 5 thus provides the background to the process engineering aspects of biochemical reactors which are covered in Chapter 6. A section on specificity of enzymes, which discusses the various theories such as lock and key mechanism, induced fit model, and transition state stabilization, is included. The revised section on microbial stoichiometry now contains cell composition, yields, growth stoichiometry, elemental balance, degree of reduction, electron balance, and product stoichiometry. The detailed derivations of the equations governing enzyme kinetics are now covered in the main text of Chapter 6.

R. Ravi and R. Vinu
Department of Chemical Engineering

S.N. Gummadi
Department of Biotechnology
Indian Institute of Technology, Madras
India

CHAPTER 1

Reactor Design—General Principles

Ramamurthy Ravi

Indian Institute of Technology Madras, Chennai, India

Learning Outcomes

1. The basic objectives of reactor design, classification of reactors, and guidelines for choice of process conditions (Sections 1.1–1.3).

2. The principles of material and energy balances, steady-state and unsteady-state cases, as well as integral and differential balance equations (Section 1.4).

3. The most common forms of rate equations (kinetics) and how to obtain them from a given mechanism (Section 1.5).

4. Derivation of design equations for a batch reactor and application of them to isothermal and nonisothermal operations; various methods of obtaining kinetics from batch reactor data for liquid-phase and gas-phase reactions (Section 1.6).

5. The assumptions inherent in the plug flow reactor (PFR) and how they lead to the design equations; application to gas-phase and liquid-phase reactions; energy balance considerations for nonisothermal operation (Section 1.7).

6. Setup and application of design equations for single and multiple (in series) continuous stirred tank reactors (CSTRs); interpretation of graphical constructions; nonisothermal operations (Section 1.8).

7. Formulation and application of performance criteria to compare batch/PFR and CSTR for single reactions (Section 1.9).

8. Types of multiple reactions; performance criteria and reactor design for multiple reactions (Section 1.10).

1.1 Basic Objectives in Design of a Reactor

In chemical engineering, physical operations such as fluid flow, heat transfer, mass transfer, and separation processes play a very large part; these have been discussed in Volumes 1 and 2. In any manufacturing process where there is a chemical change taking place, however, the chemical reactor is at the heart of the plant.

Coulson and Richardson's Chemical Engineering. http://dx.doi.org/10.1016/B978-0-08-101096-9.00001-7

In size and appearance it may often seem to be one of the least impressive items of equipment, but its demands and performance are usually the most important factors in the design of the whole plant.

When a new chemical process is being developed, at least some indication of the performance of the reactor is needed before any economic assessment of the project as a whole can be made. As the project develops and its economic viability becomes established, further work is carried out on the various chemical engineering operations involved. Thus, when the stage of actually designing the reactor in detail has been reached, the project as a whole will already have acquired a fairly definite form. Among the major decisions, which will have been taken, is the rate of production of the desired product. This will have been determined from a market forecast of the demand for the product in relation to its estimated selling price. The reactants to be used to make the product and their chemical purity will have been established. The basic chemistry of the process will almost certainly have been investigated, and information about the composition of the products from the reaction, including any by-products, should be available.

On the other hand, a reactor may have to be designed as part of a modification to an existing process. Because the new reactor has then to tie in with existing units, its duties can be even more clearly specified than when the whole process is new. Naturally, in practice, detailed knowledge about the performance of the existing reactor would be incorporated in the design of the new one.

As a general statement of the basic objectives in designing a reactor, we can say therefore that the aim is to produce a *specified product* at a *given rate* from *known reactants*. In proceeding further, however, a number of important decisions must be made and there may be scope for considerable ingenuity to achieve the best result. At the outset the two most important questions to be settled include the following:

1. The type of reactor to be used and its method of operation. Will the reaction be carried out as a batch process, a continuous flow process, or possibly a hybrid of the two? Will the reactor operate isothermally, adiabatically, or in some intermediate manner?
2. The physical condition of the reactants at the inlet to the reactor. Thus, the basic processing conditions in terms of pressure, temperature, and compositions of the reactants on entry to the reactor have to be decided, if not already specified as part of the original process design.

Subsequently, the aim is to reach logical conclusions concerning the following principal features of the reactor:

1. The overall size of the reactor, its general configuration, and, more important, dimensions of any internal structures.

2. The exact composition and physical condition of the products emerging from the reactor. The composition of the products must of course lie within any limits set in the original specification of the process.
3. The temperatures prevailing within the reactor and any provisions, which must be made for heat transfer.
4. The operating pressure within the reactor and any pressure drop associated with the flow of the reaction mixture.

1.1.1 By-products and Their Economic Importance

Before taking up the design of reactors in detail, let us first consider the very important question of whether any by-products are formed in the reaction. Obviously, consumption of reactants to give unwanted, and perhaps unsaleable, by-products is wasteful and will directly affect the operating costs of the process. Apart from this, however, the nature of any by-products formed and their amounts must be known so that plant for separating and purifying the products from the reaction may be correctly designed. The appearance of unforeseen by-products on start-up of a full-scale plant can be utterly disastrous. Economically, although the cost of the reactor may sometimes not appear to be great compared with that of the associated separation equipment such as distillation columns, etc., it is the composition of the mixture of products issuing from the reactor that determines the capital and operating costs of the separation processes.

For example, in producing ethylene[1] together with several other valuable hydrocarbons such as butadiene from the thermal cracking of naphtha, the design of the whole complex plant is determined by the composition of the mixture formed in a tubular reactor in which the conditions are very carefully controlled. As we shall see later, the design of a reactor itself can affect the amount of by-products formed and therefore the size of the separation equipment required. The design of a reactor and its mode of operation can thus have profound repercussions on the remainder of the plant.

1.1.2 Preliminary Appraisal of a Reactor Project

In the following pages we shall see that reactor design involves all the basic principles of chemical engineering with the addition of chemical kinetics. Mass transfer, heat transfer, and fluid flow are all concerned and complications arise when, as so often is the case, interaction occurs between these transfer processes and the reaction itself. In designing a reactor it is essential to weigh up all the various factors involved and, by an exercise of judgment, to place them in their proper order of importance. Often the basic design of the reactor is determined by what is seen to be the most troublesome step. It may be the chemical kinetics; it may be mass transfer between phases; it may be heat transfer; or it may even be the need to ensure safe operation. For example, in oxidizing naphthalene or o-xylene to phthalic anhydride with air, the reactor must be designed so that ignitions, which are not infrequent, may be rendered harmless. The theory of reactor design is being extended rapidly and more precise methods for detailed design and optimization are being evolved. However, if the final design is to be successful, *the major decisions taken at the outset must be correct.* Initially, a careful appraisal of the basic role and functioning of the reactor is required and at this stage the application of a little chemical engineering common sense may be invaluable.

1.2 Classification of Reactors and Choice of Reactor Type
1.2.1 Homogeneous and Heterogeneous Reactors

Chemical reactors may be divided into two main categories: homogeneous and heterogeneous. In homogeneous reactors, only one phase, usually a gas or liquid, is present. If more than one reactant are involved, provision must of course be made for mixing them together to form a homogenous whole. Often, mixing the reactants is the way of starting off the reaction, although sometimes the reactants are mixed and then brought to the required temperature.

In heterogeneous reactors, two, or possibly three, phases are present; common examples being gas—liquid, gas—solid, liquid—solid, and liquid—liquid systems. In cases where one of the phases is a solid, it is quite often present as a catalyst; gas—solid catalytic reactors particularly form an important class of heterogeneous chemical reaction systems. It is worth noting that, in a heterogeneous reactor, the chemical reaction itself may be truly heterogeneous, but this is not necessarily so. In a gas—solid catalytic reactor, the reaction takes place on the surface of the solid and is thus heterogeneous. However, bubbling a gas through a liquid may serve just to dissolve the gas in the liquid where it then reacts homogeneously; the reaction is thus homogeneous but the reactor is heterogeneous in that it is required to effect contact between two phases—gas and liquid. Generally, heterogeneous reactors exhibit a greater variety of configuration and contacting pattern than homogeneous reactors. Initially, therefore, we shall be concerned mainly with the simpler homogeneous reactors, although parts of the treatment that follows can be extended to heterogeneous reactors with little modification.

1.2.2 Batch Reactors and Continuous Reactors

Another kind of classification which cuts across the homogeneous—heterogeneous division is the mode of operation—batchwise or continuous. Batchwise operation, shown in Fig. 1.1A, is familiar to anybody who has carried out small-scale preparative reactions in the laboratory. There are many situations, however, especially in large-scale operation, where considerable advantages accrue by carrying out a chemical reaction continuously in a flow reactor.

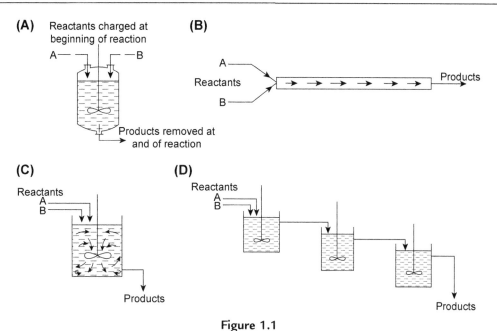

Figure 1.1

Basic types of chemical reactors. (A) Batch reactor. (B) Tubular flow reactor. (C) Continuous stirred tank reactor (CSTR) or "backmix reactor." (D) CSTRs in series as frequently used.

Fig. 1.1 illustrates the two basic types of flow reactor which may be employed. In the *tubular flow reactor* (B) the aim is to pass the reactants along a tube so that there is as little intermingling as possible between the reactants entering the tube and the products leaving at the far end. In the *continuous stirred tank reactor* (CSTR) (C) an agitator is deliberately introduced to disperse the reactants thoroughly into the reaction mixture immediately after they enter the tank. The product stream is drawn off continuously and, in the ideal state of perfect mixing, will have the same composition as the contents of the tank. In some ways, using a CSTR, or *backmix reactor* as it is sometimes called, seems a curious method of conducting a reaction because as soon as the reactants enter the tank they are mixed and a portion leaves in the product stream flowing out. To reduce this effect, it is often advantageous to employ a number of stirred tanks connected in series as shown in Fig. 1.1D.

The stirred tank reactor is by its nature well suited to liquid-phase reactions. The tubular reactor, although sometimes used for liquid-phase reactions, is the natural choice for gas-phase reactions, even on a small scale. Usually the temperature or catalyst is chosen so that the rate of reaction is high, in which case a comparatively small tubular reactor is sufficient to handle a high volumetric flow rate of gas. A few gas-phase reactions, examples being partial combustion and certain chlorinations, are carried out in reactors, which resemble the stirred tank reactor; rapid mixing is usually brought about by arranging for the gases to enter with a vigorous swirling motion instead of by mechanical means.

1.2.3 Variations in Contacting Pattern—Semibatch Operation

Another question which should be asked in assessing the most suitable type of reactor is whether there is any advantage to be gained by varying the contacting pattern. Fig. 1.2A illustrates the *semibatch* mode of operation. The reaction vessel here is essentially a batch reactor, and at the start of a batch it is charged with one of the reactants **A**. However, the second reactant **B** is not all added at once, but continuously over the period of the reaction. This is the natural and obvious way to carry out many reactions. For example, if a liquid has to be treated with a gas, perhaps in a chlorination or hydrogenation reaction, the gas is normally far too voluminous to be charged all at once to the reactor; instead it is fed continuously at the rate at which it is used up in the reaction. Another case is where the reaction is too violent if both reactants are mixed suddenly together. Organic nitration, for example, can be conveniently controlled by regulating the rate of addition of the nitrating acid. The maximum rate of addition of the second reactant in such a case will be determined by the rate of heat transfer.

A characteristic of semibatch operation is that the concentration C_B of the reactant added slowly, **B** in Fig. 1.2, is low throughout the course of the reaction. This may be an advantage if more than one reaction is possible, and if the desired reaction is favored by a low value of C_B. Thus, the semibatch method may be chosen for a further reason, that of improving the yield of the desired product, as shown in Section 1.10.4.

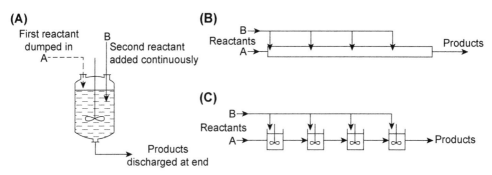

Figure 1.2

Examples of possible variations in reactant contacting pattern. (A) Semibatch operation. (B) Tubular reactor with divided feed. (C) Stirred tank reactors with divided feed (in each case the concentration of **B**, C_B, is low throughout).

Summarizing, a semibatch reactor may be chosen:

1. to react a gas with a liquid,
2. to control a highly exothermic reaction, and
3. to improve product yield in suitable circumstances.

In semibatch operation, when the initial charge of **A** has been consumed, the flow of **B** is interrupted, the products are discharged, and the cycle will begin again with a fresh charge of **A**. If required, however, the advantages of semibatch operation may be retained, but the reactor system is designed for continuous flow of both reactants. In the tubular flow version (Fig. 1.2B) and the stirred tank version (Fig. 1.2C), the feed of **B** is divided between several points. These are known as *cross-flow* reactors. In both cases C_B is low throughout.

1.2.4 Influence of Heat of Reaction on Reactor Type

Associated with every chemical change there is a heat of reaction (or more precisely, an enthalpy change due to reaction), and only in a few cases it is so small that it can be neglected. The magnitude of the heat of reaction often has a major influence on the design of a reactor. With a strongly exothermic reaction, for example, a substantial rise in temperature of the reaction mixture will take place unless provision is made for heat to be transferred as the reaction proceeds. It is important to try to appreciate clearly the relation between the enthalpy of reaction, the heat transferred, and the temperature change of the reaction mixture; quantitatively this is expressed by an energy balance (Section 1.4). If the temperature of the reaction mixture is to remain constant (isothermal operation), the heat equivalent to the heat of reaction at the operating temperature must be transferred to or from the reactor. If no heat is transferred (adiabatic operation), the temperature of the reaction mixture will rise or fall as the reaction proceeds. In practice, it may be most convenient to adopt a policy intermediate between these two extremes; in the case of a strongly exothermic reaction, some heat transfer from the reactor may be necessary to keep the reaction under control, but a moderate temperature rise may be quite acceptable, especially if strictly isothermal operation would involve an elaborate and costly control scheme.

In setting out to design a reactor, therefore, two very important questions to ask include the following:

1. What is the heat of reaction?
2. What is the acceptable range over which the temperature of the reaction mixture may be permitted to vary?

The answers to these questions may well dominate the whole design. Usually, the temperature range can only be roughly specified; often the lower temperature limit is determined by the slowing down of the reaction and the upper temperature limit by the onset of undesirable side reactions.

1.2.4.1 Adiabatic Reactors

If it is feasible, adiabatic operation is to be preferred for simplicity of design. Fig. 1.3 shows the reactor section of a plant for the catalytic reforming of petroleum naphtha; this is an important process for improving the octane number of gasoline. The reforming reactions are mostly endothermic so that in adiabatic operation the temperature would fall during the course of the reaction. If the reactor were made as one single unit, this temperature fall would be too large, i.e., either the temperature at the inlet would be too high and undesired reactions would occur, or the reaction would be incomplete because the temperature near the outlet would be too low. The problem is conveniently solved by dividing the reactor into three sections. Heat is supplied externally between the sections, and the intermediate temperatures are raised so that each section of the reactor will operate adiabatically. Dividing the reactor into sections also has the advantage that the intermediate temperature can be adjusted independently of the inlet temperature; thus an optimum temperature distribution can be achieved. In this example, we can see that the furnaces where heat is transferred and the catalytic reactors are quite separate units, each designed specifically for one function. This separation of function generally provides ease of control, flexibility of operation, and often leads to a good overall engineering design.

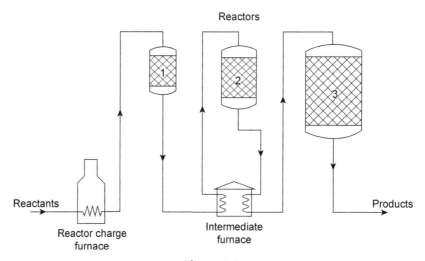

Figure 1.3

Reactor system of a petroleum naphtha catalytic reforming plant. (The reactor is divided into three units each of which operates *adiabatically*, the heat required being supplied at intermediate stages via an external furnace).

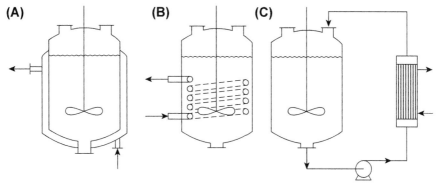

Figure 1.4
Batch reactors showing different methods of heating or cooling. (A) Jacket. (B) Internal coils. (C) External heat exchangers.

1.2.4.2 Reactors With Heat Transfer

If the reactor does not operate adiabatically, then its design must include provision for heat transfer. Fig. 1.4 shows some of the ways in which the contents of a batch reactor may be heated or cooled. In *A* and *B*, the jacket and the coils form part of the reactor itself, whereas in *C*, an external heat exchanger is used with a recirculating pump. If one of the constituents of the reaction mixture, possibly a solvent, is volatile at the operating temperature, the external heat exchanger may be a reflux condenser, just as in the laboratory.

Fig. 1.5 shows ways of designing tubular reactors to include heat transfer. If the amount of heat to be transferred is large, then the ratio of heat transfer surface to reactor volume will be large, and the reactor will look very much like a heat exchanger as in Fig. 1.5B. If the reaction has to be carried out at a high temperature and is strongly endothermic (for example, the production of ethylene by the thermal cracking of naphtha or

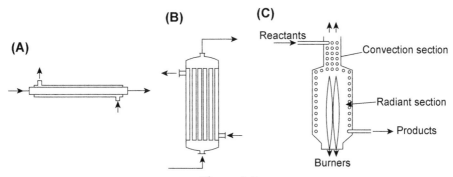

Figure 1.5
Methods of heat transfer to tubular reactors. (A) Jacketed pipe. (B) Multitube reactor (tubes in parallel). (C) Pipe furnace (pipes mainly in series although some pipe runs may be in parallel).

ethane—see also Section 1.7.1, Example 1.5), the reactor will be directly fired by the combustion of oil or gas and will look like a pipe furnace (Fig. 1.5C).

1.2.4.3 Autothermal Reactor Operation

If a reaction requires a relatively high temperature before it will proceed at a reasonable rate, the products of the reaction will leave the reactor at a high temperature and, in the interests of economy, heat will normally be recovered from them. Since heat must be supplied to the reactants to raise them to the reaction temperature, a common arrangement is to use the hot products to heat the incoming feed as shown in Fig. 1.6A. If the reaction is sufficiently exothermic, enough heat will be produced in the reaction to overcome any losses in the system and to provide the necessary temperature difference in the heat exchanger. The term *autothermal* is used to describe such a system which is completely self-supporting in its thermal energy requirements.

The essential feature of an autothermal reactor system is the feedback of reaction heat to raise the temperature and hence the reaction rate of the incoming reactant stream. Fig. 1.6 shows a number of ways in which this can occur. With a tubular reactor the feedback may be achieved by external heat exchange, as in the reactor shown in Fig. 1.6A, or by internal heat exchange as in Fig. 1.6B. Both of these are catalytic reactors; their thermal characteristics are discussed in more detail in Chapter 3, Section 3.6.2. Being catalytic the reaction can only take place in that part of the reactor which holds the catalyst, so the temperature profile has the form indicated alongside the reactor. Fig. 1.6C shows a CSTR in which the entering cold feed immediately mixes with a large volume of hot products and rapid reaction occurs. The combustion chamber of a liquid-fueled rocket motor is a reactor of this type, the products being hot gases which are ejected at high speed. Fig. 1.6D shows another type of combustion process in which a laminar flame of conical shape is stabilized at the orifice of a simple gas burner. In this case the feedback of combustion heat occurs by transfer upstream in a direction opposite to the flow of the cold reaction mixture.

Another feature of the autothermal system is that, although ultimately it is self-supporting, an external source of heat is required to start it up. The reaction has to be ignited by raising some of the reactants to a temperature sufficiently high for the reaction to commence. Moreover, a stable operating state may be obtainable only over a limited range of operating conditions. This question of stability is discussed further in connection with autothermal operation of a CSTR (Section 1.8.4).

1.3 Choice of Process Conditions

The choice of temperature, pressure, reactant feed rates, and compositions at the inlet to the reactor is closely bound up with the basic design of the process as a whole. In arriving

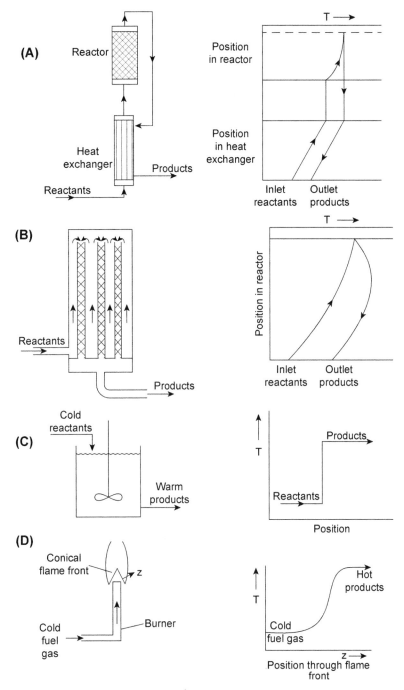

Figure 1.6

Autothermal reactor operation. (A) Tubular reactor with external heat exchange. (B) Tubular reactor with internal heat exchange. (C) CSTR. (D) Gas burner in a combustion process.

at specifications for these quantities, the engineer is guided by knowledge available on the fundamental physical chemistry of the reaction. Usually he or she will also have results of laboratory experiments giving the fraction of the reactants converted and the products formed under various conditions. Sometimes he or she may have the benefit of highly detailed information on the performance of the process from a pilot plant, or even a large-scale plant. Although such direct experience of reactor conditions may be invaluable in particular cases, we shall here be concerned primarily with design methods based upon fundamental physicochemical principles.

1.3.1 Chemical Equilibria and Chemical Kinetics

The two basic principles involved in choosing conditions for carrying out a reaction are thermodynamics, under the heading of chemical equilibrium, and chemical kinetics. Strictly speaking, every chemical reaction is reversible and, no matter how fast a reaction takes place, it cannot proceed beyond the point of chemical equilibrium in the reaction mixture at the particular temperature and pressure concerned. Thus, under any prescribed conditions, the principle of chemical equilibrium, through the equilibrium constant, determines *how far* the reaction can possibly proceed given sufficient time for equilibrium to be reached. On the other hand, the principle of chemical kinetics determines at what *rate* the reaction will proceed towards this maximum extent. If the equilibrium constant is very large, then for all practical purposes the reaction may be said to be *irreversible*. However, even when a reaction is claimed to be *irreversible* an engineer would be very unwise not to calculate the equilibrium constant and check the position of equilibrium, especially if high conversions are required.

In deciding process conditions, the two principles of thermodynamic equilibrium and kinetics need to be considered together; indeed, any complete rate equation for a reversible reaction will include the equilibrium constant or its equivalent (see Section 1.5.4), but complete rate equations are not always available to the engineer. The first question to ask is in what temperature range will the chemical reaction take place at a reasonable rate (in the presence, of course, of any catalyst, which may have been developed for the reaction)? The next step is to calculate values of the equilibrium constant in this temperature range using the principles of chemical thermodynamics. (Such methods are beyond the scope of this chapter and any reader unfamiliar with this subject should consult a standard textbook.[2]) The activity-based equilibrium constant K_a of a reaction depends only on the temperature as indicated by the relation:

$$\frac{d \ln K_a}{dT} = \frac{\Delta h_s^{rxn}}{\mathbf{R}T^2} \tag{1.1}$$

where Δh_s^{rxn} is the standard enthalpy change of reaction or more commonly known as the *standard heat of reaction*. If the reaction mixture is an ideal gas mixture, then K_a may be replaced by K_p, the partial pressure–based equilibrium constant, in Eq. (1.1). The equilibrium constant is then used to determine the limit to which the reaction can proceed under the conditions of temperature, pressure, and reactant compositions which appear to be most suitable.

1.3.2 Calculation of Equilibrium Conversion

Whereas the equilibrium constant itself depends on the temperature only, the conversion at equilibrium depends on the composition of the original reaction mixture and, in general, on the pressure. If the equilibrium constant is very high, the reaction may be treated as being irreversible. If the equilibrium constant is low, however, it may be possible to obtain acceptable conversions only by using high or low pressures. Two important examples are the following reactions:

$$C_2H_4 + H_2O \rightleftharpoons C_2H_5OH$$
$$N_2 + 3H_2 \rightleftharpoons 2NH_3$$

both of which involve a decrease in the number of moles as the reaction proceeds, and therefore high pressures are used to obtain satisfactory equilibrium conversions.

Thus, in those cases in which reversibility of the reaction imposes a serious limitation, the equilibrium conversion must be calculated in order that the most advantageous conditions to be employed in the reactors may be chosen; this may be seen in detail in the following example of the styrene process. A study of the design of this process is also very instructive in showing how the basic features of the reaction, namely equilibrium, kinetics, and suppression of by-products, have all been satisfied in quite a clever way by using steam as a diluent.

Example 1.1

A Process for the Manufacture of Styrene by the Dehydrogenation of Ethylbenzene

Let us suppose that we are setting out from first principles to investigate the dehydrogenation of ethylbenzene, which is a well-established process for manufacturing styrene:

$$C_6H_5 \cdot CH_2 \cdot CH_3 = C_6H_5 \cdot CH{:}CH_2 + H_2$$

There is a catalyst available, which will give a suitable rate of reaction at 560°C. At this temperature the pressure-based equilibrium constant for the reaction above is

$$\frac{P_{St} \times P_H}{P_{Et}} = K_p = 100 \text{ mbar} = 10^4 \text{ N/m}^2 \tag{A}$$

where P_{Et}, P_{St}, and P_H are the partial pressures of ethylbenzene, styrene, and hydrogen, respectively.

Part (i)

Feed pure ethylbenzene: If a feed of pure ethylbenzene is used at 1 bar pressure, determine the fractional conversion at equilibrium.

Solution

This calculation requires not only the use of the equilibrium constant, but also a material balance over the reactor. To avoid confusion, it is as well to set out this material balance quite clearly even in this comparatively simple case.

First it is necessary to choose a basis; let this be 1 mol of ethylbenzene fed into the reactor: a fraction α_e of this will be converted at equilibrium. Then, from the above stoichiometric equation, α_e mole styrene and α_e mole hydrogen are formed, and $(1 - \alpha_e)$ mole ethylbenzene remains unconverted. Let the total pressure at the outlet of the reactor be P which we shall later set equal to 1 bar.

$$C_6H_5 \cdot C_2H_5 \rightarrow \boxed{\text{REACTOR}} \rightarrow \begin{matrix} C_6H_5 \cdot C_2H_5 \\ C_6H_5 \cdot C_2H_3 \\ H_2 \end{matrix}$$

Temperature 560°C = 833 K

Pressure P (1 bar = 1.0×10^5 N/m²)

	In		Out		
			a	*b*	*c*
	Mole		Mole	Mole fraction	Partial pressure
$C_6H_5 \cdot C_2H_5$	1		$1 - \alpha_e$	$\frac{1-\alpha_e}{1+\alpha_e}$	$\frac{1-\alpha_e}{1+\alpha_e}P$
$C_6H_5 \cdot C_2H_3$	—		α_e	$\frac{\alpha_e}{1+\alpha_e}$	$\frac{\alpha_e}{1+\alpha_e}P$
H_2	—		α_e	$\frac{\alpha_e}{1+\alpha_e}$	$\frac{\alpha_e}{1+\alpha_e}P$
		Total	$1 + \alpha_e$		

Since for 1 mol of ethylbenzene entering, the total number of moles increases to $1 + \alpha_e$, the mole fractions of the various species in the reaction mixture at the reactor outlet are shown in column *b* above. At a total pressure P, the partial pressures are given in column *c*. If the re-action mixture is at chemical equilibrium, these partial pressures must satisfy Eq. (A) above:

$$K_p = \frac{P_{St} \times P_H}{P_{Et}} = \frac{\dfrac{\alpha_e}{(1+\alpha_e)}P \dfrac{\alpha_e}{(1+\alpha_e)}P}{\dfrac{(1-\alpha_e)}{(1+\alpha_e)}P} = \frac{\alpha_e^2}{1-\alpha_e^2}P$$

i.e.,

$$\frac{\alpha_e^2}{1-\alpha_e^2}P = 1.0 \times 10^4 \text{ N/m}^2 \tag{B}$$

Thus, when $P = 1$ bar, $\alpha_e = 0.30$; i.e., the maximum possible conversion using pure ethylbenzene at 1 bar is only 30%; this is not very satisfactory (although it is possible in some processes to operate at low conversions by separating and recycling reactants). Ways of improving this figure are now sought.

Note that Eq. (B) shows that as P decreases, α_e increases; this is the quantitative expression of Le Chatelier's principle that, because the total number of moles increases in the reaction, the decomposition of ethylbenzene is favored by a reduction in pressure. There are, however, disadvantages in operating such a process at subatmospheric pressures. One disadvantage is that any ingress of air through leaks might result in ignition. A better solution in this instance is to reduce the partial pressure by diluting the ethylbenzene with an inert gas, while maintaining the total pressure slightly in excess of atmospheric. The inert gas most suitable for this process is steam; one reason for this is that it can be condensed easily in contrast to a gas such as nitrogen, which would introduce greater problems in separation.

Part (ii)
Feed ethylbenzene with steam: If the feed to the process consists of ethylbenzene diluted with steam in the ratio 15 mol steam:1 mol ethylbenzene, determine the new fractional conversion at equilibrium α'_e.

Solution
Again we set out the material balance in full, the basis being 1 mole ethylbenzene into the reactor.

$C_6H_5 \cdot C_2H_5$ $C_6H_5 \cdot C_2H_5$

H_2O $\rightarrow \boxed{\text{REACTOR}} \rightarrow$ $C_6H_5 \cdot C_2H_3$ Temperature 560°C = 833 K

                   H_2 Pressure P (1 bar = 1.0×10^5 N/m²)

                   H_2O

	In		Out		
			a	*b*	*c*
	Mole		Mole	Mole fraction	Partial pressure
$C_6H_5 \cdot C_2H_5$	1		$1 - \alpha'_e$	$\frac{1-\alpha'_e}{16+\alpha'_e}$	$\frac{1-\alpha'_e}{16+\alpha'_e}P$
$C_6H_5 \cdot C_2H_3$	—		α'_e	$\frac{\alpha'_e}{16+\alpha'_e}$	$\frac{\alpha'_e}{16+\alpha'_e}P$
H_2	—		α'_e	$\frac{\alpha'_e}{16+\alpha'_e}$	$\frac{\alpha'_e}{16+\alpha'_e}P$
H_2O	15		15	$\frac{15}{16+\alpha'_e}$	
		Total	$16 + \alpha'_e$		

i.e.,

$$K_p = \frac{P_{St} \times P_H}{P_{Et}} = \frac{\dfrac{\alpha'_e}{(16 + \alpha'_e)}P \dfrac{\alpha'_e}{(16 + \alpha'_e)}P}{\dfrac{(1 - \alpha'_e)}{(16 + \alpha'_e)}P}$$

$$= \frac{\alpha'^2_e}{(16 + \alpha'_e)(1 - \alpha'_e)}P \qquad\qquad (C)$$

$$\frac{\alpha'^2_e}{(16 + \alpha'_e)(1 - \alpha'_e)}P = 1.0 \times 10^4$$

Thus when $P = 1$ bar, $\alpha'_e = 0.70$, i.e., the maximum possible conversion has now been raised to 70%. Inspection of Eq. (C) shows that the equilibrium conversion increases as the ratio of steam to ethylbenzene increases. However, as more steam is used, its cost increases and offsets the value of the increase in ethylbenzene conversion. The optimum steam:ethylbenzene ratio is thus determined by an economic balance.

Part (iii)
Final choice of reaction conditions in the styrene process.

Solution
The use of steam has a number of other advantages in the styrene process. The most important of these is that it acts as a source of internal heat supply so that the reactor can be operated adiabatically. The dehydrogenation reaction is strongly endothermic, the standard heat of reaction at 560°C being $\Delta h_s^{rxn} = 125{,}000$ kJ/kmol. It is instructive to look closely at the conditions that were originally worked out for this process (Fig. 1.7). Most of the steam, 90% of the total used, is heated separately from the ethylbenzene stream, and to a higher temperature (710°C) than is required at the inlet to the reactor. The ethylbenzene is heated in the heat exchangers to only 520°C and is then rapidly mixed with the hotter steam to give a temperature of 630°C at the inlet to the catalyst bed. If the ethylbenzene were heated to 630°C more slowly by normal heat exchange decomposition and coking of the heat transfer surfaces would tend to occur. Moreover, the tubes of this heat exchanger would have to be made of a more expensive alloy to resist the more severe working conditions. To help avoid coking, 10% of the steam used is passed through the heat exchanger with the ethylbenzene. The presence of a large proportion of steam in the reactor also prevents coke deposition on the catalyst. By examining the equilibrium constant of reactions involving carbon such as

$$C_6H_5 \cdot CH_2 \cdot CH_3 \rightleftharpoons 8C + 5H_2$$
$$C + H_2O \rightleftharpoons CO + H_2$$

it may be shown that coke formation is not possible at high steam:ethylbenzene ratios. The styrene process operates with a fractional conversion of ethylbenzene per pass of 0.40 compared with the equilibrium conversion of 0.70. This actual conversion of 0.40 is determined by the *rate* of the reaction over the catalyst at the temperature prevailing in the reactor. (Adiabatic operation means that the temperature falls with increasing conversion and the reaction tends to be quenched at the outlet.) The unreacted ethylbenzene is separated and recycled to the reactor. The overall yield in the process, i.e., moles of ethylbenzene transformed into styrene per mole of ethylbenzene supplied, is 0.90, the remaining 0.10 being consumed in unwanted side reactions. Notice that the conversion per pass could be increased by increasing the temperature at the inlet to the catalyst bed beyond 630°C, but the undesirable side reactions would increase and the overall yield of the process would fall. The figure of 630°C for the inlet temperature is thus determined by an economic balance between the cost of separating unreacted ethylbenzene (which is high if the inlet temperature and conversion per pass are low) and the cost of ethylbenzene consumed in wasteful side reactions (which is high if the inlet temperature is high).

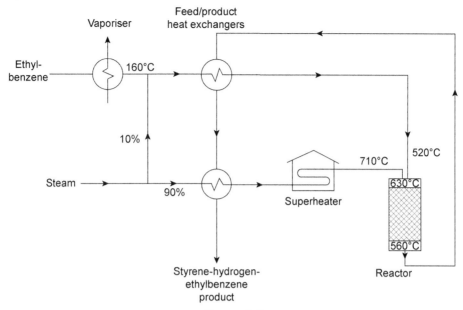

Figure 1.7

A process for styrene from ethylbenzene using 15 mol steam: 1 mol ethylbenzene, operating
pressure 1 bar, conversion per pass 0.40, overall relative yield 0.90.

1.3.3 Ultimate Choice of Reactor Conditions

The use of steam in the styrene process mentioned above is an example of how an
engineer can exercise a degree of ingenuity in reactor design. The advantages conferred by
the steam may be summarized as follows:

1. it lowers the partial pressure of the ethylbenzene without the need to operate at subatmospheric pressures;
2. it provides an internal heat source for the endothermic heat of reaction, making adiabatic operation possible; and
3. it prevents coke formation on the catalyst and coking problems in the ethylbenzene
   heaters.

As the styrene process shows, it is not generally feasible to operate a reactor with a
conversion per pass equal to the equilibrium conversion. The rate of a chemical reaction
decreases as equilibrium is approached, so that the equilibrium conversion can only be
attained if either the reactor is very large or the reaction is unusually fast. The size of
reactor required to give any particular conversion, which of course cannot exceed the
maximum conversion predicted from the equilibrium constant, is calculated from the

kinetics of the reaction. For this purpose we need quantitative data on the rate of reaction, and the rate equations which describe the kinetics are considered in Section 1.5.

If there are two or more reactants involved in the reaction, both can be converted completely in a single pass only if they are fed to the reactor in the stoichiometric proportion. In many cases, the stoichiometric ratio of reactants may be the best, but in some instances, where one reactant (especially water or air) is very much cheaper than the other, it may be economically advantageous to use it in excess. For a given size of reactor, the object is to increase the conversion of the more costly reactant, possibly at the expense of a substantial decrease in the fraction of the cheaper reactant converted. Examination of the kinetics of the reaction is required to determine whether this can be achieved and calculate quantitatively the effects of varying the reactant ratio. Another and perhaps more common reason for departing from the stoichiometric proportions of reactants is to minimize the amount of by-products formed. This question is discussed further in Section 1.10.4.

Ultimately, the final choice of the temperature, pressure, reactant ratio, and conversion at which the reactor will operate depends on an assessment of the overall economics of the process. This will take into account the cost of the reactants, the cost of separating the products, and the cost associated with any recycle streams. It should include all the various operating costs and capital costs of reactor and plant. In the course of making this economic assessment, a whole series of calculations of operating conditions, final conversion, and reactor size may be performed with the aid of a computer, provided that the data are available. Each of these sets of conditions may be technically feasible, but the one chosen will be that which gives the maximum profitability for the project as a whole.

1.4 Material and Energy Balances

The ability to write down material and energy balance equations for a particular reactor configuration is the first step toward reactor analysis and design. It is important to be familiar with both an overall balance for the reactor as a whole and also a local balance valid at each point in the reactor. The latter is required in cases such as the plug flow reactor (PFR) where concentrations and/or temperature vary with position within the reactor.

The treatment of material and energy balances adopted here is different from those in the standard reaction engineering textbooks and is more akin to what is found in books on transport phenomena.[3] Reaction engineering textbooks can afford a simplified treatment because the systems they primarily deal with are ideal reactors. There is an attempt here to show that the fundamental forms of the equations found in the latter[3] provide a secure starting point on two counts: (1) the simplifications that lead to the final equations are made transparent and (2) a pointer to handling more complicated reactor systems is provided. But there is a price to be paid. The reader has to become familiar with a few simple vector operations such as the scalar or dot product, gradient, and divergence. These are explained briefly wherever they occur in this chapter. For a more detailed treatment, the reader may consult Bird et al.[3]

We first consider the simpler case of material balances and then undertake a derivation of the energy balance equations.

1.4.1 Material Balance and the Concept of Rate of Generation of a Species

We start with the material balance for a fixed control volume, V_c, through which a single component fluid flows. The mass balance can be written as

$$\text{Rate of accumulation} \ = \ \text{input rate} \ - \ \text{output rate.} \tag{1.2}$$

A very general representation of the above equation can be given within the framework of continuum mechanics:

$$\frac{dm}{dt} = \frac{d}{dt} \int_{V_c} \rho \, dV = - \int_{A_c} \rho \mathbf{v} \cdot \mathbf{n} \, dA \tag{1.3}$$

where $m = \int_{V_c} \rho dV$ represents the mass of fluid inside the control volume at any time, t, with $\rho = \rho(\mathbf{x},t)$ denoting the fluid density at a location indicated by the position vector $\mathbf{x}$ at an instant of time t. The quantity $\mathbf{v}$ is the velocity vector, whereas $\mathbf{n}$ is the *outward* unit normal vector.

The area integral in the equation has the following interpretation: the quantity $\rho \mathbf{v}$ is a vector and has the unit of mass/area/time. Thus it is denoted as the mass flux vector. Of course, mass itself is a scalar quantity and the term $\rho \mathbf{v} \cdot \mathbf{n}$ represents the flux of mass *out* of the control volume at a particular location on the surface enclosing the control volume where the fluid density is ρ, the velocity is $\mathbf{v}$, and the unit *outward* normal is $\mathbf{n}$. The quantity $\rho \mathbf{v} \cdot \mathbf{n}$ is the scalar dot product of the vector $\rho \mathbf{v}$ with the vector $\mathbf{n}$ and is given by

$$\rho \mathbf{v} \cdot \mathbf{n} = \rho v_x n_x + \rho v_y n_y + \rho v_z n_z$$

in the familiar rectilinear Cartesian coordinate system, where (v_x, v_y, v_z) and (n_x, n_y, n_z) are the components of the vector $\mathbf{v}$ and $\mathbf{n}$, respectively. Hence the integral $\int_{A_c} \rho \mathbf{v} \cdot \mathbf{n} \, dA$ represents the net *efflux* of mass from the control volume whereas $- \int_{A_c} \rho \mathbf{v} \cdot \mathbf{n} \, dA$ represents the net influx (in−out) of mass into the control volume with A_c representing the area enclosing the control volume. Thus Eq. (1.3) may be written in a more familiar form as

$$\frac{dm}{dt} = m_{r,in} - m_{r,out}$$

where $m_{r,in}$ and $m_{r,out}$ are, respectively, the rate at which mass flows in and out of the control volume.

An extension of the above equations to mixtures (having n components, say) would involve two factors: (1) a mass balance for each of the n components, (2) generation (or disappearance) of mass of a component due to chemical reactions. Thus Eq. (1.2) is to be modified as

(Rate of accumulation)$_i$ = (input rate)$_i$ − (output rate)$_i$ + (generation rate)$_i$, $i = 1, ..., n$.

Correspondingly, Eq. (1.3) may be generalized as

$$\frac{dm_i}{dt} = \frac{d}{dt} \int_{V_c} \rho_i dV = - \int_{A_c} \rho_i \mathbf{v}_i \cdot \mathbf{n} \, dA + \int_{V_c} R_i \, dV, i = 1, ..., n \tag{1.4}$$

where ρ_i is the mass density of species i and $\mathbf{v}_i$ is the velocity of species i. The quantity R_i represents the net rate of generation per unit volume of species i due to chemical reactions. It is expressed as mass/volume/time for a homogeneous reaction, which is our focus in this chapter. Thus the integral $\int_{V_c} R_i \, dV$ is the net rate of generation of species i due to chemical reactions in the control volume. If the quantity R_i is negative, then there is a net disappearance of species i due to chemical reactions. This means that the rate at which species i is being produced by reactions is *less* than the rate at which it is being consumed. The other two terms in Eq. (1.4) have an interpretation similar to that of Eq. (1.3), except now that they refer to a particular component i. Thus, $- \int_{A_c} \rho_i \mathbf{v}_i \cdot \mathbf{n} \, dA$ is the *net* influx of species i into the control volume and $\int_{V_c} \rho_i \, dV = m_i$ is the mass of species i in the control volume at time t.

Summing all the equations in (1.4), we obtain

$$\frac{d \sum_{i=1}^{n} m_i}{dt} = - \int_{A_c} \left(\sum_{i=1}^{n} \rho_i \mathbf{v}_i \right) \cdot \mathbf{n} \, dA + \int_{V_c} \left(\sum_{i=1}^{n} R_i \right) dV. \tag{1.5}$$

We denote $m = \sum_{j=1}^{n} m_j$ as the total mass of the mixture in the control volume, $\rho = \sum_{j=1}^{n} \rho_j$ as the total local density of the mixture, and $\mathbf{v} = \left(\sum_{i=1}^{n} \rho_i \mathbf{v}_i \right) \big/ \rho$ as the mass average velocity of the mixture. Further, we note that

$$\sum_{i=1}^{n} R_i = 0. \tag{1.6}$$

This expresses the fact while one species may be converted to another through chemical reactions, there can be no net creation or destruction of mass due to reactions. Thus Eq. (1.5) may be written as

$$\frac{dm}{dt} = - \int_{A_c} \rho \mathbf{v} \cdot \mathbf{n} \, dA \tag{1.7}$$

which looks identical to Eq. (1.3), but the quantities in Eq. (1.7) refer to mixture quantities. Dividing each of the equations in Eq. (1.4) by the molecular weight of the corresponding species i, we obtain the molar analog of Eq. (1.4):

$$\frac{dn_i}{dt} = \frac{d}{dt} \int_{V_c} C_i \, dV = - \int_{A_c} C_i \mathbf{v}_i \cdot \mathbf{n} \, dA + \int_{V_c} r_i dV, i = 1, \dots, n \qquad (1.8)$$

where n_i is the number of moles of component i within the control volume at time t, C_i is the molar concentration of species i, and r_i is the net *molar* rate of generation of species i expressed as moles/volume/time. Since $-\int_{A_c} C_i \mathbf{v}_i \cdot \mathbf{n} \, dA$ represents the *net* molar rate of species i entering the reactor, Eq. (1.8) may be written as

$$\frac{dn_i}{dt} = F_{i,in} - F_{i,out} + \int_{V_c} r_i \, dV, i = 1, \dots, n \qquad (1.9)$$

where $F_{i,in}$ and $F_{i,out}$ are the inlet and outlet molar rates, respectively, of species i. In the most common case, there is one inlet and one outlet to a reactor. The quantity r_i occupies a central place in reaction engineering and its determination and representation will be discussed in detail in Section 1.5. Often $-r_{A}$, which represents the rate of disappearance of a component A, is used to represent the rate of a reaction in which A is a reactant.

Eqs. (1.4), (1.8), and (1.9) are sufficient for the analysis of the well-mixed batch and CSTRs. For a *batch reactor*, the flow terms in Eq. (1.9) are zero and hence the rate of accumulation of species i equals its rate of generation by reaction. For a CSTR, if operating in a *steady state,* the *rate of accumulation* term (left-hand side of Eq. 1.9) is zero, and hence the *rate of generation of species i by reaction* is just balanced by the difference between outflow and inflow.

For unsteady state operation of a flow reactor, it is important to appreciate the distinction between the *reaction* term and the *accumulation* term, which are equal for a batch reactor. Transient operation of a flow reactor occurs during start-up and in response to disturbances in the operating conditions. The nature of transients induced by disturbances and the differences between the reaction and accumulation terms can best be visualized for the case of a CSTR (Fig. 1.8). In Fig. 1.8A, the reactor is operating in a steady state. In Fig. 1.8B, it is subject to an increase in the input of reactant owing to a disturbance in the feed composition. This results in a rise in the concentration of the reactant within the reaction vessel corresponding to the *accumulation* term which is quite distinct from, and additional to, the *reactant removal by the reaction* term. Fig. 1.8C shows another kind of transient which will cause compositions in the reactor to change, namely a change in the volume of reaction mixture contained in the reactor. Other variables which must be controlled, apart from feed composition and flow rates in and out, are temperature and, for gas reactions particularly, pressure. Variations of any of these quantities with time will cause a change in the composition levels of the reactant in the reactor, and these will

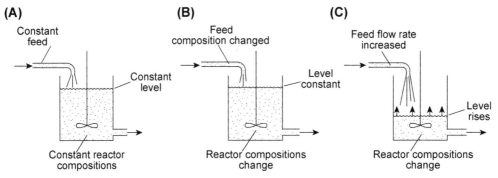

Figure 1.8

Continuous stirred tank reactor showing steady-state operation (A), and two modes of unsteady state operation (B) and (C).

appear in the *accumulation* term in the material balance. Of course, since the reaction rate is also affected by the composition levels in the reactor, the reaction term in Eq. (1.9) will also be affected.

For the PFR, where there is a *continuous* variation of composition along the length of the reactor, a local *continuity* equation is required. Towards this end, we use the divergence theorem[3] given by

$$\int_{A_c} \boldsymbol{\rho}_i \mathbf{v}_i \cdot \mathbf{n} \, dA = \int_{V_c} \nabla \cdot \boldsymbol{\rho}_i \mathbf{v}_i \, dV \tag{1.10}$$

where "$\nabla \cdot$" is the divergence operator. [In rectilinear Cartesian coordinate system, the divergence of vector $\mathbf{a}$, with components a_x, a_y, and a_z, is given by

$$\nabla \cdot \mathbf{a} = \frac{\partial a_x}{\partial x} + \frac{\partial a_y}{\partial y} + \frac{\partial a_z}{\partial z}$$

whereas in the cylindrical coordinate system, defined by the coordinates (r_c, ϕ, z),

$$\nabla \cdot \mathbf{a} = \frac{1}{r_c} \frac{\partial}{\partial r_c} (r_c \, a_{r_c}) + \frac{1}{r_c} \frac{\partial a_\phi}{\partial \phi} + \frac{\partial a_z}{\partial z},$$

where (a_{r_c}, a_ϕ, a_z) are the components of the vector $\mathbf{a}$ in the cylindrical coordinates.]

Using Eq. (1.10) in Eq. (1.4), and bringing all the terms into a single volume integral, we obtain

$$\int_{V_c} \left[\frac{\partial \boldsymbol{\rho}_i}{\partial t} + \nabla \cdot \boldsymbol{\rho}_i \mathbf{v}_i - R_i \right] dV = 0, i = 1, \ldots, n. \tag{1.11}$$

It must be noted that the time derivative operator can be taken inside the integral in Eq. (1.4) because the control volume is fixed in space. The vanishing of an integral does not necessarily mean vanishing of the integrand. But in this case, using standard limiting arguments, we may conclude that

$$\frac{\partial \boldsymbol{\rho}_i}{\partial t} + \nabla \cdot \boldsymbol{\rho}_i \mathbf{v}_i = R_i, i = 1, \ldots, n \tag{1.12}$$

While Eqs. 1.3, 1.4, 1.8, and 1.9 are *global* balances valid for the system as a whole, Eq. (1.12) is a local balance equation, valid at every point within the system. Summing Eq. (1.12), we obtain the overall mixture continuity equation:

$$\frac{\partial \boldsymbol{\rho}}{\partial t} + \nabla \cdot \boldsymbol{\rho} \mathbf{v} = 0, \tag{1.13}$$

where Eq. (1.6) has been used. The molar analog of Eq. (1.12) is

$$\frac{\partial C_i}{\partial t} + \nabla \cdot C_i \mathbf{v}_i = r_i, \quad i = 1, \ldots, n. \tag{1.14}$$

1.4.2 Energy Balance

The earliest instance of an energy balance came out of classical thermodynamics through the first law of thermodynamics for a closed system containing a homogeneous mass of a single component fluid. It may be expressed as

$$\frac{dE_I}{dt} = Q_r + W_r$$

where E_I is the internal energy of the mass of fluid in the system, Q_r is the rate at which heat is added to the system, and W_r is the rate at which work is done on the system. To extend this equation to open systems would require taking into account the kinetic energy as well as the flow of energy in and out of the system. A general balance for a fixed control volume may be expressed as

$$\frac{d(E_I + E_K)}{dt} = \text{net influx } (E_I + E_K) + Q_r + W_r \tag{1.15}$$

where E_K refers to the kinetic energy. The heating rate can be divided into two parts, one arising out of transfer of heat by conduction across the surface enclosing the control volume and another by radiation. In most chemical engineering applications, the effect of radiation can be neglected. The rate of work done can also be classified into two types, one termed as "flow work" and another referred to as "shaft work." The former is the work done on the fluid to cause its flow through the control volume, whereas the latter arises out of work done by stirrer, mixer, etc. The flow work can be further divided into two parts: work due to body forces and work due to contact force or stress force. The work done by the contact force consists in turn of work due to pressure force and that due to shear force. If the body forces are conservative, that is, if they are derivable from a time-independent potential, then the potential energy term can be added to the internal and kinetic energy in Eq. 1.15.[3] Similarly the work due to the pressure force can be combined with the influx of internal energy to yield the influx of enthalpy. In summary, with these modifications, Eq. (1.15) takes the form

$$\frac{d(E_I + E_K + E_P)}{dt} = \text{net influx } (H + E_K + E_P) + Q_r + W_{r,shear} + W_{r,shaft} \tag{1.16}$$

where H refers to the enthalpy and E_P to the potential energy.

We first deal with simpler forms of the above equation because in reaction engineering, usually, kinetic energy and potential energy are neglected when compared to the internal energy or enthalpy. The shear stress work term is also neglected. Further, following the continuum hypothesis, we may express the total internal energy of the fluid within the control volume as

$$E_I = \int_{V_c} C\tilde{\varepsilon}dV$$

where C is the molar concentration and $\tilde{\varepsilon}$ is the specific internal energy per mole. Further, the net influx of enthalpy may be written as

$$\text{Net influx of enthalpy into control volume} = -\int_{A_c} C\tilde{h}\mathbf{v}\cdot\mathbf{n}dA$$

where $\tilde{h}$ is the specific enthalpy per mole. With the above approximations and representations, Eq. (1.16) becomes

$$\frac{d}{dt}\int_{V_c} C\tilde{\varepsilon}dV = -\int_{A_c} C\tilde{h}\mathbf{v}\cdot\mathbf{n}\, dA + Q_r + W_{r,shaft} \tag{1.17}$$

All of the above equations apply directly to a fluid with only one component. For a mixture, which is what we have to deal with in reaction engineering, it appears the convention to adopt the same form as the above equations except that the quantities in the equation are to be regarded as corresponding to the mixture. The simplest representation of mixture properties is in terms of the pure component properties and is obtained by neglecting the so-called change in properties associated with mixing. Thus, to get the mixture equations, we apply the following rules:

$$\tilde{\varepsilon} \rightarrow \sum_{i=1}^{n} x_i\tilde{\varepsilon}_i, \; C\tilde{h}\mathbf{v} \rightarrow \sum_{i=1}^{n} C_i\tilde{h}_i\mathbf{v}_i$$

where x_i is the mole fraction of species i in the mixture, $\widetilde{\varepsilon}_i$ and $\widetilde{h}_i$ are the pure component-specific internal energy and enthalpy of species i at the temperature and pressure of the mixture. Thus Eq. (1.17) may be written as

$$\frac{d}{dt} \int_{V_c} \left(\sum_{i=1}^{n} C_i \widetilde{\varepsilon}_i \right) dV = - \int_{A_c} \sum_{i=1}^{n} C_i \widetilde{h}_i \mathbf{v}_i \cdot \mathbf{n} dA + Q_r + W_{r,shaft}. \tag{1.18}$$

At steady state, we obtain

$$- \int_{A_c} \sum_{i=1}^{n} C_i \widetilde{h}_i \mathbf{v}_i \cdot \mathbf{n} dA + Q_r + W_{r,shaft} = 0. \tag{1.19}$$

In the most common case of one inlet and one outlet and when $\widetilde{h}_i$ is independent of position on the control surface at the inlet and outlet, we may write Eq. (1.19) as

$$\sum_{i=1}^{n} F_{i,in} \widetilde{h}_{i,in} - \sum_{i=1}^{n} F_{i,out} \widetilde{h}_{i,out} + Q_r + W_{r,shaft} = 0 \tag{1.20}$$

We note that the *magnitude* of $\int C_i \mathbf{v}_i \cdot \mathbf{n} \, dA$ evaluated at the inlet will equal $F_{i,in}$, whereas that evaluated at the outlet will yield $F_{i,out}$. Eq. (1.20) is a convenient starting point for analyzing the nonisothermal operation of a CSTR under steady-state conditions (Section 1.8.4). The developments leading from Eqs. (1.15) to (1.20) serve to bring out the generalizations and simplifications inherent in Eq. (1.20).

To develop further the unsteady state equation, we express the accumulation term too in terms of enthalpy. Noting that

$$\widetilde{h}_i = \widetilde{\varepsilon}_i + P\widetilde{v}_i \tag{1.21}$$

where $\widetilde{v}_i$ is the molar specific volume of pure component i at the temperature (T) and pressure (P) of the mixture, we obtain

$$\sum_{i=1}^{n} C_i \widetilde{\varepsilon}_i = \sum_{i=1}^{n} C_i \widetilde{h}_i - P \sum_{i=1}^{n} C_i \widetilde{v}_i$$

However,

$$\sum_{i=1}^{n} C_i \widetilde{v}_i = C \sum_{i=1}^{n} x_i \widetilde{v}_i = C\widetilde{v} = 1 \tag{1.22}$$

where $\widetilde{v}$ is the molar volume of the mixture (volume/mole) and is the inverse of C, the molar concentration of the mixture (mole/volume). In equating $\sum_{i=1}^{n} x_i \widetilde{v}_i$ to $\widetilde{v}$ in

Eq. (1.22), we follow the spirit of our earlier stated assumption that change in properties on mixing is neglected. Hence from Eqs. (1.21) and (1.22),

$$\sum_{i=1}^{n} C_i \widetilde{\varepsilon}_i = \sum_{i=1}^{n} C_i \widetilde{h}_i - P \tag{1.23}$$

Therefore,

$$\frac{\partial}{\partial t}\left(\sum_{i=1}^{n} C_i \widetilde{\varepsilon}_i\right) = \frac{\partial}{\partial t}\left(\sum_{i=1}^{n} C_i \widetilde{h}_i\right) - \frac{\partial P}{\partial t} = \sum_{i=1}^{n}\left(C_i \frac{\partial \widetilde{h}_i}{\partial t} + \widetilde{h}_i \frac{\partial C_i}{\partial t}\right) - \frac{\partial P}{\partial t} \tag{1.24}$$

Further, $\widetilde{h}_i$ is a function of P and T so that

$$\frac{\partial \widetilde{h}_i}{\partial t} = \left(\frac{\partial \widetilde{h}_i}{\partial P}\right)_T \frac{\partial P}{\partial t} + \left(\frac{\partial \widetilde{h}_i}{\partial T}\right)_P \frac{\partial T}{\partial t}. \tag{1.25}$$

The derivative $\left(\partial \widetilde{h}_i / \partial P\right)_T$ may be obtained from an equation of state for the component i and is usually neglected for liquid-phase reactions. The derivative $\left(\partial \widetilde{h}_i / \partial T\right)_P$ is the (molar) specific heat at constant pressure denoted by $\widetilde{c}_{Pi}$. Taking into account these definitions and using Eqs. (1.24) and (1.25) in Eq. (1.18), we obtain

$$\int_{V_c} \frac{\partial P}{\partial t}\left(\sum_{i=1}^{n} C_i \left(\frac{\partial \widetilde{h}_i}{\partial P}\right)_T - 1\right) dV + \int_{V_c}\left(\sum_{i=1}^{n} C_i \widetilde{c}_{Pi}\right)\frac{\partial T}{\partial t} dV + \int_{V_c}\left(\sum_{i=1}^{n} \widetilde{h}_i \frac{\partial C_i}{\partial t} dV\right)$$

$$= -\int_{A_c} \sum_{i=1}^{n} C_i \widetilde{h}_i \, \mathbf{v}_i \cdot \mathbf{n} \, dA + Q_r + W_{r,shaft}. \tag{1.26}$$

Eq. (1.26) is adequate for analyzing the nonisothermal operation of a batch reactor (Sections 1.6.4 and 1.6.5). As we have remarked above, Eq. (1.20) is a good starting point for analyzing a CSTR operating at steady state. For the PFR, we need a local balance. A simplified set of equations useful for applying directly to a CSTR and a PFR operating at steady state is presented in the Appendix at the end of this chapter.

1.5 Chemical Kinetics and Rate Equations

When a homogeneous mixture of reactants is passed into a reactor, either batch or tubular, the concentrations of the reactants fall as the reaction proceeds. Experimentally it has been found that, in general, the rate of the reaction decreases as the concentrations of the reactants decrease. To calculate the size of the reactor required to manufacture a particular product at a desired overall rate of production, the design engineer therefore needs to know how the rate of reaction at any time or at any point in the reactor depends on the concentrations of the reactants. Since the reaction rate varies also with temperature, generally increasing rapidly with increasing temperature, a *rate equation,* expressing the rate of formation of a species *i* (the quantity r_i in Eqs. 1.8, 1.9, and 1.14) as a function of concentrations and temperature, is required to design a reactor.

1.5.1 Definition of Order of Reaction and Rate Constant

For concreteness, let us consider a homogeneous irreversible reaction:

$$a\mathbf{A} + b\mathbf{B} \rightarrow c\mathbf{C} + d\mathbf{D} \tag{1.27}$$

where **A** and **B** are reactants, **C** and **D** are products. The above equation represents the fact that *a* moles of **A** and *b* moles of **B** react to yield *c* moles of **C** and *d* moles of **D**. A fundamental relation between the rates of generation of the species involved in the above reaction is given by

$$\frac{r_A}{v_A} = \frac{r_B}{v_B} = \frac{r_C}{v_C} = \frac{r_D}{v_D} \tag{1.28}$$

where v_A, v_B, v_c, and v_D are the corresponding *stoichiometric* coefficients in the reaction (1.27). In the reaction *as written,*

$$v_A = -a, \quad v_B = -b, \quad v_C = c, \quad v_D = d.$$

That is, as per convention, the stoichiometric coefficients of the components that appear on the reactant side are assigned a negative sign. It must be noted that Eq. (1.27) is not a unique representation of the stoichiometry of the reaction. For instance, it can also be represented by

$$\alpha a\mathbf{A} + \alpha b\mathbf{B} \rightarrow \alpha c\mathbf{C} + \alpha d\mathbf{D}$$

in which case

$$v_A = -\alpha a, v_B = -\alpha b, v_C = \alpha c, v_D = \alpha d.$$

However, as may be seen from Eq. (1.28), the relationship between the rates of generation of the various species remains unaltered. Eq. (1.28) may be more generally represented by

$$\frac{r_i}{v_i} = \frac{r_j}{v_j} \tag{1.29}$$

for any two species *i* and *j* involved in a single reaction. Eq. (1.29) suggests a definition of a *reaction rate, r,* which is independent of the species, namely,

$$r \equiv \frac{r_i}{v_i}, \tag{1.30}$$

where i is any species involved in the reaction. Following convention in chemical engineering, we prefer to deal with a particular rate r_i than r.

At constant temperature, the rate of generation of a species A, r_A, or equivalently the rate of disappearance of A, $-r_A$, in a reaction such as 1.27, is a function of the concentrations of the reactants. Experimentally, it has been found that often (but not always) the function has the mathematical form

$$-r_A = kC_A^p C_B^q \tag{1.31}$$

C_A is the molar concentration of **A** and C_B the molar concentration of **B**. The exponents p and q in this expression are quite often (but not necessarily) whole numbers. When the functional relationship has the form of Eq. (1.31), the reaction is said to be of order p with respect to reactant **A**, q with respect to **B**. The *overall* order of the reaction is $(p + q)$.

The coefficient k in Eq. (1.31) is referred to as the *rate constant* of the reaction. Its dimensions depend on the exponents p and q (i.e., on the order of the reaction); the units in which it is to be expressed may be inferred from the defining Eq. (1.31). For example, if a reaction

$$\textbf{A} \rightarrow \text{Products}$$

behaves as a simple first-order reaction, it has a rate equation

$$-r_A = k_1 C_A \tag{1.32}$$

If $-r_A$ is measured in units of kmol/m^3/s and the concentration C_A in kmol/m^3, then k_1 has the unit s^{-1}. On the other hand, if the reaction above behaved as a second-order reaction with a rate equation

$$-r_A = k_2\, C_A{}^2 \tag{1.33}$$

the unit of this rate constant, with $-r_A$ in kmol/m^3 s and C_A in kmol/m^3, is m^3(kmol)$^{-1}$ s^{-1}. A possible source of confusion is that in some instances in the chemical literature, the rate equation, for, say, a second-order gas-phase reaction may be written as $-r_A = k_p P_A^2$, where P_A is the partial pressure of **A** and may be measured in N/m^2, bar, or even in mm Hg. This form of expression results in rather confusing hybrid units for k_p and is not to be recommended.

If a large excess of one or more of the reactants is used, such that the concentration of that reactant changes hardly at all during the course of the reaction, the effective order of the reaction is reduced. Thus, if in carrying out a reaction which is normally second order with a rate equation $-r_A = k_2 C_A C_B$ an excess of **B** is used, then C_B remains approximately constant and almost equal to the initial value C_{B0}. The rate equation may then be written $-r_A = k_1' C_A$, where $k_1' = k_2 C_{B0}$ is an *effective* rate constant and the reaction is now said to be *pseudo—first-order.*

1.5.2 Influence of Temperature: Activation Energy

Experimentally, the influence of temperature on the rate constant of a reaction is well represented by the original equation of Arrhenius:

$$k = A \exp(-\mathbf{E}/\mathbf{R}T) \tag{1.34}$$

where T is the absolute temperature and $\mathbf{R}$ the universal gas constant. In this equation, $\mathbf{E}$ is termed as the *activation energy,* and A the *frequency factor.* There are theoretical reasons to suppose that temperature dependence should be more exactly described by an equation of the form $k = A'T^m \exp(-\mathbf{E}/\mathbf{R}T)$, with m usually in the range 0–2. However, the influence of the exponential term in Eq. (1.34) is in practice so strong as to mask any variation in A with temperature, and the simple form of the relationship, Eq. (1.34), is therefore quite adequate. The quantity $\mathbf{E}$ is called the *activation energy* because in the molecular theory of chemical kinetics it is associated with an energy barrier that the reactants must surmount to form an activated complex in the transition state. Similarly, A is associated with the frequency with which the activated complex breaks down into products or, in terms of the simple collision theory, it is associated with the frequency of collisions.

Values of the activation energy $\mathbf{E}$ are in J/kmol in the SI system but are usually quoted in kJ/kmol (or J/mol); using these values, $\mathbf{R}$ must then be expressed as kJ/kmol K. For most reactions the activation energy lies in the range 50,000–250,000 kJ/kmol, which implies a very rapid increase in rate constant with temperature. Thus, for a reaction which is occurring at a temperature in the region of 100°C and has an activation energy of 100,000 kJ/kmol, the reaction rate will be doubled for a temperature rise of only 10°C.

Thus, the complete rate equation for an irreversible reaction such as 1.27 normally has the form

$$-r_A = A \exp(-\mathbf{E}/\mathbf{R}T)C_A^p C_B^q \tag{1.35}$$

Unfortunately, the exponential temperature term $\exp(-\mathbf{E}/\mathbf{R}T)$ is rather troublesome to handle mathematically, both by analytical methods and numerical techniques. In reactor design, this means that calculations for reactors, which are not operated isothermally tend to become complicated. In a few cases, useful results can be obtained by abandoning the exponential term altogether and substituting a linear variation of reaction rate with temperature, but this approach is quite inadequate unless the temperature range is very small.

1.5.3 Rate Equations and Reaction Mechanism

One of the reasons why chemical kinetics is an important branch of physical chemistry is that the rate of a chemical reaction may be a significant guide to its mechanism. The engineer concerned with reactor design and development is not interested in reaction mechanism per se, but should be aware that an insight into the mechanism of the reaction can provide a valuable clue to the kind of rate equation to be used in a design problem. In the present chapter, it will be possible to make only a few observations on the subject, and for further information the excellent text of Moore and Pearson[4] should be consulted.

The first point that must be made is that the *overall* stoichiometry of a reaction is *no guide whatsoever* to its rate equation or to the mechanism of reaction. A stoichiometric equation is no more than a material balance; thus the reaction

$$KClO_3 + 6FeSO_4 + 3H_2SO_4 \rightarrow KCl + 3Fe_2(SO_4)_3 + 3H_2O$$

is in fact second order in dilute solution with the rate of reaction proportional to the concentrations of ClO_3^- and Fe^{2+} ions. In the general case the stoichiometric coefficients ν_A, ν_B, ν_C are not necessarily related to the orders p, q, r for the reaction.

However, if it is known from kinetic or other evidence that a reaction, $\mathbf{M} + \mathbf{N} \rightarrow$ Product, is a simple *elementary reaction,* i.e., if it is known that its mechanism is simply the interaction between a molecule of $\mathbf{M}$ and a molecule of $\mathbf{N}$, then the molecular theory of reaction rates predicts that the rate of this elementary step is proportional to the concentration of species $\mathbf{M}$ and the concentration of species $\mathbf{N}$, i.e., it is second order overall. The reaction is also said to be *bimolecular* since two molecules are involved in the actual chemical transformation. More generally, we regard a reaction step as elementary *as written* if the order corresponding to each of the reactants equals its stoichiometric coefficient.

Thus, the reaction between H_2 and I_2 is known to occur by an elementary bimolecular reaction:

$$H_2 + I_2 \rightarrow 2HI$$

and the rate of the forward reaction corresponds to the equation:

$$-r_{I_2} = k_f C_{H_2} C_{I_2}$$

For many years the hydrogen—iodine reaction was quoted in textbooks as being virtually the only known example of a simple bimolecular reaction. There is now evidence[5] that in parallel to the main bimolecular transformation, some additional reactions involving iodine atoms do occur.

Whereas in the hydrogen—iodine reaction, atomic iodine plays only a minor part, in the reaction between hydrogen and bromine, bromine and hydrogen atoms are the principal intermediates in the overall transformation.

The kinetics of the reaction is quite different from that of the hydrogen—iodine reaction, although the stoichiometric equation:

$$H_2 + Br_2 \rightarrow 2HBr$$

looks similar. The reaction actually has a chain mechanism consisting of the elementary steps:

$$Br_2 \rightleftharpoons 2Br\cdot \quad \textit{chain initiaton and termination}$$

$$\left.\begin{array}{l} Br\cdot + H_2 \rightleftharpoons HBr + H\cdot \\ H\cdot + Br_2 \rightarrow HBr + Br\cdot \end{array}\right\} \textit{chain propagation}$$

The rate of the last reaction, for example, is proportional to the concentration of **H·** and the concentration of Br_2, i.e., it is second order. When the rates of these elementary steps are combined into an overall rate equation, this becomes

$$r_{HBr} = \frac{k' C_{H_2} C_{Br_2}^{1/2}}{1 + k'' \dfrac{C_{HBr}}{C_{Br_2}}} \tag{1.36}$$

where k' and k'' are constants, which are combinations of the rate constants of the elementary steps. This rate equation has a different form from the usual type given by Eq. (1.31) and cannot therefore be said to have any order because the definition of order applies only to the usual form.

A basic principle used to obtain expressions such as 1.36 from a mechanism is the **pseudo—steady-state approximation (PSSA)**. According to this principle, the net rate of formation of any reactive intermediate is taken to be zero. The rationale is that such a species reacts as quickly as it is formed. For the HBr formation mechanism, the principle would apply to the radicals H• and Br•. Thus,

$$r_{Br\cdot} = r_{H\cdot} = 0. \tag{1.37}$$

Further, since each of the reactions in the mechanism is an elementary reaction,

$$r_{Br\cdot} = 2k_1 C_{Br_2} - k_{-1}(C_{Br\cdot})^2 - k_2 C_{Br\cdot} C_{H_2} + k_{-2} C_{H\cdot} C_{HBr} + k_3 C_{H\cdot} C_{Br_2} = 0 \tag{1.38}$$

and

$$r_{H\cdot} = k_2 C_{Br\cdot} C_{H_2} - k_{-2} C_{H\cdot} C_{HBr} - k_3 C_{H\cdot} C_{Br_2} = 0 \tag{1.39}$$

The factor 2 in the first term on the right-hand side of the equation indicates that k_1 represents the rate constant for the disappearance of Br_2 through the forward reaction in the initiation step. That is, $-r_{Br_2} = k_1 C_{Br_2}$. From Eqs. (1.38) and (1.39), one obtains

$$2k_1 C_{Br_2} - k_{-1}(C_{Br\cdot})^2 = 0 \tag{1.40}$$

This implies that the first step is essentially in equilibrium. This is sometimes invoked as an independent principle: a particular elementary step in a mechanism is in equilibrium and is referred to as the *quasi-equilibrium* assumption. From Eq. (1.40), we get an expression for $C_{Br·}$:

$$C_{Br·} = \left(2\frac{k_1}{k_{-1}}C_{Br_2}\right)^{1/2} \tag{1.41}$$

Substituting in Eq. (1.39), we get an expression for $C_{H·}$:

$$C_{H·} = \frac{k_2 C_{H_2}}{(k_3 C_{Br_2} + k_{-2}C_{HBr})}\left(2\frac{k_1}{k_{-1}}C_{Br_2}\right)^{1/2}. \tag{1.42}$$

The final rate expression is obtained by noting that

$$r_{HBr} = k_3 C_{H·}C_{Br_2} + k_2 C_{Br·}C_{H_2} - k_{-2}C_{H·}C_{HBr}.$$

Using Eq. (1.39), we obtain

$$r_{HBr} = 2k_3 C_{H·}C_{Br_2}. \tag{1.43}$$

Substituting for $C_{H·}$ from Eq. (1.42) and rearranging, we get

$$r_{HBr} = \frac{2k_2\left(\dfrac{2k_1}{k_{-1}}\right)^{1/2} C_{H_2}(C_{Br_2})^{1/2}}{\left(1 + \dfrac{k_{-2}}{k_3}\dfrac{C_{HBr}}{C_{Br_2}}\right)} \tag{1.44}$$

Comparing with Eq. (1.36), we note that

$$k' = 2k_2\left(\frac{2k_1}{k_{-1}}\right)^{1/2}, \quad k'' = \frac{k_{-2}}{k_3}. \tag{1.45}$$

We shall find that the rate equations of gas–solid heterogeneous catalytic reactions (Chapter 3) also do not, in general, have the same form as Eq. (1.31).

However, many reactions, although their mechanism may be quite complex, do conform to simple first- or second-order rate equations. This is because the rate of the overall reaction is limited by just one of the elementary reactions that is then said to be rate determining. The kinetics of the overall reaction thus reflects the kinetics of this particular step. An example is the pyrolysis of ethane,[5] which is important industrially as a source of ethylene[1] (see also Section 1.7.1; Example 1.5). The main overall reaction is

$$C_2H_6 \rightarrow C_2H_4 + H_2$$

Although there are complications concerning this reaction, under most circumstances it is first order, the kinetics being largely determined by the first step in a chain mechanism:

$$C_2H_6 \rightarrow 2CH_3 \cdot$$

which is followed by the much faster reactions:

$$\left. \begin{array}{rcl} CH_3 \cdot + C_2H_6 & \rightarrow & C_2H_5 \cdot + CH_4 \\ C_2H_5 \cdot & \rightarrow & C_2H_4 + H \cdot \\ H \cdot + C_2H_6 & \rightarrow & C_2H_5 \cdot + H_2 \end{array} \right\} \text{ chain propagation}$$

Eventually the reaction chains are broken by termination reactions. Other free radical reactions also take place to a lesser extent leading to the formation of CH_4 and some higher hydrocarbons among the products.

1.5.4 Reversible Reactions

For reactions that do not proceed virtually to completion, it is necessary to include the kinetics of the reverse reaction, or the equilibrium constant, in the rate equation.

The equilibrium state in a chemical reaction can be considered from two distinct points of view. The first is from the standpoint of classical thermodynamics and leads to relationships between the equilibrium constant and thermodynamic quantities such as standard free energy change and heat of reaction, from which we can very usefully calculate equilibrium conversion (Example 1.1 in Section 1.3.2). The second is a kinetic viewpoint, in which the state of chemical equilibrium is regarded as a dynamic balance between forward and reverse reactions; at equilibrium the rates of the forward reactions and reverse reactions are just equal to each other, making the net rate of transformation zero.

Consider a reversible reaction:

$$\mathbf{A} + \mathbf{B} \underset{k_r}{\overset{k_f}{\rightleftharpoons}} \mathbf{M} + \mathbf{N}$$

which is second order overall in each direction and first order with respect to each species. The hydrolysis of an ester such as ethyl acetate is an example:

$$CH_3COOC_2H_5 + NaOH \rightleftharpoons CH_3COONa + C_2H_5OH$$

The net rate of disappearance of species A is given by

$$-r_A = k_f C_A C_B - k_r C_M C_N \tag{1.46}$$

At equilibrium, when $C_A = C_{Ae}$, etc., r_A is zero and we have

$$k_f C_{Ae} C_{Be} = k_r C_{Me} C_{Ne}$$

or

$$\frac{C_{Me} C_{Ne}}{C_{Ae} C_{Be}} = \frac{k_f}{k_r} \tag{1.47}$$

But $C_{Me} C_{Ne}/C_{Ae} C_{Be}$ is the concentration-based equilibrium constant K_c and hence $k_f/k_r = K_c$. Often it is convenient to substitute for k_r in Eq. (1.46) so that we have a typical example of a *rate equation for a reversible reaction*:

$$-r_A = k_f \left(C_A C_B - \frac{C_M C_N}{K_c} \right) \tag{1.48}$$

We see from the above example that the forward and reverse rate constants are not completely independent, but are related by the equilibrium constant, which in turn is related to the thermodynamic free energy change, etc. More detailed examination of the kinds of kinetic equations that might be used to describe the forward and reverse reactions shows that, to be consistent with the thermodynamic equilibrium constant, the form of the rate equation for the reverse reaction cannot be completely independent of the forward rate equation. A good example is the formation of phosgene:

$$CO + Cl_2 \rightleftharpoons CO\,Cl_2$$

The rate of the forward reaction is given by $r_f = k_f C_{CO} C_{Cl_2}^{3/2}$. This rate equation indicates that the chlorine concentration must also appear in the reverse rate equation. Let this be $r_r = k_r C_{COCl_2}^p C_{Cl_2}^q C_{CO}^s$; then at equilibrium, when $r_f = r_r$, we must have

$$k_f \left(C_{CO} C_{Cl_2}^{3/2} \right)_{eq} = k_r \left(C_{COCl_2}^p C_{Cl_2}^q C_{CO}^s \right)_{eq}$$

or

$$\frac{k_f}{k_r} = K_c = \left(C_{COCl_2}^p C_{Cl_2}^{(q-3/2)} C_{CO}^{(s-1)} \right)_{eq} \tag{1.49}$$

But we know from the thermodynamic equilibrium constant that

$$K_c = \left(\frac{C_{COCl_2}}{C_{CO} C_{Cl_2}} \right)_{eq} \tag{1.50}$$

Therefore it follows that, for consistency with the thermodynamic concept of equilibrium, $p = 1$, $q = \frac{1}{2}$, and $s = 0$. The complete rate equation is therefore

$$-r_{CO} = k_f C_{CO} C_{Cl_2}^{3/2} - k_r C_{COCl_2} C_{Cl_2}^{1/2}$$

or

$$-r_{CO} = k_f \left(C_{CO} C_{Cl_2}^{3/2} - \frac{C_{COCl_2} C_{Cl_2}^{1/2}}{K_c} \right) \tag{1.51}$$

A more detailed treatment of the implications of thermodynamic consistency for reaction kinetics may be found in Denbigh.[2]

1.5.5 Experimental Determination of Kinetic Constants

The interpretation of laboratory-scale experiments to determine order and rate constant is another subject which is considered at length in physical chemistry texts.[4,5] Essentially, it is a process of fitting a rate equation of the general form given by Eq. (1.31) to a set of numerical data. In a batch reactor, the experiments that are carried out to obtain the kinetic constants may be of two kinds, depending on whether the rate equation is to be used in *differential* form or in its *integrated* form. If the differential form is to be used then experiments must be designed so that the rate of disappearance of reactant A, $-r_A$, can be measured without its concentration changing appreciably. With batch or tubular reactors, this has the disadvantage in practice that very accurate measurements of C_A must be made so that, when differences in concentration ΔC_A are taken to evaluate r_A, the difference may be obtained with sufficient accuracy. CSTRs do not suffer from this disadvantage; by operating in the steady state, steady concentrations of the reactants are maintained and the rate of reaction is determined readily.

If the rate equation is to be employed in its integrated form, the problem of determining kinetic constants from experimental data from batch or tubular reactors is in many ways equivalent to taking the design equations and working backwards. Experiments using the various reactors to determine kinetic constants are discussed in the sections describing these reactors (Sections 1.6.6, 1.7.4 and 1.8.5).

Unfortunately, many of the chemical processes that are important industrially are quite complex. A complete description of the kinetics of a process, including by-product formation as well as the main chemical reaction, may involve several individual reactions, some occurring simultaneously and some proceeding in a consecutive manner. Often the results of laboratory experiments in such cases are ambiguous and, even if complete elucidation of such a complex reaction pattern is possible, it may take several man-years of experimental effort. Whereas ideally the design engineer would like to have a complete set of rate equations for all the reactions involved in a process, in practice the data available to him often fall far short of this.

1.6 Batch Reactors

Whereas continuous flow reactors are likely to be most economical for large-scale production, the very real advantages of batch reactors, especially for smaller scale production, should not be overlooked. Small batch reactors generally require less auxiliary equipment, such as pumps, and their control systems are less elaborate and costly than those for continuous reactors, although manpower needs are greater. However, large batch reactors may sometimes be fitted with highly complex control systems. A big advantage of batch reactors in the dyestuff, fine chemical and pharmaceutical industries is their versatility. A corrosion-resistant batch reactor such as an enamel or rubber-lined jacketed vessel (Fig. 1.4A) or a stainless steel vessel with heating and cooling coils (Fig. 1.4B) can be used for a wide variety of similar kinds of reaction. Sometimes only a few batches per year are required to meet the demand for an unusual product. In some processes, such as polymerizations and fermentations, batch reactors are traditionally preferred because the interval between batches provides an opportunity to clean the system thoroughly and ensure that no deleterious intermediates such as foreign bacteria build up and spoil the product. Moreover, it must not be forgotten that a squat tank is the most economical shape for holding a given volume of liquid, and for slow reactions a tubular flow reactor with a diameter sufficiently small to prevent backmixing would be more costly than a simple batch reactor. Although at present we are concerned mainly with homogeneous reactions, we should note that the batch reactor has many advantages for heterogeneous reactions; the agitator can be designed to suspend solids in the liquid and to disperse a second immiscible liquid or a gas.

In calculating the volume required for a batch reactor, we shall be specifying the volume of liquid which must be processed. In designing the vessel itself the heights should be increased by about 10% to allow freeboard for waves and disturbances on the surface of the liquid; additional freeboard may have to be provided if foaming is anticipated.

1.6.1 Calculation of Reaction Time: Basic Design Equation

Calculation of the time required to reach a particular conversion is the main objective in the design of batch reactors. Knowing the amount of reactant converted, i.e., the amount of the desired product formed per unit volume in this reaction time, the volume of reactor required for a given production rate can be found by simple scale-up as shown in the example on ethyl acetate.

The reaction time t_r is determined by applying the general material balance Eq. (1.9). In the most general case, when the volume of the reaction mixture is not constant throughout the reaction, it is convenient to make the material balance over the whole volume of the fluid in the reactor V_b. For a batch reactor, the flow terms are zero. Hence, applying Eq. (1.9) for the reactant **A** and assuming a uniformly mixed batch reactor (which implies that $\int_{V_b} r_A dV = r_A V_b$), we obtain

$$\frac{dn_A}{dt} = r_A V_b \tag{1.52}$$

If n_{A0} moles are charged initially, the number of moles remaining at time t, n_A, may be expressed in terms of the fraction of **A** converted, α_A. The fractional conversion of A is defined as

$$\alpha_A \equiv 1 - \frac{n_A}{n_{A0}}. \tag{1.53}$$

Thus $n_A = n_{A0}(1 - \alpha_A)$ and hence

$$\frac{dn_A}{dt} = \frac{d}{dt}[n_{A0}(1 - \alpha_A)] = -n_{A0}\frac{d\alpha_A}{dt}$$

Substituting into Eq. (1.52), we obtain

$$-n_{A0}\frac{d\alpha_A}{dt} = r_A V_b$$

Then, integrating over the period of the reaction (t_r) to a final conversion α_{Af}, we obtain the basic design equation:

$$t_r = n_{A0} \int_0^{\alpha_{Af}} \frac{d\alpha_A}{(-r_A)V_b} \tag{1.54}$$

For many liquid-phase reactions, it is reasonable to neglect any change in volume of the reaction mixture. It can be easily seen from Eq. (1.7) that a sufficient condition for this assumption to be valid is that the density of the reaction mixture is independent of composition. From Eq. (1.7), with the flow term equal to zero, we get

$$\frac{dm}{dt} = \frac{d(\rho V_b)}{dt} = 0.$$

Thus if ρ remains constant then so does V_b. Eq. (1.54) then becomes

$$t_r = \frac{n_{A0}}{V_b} \int_0^{\alpha_{Af}} \frac{d\alpha_A}{-r_A} = C_{A0} \int_0^{\alpha_{Af}} \frac{d\alpha_A}{-r_A} \tag{1.55}$$

where C_{A0} is the initial concentration of A. For a constant-volume batch reactor, it is sometimes more convenient to deal directly with concentration than conversion. For a constant V_b, we may write Eq. (1.52) as

$$\frac{d(n_A/V_b)}{dt} = \frac{dC_A}{dt} = r_A \tag{1.56}$$

Integrating Eq. (1.56), we obtain the reaction time as

$$t_r = \int_{C_{A0}}^{C_{Af}} \frac{dC_A}{r_A} = \int_{C_{Af}}^{C_{A0}} \frac{dC_A}{-r_A} \tag{1.57}$$

where C_{Af} is the final concentration. Another measure of conversion that is useful in dealing with constant-volume batch reactors is the quantity χ defined as

$$\chi \equiv C_{A0} - C_A = C_{A0}\alpha_A \tag{1.58}$$

This quantity represents the number of moles of A converted per unit volume of the reactor.

Eqs. (1.52)−(1.57) focus on a particular species, usually a reactant in the reaction. The number of moles of any other species (reactant or product) can be determined using Eq. (1.29). Thus for a species i, we have, analogous to Eq. (1.52),

$$\frac{dn_i}{dt} = r_i V_b \tag{1.59}$$

Dividing Eq. (1.59) by Eq. (1.52) and using Eq. (1.29), we obtain

$$\frac{r_i}{r_A} = \frac{dn_i}{dn_A} = \frac{\nu_i}{\nu_A} \tag{1.60}$$

Integrating Eq. (1.60), we obtain

$$n_i - n_{i0} = \frac{\nu_i}{\nu_A}(n_A - n_{A0}) = \frac{\nu_i}{-\nu_A}(n_{A0} - n_A) \tag{1.61}$$

For a constant-volume batch reactor, we get, on dividing Eq. (1.61) by the volume of the reaction mixture,

$$C_i - C_{i0} = \frac{\nu_i}{-\nu_A}(C_{A0} - C_A) \tag{1.62}$$

The equations that result from the integration of Eq. (1.55) or 1.57 for reactions of various orders are discussed in considerable detail in most texts dealing with the physicochemical aspects of chemical kinetics.[4,5] The integrated forms of Eq. (1.55) for a variety of the simple rate equations are shown in Table 1.1 in terms of the quantity χ defined by Eq. (1.58). The integrated forms can be easily verified by the reader if desired.

One particular point of interest is the expression for the *half-life* of a reaction $t_{1/2}$; this is the time required for one half of the reactant in question to disappear. A first-order reaction is unique in that the *half-life* is independent of the initial concentration of the reactant. This characteristic is sometimes used as a test of whether a reaction really is first order. Also since $t_{1/2} = \frac{1}{k_1}\ln 2$, a first-order rate constant can be readily converted into a *half-life* which one can easily remember as characteristic of the reaction (see Section 1.6.6 for a more detailed discussion on half-life).

Table 1.1: Rate equations for constant-volume batch reactors.

Reaction Type	Rate Equation	Integrated Form
Irreversible Reactions		
First order $A \rightarrow$ products	$\frac{d\chi}{dt} = k_1(C_{A0} - \chi)$	$t = \frac{1}{k_1} \ln \frac{C_{A0}}{(C_{A0} - \chi)}$ $t_{1/2} = \frac{\ln 2}{k_1}$
Second order $A + B \rightarrow$ products	$\frac{d\chi}{dt} = k_2(C_{A0} - \chi)(C_{B0} - \chi)$	$t = \frac{1}{k_2(C_{B0} - C_{A0})} \ln \frac{C_{A0}(C_{B0} - \chi)}{C_{B0}(C_{A0} - \chi)}$ $C_{A0} \neq C_{B0}$
$2A \rightarrow$ products	$\frac{d\chi}{dt} = k_2(C_{A0} - \chi)^2$	$t = \frac{1}{k_2}\left(\frac{1}{C_{A0} - \chi} - \frac{1}{C_{A0}}\right) t_{1/2} = \frac{1}{k_2 C_{A0}}$
Order p, one reactant $A \rightarrow$ products	$\frac{d\chi}{dt} = k(C_{A0} - \chi)^p$	$t = \frac{1}{k(p-1)}\left[\frac{1}{(C_{A0} - \chi)^{p-1}} - \frac{1}{C_{A0}^{p-1}}\right]$ $t_{1/2} = \frac{2^{p-1} - 1}{k(p-1)C_{A0}^{p-1}} \quad p \neq 1$
Reversible Reactions		
First order Both directions $A \underset{k_r}{\overset{k_f}{\rightleftharpoons}} M$	$\frac{d\chi}{dt} = k_f(C_{A0} - \chi) - k_r(C_{M0} + \chi)$	$t = \frac{1}{(k_f + k_r)} \ln \frac{k_f C_{A0} - k_r C_{M0}}{k_f(C_{A0} - \chi) - k_r(C_{M0} + \chi)}$ or since $K_c = k_f/k_r$ $t = \frac{K_c}{k_f(1 + K_c)} \ln \frac{K_c C_{A0} - C_{M0}}{K_c(C_{A0} - \chi) - (C_{M0} + \chi)}$ If $C_{M0} = 0$ $t = \frac{1}{(k_f + k_r)} \ln \frac{\chi_e}{(\chi_e - \chi)}$
Second order Both directions $A + B \underset{k_r}{\overset{k_f}{\rightleftharpoons}} M + N$	If $C_{A0} = C_{B0}$ and $C_{M0} = C_{N0} = 0$ $\frac{d\chi}{dt} = k_f(C_{A0} - \chi)^2 - k_r \chi^2$	$t = \frac{\sqrt{K_c}}{2k_f C_{A0}} \ln \frac{C_{A0} + \chi[1/(\sqrt{K_c}) - 1]}{C_{A0} - \chi[1/(\sqrt{K_c}) + 1]}$

A further point of interest about the equations shown in Table 1.1 is to compare the shapes of graphs of fractional conversion or moles converted versus time for reactions of different orders p. Fig. 1.9 shows a comparison between first- and second-order reactions involving a single reactant only, together with the straight line for a zero-order reaction. The rate constants have been taken so that the curves coincide at 50% conversion. It may be seen that the rate of the second-order reaction is high at first but falls rapidly with increasing time and, compared with first-order reactions, longer reaction times are required for high conversions. The zero-order reaction is the only one where the reaction rate does not decrease with increasing conversion. Many biological systems have apparent reaction orders between 0 and 1 and will have a behavior intermediate between the curves shown.

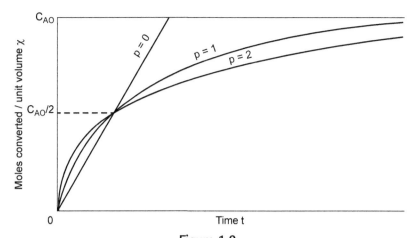

Figure 1.9
Batch reactions at constant volume: comparison of curves for zero-, first-, and second-order reactions.

1.6.2 Reaction Time—Isothermal Operation

If the reactor is to be operated isothermally, the rate of disappearance of A, $-r_A$, can be expressed as a function of concentrations only, and the integration in Eq. (1.55) or 1.57 is carried out. We now consider an example with a rather complicated rate equation involving a reversible reaction and show also how the volume of the batch reactor required to meet a particular production requirement is calculated.

Example 1.2
Production of Ethyl Acetate in a Batch Reactor

Ethyl acetate is to be manufactured by the esterification of acetic acid with ethanol in an isothermal batch reactor. A production rate of 10 tonne/day of ethyl acetate is required.

$$CH_3 \cdot COOH + C_2H_5OH \rightleftharpoons CH_3 \cdot COOC_2H_5 + H_2O$$

$$A \qquad\quad B \qquad\qquad M \qquad\quad N$$

The reactor will be charged with a mixture containing 500 kg/m^3 ethanol and 250 kg/m^3 acetic acid, the remainder being water, and a small quantity of hydrochloric acid to act as a catalyst. The density of this mixture is 1045 kg/m^3 which will be assumed constant throughout the reaction. The reaction is reversible with a rate equation that, over the concentration range of interest, can be written as

$$-r_A = k_f C_A C_B - k_r C_M C_N$$

At the operating temperature of 100°C the rate constants have the following values:

$$k_f = 8.0 \times 10^{-6} \text{ m}^3/\text{kmol/s}$$
$$k_r = 2.7 \times 10^{-6} \text{ m}^3/\text{kmol/s}$$

The reaction mixture will be discharged when the conversion of the acetic acid is 30%. A time of 30 min is required between batches for discharging, cleaning, and recharging. Determine the volume of the reactor required.

Solution

After a time t, if α_A is the fractional conversion of acetic acid (**A**), its concentration will be $C_{A0}(1 - \alpha_A)$ where C_{A0} is the initial concentration (see Eq. 1.58). Thus, at time t, $C_{A0}\alpha_A$ kmol/m^3 of acetic acid has reacted. From the stoichiometry of the reaction, and as per Eq. 1.62, $C_{A0}\alpha_A$ kmol/m^3 of ethanol also will have reacted and the same number of moles of ester and water will have been formed. The rate equation may thus be written as

$$-r_A = k_f C_{A0}(1 - \alpha_A)(C_{B0} - C_{A0}\alpha_A) - k_r(C_{M0} + C_{A0}\alpha_A)(C_{N0} + C_{A0}\alpha_A)$$

which can be rewritten as

$$-r_A = k_f C_{A0}^2(1 - \alpha_A)(C_{B0}/C_{A0} - \alpha_A) - k_r C_{A0}^2(C_{M0}/C_{A0} + \alpha_A)(C_{N0}/C_{A0} + \alpha_A)$$

From the original composition of the mixture, its density and the molecular weights of acetic acid, ethanol, and water which are 60, 46, and 18, respectively, $C_{A0} = 4.2$ kmol/m^3; $C_{B0} = 10.9$ kmol/m^3; $C_{M0} = 0$ kmol/m^3, $C_{N0} = 16.4$ kmol/m^3. Thus, from Eq. (1.55), t_r in seconds is given by

$$t_r = \frac{1}{4.2} \int_0^{\alpha_{Af}} \frac{d\alpha_A}{8.0 \times 10^{-6}(1 - \alpha_A)(10.9/4.2 - \alpha_A) - 2.7 \times 10^{-6}\alpha_A(16.4/4.2 + \alpha_A)}$$

This integral may be evaluated either by splitting into partial fractions, or by graphical or numerical means. Using the method of partial fractions, we obtain after some fairly lengthy manipulation, for a final conversion of 0.3,

$$t_r = 7165.32 \left[\ln \frac{6.845 - \alpha_A}{0.575 - \alpha_A} \right]_0^{0.3} = 4940 \text{ s}.$$

Thus 1 m^3 of reactor volume produces 1.26 kmol of ethyl acetate (molecular weight 88) in a total batch time of 6740 s, i.e., in 4940 s reaction time and 1800 s shutdown time. This is an average production rate of

$$1.26 \times 88 \times \frac{24 \times 60^2}{6740}$$

i.e., 1420 kg/day per m^3 of reactor volume. Because the required production rate is 10,000 kg/day the required reactor volume is $10,000/1420 = \underline{\underline{7.1 \text{ m}^3}}$.

The above example is useful also in directing attention to an important point concerning reversible reactions in general. A reversible reaction will not normally go to completion, but will slow down as equilibrium is approached. This progress towards equilibrium can, however, sometimes be disturbed by continuously removing one or more of the products as

formed. In the actual manufacture of ethyl acetate, the ester is removed as the reaction proceeds by distilling off a ternary azeotrope of molar composition ethyl acetate 60.1%, ethanol
12.4%, and water 27.5%. The net rate of reaction is thereby increased as the rate equation
above shows; because C_M is always small the term for the rate of the reverse reaction $k_r C_M C_N$
is always small and the net rate of reaction is virtually equal to the rate of the forward reaction above, i.e., $k_f C_A C_B$.

1.6.3 Maximum Production Rate

For most reactions, the rate decreases as the reaction proceeds (important exceptions being
a number of biological reactions that are autocatalytic). For a reaction with no volume
change, from Eqs. (1.56) and (1.58), we get

$$-r_A = -\frac{dC_A}{dt} = \frac{d\chi}{dt}.$$

Thus the rate of disappearance of the reactant A, $-r_A$, is represented by the slope of the
curve of χ (moles converted per unit volume) versus time (Fig. 1.10), which decreases
steadily with increasing time. The maximum reaction rate occurs at zero time, and, if our
sole concern were to obtain maximum output from the reactor and the shutdown time were
zero, it appears that the best course would be to discharge the reactor after only a short
reaction time t_r and refill with fresh reactants. It would then be necessary, of course, to
separate a large amount of reactant from a small amount of product. However, if the
shutdown time is appreciable and has a value t_s, then as we have seen in the example on
ethyl acetate above, the average production rate per unit volume is

$$\frac{\chi}{t_r + t_s}$$

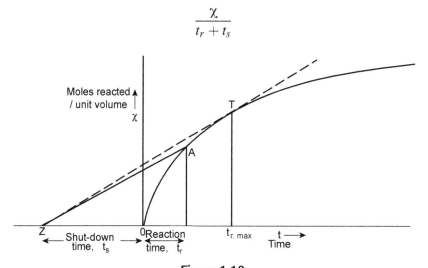

Figure 1.10
Maximum production rate in a batch reactor with a shutdown time t_s.

The maximum production rate is therefore given by the maximum value of

$$\frac{\chi}{t_r + t_s}$$

This maximum can be most conveniently found graphically (Fig. 1.10). The average production rate is given by the slope of the line ZA; this is obviously maximum when the line is tangent to the curve of χ versus t, i.e., ZT as shown. The reaction time obtained $t_{r\ max}$ is not necessarily the optimum for the process as a whole, however. This will depend in addition upon the costs involved in feed preparation, separation of the products, and storage.

1.6.4 Reaction Time—Nonisothermal Operation

If the temperature is not constant but varies during the course of the reaction, then the rate of disappearance of A, $-r_A$, in Eq. (1.55) or 1.57 will be a function of temperature as well as concentration (see Eq. (1.35), for instance) and hence the integrals in those equations cannot be directly evaluated. The temperature at any stage is determined by an integral energy balance, the general form of which is given by Eq. (1.26). Since there is no material flow into or out of a batch reactor during reaction, the enthalpy changes associated with such flows in continuous reactors are absent. However, there may be a flow of heat to or from the reactor by heat transfer using the type of equipment shown in Fig. 1.4. In the case of a jacketed vessel or one with an internal coil, the heat transfer coefficient will be largely dependent on the agitator speed, which is usually held constant. Thus, assuming that the viscosity of the liquid does not change appreciably, which is reasonable in many cases (except for some polymerizations), the heat transfer coefficient may be taken as constant. If heating is effected by condensing saturated steam at constant pressure, as in Fig. 1.11B, the temperature on the coil side T_c is constant. If cooling is carried out with water (Fig. 1.11C), the rise in temperature of the water may be small if

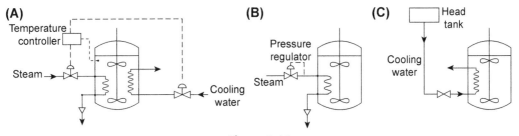

Figure 1.11
Methods of operating batch reactors. (A) Isothermal operation of an exothermic reaction; heating to give required initial temperature, cooling to remove heat of reaction; (B) and (C) nonisothermal operation; simple schemes.

the flow rate is large, and T_c is again taken as a constant. Thus, we may write the rate of heat transfer to cooling coils of area A_t as

$$Q_r = -UA_t(T - Tc) \tag{1.63}$$

where T is the temperature of the reaction mixture. The negative sign is because by the convention that we have adopted, Q_r represents the rate of heat transfer *to* the reactor. The energy balance taken over the whole reactor is represented by Eq. (1.26). Given that we are considering a well-mixed batch reactor in which there is no variation with position of temperature and concentrations, and further neglecting shaft work and pressure variation with time, we get, from Eq. (1.26),

$$\left(\sum_{i=1}^{n} C_i \widetilde{c}_{Pi} \right) \frac{dT}{dt} V_b + \left(\sum_{i=1}^{n} \widetilde{h}_i \frac{dC_i}{dt} \right) V_b = Q_r.$$

Noting that $n_i = C_i V_b$, we rewrite the above equation as

$$\left(\sum_{i=1}^{n} n_i \widetilde{c}_{Pi} \right) \frac{dT}{dt} + \left(\sum_{i=1}^{n} \widetilde{h}_i \frac{dn_i}{dt} \right) = Q_r.$$

Substituting for Q_r from Eq. (1.63) and for dn_i/dt from Eq. (1.59), we get

$$\left(\sum_{i=1}^{n} n_i \widetilde{c}_{Pi} \right) \frac{dT}{dt} = -\left(\sum_{i=1}^{n} r_i \widetilde{h}_i \right) V_b - UA_t(T - Tc). \tag{1.64}$$

The term $\sum_{i=1}^{n} r_i \widetilde{h}_i$ can be written in terms of the rate of generation r_A of a single species A using Eq. (1.60) as follows:

$$\sum_{i=1}^{n} r_i \widetilde{h}_i = \sum_{i=1}^{n} \frac{r_A}{\nu_A} \nu_i \widetilde{h}_i = \frac{-r_A}{-\nu_A} \sum_{i=1}^{n} \nu_i \widetilde{h}_i \tag{1.65}$$

We note that the quantity $\sum_{i=1}^{n} \nu_i \widetilde{h}_i$ may be regarded as the enthalpy change associated with reaction. It is commonly referred to as the "heat of reaction" and denoted by Δh^{rxn}. The quantity $\Delta h^{rxn}/(-\nu_A)$ may be regarded as the enthalpy change of reaction per mole of A and is denoted by Δh_A^{rxn}. Hence

$$\Delta h_A^{rxn} = \frac{\Delta h^{rxn}}{-\nu_A} = \frac{1}{-\nu_A} \sum_{i=1}^{n} \nu_i \widetilde{h}_i \tag{1.66}$$

[Note: The definition of Δh_A^{rxn} involves a sum, yet use of "Δ" indicates a difference. This apparent contradiction is resolved if the sign convention concerning the stoichiometric coefficients is considered. Thus, for a reaction such as 1.27, Δh_A^{rxn} will be given by

$$\Delta h_A^{rxn} = \frac{1}{a} \left(c\widetilde{h}_C + d\widetilde{h}_D - a\widetilde{h}_A - b\widetilde{h}_B \right).$$

Further, the quantities Δh^{rxn} or Δh_A^{rxn} involve pure component enthalpies evaluated at the temperature and pressure of the reaction. On the other hand, the "standard" enthalpy change referred to in thermodynamics textbooks are at the standard pressure (usually 1 bar). Further, the standard state may be either the ideal gas associated with a given component or any other suitably specified standard state. Hence care must be exercised in adapting the values given in the tables in thermodynamics textbooks].

In view of Eqs. (1.65) and (1.66), the energy balance Eq. (1.64) can be written as

$$\left(\sum_{i=1}^{n} n_i \widetilde{c}_{Pi} \right) \frac{dT}{dt} = \Delta h_A^{rxn} r_A V_b - UA_t(T - Tc) \tag{1.67}$$

Finding the time required for a particular conversion involves the solution of two simultaneous equations, i.e., 1.52 or 1.56 for the material balance and 1.67 for the energy balance. Generally, a solution in analytical form is unobtainable and numerical methods must be used. Taking, for example, a first-order reaction in a constant-volume batch reactor, we have for the material balance

$$r_A = \frac{dC_A}{dt} = -A \exp\left(\frac{-E}{RT} \right) C_A$$

and for the energy balance

$$\frac{dT}{dt} = \frac{\left(-\Delta h_A^{rxn} \right) \cdot A \exp\left(\frac{-E}{RT} \right) C_A \cdot V_b - UA_t(T - Tc)}{\left(\sum_{i=1}^{n} n_i \widetilde{c}_{Pi} \right)}$$

These equations represent a system of two coupled first-order differential equations in C_A and T with t as the independent variable. The solution will be as (1) $C_A(t)$ and (2) $T(t)$.

A typical requirement is that the temperature shall not rise above T_{mx} to avoid by-products or hazardous operation. The forms of the solutions obtained are sketched in Fig. 1.12 where instead of the concentration, the variable χ as defined by Eq. (1.58) is plotted versus time.

[*Note*: Typically, the mixture heat capacity $\sum_{i=1}^{n} n_i \widetilde{c}_{Pi}$ or the specific heat $\sum_{i=1}^{n} C_i \widetilde{c}_{Pi}$ is a weak function of temperature or of the composition of the reaction mixture and may be assumed to be a constant. In some cases, the temperature dependence of Δh_A^{rxn} has to be taken into account. This may be done through Eq. (A.4) of the Appendix at the end of this chapter.]

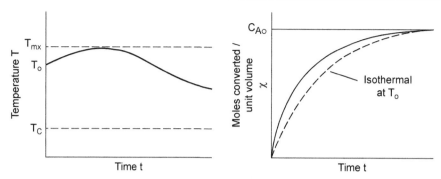

Figure 1.12

Nonisothermal batch reactor. Typical curves for an exothermic reaction with just sufficient cooling (constant U and T_c) to prevent temperature rising above T_{mx}.

1.6.5 Adiabatic Operation

If the reaction is carried out adiabatically (i.e., without heat transfer, so that $Q_r = 0$), the energy balance shows that the temperature at any stage in the reaction can be expressed in terms of the conversion only. This is because, however, fast or slow the reaction, the enthalpy change due to reaction is retained as sensible heat in the reactor. Thus, for reaction at constant volume, substituting for r_A from Eq. (1.56) in Eq. (1.67), discarding the heat transfer term and rewriting n_i as $C_i V_b$, we get

$$\left(\sum_{i=1}^{n} C_i \tilde{c}_{Pi} \right) \frac{dT}{dt} V_b = \Delta h_A^{rxn} \frac{dC_A}{dt} V_b.$$

Thus

$$\frac{dT}{dC_A} = \frac{\Delta h_A^{rxn}}{\sum_{i=1}^{n} C_i \tilde{c}_{Pi}}. \tag{1.68}$$

Eq. (1.68) may be solved to give the temperature as a function of C_A. Usually the change in temperature $(T - T_0)$, where T_0 is the initial temperature, is proportional to C_A because $\sum_{i=1}^{n} C_i \tilde{c}_{Pi}$, the mixture-specific heat, does not vary appreciably with temperature or conversion. The appropriate values of the rate constant are then used to carry out the integration of Eq. (1.55) or 1.57 numerically as shown in the following example.

Example 1.3
Adiabatic Batch Reactor

Acetic anhydride is hydrolyzed by water in accordance with the equation:

$$(CH_3 \cdot CO)_2O + H_2O \rightleftharpoons 2CH_3 \cdot COOH$$

In a dilute aqueous solution where a large excess of water is present, the reaction is irreversible and pseudo–first-order with respect to the acetic anhydride. The variation of the pseudo–first-order rate constant with temperature is as follows:

Temperature	(°C)	15	20	25	30
	(K)	288	293	298	303
Rate constant	(s^{-1})	0.00134	0.00188	0.00263	0.00351

A batch reactor for carrying out the hydrolysis is charged with an anhydride solution containing 0.30 kmol/m^3 at 15°C. The specific heat and density of the reaction mixture are 3.8 kJ/kg/K and 1070 kg/m^3, respectively, and may be taken as constant throughout the course of the reaction. The reaction is exothermic, the heat of reaction per kilomoles of anhydride being $-210,000$ kJ/kmol. If the reactor is operated adiabatically, estimate the time required for the hydrolysis of 80% of the anhydride.

Solution

For the purposes of this example we shall neglect the heat capacity of the reaction vessel. Since the anticipated temperature rise is small, the enthalpy change of reaction will be taken as independent of temperature. Further, from the given data, we have

$$\sum_{i=1}^{n} C_i \tilde{c}_{Pi} = 3.8 \times 1070 = 4066 \text{ kJ/m}^3/\text{K}.$$

Hence, from Eq. (1.68),

$$\frac{dT}{dC_A} = \frac{\Delta h_A^{rxn}}{\sum\limits_{i=1}^{n} C_i \tilde{c}_{Pi}} = \frac{-210000}{4066} = -51.65 \text{ m}^3\text{K/kmol}.$$

Integrating, we get

$$T - T_0 = -51.65(C_A - C_{A0}) = 51.65(C_{A0} - C_A) = 51.65 C_{A0}\alpha_A = 15.5\alpha_A.$$

Thus, if the reaction went to completion ($\alpha_A = 1$), the adiabatic temperature rise would be 15.5°C.

From the given data of the rate constant (k_1, pseudo–first-order) versus temperature, we obtain a fit to the Arrhenius form and obtain the following values:

$$A = 4.15 \times 10^5 \text{s}^{-1}, E = 4.68 \times 10^4 \text{ J/mol}$$

Substituting the required numerical values into Eq. (1.55), with $\mathbf{R} = 8.314$ J/mol, we get

$$t_r = C_{A0} \int_0^{\alpha_{Af}} \frac{d\alpha_A}{-r_A} = \int_0^{0.8} \frac{d\alpha_A}{4.15 \times 10^5 \exp\left[\dfrac{-4.68 \times 10^4}{8.314 \times (288.15 + 15.5\alpha_A)}\right](1 - \alpha_A)}$$

Carrying out the integration numerically, we obtain $t_r \approx 730$ s ≈ 12 min.

1.6.6 Kinetics From Batch Reactor Data

As pointed out in Section 1.5.5, two methods of analysis are adopted in determining kinetics from batch reactor data, the differential method and the integral method. We explain the two methods through examples.

1.6.6.1 Differential Method

Consider an irreversible, liquid-phase reaction of the form 1.27. We assume that the rate of disappearance of A can be expressed as Eq. (1.31). Combining Eq. (1.56), which is valid for a constant-volume batch reactor, with Eq. (1.31), we write

$$-\frac{dC_A}{dt} = kC_A^p C_B^q.$$
(1.69)

Further, it must be noted that C_B is not independent of C_A but given, on account of stoichiometry (Eq. 1.62) by

$$C_B = C_{B0} - \frac{b}{a}(C_{A0} - C_A).$$
(1.70)

A conventional approach to determining kinetics for this reaction is to conduct two types of experiments, one in which the initial amounts of A and B are in the stoichiometric ratio, i.e.,

$$\frac{C_{B0}}{C_{A0}} = \frac{b}{a}$$
(1.71)

and another in which the initial amount of B is much in excess of A, i.e.,

$$C_{B0} \gg \frac{b}{a} C_{A0}.$$
(1.72)

In each experiment, the data is collected as $C_A(t)$, i.e., the concentration of A as a function of time. For the first type of experiment, Eqs. (1.70) and (1.71) imply

$$C_B = \frac{b}{a} C_A.$$
(1.73)

Thus Eq. (1.69) may be written as

$$\ln\left(-\frac{dC_A}{dt}\right) = \ln k + (p+q)\ln C_A + q\ln\frac{b}{a}$$

From the above equation, it is clear that if $\ln(-dC_A/dt)$ is plotted versus $\ln C_A$, we will get a straight line of slope $(p+q)$ which is the overall order of the reaction.

In the second type of experiment characterized by Eq. (1.72), $C_B \simeq C_{B0}$ and hence Eq. (1.69) may be written as

$$\ln\left(-\frac{dC_A}{dt}\right) = \ln k + p \ln C_A + q \ln C_{B0}.$$

Thus, if we plot $\ln(-dC_A/dt)$ versus $\ln C_A$ using the data of this experiment, then the result will be a straight line of slope p. Because $(p + q)$ is known from the first experiment, q can also be found. Then, k can be found from the value of the intercept in either of the plots. Thus, all the parameters in the rate expression are determined.

Alternately, one could perform one set of experiments for an initial condition different from that of Eq. (1.71) or 1.72 and collect C_A (t). Then the parameters k, p, and q could be determined by a parameter estimation method such as least squares method. Also, implicit in the above discussion is the assumption that all the experiments are performed at a single temperature. If one wants to determine the temperature dependence of the rate constant as per Eq. (1.34), one needs to conduct the set of experiments at different temperatures, plot $\ln k$ versus $1/T$ to determine the activation energy and the frequency factor (see Example 1.3).

1.6.6.2 Integral Method

We illustrate this method with a simpler reaction of the form

$$A \rightarrow Products \tag{1.74}$$

and assume that the rate equation is of the form

$$-r_A = kC_A^p \tag{1.75}$$

Combining with Eq. (1.56) for a constant-volume batch reactor, we obtain

$$-\frac{dC_A}{dt} = kC_A^p. \tag{1.76}$$

Integrating Eq. (1.76), we get

$$C_A^{1-p} = C_{A0}^{1-p} - (1 - p)kt, \quad p \neq 1; \quad C_A = C_{A0}e^{-kt}, \quad p = 1. \tag{1.77}$$

In the integral method, we have to assume a value of the order p. For $p \neq 1$, we plot C_A^{1-p} versus t. If the assumed order is correct, then the plot will be a straight line of slope $(p - 1)k$ from which k may be found. For orders greater than one, clearly, the slope would be positive, whereas for orders less than one, it would be negative. If the plot is not a straight line for the chosen value of p, then another value of p is chosen and the procedure continued until a satisfactory fit is obtained. For $p = 1$, a plot of $\ln C_A$ versus time will yield the rate constant. The method can be easily extended to more complicated reactions of the form 1.31.

1.6.6.3 Differential Versus Integral Method: Comparison

The differential method directly gives the kinetic parameters, whereas an iterative procedure is required when using the integral method. However, the differential method requires the evaluation of derivatives (dC_A/dt) from experimental data and this can be inaccurate. A compromise can be found by first using the differential method to get an estimate of the order(s) of the reaction and then using the integral method to fine tune the parameters.

1.6.6.4 Fractional Life Method

For reactions of the form 1.74 and whose kinetics is described by Eq. (1.75), the time required for a fractional conversion of A of α is given by

$$t_\alpha = \frac{1}{k} \frac{C_{A0}^{1-p}\left[1 - (1-\alpha)^{1-p}\right]}{(1-p)}, p \neq 1; \quad t_\alpha = -\frac{1}{k}\ln(1-\alpha), p = 1 \qquad (1.78)$$

The above equations suggest a way of determining the kinetic parameters by measuring the time required for a certain degree of conversion, α, as a function of the initial concentration of the reactant C_{A0}. Typically, $\alpha = 1/2$ is chosen resulting in the method of half-time. It can be easily shown from the above equations that for $p \neq 1$, a plot of $\ln t_{1/2}$ versus $\ln C_{A0}$ would yield a straight line of slope $(1 - p)$. For a first-order reaction, as pointed out in Section 1.6.1, $t_{1/2}$ is independent of C_{A0} and is given by $\ln 2/k$ from Eq. (1.78).

The advantage of the half-life method over the integral method is that the iterative method for determining the order of the reaction is avoided. However, several experimental runs must be conducted, each with a different C_{A0}. To get an estimate of the order from a single run, it is useful to look at the ratios of time required for two different degrees of fractional conversion, say, $t_{\alpha'}/t_{\alpha''}$. For instance, α' could be ½ and α'' could be 1/3 or ¼. These ratios are tabulated for some common reactions[6] from which an estimate of the order of the reaction can be obtained.

A characteristic feature implied by Eq. (1.78) is that reactions for which $p < 1$ go to completion in a finite time while those for which $p \geq 1$ go to completion only asymptotically, i.e., as $t \to \infty$.

1.6.6.5 Kinetics of Gas-Phase Reactions From Pressure Measurements

Often it is not feasible to measure concentrations directly. Typically, for liquid-phase reactions, a property such as conductivity is monitored and correlated with the concentration. For gas-phase reactions, pressure measurement offers a convenient way to monitor the progress of the reaction. We consider a general reaction in which species A is one of the reactants. Then, summing over all the species in Eq. (1.61) leads to

$$n_T(t) = \sum_{i=1}^{n} n_i(t) = \sum_{i=1}^{n} n_{i0} + \frac{[n_{A0} - n_A(t)]}{-\nu_A} \sum_{i=1}^{n} \nu_i = n_{T0} + \frac{[n_{A0} - n_A(t)]}{-\nu_A} \sum_{i=1}^{n} \nu_i$$

where $n_T(t)$ is the total number of moles at time t and $n_{T0} = \sum_{i=1}^{n} n_{i0}$ is the initial total number of moles. Using Eq. (1.53), we obtain

$$n_T(t) = n_{T0} + \frac{n_{A0}\alpha_A}{-\nu_A} \sum_{i=1}^{n} \nu_i = n_{T0} \left[1 + y_{A0} \frac{\sum_{i=1}^{n} \nu_i}{(-\nu_A)} \alpha_A \right] \tag{1.79}$$

where $y_{A0} = n_{A0}/n_{T0}$ is the mole fraction of A in the initial reaction mixture. Defining the "expansion" factor ε_A as

$$\varepsilon_A \equiv y_{A0} \frac{\sum_{i=1}^{n} \nu_i}{(-\nu_A)}, \tag{1.80}$$

we may write Eq. (1.79) as

$$n_T(t) = n_{T0}[1 + \varepsilon_A \alpha_A] \tag{1.81}$$

Assuming an ideal gas mixture for which

$$P = \frac{n_T RT}{V}, \tag{1.82}$$

we can write Eq. (1.81) as

$$P(t) = P_0[1 + \varepsilon_A \alpha_A] \tag{1.83}$$

for an isothermal constant-volume batch reactor. It is to be noted that for a reaction of the form 1.27,

$$\varepsilon_A = y_{A0} \frac{(d+c) - (b+a)}{a} \tag{1.84}$$

with the assumed sign convention for the stoichiometric coefficients.

Example 1.4

We now consider a reaction

$$A \rightarrow B + 2C$$

for which we look for a rate expression of the form 1.75. We assume that only A is present at the beginning of the reaction. Thus $y_{A0} = 1$. For the above reaction conditions

$$\varepsilon_A = 1. \frac{(2+1) - (1)}{1} = 2$$

from Eq. (1.84) and hence

$$\alpha_A = \frac{1}{2}\left(\frac{P}{P_0} - 1\right)$$

from Eq. (1.83). Thus

$$C_A = C_{A0}(1 - \alpha_A) = C_{A0}\frac{3P_0 - P}{2P_0}.$$

Further, $C_{A0} = N_{A0}/V = N_{T0}/V = P_0/\mathbf{R}T$. Hence

$$C_A = \frac{3P_0 - P}{2\mathbf{R}T}, \quad \frac{dC_A}{dt} = -\frac{1}{2\mathbf{R}T}\frac{dP}{dt}.$$

From Eq. (1.76),

$$\frac{dP}{dt} = k(2\mathbf{R}T)^{1-p}(3P_0 - P)^p.$$

Thus if one were to adopt the differential method a plot of $\ln\frac{dP}{dt}$ versus $\ln(3P_0 - P)$ would yield the slope as the order p. If we adopt the integral method, we have to choose a value of p. Suppose $p = 1$. Then

$$\frac{dP}{dt} = k(3P_0 - P).$$

Integrating, we get

$$\ln(2P_0) - \ln(3P_0 - P) = kt.$$

Thus a plot of $\ln(3P_0 - P)$ versus time would yield a straight line if the assumed order is correct. Then the slope would give $-k$. Otherwise, another choice of p must be made and the procedure continued until a proper fit to the pressure versus time data is found.

1.7 Tubular Flow Reactors

The tubular flow reactor (Fig. 1.1B) is chosen when it is desired to operate the reactor continuously but without backmixing of reactants and products. In the case of an *ideal* tubular reactor or PFR, the reaction mixture passes through in a state of *plug flow* which, as the name suggests, means that the fluid moves like a solid plug or piston. Furthermore, in the ideal reactor it is assumed that not only the local mass flux but also the fluid properties, temperature, pressure, and compositions are uniform across any section normal to the fluid motion. Of course the compositions, and possibly the temperature and pressure also, change between inlet and outlet of the reactor in the longitudinal direction. In the elementary treatment of tubular reactors, *longitudinal dispersion*, i.e., mixing by diffusion and other processes in the direction of flow, is also neglected.

Thus, in the idealized tubular reactor all elements of fluid take the same time to pass through the reactor and experience the same sequence of temperature, pressure, and composition changes. In calculating the size of such a reactor, we are concerned with its volume only; its shape does not affect the reaction so long as plug flow occurs.

The flow pattern of the fluid is, however, only one of the criteria which determine the shape eventually chosen for a tubular reactor. The factors which must be taken into account are the following:

1. whether plug flow can be attained,
2. heat transfer requirements,
3. pressure drop in the reactor,
4. support of catalyst, if present, and
5. ease and cheapness of construction.

Fig. 1.13 shows various configurations which might be chosen. One of the cheapest ways of enclosing a given volume is to use a cylinder of height approximately equal to its diameter. In Fig. 1.13A the reactor is a simple cylinder of this kind. Without packing, however, swirling motions in the fluid would cause serious departures from plug flow. With packing in the vessel, such movements are damped out and the simple cylinder is then quite suitable for catalytic reactions where no heat transfer is required. If pressure drop is a problem, the depth of the cylinder may be reduced and its diameter increased as in Fig. 1.13B; to avoid serious departures from plug flow in such circumstances, the catalyst must be uniformly distributed and baffles are often used near the inlet and outlet.

When heat transfer to the reactor is required, a configuration with a high surface to volume ratio is employed. In the reactors shown in Fig. 1.13C and D the reaction volume is made up of a number of tubes. In *c* they are arranged in parallel, whereas in *d* they are in series. The parallel arrangement gives a lower velocity of the fluid in the tubes, which in turn results in a lower pressure drop, but also a lower heat transfer coefficient (which affects the temperature of the reactant mixture and must be taken into account in calculating the reactor volume). The parallel arrangement is very suitable if a second fluid outside the tubes is used for heat transfer; parallel tubes can be arranged between tube sheets in a compact bundle fitted into a shell, as in a shell and tube heat exchanger. On the other hand, with tubes in series, a high fluid velocity is obtained inside the tubes and a higher heat transfer coefficient results. The series arrangement is therefore often the more suitable if heat transfer is by radiation, when the high heat transfer coefficient helps to prevent overheating of the tubes and coke formation in the case of organic materials.

In practice, there is always some degree of departure from the ideal plug flow condition of uniform velocity, temperature, and composition profiles. If the reactor is not packed and the flow is turbulent, the velocity profile is reasonably flat in the region of the turbulent

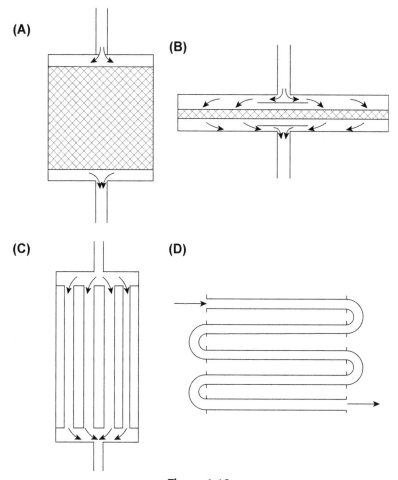

Figure 1.13

Various configurations for tubular reactors. (A) Simple cylindrical shell: suitable only if packed with catalyst. (B) Shallow cylinder giving low pressure drop through catalyst bed. (C) Tubes in parallel: relatively low tube velocity. (D) Tubes in series: high tube velocity.

core , but in laminar flow, the velocity profile is parabolic. More serious, however, than departures from a uniform velocity profile are departures from a uniform temperature profile. If there are variations in temperature across the reactor, there will be local variations in reaction rate and therefore in the composition of the reaction mixture. These transverse variations in temperature may be particularly serious in the case of strongly exothermic catalytic reactions which are cooled at the wall (Chapter 3, Section 3.6.1). An excellent discussion on how deviations from plug flow arise is given by Denbigh and Turner.[7]

1.7.1 Basic Design Equations for a Tubular Reactor

The basic equation for a tubular reactor is obtained by applying the general material balance Eq. (1.14), with the plug flow assumptions. For concreteness, we assume a tubular reactor of constant cross section whose cross-sectional area is given by A_C. Based on the discussion in the preceding section, we may summarize the key features of the PFR as follows:

1. Fluid properties (compositions, temperature, pressure, etc.) are uniform across any section normal to the fluid motion.
2. Mixing by diffusion and other processes in the direction of flow is neglected.
3. Flat velocity profile.

These assumptions may be summarized as follows:

$$C_j = C_j(z), \mathbf{v}_j = v(z)\, \mathbf{e}_z, j = 1, ..., n \qquad (1.85)$$

where the z is the direction of flow and $\mathbf{e}_z$ is the unit vector in that direction. The equality of all the species velocities signifies absence of diffusion. Then, at steady state, the continuity Eq. (1.14) becomes

$$\frac{d}{dz}(C_j v) = r_j \qquad (1.86)$$

We note that the molar flow rate of species j entering the reactor at a distance z along the reactor is given by

$$F_j(z) = -\int_A C_j \mathbf{v}_j \cdot \mathbf{n}\, dA = -\int_A C_j(z) v(z)\, \mathbf{e}_z \cdot (-\mathbf{e}_z)\, dA$$

where the integral is over the cross section of the reactor at a particular position along the length of the reactor. The second equality follows from the plug flow assumptions 1.85 and the fact that the outward normal at any location is opposite to the direction of flow. Given that the integrand is independent of position on the cross section, we get

$$F_j(z) = C_j(z) v(z) A_C. \qquad (1.87)$$

Multiplying and dividing the left-hand side of Eq. (1.86) by A_C, using Eq. (1.87) and noting that $(d/dz) \div A_C = d/dV$, we get the well-known PFR balance equation:

$$\frac{dF_j}{dV} = r_j \qquad (1.88)$$

Typically, for a single reaction, Eq. (1.88) is written for a reactant A. Thus,

$$\frac{dF_A}{dV} = r_A \qquad (1.89)$$

It is convenient to introduce a conversion for flow reactors defined as

$$\alpha_A \equiv 1 - \frac{F_A}{F_{A,in}} \tag{1.90}$$

where $F_{A,in}$ is the molar flow rate of A at the inlet to the reactor. Thus

$$F_A = F_{A,in}(1 - \alpha_A) \tag{1.91}$$

and Eq. (1.89) becomes

$$F_{A,in}\frac{d\alpha_A}{dV} = -r_A \tag{1.92}$$

Integrating, we obtain

$$V_t = F_{A,in} \int_0^{\alpha_{A,out}} \frac{d\alpha_A}{-r_A} \tag{1.93}$$

where $\alpha_{A,out}$ is the required conversion of A to be achieved by the reactor and V_t the volume of the tubular reactor required to achieve the specified conversion.

The conventional way of deriving Eq. (1.93) involves writing the material balance with respect to a reactant **A** over a differential element of volume δV_t (Fig. 1.14). The fractional conversion of **A** in the mixture entering the element is α_A and leaving it is $(\alpha_A + \delta\alpha_A)$. If $F_{A,in}$ is the feed rate of **A** into the reactor (moles per unit time) the material balance over δV_t gives

$$\underset{\text{Inflow}}{F_{A,in}(1-\alpha_A)} - \underset{\text{Outflow}}{F_{A,in}(1-\alpha_A - \delta\alpha_A)} + \underset{\text{Reaction}}{r_A \delta V_t} = 0 \quad \text{(in steady state)}$$

from which Eq. (1.92) can be obtained. The somewhat long-winded procedure that we have adopted brings out more explicitly how the plug flow assumptions lead to the design equation for an ideal tubular reactor.

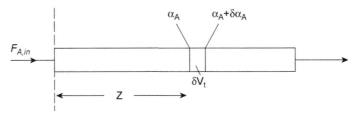

Figure 1.14
Differential element of a tubular reactor.

Dividing Eq. (1.88) by Eq. (1.89) and then using stoichiometry (Eq. 1.29), we get

$$\frac{dF_j}{dF_A} = \frac{r_j}{r_A} = \frac{\nu_j}{\nu_A}. \tag{1.94}$$

On integrating and using the definition for conversion, we get

$$F_j = F_{j,in} + \frac{\nu_j}{(-\nu_A)}\left(F_{A,in} - F_A\right) = F_{A,in}\left[\frac{F_{j,in}}{F_{A,in}} + \frac{\nu_j}{(-\nu_A)}\alpha_A\right]. \tag{1.95}$$

It may be noted that if we substitute $j = A$ in the above equation, we recover the equation for A, namely, Eq. (1.91). To carry out the integration in Eq. (1.93), we need to express $-r_A$ in terms of α_A. Typically, $-r_A$ is expressed in terms of the concentrations of the reacting species.

Thus we need to express C_j in terms of α_A. From Eq. (1.87), we note that

$$C_j = F_j/q_r \tag{1.96}$$

where $q_r = \nu A_C$ is the volumetric flow rate. Thus, we need to express F_j and q_r as functions of α_A. Eq. (1.95) gives the appropriate expression for F_j.

We first consider the special case when the velocity v and hence q_r are independent of z. As noted for the batch reactor, for the special case to be valid, a sufficient condition is constant mass density of the reaction mixture. This may be seen from Eq. (1.13) that reduces, at constant ρ, to

$$\nabla \cdot \mathbf{v} = \frac{dv}{dz} = 0$$

for the PFR. Thus v and hence q_r are independent of z. Then Eq. (1.89) may be written as

$$q_r \frac{dc_A}{dV} = r_A \tag{1.97}$$

where $F_A = C_A q_r$ (Eq. 1.96 for $j = A$) has been used. We may define a quantity called the space time of the reactor as

$$\tau \equiv V_t/q_r \tag{1.98}$$

in terms of which Eq. (1.97) can be written as

$$\frac{dc_A}{d\tau} = r_A \tag{1.99}$$

Integrating the above equation, we obtain an expression for the space time of the reactor from which the required volume can be found:

$$\tau = \frac{V_t}{q_r} = \int_{C_{A,in}}^{C_{A,out}} \frac{dC_A}{r_A} = \int_{C_{A,out}}^{C_{A,in}} \frac{dC_A}{-r_A} \tag{1.100}$$

The concentration of any species can then be written, from Eqs. (1.95) and (1.96), as

$$C_j = \frac{F_j}{q_r} = \frac{F_{A,in}}{q_r}\left[\frac{F_{j,in}}{F_{A,in}} + \frac{v_j}{(-v_A)}\alpha_A\right] = C_{A,in}\left[\frac{C_{j,in}}{C_{A,in}} + \frac{v_j}{(-v_A)}\,\alpha_A\right] \qquad (1.101)$$

Eq. (1.99) has a striking similarity to Eq. (1.56) for a constant-volume batch reactor with τ replacing t. Thus all the design equations derived for the constant-volume batch reactor apply for the PFR with constant q_r with the time (t) in those equations replaced by τ. This case applies to liquid-phase reactions where volumetric changes can be neglected and gives rise to the following physical picture of a PFR. We may imagine that a small volume of reaction mixture is encapsulated by a membrane (Fig. 1.15) in which it is free to expand or contract at constant pressure. It will behave as a miniature batch reactor, spending a time said to be the *residence time,* in the reactor, and emerging with the conversion $\alpha_{A,out}$. If expansion or contraction of a volume element does not occur, then the residence time equals the space time defined by Eq. (1.98).

The other case commonly considered is a gas-phase reaction where the reaction mixture can be approximated as an ideal gas mixture. For such a mixture, a local equation of state appropriate for a flow system (analogous to Eq. 1.82) is

$$P = C_t\,\mathbf{R}\,T \qquad (1.102)$$

where $C_t = \sum_{j=1}^{n} C_j$ is the total concentration of the mixture. Noting Eq. (1.96), we get

$$C_t = \sum_{j=1}^{n} C_j = \frac{1}{q_r}\sum_{j=1}^{n} F_j = \frac{F_t}{q_r} \qquad (1.103)$$

where $F_t = \sum_{j=1}^{n} F_j$ is the total molar flow rate at any location along the length of the reactor. We now apply Eq. (1.103) at two locations, one at the inlet and another at any point along the reactor. Thus we get

$$\frac{C_t}{C_{t,in}} = \frac{F_t}{F_{t,in}}\cdot\frac{q_{r,in}}{q_r} \qquad (1.104)$$

Similarly, from Eq. (1.102), we get

$$\frac{C_t}{C_{t,in}} = \frac{P}{P_{in}}\frac{T_{in}}{T} \qquad (1.105)$$

Volume charged/unit time

Figure 1.15

Tubular reactor: residence time $\tau = V_t/q_r$ if volumetric flow rate constant throughout reactor.

Combining the above two equations, we get

$$\frac{q_r}{q_{r,in}} = \frac{F_t}{F_{t,in}} \cdot \frac{P_{in}}{P} \cdot \frac{T}{T_{in}} \tag{1.106}$$

Rewriting F_j from Eq. (1.95) as

$$F_j = F_{j,in} + F_{A,in}\frac{\nu_j}{(-\nu_A)}\alpha_A \tag{1.107}$$

we get, on summing over all components,

$$F_t = F_{t,in} + F_{A,in}\frac{\sum_{j=1}^{n}\nu_j}{(-\nu_A)}\alpha_A = F_{t,in}\left[1 + y_{A,in}\frac{\sum_{j=1}^{n}\nu_j}{(-\nu_A)}\alpha_A\right] = F_{t,in}(1 + \varepsilon_A\alpha_A) \tag{1.108}$$

where $y_{A,in} = F_{A,in}/F_{t,in}$ is the mole fraction of A in the inlet stream and

$$\varepsilon_A = y_{A,in}\frac{\sum_{j=1}^{n}\nu_j}{(-\nu_A)} \tag{1.109}$$

is defined analogous to that of Eq. (1.80) for a batch reactor. From Eqs. (1.106) and (1.108), we get

$$\frac{q_r}{q_{r,in}} = (1 + \varepsilon_A\alpha_A)\cdot\frac{P_{in}}{P}\cdot\frac{T}{T_{in}} \tag{1.110}$$

Thus, we get

$$C_j = \frac{F_j}{q_r} = \frac{F_{A,in}\left[\dfrac{F_{j,in}}{F_{A,in}} + \dfrac{\nu_j}{(-\nu_A)}\alpha_A\right]}{q_{r,in}(1 + \varepsilon_A\alpha_A)\cdot\dfrac{P_{in}}{P}\cdot\dfrac{T}{T_{in}}} = C_{A,in}\frac{\left[\dfrac{C_{j,in}}{C_{A,in}} + \dfrac{\nu_j}{(-\nu_A)}\alpha_A\right]}{(1 + \varepsilon_A\alpha_A)}\cdot\frac{P}{P_{in}}\cdot\frac{T_{in}}{T} \tag{1.111}$$

which is to be compared to Eq. (1.101) for the constant volumetric flow rate case.

For many tubular reactors, the pressure drop due to flow of the reaction mixture is relatively small, so that the reactor operates at almost constant pressure. If further, the reactor is operated isothermally, then the above equations simplify to

$$\frac{q_r}{q_{r,in}} = (1 + \varepsilon_A\alpha_A) \tag{1.112}$$

and

$$C_j = C_{A,in} \frac{\left[\dfrac{C_{j,in}}{C_{A,in}} + \dfrac{\nu_j}{(-\nu_A)} \alpha_A \right]}{(1 + \varepsilon_A \alpha_A)}. \tag{1.113}$$

For $j = A$, we get

$$C_A = C_{A,in} \frac{(1 - \alpha_A)}{(1 + \varepsilon_A \alpha_A)}. \tag{1.114}$$

We note that for gas-phase reactions where there is no net change in total number of moles between the reactant and product side, $\varepsilon_A = 0$. Hence the volumetric flow rate remains constant across the reactor, and we recover the equations for the constant mass density case applicable to liquid-phase reactions.

We now illustrate the above concepts with an isothermal gas-phase reaction in which expansion of a volume element does occur.

Example 1.5

Production of Ethylene by Pyrolysis of Ethane in an Isothermal Tubular Reactor

Ethylene is manufactured on a very large scale[1] by the thermal cracking of ethane in the gas phase:

$$C_2H_6 \rightleftharpoons C_2H_4 + H_2$$

Significant amounts of CH_4 and C_2H_2 are also formed but will be ignored for the purposes of this example. The ethane is diluted with steam and passed through a tubular furnace. Steam is used for reasons very similar to those in the case of ethylbenzene pyrolysis (Section 1.3.2, Example 1.1): in particular, it reduces the amounts of undesired by-products. The economic optimum proportion of steam is, however, rather less than in the case of ethylbenzene. We will suppose that the reaction is to be carried out in an isothermal tubular reactor that will be maintained at 900°C. Ethane will be supplied to the reactor at a rate of 20 tonne/h; it will be diluted with steam in the ratio 0.5 mol steam:1 mol ethane. The required fractional conversion of ethane is 0.6 (the conversion per pass is relatively low to reduce by-product formation; unconverted ethane is separated and recycled). The operating pressure is 1.4 bar total and will be assumed constant, i.e., the pressure drop through the reactor will be neglected.

Laboratory experiments, confirmed by data from large-scale operations, have shown that ethane decomposition is a homogeneous first-order reaction, the rate constant (s^{-1}) being given by the equation[8]

$$k_1 = 1.535 \times 10^{14} \exp\left(-294,000/RT\right), \quad R \text{ in J/mol/K}.$$

We are required to determine the volume of the reactor.

Solution

The ethane decomposition reaction is in fact reversible, but in the first instance, to avoid undue complication, we shall neglect the reverse reaction; a more complete and satisfactory treatment is given below. For a simple first-order reaction, the rate equation is

$$-r_A \quad = \quad k_1 C_A$$
$$\text{kmol/m}^3\text{s} \quad \text{s}^{-1}\text{kmol/m}^3$$

These units for r_A and C_A will eventually lead to the unit m^3 for the volume of the reactor when we come to use Eq. (1.93). When we substitute for $-r_A$ in Eq. (1.93), however, to integrate, we must express C_A in terms of α_A, where the reactant **A** is C_2H_6. We can directly use Eq. (1.114). For this we note

$$y_{A,in} = \frac{1}{1+0.5} = 2/3, \sum_{j=1}^{C} \nu_j = 1+1-1 = 1, -\nu_A = 1, \varepsilon_A = 2/3.$$

Hence,

$$C_A = C_{A,in}\frac{(1-\alpha_A)}{\left(1+\dfrac{2}{3}\alpha_A\right)}.$$

To find $C_{A,in}$, we note that

$$C_{A,in} = C_t\, y_{A,in} = \frac{P}{RT}y_{A,in} = \frac{1.4 \times 10^5}{8.314 \times 10^3 \times 1173}\cdot\frac{2}{3} = 9.57 \times 10^{-3}\,\text{kmol/m}^3.$$

At 900°C (1173 K) the value of k_1 is 12.8 s^{-1}. The feed rate of ethane (molecular weight 30) is to be 20 tonne/h, which is equivalent to $F_{A,in} = 0.185$ kmol/s. Hence, to find the volume of the reactor, using the basic design Eq. (1.93) and the first-order rate equation:

$$V_t = F_{A,in}\int_0^{\alpha_{A,out}}\frac{d\alpha_A}{-r_A} = \frac{F_{A,in}}{k_1 C_{A,in}}\int_0^{0.6}\frac{\left(1+\dfrac{2}{3}\alpha_A\right)d\alpha_A}{(1-\alpha_A)}.$$

The integral can be evaluated as

$$\int_0^{0.6}\frac{\left(1+\dfrac{2}{3}\,\alpha_A\right)d\,\alpha_A}{(1-\alpha_A)} = \left[-\frac{2}{3}\alpha_A - \frac{5}{3}\ln(1-\alpha_A)\right]_0^{0.6} = 1.1272$$

Introducing the numerical values:

$$V_t = \frac{0.185 \times 1.1272}{12.8 \times 9.57 \times 10^{-3}} = 1.7\text{m}^3.$$

The pyrolysis reaction is strongly endothermic so that one of the main problems in designing the reactor is to provide for sufficiently high rates of heat transfer. The volume calculated above would be made up of a series of tubes, probably in the range 50–150 mm diameter, arranged in a furnace similar to that in Fig. 1.5C. (For further details of ethylene plants. see Miller.[1])

Calculation with reversible reaction. At 900°C the equilibrium constant K_p for ethane decomposition $P_{C2H4}P_{H2}/P_{C2H6}$ is 3.2 bar; using the method described in Example 1.1 the equilibrium conversion of ethane under the conditions above (i.e., 1.4 bar, 0.5 kmol steam added) is 0.86. This shows that the influence of the reverse reaction is appreciable.

The rate equation for a reversible reaction $A \rightleftharpoons M + N$, which is first order in the forward direction and first order with respect to each of M and N in the reverse direction, may be written in terms of the equilibrium constant K_C:

$$-r_A = k_1 \left[C_A - \frac{C_M C_N}{K_C} \right]$$

Since $K_C = (C_M C_N / C_A)_e$ and $C_i = C_t y_i = P y_i / RT = P_i / RT$ from Eq. (1.102), we obtain

$$K_P = (P_M P_N / P_A)_e = K_C RT$$

where P_i's are the partial pressures of the components. Thus

$$-r_A = k_1 \left[C_A - \frac{C_M C_N}{K_P} \cdot RT \right]$$

Further, $\nu_M = \nu_N = 1, C_{M,in} = C_{N,in} = 0$. Hence, using Eq. (1.113),

$$C_M = C_N = C_{A,in} \frac{\alpha_A}{\left(1 + \frac{2}{3}\alpha_A\right)}.$$

Substituting into Eq. (1.93), we get

$$V_t = F_{A,in} \int_0^{\alpha_{A,out}} \frac{d\alpha_A}{-r_A} = \frac{F_{A,in}}{k_1 C_{A,in}} \int_0^{0.6} \frac{\left(1 + \frac{2}{3}\alpha_A\right)^2 d\alpha_A}{(1 - \alpha_A)\left(1 + \frac{2}{3}\alpha_A\right) - \frac{RT}{K_p} C_{A,in}\alpha_A^2}.$$

Introducing the numerical values,

$$\frac{RT}{K_p} C_{A,in} = \frac{8.314 \times 10^3 \times 1173}{3.2 \times 10^5} \times 9.57 \times 10^{-3} = 0.292.$$

The integral can then be evaluated to be 1.22. Hence

$$V_t = \frac{F_{A,in}}{k_1 C_{A,in}} \times 1.22 = \frac{0.185}{12.8 \times 9.57 \times 10^{-3}} \times 1.22 = 1.51 \times 1.22 = 1.84 \text{ m}^3.$$

which may be compared with the previous result above (1.7 m³).

1.7.2 Tubular Reactors—Nonisothermal Operation

In designing and operating a tubular reactor when the enthalpy change of reaction or "heat of reaction" is appreciable, strictly isothermal operation is rarely achieved and usually is not economically justifiable, although the aim may be to maintain the local temperatures within fairly narrow limits. On the assumption of plug flow, the rate of temperature rise or fall along the reactor dT/dz is determined by an energy balance equation. We refer the reader to the Appendix at the end of this chapter for details of the derivation but here give the relevant equations. From Eq. (A.12), we can write the equation for the temperature gradient as

$$\frac{dT}{dz} = \frac{r_A A_C \Delta h_A^{rxn}(T) + 2\pi R q_0(z)}{\sum_{i=1}^{n} F_{i,in} \tilde{c}_{p,i} + F_{A,in} \alpha_A \Delta \tilde{c}_p}. \tag{1.115}$$

where $A_C = \pi R^2$ is the cross-sectional area of the tubular reactor, $\Delta h_A^{rxn}(T)$ is the heat of reaction as defined by Eq. (1.66), $\Delta \tilde{c}_p$ is similarly defined as in Eq. (A.5), and $q_0(z)$ is the external heat flux supplied to the PFR. Eq. (1.115) is valid for constant specific heats and negligible shaft work. This equation is to be solved along with the material balance Eq. (1.92) written in terms of the axial distance z as

$$\frac{d\alpha_A}{dz} = \frac{(-r_A)A_C}{F_{A,in}}. \tag{1.116}$$

The design engineer can arrange for the heat flux $q_0(z)$ to vary with position in the reactor according to the requirements of the reaction. Consider, for example, the pyrolysis of ethane for which a reactor similar to that shown in Fig. 1.5C might be used. The cool feed enters the convection section, the duty of which is to heat the reactant stream to the reaction temperature. As the required reaction temperature is approached, the reaction rate increases, and a high rate of heat transfer to the fluid stream is required to offset the large endothermic heat of reaction. This high heat flux at a high temperature is effected in the radiant section of the furnace. Detailed numerical computations are usually made by splitting up the reactor in the furnace into a convenient number of sections.

From a general point of view, three types of expression for the heat transfer term can be distinguished:

1. *Adiabatic operation*: $q_0(z) = 0$. The heat released in the reaction is retained in the reaction mixture so that the temperature rise along the reactor parallels the extent of the conversion α. The material balance and heat balance equations can be solved in a manner similar to that used in the example on an adiabatic batch reactor (Section 1.6.5, Example 1.3). Adiabatic operation is important in heterogeneous tubular reactors and is considered further under that heading in Chapter 3.
2. *Constant heat transfer coefficient*: This case is again similar to the one for batch reactors (Section 1.6.4) and is also considered further in Chapter 3, under heterogeneous reactors.
3. *Constant heat flux*: If part of the tubular reactor is situated in the radiant section of a furnace, as in Fig. 1.5C, and the reaction mixture is at a temperature considerably lower

than that of the furnace walls, heat transfer to the reactor occurs mainly by radiation, and the rate will be virtually independent of the actual temperature of the reaction mixture. For this part of the reactor $q_0(z)$ may be considered to be virtually constant.

1.7.3 Pressure Drop in Tubular Reactors

For a homogeneous tubular reactor, the pressure drop corresponding to the desired flow rate is often relatively small and does not usually impose any serious limitation on the conditions of operation. The pressure drop must, of course, be calculated as part of the design so that ancillary equipment may be specified. Only for gases at low pressures or, liquids of high viscosity, e.g., polymers, the pressure drop is likely to have a major influence on the design.

In heterogeneous systems, however, the question of pressure drop may be more serious. If the reaction system is a two-phase mixture of liquid and gas, or if the gas flows through a deep bed of small particles, the pressure drop should be checked at an early stage in the design so that its influence can be assessed. The methods of calculating such pressure drops are much the same as those for flow without reaction.

1.7.4 Kinetic Data From Tubular Reactors

In the laboratory, tubular reactors are very convenient for gas-phase reactions, and for any reaction, which is so fast that it is impractical to follow it batchwise. Measurements are usually made when the reactor is operating in a steady state, so that the conversion at the outlet or at any intermediate point does not change with time. For fast reactions particularly, a physical method of determining the conversion, such as ultraviolet or infrared absorption, is preferred to avoid disturbing the reaction. The conversion obtained at the outlet is regulated by changing either the flow rate or the volume of the reactor.

The reactor may be set up either as a differential reactor, in which case concentrations are measured over a segment of the reactor with only a small change in conversion, or as an integral reactor with an appreciable change in conversion. When the integral method is used for gas-phase reactions in particular, the pressure drop should be small; when there is a change in the number of moles between reactants and products, integrated forms of Eq. (1.93), which allow for constant pressure expansion or contraction must be used for interpretation of the results. Thus, for an irreversible first-order reaction of the kind $\mathbf{A} \rightarrow v_M\mathbf{M}$, using a feed of pure $\mathbf{A}$ the integrated form of Eq. (1.93), assuming plug flow, is

$$V_t = \frac{F_{A,in}}{k_1 C_{A,in}} \left[v_M \ln \frac{1}{\left(1 - \alpha_{A,out}\right)} - (v_M - 1)\alpha_{A,out} \right]$$

To obtain basic kinetic data, laboratory tubular reactors are usually operated as closely as possible to isothermal conditions. If, however, a full-scale tubular reactor is to be operated adiabatically, it may be desirable to obtain data on the small scale adiabatically. If the small reactor has the same length as the full-scale version but a reduced cross section, it may be regarded as a longitudinal element of the large reactor and, assuming plug flow applies to both, scaling up to the large reactor is simply a matter of increasing the cross-sectional area in proportion to the feed rate. One of the problems in operating a small reactor adiabatically, especially at high temperatures, is to prevent heat loss because the surface to volume ratio is large. This difficulty may be overcome by using electrical heating elements wound in several sections along the tube, each controlled by a servosystem with thermocouples, which senses the temperature difference between the reaction mixture and the outside of the tube (Fig. 1.16). In this way, the temperature of the heating jacket follows exactly the adiabatic temperature path of the reaction, and no heat is lost or gained by the reaction mixture itself. This type of reactor is particularly valuable in developing heterogeneous packed-bed reactors for which it is still reasonable to assume the plug flow model even for large diameters.

1.8 Continuous Stirred Tank Reactors

The stirred tank reactor in the form of either a single tank, or more often a series of tanks (Fig. 1.1D), is particularly suitable for liquid-phase reactions and is widely used in the organic chemicals industry for medium- and large-scale production. It can form a unit in a continuous process, giving consistent product quality, ease of automatic control, and low

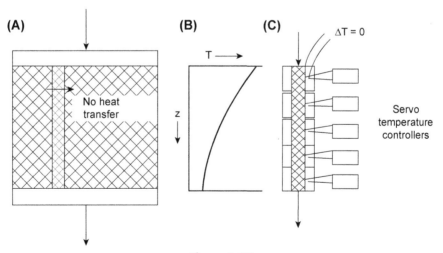

Figure 1.16

Laboratory-scale reproduction of adiabatic tubular reactor temperature profile. (A) Large-scale reactor. (B) Adiabatic temperature profile, endothermic reaction. (C) Laboratory-scale reactor.

manpower requirements. Although, as we shall see below, the volume of a stirred tank reactor must be larger than that of a plug flow tubular reactor for the same production rate, this is of little importance because large volume tanks are relatively cheap to construct. If the reactor has to be cleaned periodically, as happens sometimes in polymerizations or in plant used for manufacturing a variety of products, the open structure of a tank is an advantage.

In a stirred tank reactor, the reactants are diluted immediately on entering the tank; in many cases this favors the desired reaction and suppresses the formation of by-products. Because fresh reactants are rapidly mixed into a large volume, the temperature of the tank is readily controlled, and hot spots are much less likely to occur than in tubular reactors. Moreover, if a series of stirred tanks is used, it is relatively easy to hold each tank at a different temperature so that an optimum temperature sequence can be attained.

1.8.1 Assumption of Ideal Mixing: Residence Time

In the theory of CSTRs, an important basic assumption is that the contents of each tank are *well mixed*. This means that the compositions in the tank are everywhere uniform and that the product stream leaving the tank has the same composition as the mixture within the tank. This assumption is reasonably well borne out in practice unless the tank is exceptionally large, the stirrer inadequate, or the reaction mixture very viscous.

In the treatment that follows, it will be assumed that the mass density of the reaction mixture is constant throughout a series of stirred tanks. Thus, at steady state, if the volumetric feed rate is q_r, then the rate of outflow from each tank will also be q_r. This can be easily seen from the overall mixture balance Eq. (1.7). Applying it to a CSTR at steady state leads to

$$\int_{A_c} \rho \mathbf{v} \cdot \mathbf{n} d\, A = 0.$$

If ρ is a constant, then the above equation leads to $\int_{A_c} \mathbf{v} \cdot \mathbf{n} d\, A = 0$, which may be interpreted as implying that the net volumetric flow rate across the CSTR is zero or equivalently $q_{r,in} = q_{r,out}$. Material balances may then be written on a volume basis, and this considerably simplifies the treatment. In practice, the constancy of the density of the mixture is a reasonable assumption for liquids, and any correction which may need to be applied is likely to be small.

The space time for a CSTR of volume V_C may be defined as V_C/q_r in just the same way as for a tubular reactor. However, in a homogeneous reaction mixture, it is not possible to identify particular elements of fluid as having any particular residence time, because there is complete mixing on a molecular scale. If the feed consists of a suspension of particles, it may be shown that, although there is a distribution of residence times among the individual particles, the mean residence time does correspond to V_C/q_r if the system is ideally mixed.

1.8.2 Design Equations for Continuous Stirred Tank Reactors

When a series of stirred tanks is used as a chemical reactor, and the reactants are fed at a constant rate, eventually the system reaches a steady state such that the concentrations in the individual tanks, although different, do not vary with time. When the general material balance Eq. (1.9) is applied, the accumulation term is therefore zero. Considering first a single reactor, Eq. (1.9) reduces to

$$F_{j,in} - F_{j,out} + r_j V_C = 0, j = 1, \ldots, n \tag{1.117}$$

For a single reaction, the above equation is typically written for one of the reactants A. Thus,

$$F_{A,in} - F_{A,out} + r_A V_C = 0. \tag{1.118}$$

Thus, for a single CSTR,

$$V_C = \frac{F_{A,in} - F_{A,out}}{-r_A} = \frac{F_{A,in}\alpha_{A,out}}{-r_A}, \tag{1.119}$$

where the definition of conversion, Eq. (1.90), appropriate for flow reactors, has been used. Eq. (1.119) is the counterpart of Eq. (1.93) for a tubular reactor.

Writing Eq. (1.117) for an arbitrary species j involved in the reaction and using the implication of stoichiometry (Eq. 1.29), we obtain

$$F_{j,out} = F_{j,in} + r_j V_C = F_{j,in} + \frac{\nu_j}{\nu_A} r_A V_C.$$

Substituting for $r_A V_C$ from Eq. (1.118), we get

$$F_{j,out} = F_{j,in} + \frac{\nu_j}{(-\nu_A)}\left(F_{A,in} - F_{A,out}\right).$$

Applying the definition of conversion, Eq. (1.90), leads to

$$F_{j,out} = F_{j,in} + F_{A,in}\frac{\nu_j}{(-\nu_A)}\alpha_{A,out} \tag{1.120}$$

Eq. (1.120) is analogous to Eq. (1.107) for the PFR. The difference is that Eq. (1.107) represents the continuous variation of F_j within the PFR, whereas Eq. (1.120) applies to the exit of the CSTR.

We now consider a system of CSTRs in series and the most general case in which the mass density of the mixture is not necessarily constant. The material balance on the reactant **A** is made on the basis of $F_{A,in}$ moles of **A** per unit time fed to the first tank. Then a material balance for the rth tank of volume V_{Cr} (Fig. 1.17) is, in the steady state:

$$F_{A,in}(1 - \alpha_{Ar-1}) - F_{A,in}(1 - \alpha_{Ar}) + r_{Ar}V_{Cr} = 0,$$

where α_{Ar-1} is the fractional conversion of **A** in the mixture leaving tank $r - 1$ and entering tank r and α_{Ar} is the fractional conversion of **A** in the mixture leaving tank r.

Assuming that the contents of the tank are well mixed, α_{Ar} is also the fractional conversion of the reactant A in tank r. The volume is given by

$$V_{Cr} = F_{A,in}\frac{\alpha_{Ar} - \alpha_{Ar-1}}{-r_{Ar}} \tag{1.121}$$

Stirred tanks are usually employed for reactions in liquids and, in most cases, the mass density of the reaction mixture may be assumed constant. Material balances may then be taken on the basis of the volume rate of flow q_r which is constant throughout the system of tanks. The material balance on A over tank r may thus be written, in the steady state:

$$q_r C_{Ar-1} - q_r C_{Ar} + r_{Ar}V_{Cr} = 0 \tag{1.122}$$

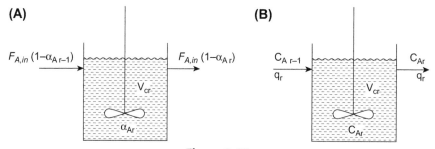

Figure 1.17
Continuous stirred-tank reactor: material balance over rth tank in steady state (A) General case; feed to first tank $F_{A,in}$ (B) Constant volumetric flowrate q_r.

where C_{Ar-1} is the concentration of **A** in the liquid entering tank r from tank $r-1$ and C_{Ar} is the concentration of **A** in the liquid leaving tank r. [*Note*: The subscript r in q_r is used to indicate the *rate* of flow and does not refer to the rth reactor].

$$V_{Cr} = q_r \frac{C_{Ar-1} - C_{Ar}}{-r_{Ar}} \tag{1.123}$$

Eq. (1.123) may be written in terms of the residence time $\tau_r = V_{Cr}/q_r$ as

$$\tau_r = \frac{C_{Ar-1} - C_{Ar}}{-r_{Ar}} \tag{1.124}$$

Eq. (1.124) may be compared with Eq. (1.100) for a tubular reactor. The difference between them is that, whereas 1.100 is an integral equation, 1.124 is a simple algebraic equation. If the reactor system consists of only one or two tanks the equations are fairly simple to solve. If a large number of tanks are employed, the equations whose general form is given by 1.124 constitute a set of finite difference equations and must be solved accordingly.

To proceed with a solution, a rate equation is required for $-r_{Ar}$. Allowance must be made for the fact that the rate constant will be a function of temperature and may therefore be different for each tank. The temperature distribution will depend on the energy balance for each tank, and this will be affected by the amount of heating or cooling that is carried out. The example that follows concerns an isothermal system of two tanks with two reactants, one of which is in considerable excess.

Example 1.6
A Two-Stage Continuous Stirred Tank Reactor

A solution of an ester $R \cdot COOR'$ is to be hydrolyzed with an excess of caustic soda solution. Two stirred tanks of equal size will be used. The ester and caustic soda solutions flow separately into the first tank at rates of 0.004 and 0.001 m^3/s and with concentrations of 0.02 and 1.0 $kmol/m^3$, respectively. The reaction:

$$R \cdot COOR' + NaOH \rightarrow R \cdot COONa + R'OH$$

is second order with a rate constant of 0.033 $m^3/kmol/s$ at the temperature at which both tanks operate. Determine the volume of the tanks required to effect 95% conversion of the ester ($\alpha_{A_2} = 0.95$).

Solution
Although the solutions are fed separately to the first tank, we may for the purpose of argument consider them to be mixed together just prior to entering the tank as shown in

Fig. 1.18. If the ester is denoted by **A**, the caustic soda by **B**, and α_A is the degree of conversion of ester, then Eq. (1.121) written for the two reactors are

$$V_{C1} = F_{A,in}\frac{\alpha_{A1}}{-r_{A1}} \tag{A}$$

$$V_{C2} = F_{A,in}\frac{\alpha_{A2} - \alpha_{A1}}{-r_{A2}} \tag{B}$$

Further,

$$-r_A = k_2 C_A C_B = k_2 C_{A,in}(1 - \alpha_A)(C_{B,in} - C_{A,in}\alpha_A) = k_2 C_{A,in}^2(1 - \alpha_A)(C_{B,in}/C_{A,in} - \alpha_A)$$

For $-r_{A1}$, we need to substitute $\alpha_A = \alpha_{A1}$ in the above equation and for $-r_{A2}$, we need to substitute $\alpha_A = \alpha_{A2}$. Dividing Eq. (A) by Eq. (B) and noting that $V_{C1} = V_{C2}$, we obtain

$$\frac{\alpha_{A2} - \alpha_{A1}}{-r_{A2}} = \frac{\alpha_{A1}}{-r_{A1}}.$$

Substituting for $-r_{A1}$ and $-r_{A2}$, and rearranging, we obtain

$$\alpha_{A1}(1 - \alpha_{A2})(C_{B,in}/C_{A,in} - \alpha_{A2}) = (\alpha_{A2} - \alpha_{A1})(1 - \alpha_{A1})(C_{B,in}/C_{A,in} - \alpha_{A1}) \tag{C}$$

Considering numerical values, the total volume flow rate $q_r = 0.005$ m^3/s.

$$C_{A,in} = C_{A0} = 0.02 \times \frac{0.004}{0.005} = 0.016 \text{ kmol/m}^3$$

$$C_{B,in} = C_{B0} = 1.0 \times \frac{0.001}{0.005} = 0.20 \text{ kmol/m}^3$$

$$\alpha_{A2} = 0.95.$$

Substituting these values into Eq. (C), we get

$$0.5725 \times \alpha_{A1} = (0.95 - \alpha_{A1})(1 - \alpha_{A1})(12.5 - \alpha_{A1}).$$

One may solve the cubic equation by trial and error or by a cubic equation solver.

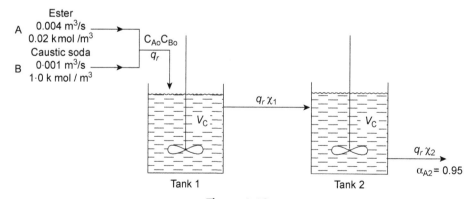

Figure 1.18
Continuous stirred tank reactor: worked example.

Alternately, noting that $\alpha_{A1} < 0.95$, we see that $\alpha_{A1} \ll 12.5$. Thus we get a simplified (approximate) quadratic equation:

$$0.5725 \times \alpha_{A1} = (0.95 - \alpha_{A1})(1 - \alpha_{A1})(12.5)$$

solving which, we get $\alpha_{A1} \approx 0.7764$. The other root would be greater than 1. Using this as an initial guess, we may, by trial and error, find

$$\alpha_{A1} \approx 0.7785.$$

Substituting this value into Eq. (B), we get

$$V_{C2} = V_{C1} = 2.8 \text{ m}^3.$$

1.8.3 Graphical Methods

For second-order reactions, graphs showing the fractional conversion for various residence times and reactant feed ratios have been drawn up by Eldridge and Piret.[9] These graphs, which were prepared from numerical calculations based on Eq. (1.124), provide a convenient method for dealing with sets of equal-sized tanks of up to five in number, all at the same temperature.

A wholly graphical method arising from Eq. (1.123) or 1.124 may be used provided that the rate of disappearance of A, $-r_A$, is a function of a single variable only, either C_A or α_A. The tanks must therefore all be at the same temperature. Experimental rate data may be used directly in graphical form without the necessity of fitting a rate equation. To establish the method, Eq. (1.124) is first rearranged to give

$$-r_{Ar} = -\frac{1}{\tau_r}(C_{Ar} - C_{Ar-1}) \tag{1.125}$$

It must be noted that this equation is valid for the case where there is negligible change in the volumetric flow rate with composition of the reaction mixture. Consider a line of slope $-1/\tau_r$ passing through the point $(C_{Ar-1}, 0)$. The equation of this line is

$$y = -\frac{1}{\tau_r}(x - C_{Ar-1}) \tag{1.126}$$

From Eq. (1.125), it is clear that the point $(C_{Ar}, -r_{Ar})$ lies on this line. Suppose the kinetics is represented by an equation of the form

$$-r_A = f(C_A). \tag{1.127}$$

Then, $-r_{Ar} = f(C_{Ar})$. Thus the point $(C_{Ar}, -r_{Ar})$ lies at the intersection of the line (1.126) and the curve (1.127) which is plotted with $-r_A$ along the y-axis and C_A along the x-axis. Thus the procedure for the graphical determination of the exit concentrations of a multitank series of CSTRs may be carried out as follows: Given the inlet concentration $C_{A,in} \equiv C_{A0}$ to the first reactor and the space times of all the reactors, we start with the point $(C_{A0}, 0)$ and draw a line of slope $-1/\tau_1$. The point where it intersects the curve (1.127), which is assumed to be provided graphically, gives the concentration C_{A1}, the concentration of the stream exiting the first reactor. Then from the point $(C_{A1}, 0)$, we draw a line of slope $-1/\tau_2$ whose intersection with the curve (1.127) gives C_{A2}, and so on for the whole series as the following example shows.

Example 1.7
Graphical Construction for a Three-Stage Continuous Stirred Tank Reactor

1. **A** system of three stirred tanks is to be designed to treat a solution containing 4.0 kmol/m^3 of a reactant **A**. Experiments with a small reactor in the laboratory gave the kinetic data shown as a graph of rate of reaction versus C_A in Fig. 1.19B. If the feed rate to the reactor system is 1.2×10^{-4} m^3/s, what fractional conversion will be obtained if each of the three tanks has a volume of 0.60 m^3?
2. Calculate the volumes of the tanks required for the same overall conversion if two equal tanks are used, and if only one tank is used.

Solution

1. Referring to Fig. 1.19B, from the point O_0 representing the feed composition of 4.0 kmol/m^3, a line O_0R_1 is drawn of slope $-1.2 \times 10^{-4}/0.60 = -2 \times 10^{-4}$ s^{-1} to intersect the rate curve at R_1. This point of intersection gives the concentration of **A** in the first tank C_{A1}. A perpendicular R_1O_1 is dropped from R_1 to the C_A axis, and from the point O_1 a second line also of slope $-2 \times 10^{-4}s^{-1}$ is drawn. The construction is continued until O_3 is reached which gives the concentration of **A** leaving the last tank C_{A3}. Reading from the figure $C_{A3} = 1.23$ kmol/m^3. The fractional conversion is therefore $(4.0-1.23)/4.0 = \underline{0.69}$.
2. When the volumes of the identical tanks are unknown the graphical construction must be carried out on a trial and error basis. The procedure for the case of two tanks is shown in Fig. 1.19C; the points O_0 and O_2 are known, but the position of R_1 has to be adjusted to make O_0R_1 and O_1R_2 parallel because these lines must have the same slope if the tanks are of equal size. From the figure this slope is $1.13 \times 10^{-4}s^{-1}$. The volume of each tank must therefore be $1.2 \times 10^{-4}/1.13 \times 10^{-4} = 1.06$ m^3.

For a single tank the construction is straightforward as shown in Fig. 1.19D and the volume obtained is $\underline{3.16 \text{ m}^3}$.

It is interesting to compare the total volume required for the same duty in the three cases.

$$\text{Total volume for tanks is } 3 \times 0.60 = 1.80 \text{ m}^3$$
$$\text{for 2 tanks } 2 \times 1.06 = 2.12 \text{ m}^3$$
$$\text{for 1 tank is } 1 \times 3.16 = 3.16 \text{ m}^3$$

These results illustrate the general conclusion that, as the number of tanks is increased, the total volume required diminishes and tends in the limit to the volume of the equivalent PFR. The only exception is in the case of a zero-order reaction for which the total volume is constant and equal to that of the PFR for all configurations.

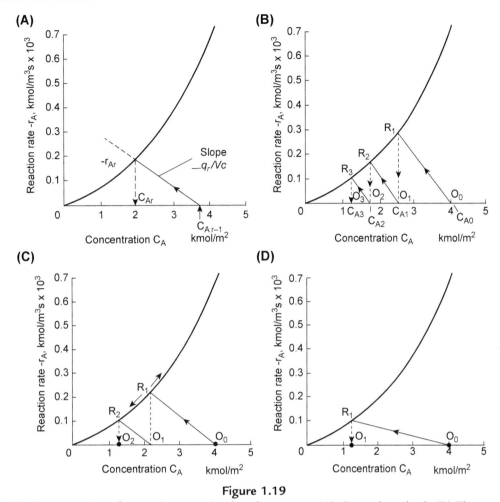

Figure 1.19

Graphical construction for continuous stirred tank reactors. (A) General method. (B) Three equal tanks, outlet concentration C_{A3} unknown. (C) Two equal tanks, volume unknown. (D) One tank, volume unknown.

1.8.4 Autothermal Operation

One of the advantages of the CSTR is the fact that it is ideally suited to autothermal operation. Feedback of the reaction heat from products to reactants is indeed a feature inherent in the operation of a CSTR consisting of a single tank only, because fresh reactants are mixed directly into the products. An important, but less obvious, point about autothermal operation is the existence of two possible stable operating conditions.

To understand how this can occur, we require an energy balance over a single tank operating at steady state. The tank is equipped with a cooling coil of area A_t through which flows a cooling medium at a temperature T_C. The reader is referred to the Appendix for a detailed derivation of the relevant equations. Under steady conditions, assumptions of negligible shaft work, negligible effect of pressure on enthalpy, and constant specific heats, we obtain the following equation based on Eqs. A.6 and A.7 of the Appendix:

$$\sum_{i=1}^{n}\left(F_{i,in}\widetilde{c}_{p,i}\right)(T - T_{in}) + F_{A,in}\alpha_{A,out}\Delta h_{A}^{rxn} + UA_{t}(T - T_{C}) = 0. \qquad (1.128)$$

It must be noted that the exit temperature T_{out} in Eq. (A.7) is replaced by T, the temperature in the reactor, in Eq. (1.128), the two temperatures being equal for a CSTR. Using the material balance Eq. (1.119) and rearranging, we obtain

$$\sum_{i=1}^{n}\left(F_{i,in}\widetilde{c}_{p,i}\right)(T - T_{in}) + UA_{t}(T - T_{C}) = (-r_{A})V_{C}\left(-\Delta h_{A}^{rxn}\right). \qquad (1.129)$$

The term for rate of energy generation due to reaction on the right-hand side of this equation varies with the temperature of operation T, as shown in diagram (A) of Fig. 1.20; as T increases, $-r_{A}$ increases rapidly at first but then tends to an upper limit as the reactant concentration in the tank approaches zero, corresponding to almost complete conversion. On the other hand, the rate of energy removal by both product outflow and heat transfer is virtually linear, as shown in diagram (B). To satisfy the energy balance equation above, the point representing the actual operating temperature must lie on both the rate of energy production curve and the rate of energy removal line, i.e., at the point of intersection as shown in (C).

In Fig. 1.20C, it may be seen how more than one stable operating temperature can sometimes occur. If the rate of energy removal is high (line 1), due to either rapid outflow or a high rate of heat transfer, there is only one point of intersection O_1, corresponding to a low operating temperature close to the reactant inlet temperature $T_{in} = T_0$ or the cooling medium temperature T_C. With a somewhat smaller flow rate or heat transfer rate (line 2), there are three points of intersection corresponding to two stable conditions of operation, O_2 at a low temperature and conversion and O_2' at a considerably higher temperature and conversion. If the reactor is started up from cold, it will settle down in the lower operating state O_2. However, if a disturbance causes the temperature to rise above the intermediate point of intersection beyond which the rate of energy production exceeds the rate of loss, the system will pass into the upper operating state O_2'. Line 3 represents an energy removal rate for which the lower operating state O_3 can only just be realized, and for line 4, only an upper operating temperature O_4' is possible.

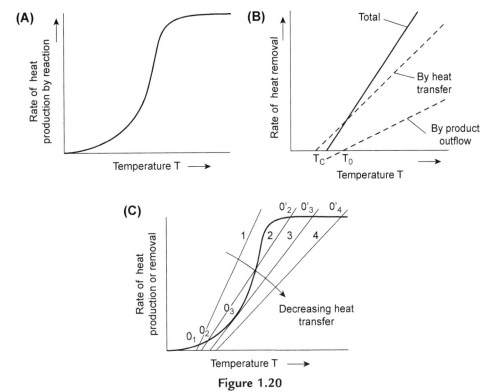

Figure 1.20

Autothermal operation of a continuous stirred tank reactor. (A) and (B) show rates of heat production and removal. (C) shows the effects of different amounts of heat transfer on possible stable operating states.

Obviously, in designing and operating a stirred tank reactor it is necessary to be aware of these different operating conditions. Further discussion of the dynamic response and control of an autothermal CSTR is given by Westerterp et al.[10]

1.8.5 Kinetic Data From Continuous Stirred Tank Reactors

For fast or moderately fast liquid-phase reactions, the stirred tank reactor can be very useful for establishing kinetic data in the laboratory. When a steady state has been reached, the composition of the reaction mixture may be determined by a physical method using a flow cell attached to the reactor outlet, as in the case of a tubular reactor. The stirred tank reactor, however, has a number of further advantages in comparison with a tubular reactor. With an appropriate ratio of reactor volume to feed rate and with good mixing, the difference in reactant composition between feed and outflow can be large, without changing the basic situation whereby all the reaction takes place at a single uniform concentration level in the reactor. Comparatively, large differences between inlet and outlet compositions are required to determine the rate of the reaction with reasonable accuracy. With a tubular reactor, on the other hand, if a large difference in reactant composition is set up across the reactor, the integral method of interpreting the results must be employed, and this may give rise to problems when dealing with complex reactions.

There is one further point of comparison. Interpretation of results from a stirred tank reactor depends on the assumption that the contents of the tank are well mixed. Interpretation of results from a tubular reactor rests on the assumption of plug flow unless the flow is laminar and is treated as such. Which of these two assumptions can be met most satisfactorily in practical experiments? Unless the viscosity of the reaction mixture is high or the reaction extremely fast, a high-speed stirrer is very effective in maintaining the contents of a stirred tank uniform. On the other hand, a tubular reactor may have to be very carefully designed if backmixing is to be completely eliminated, and in most practical situations there is an element of uncertainty about whether the plug flow assumption is valid.

In comparison with the batch reactor, the CSTR has the advantage that the rate corresponding to a particular concentration may be directly obtained from the steady-state data (as seen from Eq. (1.119) or 1.123). On the other hand, for the batch reactor, obtaining rates corresponding to a particular concentration requires differentiation of concentration with time (Eq. 1.56).

1.9 Comparison of Batch, Tubular, and Stirred Tank Reactors for a Single Reaction: Reactor Output

There are two criteria, which can be used to compare the performances of different types of reactors. The first, which is a measure of reactor productivity, is the output of product in relation to reactor size. The second, which relates to reactor selectivity, is the extent to which formation of unwanted by-products can be suppressed. When comparing reactions on the basis of output as in the present section, only one reaction need be considered, but when in the next section the question of by-product formation is taken up, more complex schemes of two or more reactions must necessarily be introduced.

In defining precisely the criterion of reactor output, it is convenient to use one particular reactant as a reference rather than the product formed. The distinction is unimportant if there is only one reaction, but is necessary if more than one reaction is involved. The *unit output*, W_A, of a reactor system may thus be defined as the moles of reactant **A** converted, per unit time, per unit volume of reaction space; in calculating this quantity it should be understood that the total moles of A converted by the whole reactor in unit time is to be divided by the total volume of the system. The unit output is therefore an average rate of reaction for the reactor as a whole and is thus distinct from the specific rate $-r_A$, which represents the local rate of reaction.

For a batch reactor, the quantity W_A, as per the definition above is

$$W_{Ab} = \frac{n_{A0}\alpha_A}{t_r V_b} = \frac{C_{A0}\alpha_A}{t_r} \tag{1.130}$$

where constant volume of the batch is assumed. For a flow reactor (CSTR or PFR),

$$W_{Af} = \frac{F_{A,in}\alpha_A}{V} = \frac{q_r C_{A,in}\alpha_A}{V} = \frac{C_{A,in}\alpha_A}{\tau} \tag{1.131}$$

where again a constant volumetric flow rate across the reactor is assumed.

We shall now proceed to compare the three basic types of reactor—batch, tubular, and stirred tank—in terms of their performance in carrying out a single first-order irreversible reaction:

$$\mathbf{A} \rightarrow \text{Products}$$

It will be assumed that there is no change in mass density and that the temperature is uniform throughout. However, it has already been shown that the conversion in a tubular reactor with plug flow is identical to that in a batch reactor irrespective of the order of the reaction, if the residence time of the tubular reactor τ is equal to reaction time of the batch reactor t_r. Thus, the comparison rests between batch and plug flow tubular reactors on the one hand, and stirred tank reactors consisting of one, two, or several tanks on the other.

1.9.1 Batch Reactor and Tubular Plug Flow Reactor

The material balance for a first-order reaction in a batch reactor is simply Eq. (1.56) applied to a first-order reaction:

$$\frac{dC_A}{dt} = -k_1 C_A$$

In terms of conversion, we get, from Eq. (1.58),

$$\frac{d\alpha_A}{dt} = k_1(1 - \alpha_A) \tag{1.132}$$

The time t_r required for a conversion α_A is given by integrating the above equation:

$$t_r = \frac{1}{k_1} \ln\left(\frac{1}{1 - \alpha_A}\right) \tag{1.133}$$

as shown in Table 1.1 (with $\chi = C_{A0}\alpha_A$). From Eq. (1.130), we obtain

$$W_{Ab} = \frac{k_1 C_{A0}\alpha_A}{\ln\left[1/(1-\alpha_A)\right]} \tag{1.134}$$

$$\frac{W_{Ab}}{k_1 C_{A0}} = \frac{\alpha_A}{\ln\left[1/(1-\alpha_A)\right]} \tag{1.135}$$

1.9.2 Continuous Stirred Tank Reactor

1.9.2.1 One Tank

Applying Eq. (1.124) to a single CSTR and first-order reaction (Fig. 1.21A) leads to

$$\tau_1 = \frac{C_{A,in} - C_{A,out}}{k_1 C_{A,out}} = \frac{C_{A,in} - C_{A,in}(1-\alpha_A)}{k_1 C_{A,in}(1-\alpha_A)} = \frac{\alpha_A}{k_1(1-\alpha_A)}.$$

Substituting into Eq. (1.131), we obtain

$$W_{A,C1} = k_1 C_{A,in}(1-\alpha_A).$$

Hence

$$\frac{W_{A,C1}}{k_1 C_{A,in}} = (1-\alpha_A) \tag{1.136}$$

1.9.2.2 Two Tanks

Applying Eq. (1.124) to each of the two tanks (Fig. 1.21B), we obtain

$$\tau_1 = \frac{C_{A,in} - C_{A1}}{k_1 C_{A1}} = \frac{\alpha_{A1}}{k_1(1-\alpha_{A1})}$$

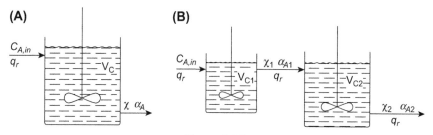

Figure 1.21

Continuous stirred tank reactors: calculation of unit output.

and

$$\tau_2 = \frac{C_{A1} - C_{A,out}}{k_1 C_{A,out}} = \frac{\alpha_{A2} - \alpha_{A1}}{k_1(1 - \alpha_{A2})}.$$

At this point there arises the question of whether the two tanks should be of the same size or different sizes. Mathematically, this question needs to be investigated with some care so that the desired objective function is correctly identified. If we wished to design a two-stage reactor we might be interested in the minimum total volume $V_C = V_{C1} + V_{C2}$ or, equivalently, total space time $\tau = \tau_1 + \tau_2$ required for a given conversion α_{A2}. This condition is met by setting

$$\left(\frac{\partial \tau}{\partial \alpha_{A1}}\right)_{\alpha_{A2}} = 0$$

This leads to

$$\alpha_{A1} = 1 - \sqrt{(1 - \alpha_{A2})}$$

and

$$\tau_1 = \tau_2 = \frac{1}{k_1}\left(\frac{1}{\sqrt{(1 - \alpha_{A2})}} - 1\right). \tag{1.137}$$

In general, it may be shown that the optimum value of the ratio τ_1/τ_2 depends on the order of reaction and is unity only for a first-order reaction. However, the convenience and reduction in costs associated with having all tanks the same size will in practice always outweigh any small increase in total volume that this may entail.

We will assume henceforth that the tanks are of equal size and space time. Thus, we get from Eqs. 1.131 and 1.137,

$$\frac{W_{A,C2}}{k_1 C_{A,in}} = \frac{\alpha_{A2}}{2}\left[\frac{(1 - \alpha_{A2})^{1/2}}{1 - (1 - \alpha_{A2})^{1/2}}\right] \tag{1.138}$$

1.9.3 Comparison of Reactors

It may be seen from Eqs. 1.135, 1.136, and 1.138 that unit output is a function of conversion. Some numerical values of the dimensionless quantity $W_A/k_1 C_{A0}$ representing

Table 1.2: Comparison of continuous stirred tank reactors and batch reactors with respect to unit output $W_A/k_1 C_{A0}$ and reactor volume (first-order reaction).

Reactor Type		Conversion			
		0.50	0.90	0.95	0.99
Batch or tubular plug flow	Unit output	0.722	0.391	0.317	0.215
CSTR one tank	Unit output	0.50	0.10	0.05	0.01
	Volume ratio CSTR/Batch	1.44	3.91	6.34	21.5
CSTR two tanks	Unit output	0.604	0.208	0.137	0.055
	Volume ratio CSTR/Batch	1.19	1.88	2.31	3.91

CSTR, continuous stirred tank reactor.

the unit output are shown in Table 1.2. Shown also in Table 1.2 are values of the following ratios for various values of the conversion:

$$\frac{\text{Unit output batch reactor}}{\text{Unit output stirred} - \text{tank reactor}} = \frac{\text{Volume stirred} - \text{tank reactor}}{\text{Volume batch reacor}}$$

For ease of comparison, we take $C_{A,in} = C_{A0}$ for flow reactors. These show that a single CSTR must always be larger than a batch or tubular plug flow reactor for the same duty, and for high conversions the stirred tank must be very much larger indeed. If two tanks are used, however, the total volume is less than that of a single tank. Although the detailed calculations for systems of three or more tanks are not given here, it can be seen in Fig. 1.22, which is based on charts prepared by Levenspiel,[11] that the total volume is progressively reduced as the number of tanks is increased. This principle is evident also in the example on stirred tank reactors solved by the graphical method used in Example 1.7, which does not refer to a first-order reaction. Calculations such as those in Table 1.2 can be extended to give results for orders of reaction both greater than and less than one. As the order of the reaction increases, so the comparison becomes even less favorable to the stirred tank reactor.

As the number of stirred tanks in a series is increased, so is the total volume of the system reduced. In the limit with an infinite number of tanks, we can expect the volume to approach that of the equivalent batch or tubular reactor because in the limiting case, plug flow is obtained. However, although the total volume of a series of tanks progressively decreases with increasing number, this does not mean that the total cost will continue to fall. The cost of a tank and its associated mixing and heat transfer equipment will be proportional to approximately the 0.6 power of its volume. When total cost is plotted against the number of tanks, as in the second curve of Fig. 1.22, the curve passes through a minimum. This usually occurs in the region of 3—6 tanks and it is most likely that a number in this range will be employed in practice.

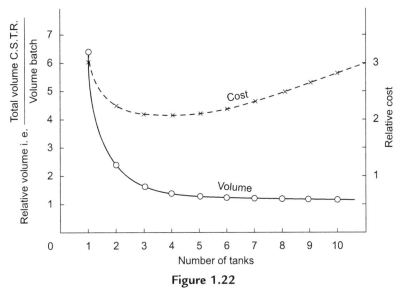

Figure 1.22

Comparison of size and cost of continuous stirred tank reactors with a batch or a tubular plug flow reactor: first-order reaction, conversion 0.95.

1.10 Comparison of Batch, Tubular, and Stirred Tank Reactors for Multiple Reactions: Reactor Yield

If more than one chemical reaction can take place in a reaction mixture, the type of reactor used may have a quite considerable effect on the products formed. The choice of operating conditions is also important, especially the temperature and the degree of conversion of the reactants. The economic importance of choosing the type of reactor, which will suppress any unwanted by-products to the greatest extent has already been stressed (Section 1.1.1).

In this section, our aim will be to take certain model reaction schemes and work out in detail the product distribution which would be obtained from each of the basic types of reactor. It is fair to say that in practice there are often difficulties in attempting to design reactors from fundamental principles when multiple reactions are involved. Information on the kinetics of the individual reactions is often incomplete, and in many instances an expensive and time-consuming laboratory investigation would be needed to fill in all the gaps. Nevertheless, the model reaction schemes examined below are valuable in indicating firstly how such limited information as may be available can be used to the best advantage, and secondly what key experiments should be undertaken in any research and development program.

1.10.1 Types of Multiple Reactions

Multiple reactions are of two basic kinds. Taking the case of one reactant only, these are the following:

1. Reactions in parallel or competing reactions of the type:

$$\mathbf{A} \rightarrow \mathbf{P} \quad \text{(desired product)}$$
$$\mathbf{A} \rightarrow \mathbf{Q} \quad \text{(unwanted product)}$$

2. Reactions in series or consecutive reactions of the type:

$$\mathbf{A} \rightarrow \mathbf{P} \rightarrow \mathbf{Q}$$

where again **P** is the desired product and **Q** the unwanted by-product.

When a second reactant **B** is involved, the situation is basically unchanged in the case of parallel reactions:

$$\mathbf{A} + \mathbf{B} \rightarrow \mathbf{P}$$
$$\mathbf{A} + \mathbf{B} \rightarrow \mathbf{Q}$$

The reactions are thus in parallel with respect to both **A** and **B**. For reactions in series, however, if the second reactant **B** participates in the reaction with the product **P** as well as with **A**, i.e., if

$$\mathbf{A} + \mathbf{B} \rightarrow \mathbf{P}$$
$$\mathbf{P} + \mathbf{B} \rightarrow \mathbf{Q}$$

then although the reactions are in series with respect to **A**, they are in parallel with respect to **B**. In these circumstances, we have

$$\mathbf{B} \, (+\, \mathbf{A} \,) \rightarrow \mathbf{P}$$
$$\mathbf{B} \, (+\, \mathbf{P} \,) \rightarrow \mathbf{Q}$$

As we shall see, however, the series character of these reactions is the more important, because **B** cannot react to give **Q** until a significant amount of **P** has been formed.

More complex reaction schemes can be regarded as combinations of these basic types of individual reaction steps.

1.10.2 Yield and Selectivity

When a mixture of reactants undergoes treatment in a reactor and more than one product is formed, part of each reactant is converted into the desired product, part is converted into undesired products, and the remainder escapes unreacted. The amount of the desired product actually obtained is therefore smaller than the amount expected had all the reactant been transformed into the desired product alone. The reaction is then said to give a certain yield of the desired product. Unfortunately, the term yield has been used by different authors for two somewhat different quantities and care must be taken to avoid confusion. Here these two usages will be distinguished by employing the terms *relative yield* and *operational yield*; in each case the amount of product formed will be expressed in terms of the stoichiometrically equivalent amount of the reactant **A** from which it was produced.

The *relative yield* Φ_A is defined by

$$\Phi_A = \frac{\text{Moles of } \mathbf{A} \text{ transformed into desired product}}{\text{Total moles of } \mathbf{A} \text{ which have reacted}} \tag{1.139}$$

The relative yield is therefore a net yield based on the amount of **A** actually consumed.

The operational yield Θ_A is defined by

$$\Theta_A = \frac{\text{Moles of } \mathbf{A} \text{ transformed into desired product}}{\text{Total moles of } \mathbf{A} \text{ fed to the reactor}} \tag{1.140}$$

It is based on the total amount of reactant **A** entering the reactor, irrespective of whether it is consumed in the reaction or passes through unchanged.

Both these quantities are fractions, and it follows from the definitions above that Φ_A always exceeds Θ_A, unless all the reactant is consumed, when they are equal.

If unreacted **A** can be recovered from the product mixture at low cost and then recycled, the relative yield is the more significant, and the reactor can probably be operated economically at quite a low conversion per pass. If it cannot be recovered and no credit can be allotted to it, the operational yield is the more relevant, and the reactor will probably have to operate at a high conversion per pass.

Another way of expressing product distribution is the *selectivity* of the desired reaction. Once more expressing the amount of product formed in terms of the amount of **A** reacted, the selectivity is defined as

$$\frac{\text{Moles of } \mathbf{A} \text{ transformed into the desired product}}{\text{Moles of } \mathbf{A} \text{ transformed into unwanted products}}$$

It is thus a product ratio and can have any value, the higher the better. It is often used to describe catalyst performance in heterogeneous reactions.

1.10.3 Reactor Type and Backmixing

When more than one reaction can occur, the extent of any backmixing of products with reactants is one of the most important factors in determining the yield of the desired product. In a well-stirred batch reactor, or in an ideal tubular reactor with plug flow, there is no backmixing, whereas in a single CSTR there is complete backmixing. Intermediate between these two extremes are systems of two or more CSTRs in series and nonideal tubular reactors in which some degree of backmixing occurs (often termed longitudinal dispersion).

Backmixing in a reactor affects the yield for two reasons. The first and most obvious reason is that the products are mixed into the reactants; this is undesirable if the required product is capable of reacting further with the reactants to give an unwanted product, as in some series reactions. The second reason is that backmixing affects the level of reactant concentration at which the reaction is carried out. If there is no backmixing, the concentration level is high at the start of the reaction and has a low value only toward the end of the reaction. With backmixing, as in a single CSTR, the concentration of reactant is low throughout. As we shall see, for some reactions, high reactant concentrations favor high yields, whereas for other reactions low concentrations are more favorable.

If two reactants are involved in a reaction, high concentrations of both, at least initially, may be obtained in a batch or tubular plug flow reactor and low concentrations of both in a single CSTR. In some circumstances, however, a high concentration of reactant **A** coupled with a low concentration of reactant **B** may be desirable. This may be achieved in a number of ways:

1. *Without recycle*: For continuous operation a cross-flow type of reactor may be used as illustrated in Fig. 1.2. The reactor can consist of either a tubular reactor with multiple injection of **B** (Fig. 1.2B), or a series of several stirred tanks with the feed of **B** divided between them (Fig. 1.2C). If a batch type of reactor were preferred, the semibatch mode of operation would be used (Fig. 1.2A). Reaction without recycle is the normal choice where the cost of separating unreacted **A** from the reaction mixture is high.
2. *With recycle*: If the cost of separating **A** is low, then a large excess of **A** can be maintained in the reactor. A single CSTR will provide a low concentration of **B**, while the large excess of **A** ensures a high concentration of **A**. Unreacted **A** is separated and recycled as shown in Fig. 1.23 (Section 1.10.5).

1.10.4 Reactions in Parallel

Let us consider the case of one reactant only but different orders of reaction for the two reaction paths. i.e.,

$$\mathbf{A} \xrightarrow{k_P} \mathbf{P} \text{ desired product}$$

$$\mathbf{A} \xrightarrow{k_Q} \mathbf{Q} \text{ unwanted product}$$

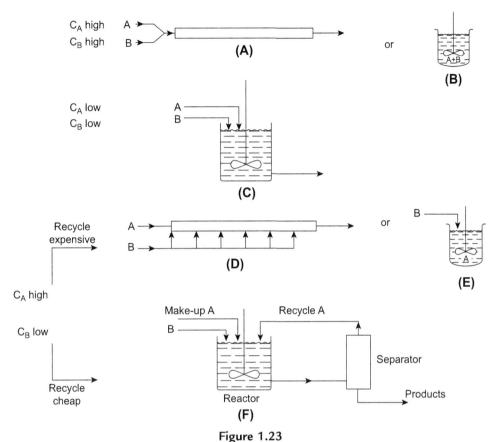

Figure 1.23
Contacting schemes to match possible concentration levels required for high relative yields in parallel reactions. (A) Plug flow tubular reactor. (B) Batch reactor. (C) Single continuous stirred tank reactor. (D) Cross-flow tubular reactor. (E) Semibatch (or fed-batch) reactor. (F) Single continuous stirred tank reactor with recycle of A. Alternatively a series of several continuous stirred tanks could be used in place of the tubular reactors in (A) and (D).

with the corresponding rate equations:

$$r_P = k_P C_A^p \tag{1.141}$$
$$r_Q = k_Q C_A^q \tag{1.142}$$

In these equations it is understood that C_A may be (1) the concentration of **A** at a particular time in a batch reactor, (2) the local concentration in a tubular reactor operating in a steady state, or (3) the concentration in a stirred tank reactor, possibly one of a series, also in a steady state. The relative yield under the circumstances may be called *the instantaneous* or *point yield* ϕ_A because C_A will change (1) with time in the batch reactor or (2) with position in the tubular reactor.

Thus

$$\phi_A = \frac{r_P}{-r_A} = \frac{r_P}{(r_P + r_Q)} = \frac{k_P C_A^p}{k_P C_A^p + k_Q C_A^q}$$

or

$$\phi_A = \left[1 + \frac{k_Q}{k_p} C_A^{(q-p)}\right]^{-1} \tag{1.143}$$

Similarly, the local selectivity is given by

$$\frac{r_P}{r_Q} = \frac{k_P}{k_Q} C_A^{(p-q)} \tag{1.144}$$

To find the overall relative yield Φ_A, i.e., the yield obtained at the end of a batch reaction or at the outlet of a tubular reactor, we note that for reactions with negligible change in mass density, for a batch reactor (see Eq. 1.56)

$$\frac{dC_j}{dt} = r_j$$

whereas for a tubular reactor (see Eq. 1.99)

$$\frac{dC_j}{d\tau} = r_j$$

In either case, we obtain

$$\frac{dC_i}{dC_j} = \frac{r_i}{r_j}$$

for any two species i and j. Thus we get

$$\frac{dC_P}{dC_A} = \frac{r_P}{r_A} = -\phi_A.$$

Integrating the above equation, we obtain

$$C_{Pf} - C_{P0} = -\int_{C_{A0}}^{C_{Af}} \phi_A dC_A = \int_{C_{Af}}^{C_{A0}} \phi_A dC_A$$

for a batch reactor and

$$C_{P,out} - C_{P,in} = -\int_{C_{A,in}}^{C_{A,out}} \phi_A dC_A = \int_{C_{A,out}}^{C_{A,in}} \phi_A dC_A$$

for a PFR. From Eq. (1.139), we get, for the overall relative yield

$$\Phi_{A,batch} = \frac{N_{Pf} - N_{P0}}{N_{A0} - N_{Af}} = \frac{C_{Pf} - C_{P0}}{C_{A0} - C_{Af}} = \frac{1}{C_{A0} - C_{Af}} \int_{C_{Af}}^{C_{A0}} \phi_A dC_A \qquad (1.145)$$

$$\Phi_{A,PFR} = \frac{F_{P,out} - F_{P,in}}{F_{A,in} - F_{A,out}} = \frac{C_{P,out} - C_{P,in}}{C_{A,in} - C_{A,out}} = \frac{1}{C_{A,in} - C_{A,out}} \int_{C_{A,out}}^{C_{A,in}} \phi_A dC_A \qquad (1.146)$$

The overall relative yield thus represents an average of the instantaneous value ϕ_A over the whole concentration range. For a stirred tank reactor consisting of a single tank in a steady state, Eqs. (1.117) and (1.118) apply. Hence

$$\Phi_{A,CSTR} = \frac{F_{P,out} - F_{P,in}}{F_{A,in} - F_{A,out}} = \frac{r_P}{-r_A} = \phi_A \qquad (1.147)$$

where r_P and $-r_A$ are evaluated at the outlet concentrations, which are equal to the corresponding reactor concentrations. Thus the overall yield is the same as the instantaneous yield evaluated at the reactor/outlet concentrations. If more than one stirred tank is used, however, an appropriate average must be taken.

For the specific kinetics represented by Eqs. (1.141) and (1.142) for which Eq. (1.143) applies, we get

$$\Phi_{A,batch} = \frac{1}{(C_{A0} - C_{Af})} \int_{C_{Af}}^{C_{A0}} \frac{dC_A}{1 + \frac{k_Q}{k_P} C_A^{(q-p)}} \qquad (1.148)$$

$$\Phi_{A,PFR} = \frac{1}{(C_{A,in} - C_{A,out})} \int_{C_{A,out}}^{C_{A,in}} \frac{dC_A}{1 + \frac{k_Q}{k_P} C_A^{(q-p)}} \qquad (1.149)$$

$$\Phi_{A,CSTR} = \frac{1}{1 + \frac{k_Q}{k_P} C_{A,out}^{q-p}}. \qquad (1.150)$$

1.10.4.1 Requirements for High Yield

1.10.4.1.1 Reactant Concentration and Reactor Type

Although Eqs. (1.148) and (1.149) gives the exact value of the final yield obtainable from a batch or tubular reactor, the nature of the conditions required for a high yield can be seen more readily from Eq. (1.143). If $p > q$, i.e., the order of the desired reaction is higher than that of the undesired reaction, a high yield ϕ_A will be obtained when C_A is high. A batch reactor, or a tubular reactor, gives a high reactant concentration at least initially and should therefore be chosen in preference to a single stirred tank reactor in which reactant concentration is low. If the stirred tank type of reactor is chosen on other grounds, it should consist of several tanks in series. In operating the reactor, any recycle streams which might dilute the reactants should be avoided. Conversely, if $p < q$, a high yield is favored by a low reactant concentration and a single stirred tank reactor is the most suitable. If a batch or tubular reactor were nevertheless chosen, dilution of the reactant by a recycle stream would be an advantage. Finally, if $p = q$, the yield will be unaffected by reactant concentration.

1.10.4.1.2 Pressure in Gas-Phase Reactions

If a high reactant concentration is required, i.e., $p > q$, the reaction should be carried out at high pressure and the presence of inert gases in the reactant stream should be avoided. Conversely, if $p < q$ and low concentrations are required, low pressures should be used.

1.10.4.1.3 Temperature of Operation

Adjusting the temperature affords a means of altering the ratio k_P/k_Q, provided that the activation energies of the two reactions are different.

Thus

$$\frac{k_P}{k_Q} = \frac{A_P}{A_Q} \exp\left[-\frac{(\mathbf{E}_P - \mathbf{E}_Q)}{\mathbf{R}T} \right] \tag{1.151}$$

Clearly, if $\mathbf{E}_P > \mathbf{E}_Q$, then k_P/k_Q increases with increase in temperature. Thus increasing the temperature will favor the formation of the desired product P. On the other hand, if $\mathbf{E}_P < \mathbf{E}_Q$, lowering the temperature will favor the formation of P relative to Q.

1.10.4.1.4 Choice of Catalyst

If a catalyst can be found which will enable the desired reaction to proceed at a satisfactory rate at a temperature which is sufficiently low for the rate of the undesired reaction to be negligible, this will usually be the best solution of all to the problem.

1.10.4.2 Yield and Reactor Output

The concentration at which the reaction is carried out affects not only the yield but also the reactor output. If a high yield is favored by a high reactant concentration, there is no conflict because the average rate of reaction and therefore the reactor output will be also high. However, if a high yield requires a low reactant concentration, the reactor output will be low. An economic optimum must be sought, balancing the cost of reactant wasted in undesired by-products against the initial cost of a larger reactor. In most cases the product distribution is the most important factor, especially (1) when raw material costs are high and (2) when the cost of equipment for the separation, purification, and recycle of the reactor products greatly exceeds the cost of the reactor.

1.10.5 Reactions in Parallel—Two Reactants

If a second reactant **B** is involved in a system of parallel reactions, then the same principles apply to **B** as to **A**. The rate equations are examined to see whether the order of the desired reaction with respect to **B** is higher or lower than that of the undesired reaction and to decide whether high or low concentrations of **B** favor a high yield of desired product.

There are three possible types of combination between the concentration levels of **A** and **B** that may be required for a high yield:

1. Both C_A and C_B high: In this case a batch or a tubular plug flow reactor is the most suitable.
2. Both C_A and C_B low: A single CSTR is the most suitable.
3. C_A high and C_B low (or C_A low and C_B high). A cross-flow reactor is the most suitable for continuous operation without recycle, and a semibatch reactor for batchwise operation. If the reactant required in high concentration can be easily recycled, a single CSTR can be used.

These ways of matching reactor characteristics to the concentration levels required are illustrated in Fig. 1.23.

Example 1.8
Two reactants undergo parallel reactions as follows:

$$\mathbf{A} \quad + \quad \mathbf{B} \quad \rightarrow \quad \mathbf{P} \text{ (desired product)}$$
$$2\mathbf{B} \quad \rightarrow \quad \mathbf{Q} \text{ (unwanted product)}$$

with the corresponding rate equations:

$$r_P = k_P C_A C_B,$$
$$r_Q = k_Q C_B^2 / 2.$$

Suggest suitable continuous contacting schemes that will give high yields of **P** (1) if the cost of separating **A** is high and recycling is not feasible, (2) if the cost of separating **A** is low and recycling can be employed. For the purpose of quantitative treatment, set $k_P = k_Q$. The desired conversion of reactant **B** is 0.95.

Solution
Inspection of the rate equations shows that with respect to **A**, the order of the desired reaction is unity, and the order of the undesired reaction is effectively zero because **A** does not participate in it. The desired reaction is therefore favored by high values of C_A. With respect

to **B**, the order of the desired reaction is unity, and the order of the unwanted reaction is two. The desired reaction is therefore favored by low values of C_B.

1. If recycling is not feasible, a cross-flow type of reactor will be the most suitable, the feed of **B** being distributed between several points along the reactor. Cross-flow reactors for this particular reaction system have been studied in considerable detail. Fig. 1.24 shows some results obtained with reactors employing five equidistant feed positions, together with the performances of a single CSTR and of a straight tubular reactor for the purpose of comparison. In case (D) the amounts of **B** fed at each point in the tubular reactor have been calculated to give the maximum yield of desired product. It may be seen from (B), however, that there is little disadvantage in having the feed of **B** distributed in equal parts. Furthermore, the five sections of the tubular reactor can be replaced by five stirred tanks, as in (C), without appreciably diminishing the yield although the total reactor volume is somewhat greater. By way of contrast, a simple tubular reactor (E) gives a substantially lower yield.

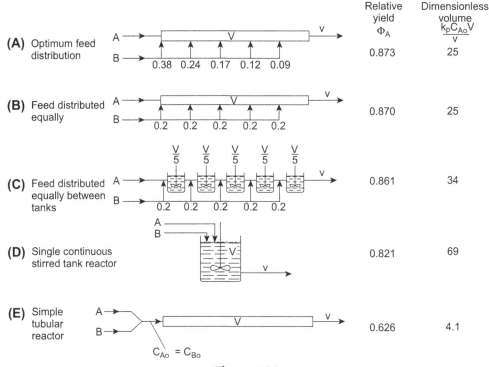

Figure 1.24

Performance of cross-flow reactors with five equidistant feed points.

Parallel reaction: $\mathbf{A} + \mathbf{B} \rightarrow \mathbf{P}$ $r_P = k_P C_A C_B$

$$2\mathbf{B} \rightarrow \mathbf{Q} \quad r_Q = k_Q C_B^2 / 2$$

with $k_P = k_Q$. Equal molar feed rates of **A** and **B**. Final conversion of **B** $= 0.95$. $v \equiv q_r$, the volumetric flow rate.

2. If **A** can be recycled, a high concentration of **A** together with a low concentration of **B** can be maintained in a single CSTR, as shown in Fig. 1.23. By suitable adjustment of the rate of recycle of **A** and the corresponding rate of outflow, the ratio of concentrations in the reactor $C_A/C_B = r'$ may be set to any desired value. The relative yield based on **B**, Φ_B, will then be given by

$$\Phi_B = \frac{k_P C_A C_B}{k_P C_A C_B + k_Q C_B^2}$$

If $k_P = k_Q$:

$$\Phi_B = \frac{C_A}{C_A + C_B} = \frac{r'}{1 + r'}$$

Even if **A** can be separated from the product mixture relatively easily as the recycle rate and hence r' is increased, the operating costs will increase and the volumes of the reactor and separator must also be increased. These costs have to be set against the cash value of the increased yield of desired product as r' is increased. Thus, the optimum setting of the recycle rate will be determined by an economic balance.

1.10.6 Reactions in Series

When reactions in series are considered, it is not possible to draw any satisfactory conclusions without working out the product distribution completely for each of the basic reactor types. The general case in which the reactions are of arbitrary order is more complex than for parallel reactions. Only the case of two first-order reactions will therefore be considered:

$$\mathbf{A \to P \to Q}$$

where **P** is the desired product and **Q** is the unwanted product. Further,

$$-r_A = k_{11} C_A; \quad r_Q = k_{12} C_P$$

1.10.6.1 Batch Reactor or Tubular Plug Flow Reactor

Let us consider unit volume of the reaction mixture in which concentrations are changing with time: this unit volume may be situated in a batch reactor or moving in plug flow in a tubular reactor. Material balances on this volume give the following equations:

$$-\frac{dC_A}{dt} = k_{11} C_A; \quad \frac{dC_P}{dt} = k_{11} C_A - k_{12} C_P; \quad \frac{dC_Q}{dt} = k_{12} C_P$$

From the first of the equations, we obtain $C_A = C_{A0}\exp(-k_{11}t)$. Substituting this expression into the second of the equations, we can solve for C_P using the method of integrating factor. If $C_P = 0$ when $t = 0$, the concentration of **P** at any time t is given by

$$C_P = C_{A0}\frac{k_{11}}{k_{12} - k_{11}}[\exp(-k_{11}t) - \exp(-k_{12}t)], k_{11} \neq k_{12}. \tag{1.152}$$

Differentiation and setting $dC_P/dt = 0$ shows that C_P passes through a maximum given by

$$\frac{C_{Pmax}}{C_{A0}} = \left(\frac{k_{11}}{k_{12}}\right)^{\frac{k_{12}}{k_{12}-k_{11}}} \tag{1.153}$$

which occurs at a time

$$t_{max} = \frac{\ln(k_{12}/k_{11})}{(k_{12} - k_{11})} \tag{1.154}$$

The relationships 1.152–1.154 are plotted in Fig. 1.25 for various values of the ratio k_{12}/k_{11}.

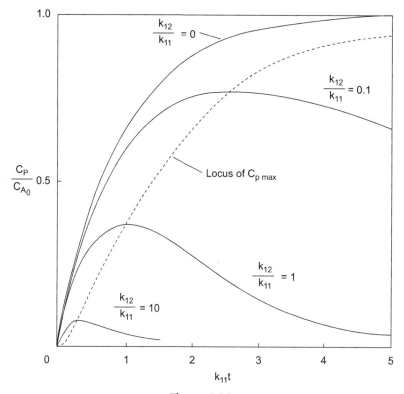

Figure 1.25

Reaction in series—batch or tubular plug flow reactor. Concentration C_P of intermediate product **P** for consecutive first-order reactions, **A** → **P** → **Q**.

Note: For $k_{11} = k_{12} \equiv k_1$, Eqs. (1.152)–(1.154) reduce to

$$C_P = C_{A0}k_1 t \exp(-k_1 t), \quad C_{P\max}/C_{A0} = 1/e, \quad t_{\max} = 1/k_1.$$

1.10.6.2 Continuous Stirred Tank Reactor—One Tank

Taking material balances in the steady state as shown in Fig. 1.26 and applying Eq. (1.122) for a single CSTR, we obtain:

1. on **A**:

$$q_r C_{A0} - q_r C_A - V_C k_{11} C_A = 0$$

2. on **P**:

$$0 - q_r C_P - V_C(k_{12}C_P - k_{11}C_A) = 0$$

hence

$$\frac{C_P}{C_{A0}} = \frac{k_{11}\tau}{(1 + k_{11}\tau)(1 + k_{12}\tau)} \tag{1.155}$$

$$\frac{C_Q}{C_{A0}} = \frac{k_{11}k_{12}\tau^2}{(1 + k_{11}\tau)(1 + k_{12}\tau)} \tag{1.156}$$

where τ is the residence time ($\tau = V_C/q_r$). C_P passes through a maximum in this case also. From $dC_P/d\tau = 0$, we obtain

$$\frac{C_{P\max}}{C_{A0}} = \frac{1}{\left[(k_{12}/k_{11})^{1/2} + 1\right]^2} \tag{1.157}$$

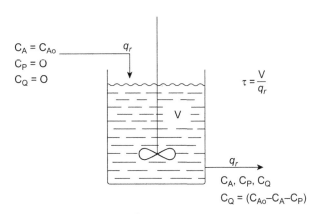

Figure 1.26

Continuous stirred tank reactor: single tank, reactions in series, **A** → **P** → **Q**.

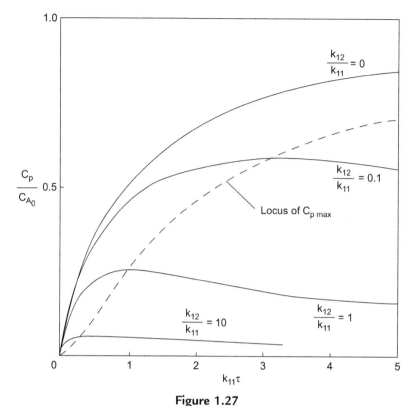

Figure 1.27
Reactions in series—single continuous stirred tank reactor. Concentration C_P of intermediate product **P** for consecutive first-order reactions, $\mathbf{A} \rightarrow \mathbf{P} \rightarrow \mathbf{Q}$.

at a residence time τ_{max} given by

$$\tau_{max} = (\boldsymbol{k}_{11}\boldsymbol{k}_{12})^{-1/2} \tag{1.158}$$

These relationships are plotted in Fig. 1.27.

1.10.6.3 Reactor Comparison and Conclusions

The curves shown in Figs. 1.25 and 1.27, which are curves of operational yield versus reduced time, can be more easily compared by plotting the relative yield of **P** against conversion of **A** as shown in Fig. 1.28. It is then apparent that the relative yield is always greater for the batch or plug flow reactor than for the single stirred tank reactor and decreases with increasing conversion. We may therefore draw the following conclusions regarding the choice of reactor type and mode of operation.

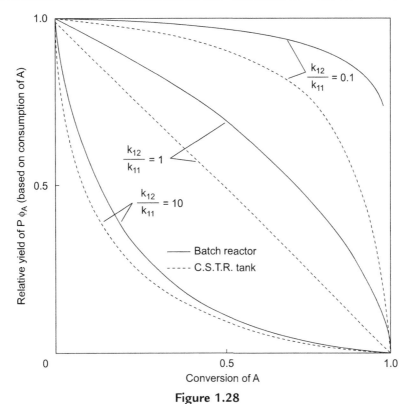

Figure 1.28

Reactions in series—comparison between batch or tubular plug flow reactor and a single continuous stirred tank reactor. Consecutive first-order reactions, **A** → **P** → **Q**.

1.10.6.3.1 Reactor Type

For the highest relative yield of **P**, a batch or tubular plug flow reactor should be chosen. If a CSTR is adopted on other grounds, several tanks should be used in series so that the behavior may approach that of a plug flow tubular reactor.

1.10.6.3.2 Conversion in Reactor

If $k_{12}/k_{11} \gg 1$, Fig. 1.28 shows that the relative yield falls sharply with increasing conversion, i.e., **P** reacts rapidly once it is formed. If possible, therefore, the reactor should be designed for a low conversion of **A** per batch or pass in a tubular reactor, with separation of **P** and recycling of the unused reactant. Product separation and recycle may be quite expensive, however, in which case we look for the conversion corresponding to the economic optimum.

1.10.6.3.3 Temperature

We may be able to exercise some control over the ratio k_{12}/k_{11} by sensible choice of the operating temperature. If $E_1 > E_2$, a high temperature should be chosen, and conversely a low temperature if $E_1 < E_2$. When a low temperature is required for the best yield, there arises the problem that reaction rates and reactor output decrease with decreasing temperature, i.e., the size of the reactor required increases. The operating temperature will thus be determined by an economic optimum. There is the further possibility of establishing a temperature variation along the reactor. For example, if for the case of $E_1 < E_2$ two stirred tanks were chosen, the temperature of the first tank could be high to give a high production rate but the second tank, in which the concentration of the product **P** would be relatively large, could be maintained at a lower temperature to avoid excessive degradation of **P** to **Q**.

1.10.6.3.4 General Conclusions

In series reactions, as the concentration of the desired intermediate **P** builds up, so the rate of degradation to the second product **Q** increases. The best course would be to remove **P** continuously as soon as it was formed by distillation, extraction, or a similar operation. If continuous removal is not feasible, the conversion attained in the reactor should be low if a high relative yield is required. As the results for the CSTR show, backmixing of a partially reacted mixture with fresh reactants should be avoided.

1.10.7 Reactions in Series—Two Reactants

The series—parallel type of reaction outlined in Section 1.10.1 is quite common among industrial processes. For example, ethylene oxide reacts with water to give monoethylene glycol, which may then react with more ethylene oxide to give diethylene glycol.

$$H_2O + C_2H_4O \rightarrow HO \cdot C_2H_4 \cdot OH$$
$$HO \cdot C_2H_4 \cdot OH + C_2H_4O \rightarrow HO \cdot (C_2H_4O)_2 \cdot H$$

i.e.,

$$A + B \rightarrow P$$
$$P + B \rightarrow Q$$

In such cases the order with respect to **B** is usually the same for both the first and the second reaction. Under these circumstances, the level of concentration of **B** at which the reactions are carried out has no effect on the relative rates of the two reactions, as may be seen by writing these as parallel reactions with respect to **B (as in Section 1.10.1)**.

These reactions will therefore behave very similarly to the reactions in series above where only one reactant was involved.

i.e.,

$$A \xrightarrow{+B} P \xrightarrow{+B} Q$$

The same general conclusions apply; since backmixing of products with reactants should be avoided, a tubular plug flow reactor or a batch reactor is preferred. However, there is one respect in which a series reaction involving a second reactant **B** does differ from simple series reaction with one reactant, even when the orders are the same. This is in the stoichiometry of the reaction; the reaction cannot proceed completely to the product **Q**, even in infinite time, if less than 2 mol of **B** per mole of **A** are supplied. Some control over the maximum extent of the reaction can therefore be achieved by choosing the appropriate ratio of **B** to **A** in the feed. For reactions which are first order in **A**, **B**, and **P**, charts[11] are available showing yields and end points reached for various feed ratios.

1.11 Appendix: Simplified Energy Balance Equations for Flow Reactors

Our starting point is Eq. (1.20), which is simplified further by neglecting shaft work:

$$\sum_{i=1}^{n} F_{i,in}\widetilde{h}_{i,in} - \sum_{i=1}^{n} F_{i,out}\widetilde{h}_{i,out} + Q_r = 0. \tag{A.1}$$

We first focus on the CSTR. Substituting for $F_{i,out}$ from Eq. (1.120), using Eq. (1.66) and rearranging, we obtain

$$\sum_{i=1}^{n} F_{i,in}\left(\widetilde{h}_{i,out} - \widetilde{h}_{i,in}\right) + F_{A,in}\alpha_{A,out}\Delta h_A^{rxn} - Q_r = 0. \tag{A.2}$$

It must be noted that Δh_A^{rxn} in Eq. (A.2) must be evaluated at the exit temperature which is equal to the temperature in the CSTR. Neglecting pressure effects on the enthalpy, we may express the difference in enthalpies as

$$\widetilde{h}_{i,out} - \widetilde{h}_{i,in} = \int_{T_{in}}^{T_{out}} \widetilde{c}_{p,i}(T)dT = \widetilde{c}_{p,i}(T_{out} - T_{in}) \tag{A.3}$$

with the second equality valid for constant specific heats. Typically, Δh_A^{rxn} is available at some reference temperature T_R. To obtain it at any other temperature, we note that

$$\widetilde{h}_i(T) = \widetilde{h}_i(T_R) + \int_{T_R}^{T} \widetilde{c}_{p,i}(T')dT'.$$

From Eq. (1.66), we may show that

$$\Delta h_A^{rxn}(T) = \Delta h_A^{rxn}(T_R) + \int_{T_R}^{T} \Delta\widetilde{c}_p(T')dT' \tag{A.4}$$

where

$$\Delta\tilde{c}_p = \frac{1}{-\nu_A}\sum_{i=1}^{n}\nu_i\tilde{c}_{p,i}. \tag{A.5}$$

When the specific heats are independent of temperature, Eq. (A.4) simplifies to

$$\Delta h_A^{rxn}(T) = \Delta h_A^{rxn}(T_R) + \Delta\tilde{c}_p(T - T_R). \tag{A.6}$$

We now write the simplified energy balance equation by substituting Eqs. A.3, A.6, and (1.63) for Q_r into Eq. (A.2):

$$\left(\sum_{i=1}^{n}F_{i,in}\tilde{c}_{p,i}\right)(T_{out} - T_{in}) + F_{A,in}\alpha_{A,out}\left[\Delta h_A^{rxn}(T_R) + \Delta\tilde{c}_p(T_{out} - T_R)\right]$$
$$+ UA_t(T_{out} - T_C) = 0 \tag{A.7}$$

Eq. (A.7) is to be solved along with the material balance Eq. (1.119). If the volume of the reactor and the inlet conditions are specified, then the exit conversion and temperature can be found. On the other hand, if the exit conversion is specified, then the volume of the CSTR and the exit temperature can be found from the equations. For adiabatic operation, Eq. (A.7) with $U = 0$ applies.

For a PFR, we apply Eq. (A.1) to an envelope including the inlet and a particular location z along the length of the reactor. Thus

$$\sum_{i=1}^{n}F_{i,in}\tilde{h}_{i,in} - \sum_{i=1}^{n}F_i(z)\tilde{h}_i(z) + Q_r(z) = 0. \tag{A.8}$$

For the first two terms involving sums in Eq. (A.8), we may follow steps that lead from Eqs. A.1−A.7. However, Q_r, which is the total heat transferred to the PFR over the chosen envelope, is to be expressed as an integral of the heat flux q_0 over the portion of the reactor between the inlet and the location z along the axis. Thus, we obtain the analog of Eq. (A.7) as

$$\left(\sum_{i=1}^{n}F_{i,in}\tilde{c}_{p,i}\right)[T(z) - T_{in}] + F_{A,in}\alpha_A(z)\left[\Delta h_A^{rxn}(T_R) + \Delta\tilde{c}_p(T(z) - T_R)\right]$$
$$- 2\pi R\int_0^z q_0(z')dz' = 0 \tag{A.9}$$

which is the simplified equation for constant specific heats. The quantity $2\pi R q_0$ is sometimes represented as dQ_r/dz. Differentiating Eq. (A.9) with respect to z, we obtain

$$\left[\sum_{i=1}^{n}F_{i,in}\tilde{c}_{p,i} + F_{A,in}\alpha_A\Delta\tilde{c}_p\right]\frac{dT}{dz} + F_{A,in}\Delta h_A^{rxn}(T)\frac{d\alpha_A}{dz} - 2\pi R q_0(z) = 0. \tag{A.10}$$

From Eq. (1.92)

$$F_{A,in}\frac{d\alpha_A}{dz} = A_C F_{A,in}\frac{d\alpha_A}{dV} = A_C(-r_A) \tag{A.11}$$

where $A_C = \pi R^2$ is the cross-sectional area of the tubular reactor. Thus Eq. (A.10) can be written as

$$\left[\sum_{i=1}^{n} F_{i,in}\widetilde{c}_{p,i} + F_{A,in}\alpha_A\Delta\widetilde{c}_p\right]\frac{dT}{dz} + (-r_A)A_C\Delta h_A^{rxn}(T) - 2\pi R q_0(z) = 0 \tag{A.12}$$

which is the required local energy balance for the PFR.

Nomenclature

		Units in SI System	Dimensions
A	Frequency factor in rate equation	*	*
A_t	Area of cooling coils	m^2	L^2
A_c	Surface area of control volume	m^2	L^2
A_C	Area of cross section of a tubular reactor	m^2	L^2
C	Molar density of reaction mixture	$kmol/m^3$	NL^{-3}
C_j	Molar concentration of species j	$kmol/m^3$	NL^{-3}
$\widetilde{c}_{p,i}$	Molar specific heat of component i	$J/kmol/K$	$MN^{-1}L^2T^{-2}\theta^{-1}$
$E_{I,K,P}$	Energy internal (I), kinetic (K), potential (P)	J	ML^2T^{-2}
F_j	Molar flow rate of species j	$kmol/s$	NT^{-1}
H	Enthalpy	J	ML^2T^{-2}
$\widetilde{h}_i$	Specific enthalpy of component i	$J/kmol$	$MN^{-1}L^2T^{-2}$
Δh_A^{rxn}	Enthalpy change of reaction per mole of **A**	$J/kmol$	$MN^{-1}L^2T^{-2}$
K_a	Activity-based equilibrium constant	*	*
K_c	Concentration-based equilibrium constant	*	*
K_p	Partial pressure—based equilibrium constant	*	*
k	Reaction rate constant	*	*
m_j	Mass of species j in a reactor or control volume	kg	M
$\mathbf{n}$	Outward unit normal vector	–	–
n_j	Moles of species j in a reactor or control volume	$kmol$	N
P	Total pressure	N/m^2	$ML^{-1}T^{-2}$
P_j	Partial pressure of species j	N/m^2	$ML^{-1}T^{-2}$
p	Order of reaction	–	–
Q_r	Rate of heat transfer to a system	W	ML^2T^{-3}
q	Order of reaction	–	–
$q_{0(z)}$	Heat flux supplied to a plug flow reactor	W/m^2	MT^{-3}
q_r	Volumetric flow rate	m^3/s	L^3T^1

Continued

		Units in SI System	Dimensions
R	Gas constant	J/kmol/K	$MN^{-1}L^2T^{-2}\theta^{-1}$
R_j	Rate (mass) of generation of species j through chemical reactions	kg/m^3/s	$ML^{-3}T^{-1}$
r_j	Rate (moles) of generation of species j through chemical reactions	kmol/m^3/s	$NL^{-3}T^{-1}$
T	Absolute temperature	K	θ
T_C	Temperature of coolant	K	θ
T_{mx}	Maximum safe operating temperature	K	θ
t	Time	s	T
t_r	Reaction time, batch reactor	s	T
t_s	Shutdown time, batch reactor	s	T
$t_{1/2}$	Half-life of reaction	s	T
U	Overall heat transfer coefficient	W/m^2/K	$MT^{-3}\theta^{-1}$
V	Volume of reaction mixture	m^3	L^3
V_c	Control volume	m^3	L^3
V_b	Volume of a batch reactor	m^3	L^3
V_C	Volume of a continuous stirred tank reactor	m^3	L^3
V_{Cr}	Volume of rth tank in a series of continuous stirred tank reactors	m^3	L^3
V_t	Volume of a tubular reactor	m^3	L^3
$\tilde{v}_i$	Molar specific volume of component i	m^3/kmol	$L^3 N^{-1}$
v	Velocity of a pure component or mass average velocity of a mixture	m/s	LT^{-1}
$\mathbf{v}_i$	Velocity of species i	m/s	LT^{-1}
W_A	Unit output of a reactor with respect to reactor **A**	kmol/m^3 s	$NL^{-3}T^{-1}$
W_{Ab}	Unit output for batch reactor	kmol/m^3 s	$NL^{-3}T^{-1}$
W_{Ac}	Unit output for a continuous stirred tank reactor	kmol/m^3 s	$NL^{-3}T^{-1}$
W_r	Rate of work done on the system	W	ML^2T^{-3}
x_i	Mole fraction of component i	—	—
y_A	Mole fraction of A	—	—
z	Distance along tubular reactor	m	L
α	Fractional conversion	—	—
α_A	Fractional conversion of reactant **A**	—	—
ρ_i	Density of species i	kg/m^3	ML^{-3}
θ	Operational yield of desired product	—	—
ν_j	Stoichiometric coefficient of component j	—	—
τ	Residence time	s	T
Φ	Relative yield of desired product, overall	—	—
ϕ	Relative yield of desired product, instantaneous	—	—
χ	Moles of reactant transformed in unit volume of reaction mixture	kmol/m^3	NL^{-3}
$\tilde{e}_i$	Specific internal energy of component i	J/kmol	$MN^{-1}L^2T^{-2}$
ε_A	Expansion factor for gas-phase reactions	—	—

Subscripts	
0	Initial condition in a batch reactor
e, eq	Equilibrium
f	Final condition in a batch reactor/forward reaction
in	Inlet condition in a flow reactor
out	Outlet condition in a flow reactor
r	Reverse reaction, rth tank in a series of continuous stirred tank reactors

*, indicates that these dimensions are dependent on order of reaction.

References

1. Miller SA, editor. *Ethylene and its industrial derivatives*. Benn; 1969.
2. Denbigh KG. *Principles of chemical equilibria*. 2nd ed. 1966. Cambridge.
3. Bird RB, Stewart WE, Lightfoot EN. *Transport phenomena*. 2nd ed. Wiley; 2002.
4. Moore JW, Pearson RG. *Kinetics and mechanism*. 3rd ed. Wiley; 1981.
5. Laidler KJ. *Chemical kinetics*. 3rd ed. Harper and Row; 1987.
6. Hill Jr CG. *An introduction to chemical engineering kinetics and reactor design*. Wiley; 1977.
7. Denbigh KG, Turner JCR. *Chemical reactor theory*. 3rd ed. 1984. Cambridge.
8. Schutt HC. Light hydrocarbon pyrolysis. *Chem Eng Prog* 1959;**55**(1):68.
9. Eldridge JW, Piret EL. Continuous flow stirred tank reactor systems. *Chem Eng Prog* 1950;**46**:290.
10. Westerterp KR, van Swaaij WPM, Beenackers AACM. *Chemical reactor design and operation*. 2nd ed. Wiley; 1984.
11. Levenspiel O. *Chemical reaction engineering*. 2nd ed. Wiley; 1972.

Further Reading

Aris R. Elementary chemical reactor analysis. In: *Butterworths reprint series*; 1989.

Barton J, Rogers R. *Chemical reaction hazards*. Rugby, UK: Institution of Chemical Engineers; 1993.

Fogler HS. *Elements of chemical reaction engineering*. 2nd ed. Prentice-Hall; 1992.

Froment GF, Bischoff KB. *Chemical reactor analysis and design*. 2nd ed. Wiley; 1989.

Holland CD, Anthony RG. *Fundamentals of chemical reaction engineering*. 2nd ed. Prentice-Hall; 1992.

Lapidus L, Amundson NR. In: *Chemical reactor theory—a review*. Prentice-Hall; 1977.

Levenspiel O. *Chemical reactor omnibook*. Corvallis, Oregon: OSU Book Stores; 1989.

Nauman EB. *Chemical reactor design*. Wiley; 1987.

Rase HF. *Chemical reactor design for process plants*, vols. 1 and 2. Wiley; 1977.

Smith JM. *Chemical engineering kinetics*. 3rd ed. McGraw-Hill; 1981.

Walas SM. Reaction kinetics for chemical engineers. In: *Butterworths reprint series*; 1989.

Walas SM. Chemical process equipment. In: *Selection and design*. Butterworths; 1988 [Chapter 17—Chemical Reactors].

CHAPTER 2

Flow Characteristics of Reactors—Flow Modeling

Ramamurthy Ravi

Indian Institute of Technology Madras, Chennai, India

Learning Outcomes

1. A clear understanding of the definitions and the physical significance of the distribution functions, $F(t)$ and $E(t)$, used to characterize nonideal flow and their determination from experiment through step and impulse responses (Section 2.1). The central role played by the convolution formula connecting the inlet and outlet tracer concentration profiles is emphasized.

2. A detailed understanding of the effects of mixing, kinetics, and nonideality on reactor conversion within the context of the zero-parameter models: the completely segregated and the maximum mixedness models (Section 2.2).

3. A systematic derivation and study of the distribution functions in the tanks-in-series model; calculation of reactor conversion from the model (Section 2.3).

4. Physical principles behind the dispersed-flow model, properties of the mathematical solutions, calculation of dispersion number from experiment, and application to calculation of reactor conversion (Section 2.4).

2.1 Nonideal Flow and Residence Time Distribution

2.1.1 Types of Nonideal Flow Patterns

So far we have developed calculation methods only for the ideal cases of plug flow in tubular reactors (Chapter 1, Section 1.7) and complete mixing in stirred tank reactors (Chapter 1, Section 1.8). In reality, the flow of fluids in reactors is rarely ideal, and although, for some reactors, design equations based on the assumption of ideal flow give acceptable results, in other cases the departures from the ideal flow state need to be taken into account. Following the development by Danckwerts[1] of the basic ideas, one of the leading contributors to the subject of nonideal flow has been Levenspiel whose papers and books,[2–4] especially *Chemical Reaction Engineering*[2] Chapters 11–16, should be consulted.

Coulson and Richardson's Chemical Engineering. http://dx.doi.org/10.1016/B978-0-08-101096-9.00002-9

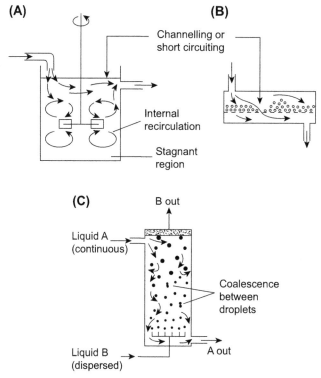

Figure 2.1

Examples of nonideal flow in chemical reactors. (A) Continuous stirred tank. (B) Gas—solid reaction with maldistribution of solid on a shallow tray. (C) Liquid—liquid reaction: Note the circulation patterns in the continuous phase **A** induced by the rising droplets of B; coalescence may occur between phase **B** droplets giving a range of sizes and upward velocities.

It is possible to distinguish various types of nonideal flow patterns in reactors (and process vessels generally), the most important being *channeling, internal recirculation*, and the presence of *stagnant regions*. These are illustrated in the examples shown in Fig. 2.1. In two-phase (and three-phase) reactors the flow patterns in one phase interact with the flow patterns in the other phase, as in the case of the liquid—liquid reactor of Fig. 2.1C. One of the major problems in the scale-up of reactors is that the flow patterns often change with a change of scale, especially in reactors involving two or more phases where flow interactions may occur. A gas—solid fluidized bed is another example of an important class of reactor whose characteristics on scale-up are difficult to predict.[4]

In this chapter, we introduce the tools for treating nonideal flow patterns in chemical reactors. The basic concepts in residence time distribution (RTD) are introduced first. Then, the distribution functions are introduced using ideas from probability theory and independent of the experimental methods commonly used to determine them. The cumulative distribution function $F(t)$ is chosen as the basis, and the RTD function $E(t)$ is shown to follow naturally from it (Section 2.1.2). Experimental methods of determining these distribution functions using a tracer are then discussed. The central result, the convolution formula relating the inlet and outlet tracer concentration for arbitrary inlet concentration profile and reactor configuration, is derived. Then the formulae for the outlet tracer concentration for the two most common experimental inputs—the impulse and step inputs—are derived. It is shown (Section 2.1.3) how the functions E and F introduced in Section 2.1.2 may be obtained from these responses. These results are then applied to obtain the well-known formulae for these functions for the ideal continuous stirred tank reactor (CSTR) and plug flow reactor (PFR) (Section 2.1.4).

In addition to the information on RTD, a model for mixing of molecules of different residence times is needed, in general, to predict conversions achievable by nonideal flow reactors. In this context, zero- and one-parameter models are introduced, and the formulae required to predict reactor conversions in each case are derived. In the case of the zero-parameter models (Section 2.2), the completely segregated and maximum mixedness models are discussed in detail. A distinguishing feature of the treatment here is the replacement of qualitative arguments on the effects of mixing and kinetics by precise results based on the work of Zwietering.[5] In the case of the one-parameter models, the well-known tanks-in-series model (Section 2.3) and the dispersion model (Section 2.4) are discussed. The chapter ends with a brief discussion of two-parameter models.

2.1.2 Residence Time Distribution: Basic Concepts and Definitions

Residence time refers to the time a fluid element spends in a reactor. Given that, typically, different fluid elements spend different times in the reactor, we have to deal with a distribution of residence times. We adopt the cumulative distribution function or the F function $F(t)$, which denotes the probability (P) that the residence time of a fluid element in the reactor, t_r, is less than or equal to t, as the basic quantity. That is,

$$F(t) \equiv P(t_r \leq t). \tag{2.1}$$

We may also interpret $F(t)$ as the fraction of molecules, entering a reactor, that would spend a time less than or equal to t in the reactor. From the basic rules of probability,[6] we can infer that the probability that the residence time lies between t_a and t_b is given by

$$P(t_a < t_r \leq t_b) = F(t_b) - F(t_a). \tag{2.2}$$

Assuming $F(t)$ to be a continuous and differentiable function, we may write Eq. (2.2) as

$$P(t_a < t_r \leq t_b) = \int_{t_a}^{t_b} F'(t)dt \tag{2.3}$$

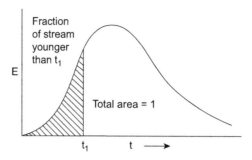

Figure 2.2
Exit age distribution function or **E**-curve; also known as the residence time distribution. The *shaded region* corresponds to $F(t_1)$.

where $F'(t) = dF/dt$. We may now define a new function

$$E(t) \equiv F'(t) \tag{2.4}$$

which may be regarded as a probability density function for the residence times and referred to as the *residence time distribution function* or the exit age distribution function. Thus Eqs. (2.2)–(2.4) can be combined and written as

$$P(t_a < t_r \le t_b) = F(t_b) - F(t_a) = \int_{t_a}^{t_b} E(t)dt \tag{2.5}$$

Eqs. (2.4) and (2.5) express the relationship between the functions $E(t)$ and $F(t)$. From Eq. (2.1), we may infer that

$$F(0) = 0, \quad F(t \to \infty) = 1. \tag{2.6}$$

With $t_a = 0$ and $t_b = \infty$ in Eq. (2.5) and the use of Eq. (2.6) leads to

$$\int_0^\infty E(t)dt = 1. \tag{2.7}$$

Eq. (2.7) reflects the fact that the probability that the residence time lies between 0 and ∞ is unity. It is common to regard $E(t)dt$ as the probability for the residence time to lie in the range $(t, t + dt)$ or, equivalently, as the fraction of molecules that would have a residence time between t and $t + dt$. Fig. 2.2 shows the relationship between $E(t)$ and $F(t)$.

2.1.3 Experimental Determination of E(t) and F(t)

2.1.3.1 The Convolution Formula

The RTD functions are usually determined by injecting a tracer of known concentration profile $C_{in}(t)$ into the reactor and measuring the tracer concentration at the outlet $C_{out}(t)$. A fundamental formula relating the inlet and outlet concentrations is given by:

$$C_{out}(t) = \int_0^t C_{in}(t - t')E(t')dt'. \tag{2.8}$$

An equivalent formula is

$$C_{out}(t) = \int_0^t C_{in}(t')E(t - t')dt'. \tag{2.9}$$

Eqs. (2.8) and (2.9) are central to the results of the subsequent sections and hence we discuss their derivations in some detail. However, the reader interested only in the application of these equations may skip the derivations. Proof of either of these formulae proceeds from the following idea. One considers the tracer molecules leaving the reactor at an instant of time t. The molar rate of the tracer at the exit is given by $q_r C_{out}(t)$ where q_r is the volumetric flow rate, *assumed constant throughout this chapter*. We can look on this rate as arising out of contributions from the tracer entering the reactor over the time range [0,t]. Consider a particular instant of time $(t - t')$ where $0 < t' < t$. The molar rate of tracer entering at this instant is $q_r C_{in}(t - t')$. If the fraction of these molecules that would have a residence time in the reactor equal to t' is $f(t')$, then $q_r C_{in}(t - t')f(t')$ may be regarded as the contribution of the tracer entering at $(t - t')$ to that leaving at t. Then the total exit molar rate of the tracer at time t may be constructed as a suitable sum or integral of the contributions arising from the inlet at each instant of time over the range [0,t]. However, because the RTD is a continuous one, we cannot assign a probability to a particular residence time but only to a range of residence times and hence cannot calculate the fractions of tracer molecules, which have a particular residence time. To implement the above idea, we follow the procedure outlined below.

We divide the time range of interest [0,t] into intervals

$$\left[t_0' = 0, t_1'\right], \left[t_1', t_2'\right], \ldots \left[t_{n-1}', t_n' = t\right]$$

with $\Delta t_i' = t_i' - t_{i-1}'$, $i = 1, \ldots n$ denoting the length of each interval. Consider the interval $\left[t_{i-1}', t_i'\right]$ and a particular instant of time ξ_i' within this interval. The inlet molar rate of the tracer at $t - \xi_i'$ is given by $q_r C_{in}\left(t - \xi_i'\right)$. The fraction of these molecules that have a residence time in the reactor between t_{i-1}' and t_i' is given by $\left[F(t_i') - F(t_{i-1}')\right]$ (see Eq. 2.2). Thus the quantity

$$Q_i = q_r C_{in}\left(t - \xi_i'\right) \cdot \left[F(t_i') - F(t_{i-1}')\right]. \tag{2.10}$$

represents the contribution of the inlet rate at $(t - \xi_i')$ to the exit rate between times $\left[t - (\xi_i' - t_{i-1}')\right]$ and $\left[t + (t_i' - \xi_i')\right]$, which corresponds to a small time interval around t

for a sufficiently fine division of the time range $[0,t]$. Using the relation 2.4 between $E(t)$ and $F(t)$, we may write Eq. (2.10) as:

$$Q_i \approx q_r C_{in}(t - \xi_i') \cdot E(\xi_i') \Delta t_i'. \tag{2.11}$$

Considering such time instants in each of the intervals and summing over their contributions to the exit rate, we obtain

$$S_n = \sum_{i=1}^{n} Q_i = q_r \sum_{i=1}^{n} C_{in}(t - \xi_i') \cdot E(\xi_i') \Delta t_i'. \tag{2.12}$$

(**Note**: To facilitate understanding, we may divide the range $[0,t]$ into equal intervals so that $\Delta t_i' = (t - 0)/n \equiv \Delta t'$ for all i and $\xi_i' = (t_{i-1}' + t_i')/2$ for all i. Then the above sum may be regarded as the contribution of the inlet rates at the time instants

$$t_0' + \Delta t'/2, t_1' + \Delta t'/2, \ldots t_{n-1}' + \Delta t'/2$$

to the exit rate over a time range $(t - \Delta t'/2, \ t + \Delta t'/2)$.

Nevertheless, in the limit of $n \to \infty$, which implies $\Delta t_i' \to 0$, one can infer that the sum in Eq. (2.12) would precisely give the exit tracer molar rate at time t, namely, $q_r C_{out}(t)$. Thus

$$\lim_{\substack{n \to \infty}} S_n = q_r C_{out}(t). \tag{2.13}$$
$$\left(\Delta t_i' \to 0\right)$$

Furthermore, from the way an integral is defined, we can easily infer from Eq. (2.12) that

$$\lim_{n \to \infty} S_n = q_r \int_0^t C_{in}(t - t')E(t')dt'. \tag{2.14}$$

From Eqs. (2.13) and (2.14), we obtain the convolution formula 2.8. To obtain Eq. (2.9), we introduce the transformation, $w = t - t'$. Then we may write Eq. (2.8) as

$$C_{out}(t) = \int_t^0 C_{in}(w)E(t - w)(-dw) = \int_0^t C_{in}(w)E(t - w)dw = \int_0^t C_{in}(t')E(t - t')dt'.$$

Thus the equivalence between Eqs. (2.8) and (2.9) is established.

2.1.3.2 Step and Impulse Responses

The two most widely used tracer experiments to determine the distribution functions involve step and impulse inputs (Fig. 2.3). In the case of the former, the input is given by

$$C_{in}^{step}(t) = C_\infty u(t) \tag{2.15}$$

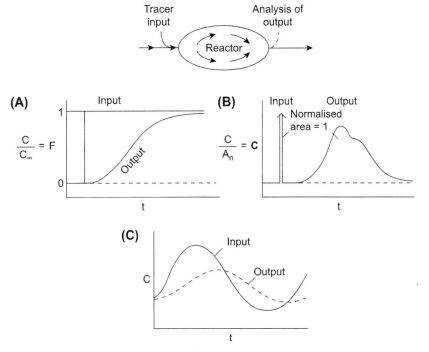

Figure 2.3
Tracer measurements; types of input signals and output responses. (A) Step input—**F**-curve.
(B) Pulse input—**C**-curve. (C) Sinusoidal input.

where $u(t)$ is the unit step function such that

$$u(t) = 0, \quad t < 0$$
$$= 1, \quad t > 0.$$

Substituting Eq. (2.15) in Eq. (2.9), we obtain

$$C_{out}^{step}(t) = C_\infty \int_0^t E(t - t')dt'.$$

Again, using the transformation $w = t - t'$, we get

$$C_{out}^{step}(t) = C_\infty \int_t^0 E(w)(-dw) = C_\infty \int_0^t E(w)(dw) = C_\infty[F(t) - F(0)] = C_\infty F(t).$$

Thus

$$F(t) = \frac{C_{out}^{step}(t)}{C_\infty}. \tag{2.16}$$

Thus, the measured outlet concentration of the tracer to a step input in tracer concentration directly yields the F function. A plot of $C_{out}^{step}(t)/C_\infty$ is referred to as the F curve.

The impulse input is best regarded as a limit of a pulse input, which is represented by

$$C_{in}(t) = C_\infty [u(t) - u(t - t_0)].$$

Thus the pulse input also involves a step in tracer concentration of magnitude C_∞ but has a finite duration t_0. The *strength* of the pulse function is given by the product $C_\infty t_0 \equiv A$. If we now progressively reduce the duration of the pulse but keeping its strength constant, then in the limit, we obtain the impulse function $\delta(t)$, which is more appropriately termed as the *Dirac delta distribution*.[2] Thus,

$$\lim_{t_0 \to 0} C_\infty [u(t) - u(t - t_0)] = A\delta(t).$$

$$C_\infty t_0 = A$$

If the volumetric flow rate of the fluid stream containing the tracer is q_r, then $n_0 = q_r A$ represents the total number of moles of the tracer present in the pulse or impulse input. Thus, the impulse input may be represented by

$$C_{in}^{imp}(t) = \frac{n_0}{q_r} \delta(t). \tag{2.17}$$

Substituting Eq. (2.17) in Eq. (2.9), we get

$$C_{out}^{imp}(t) = \frac{n_0}{q_r} \int_0^t \delta(t')E(t - t')dt'. \tag{2.18}$$

Making use of the property of the Dirac delta distribution,[2] namely,

$$\int_a^b \delta(x - x_0)f(x)dx = f(x_0), \quad a \leq x_0 \leq b,$$

we obtain, from Eq. (2.18),

$$C_{out}^{imp}(t) = \frac{n_0}{q_r} E(t),$$

from which

$$E(t) = C_{out}^{imp}(t) \frac{q_r}{n_0}. \tag{2.19}$$

Thus, the measured outlet concentration of the tracer to an impulse input directly gives the E function. A related quantity is the normalized output concentration defined by

$$\mathbf{C} \equiv \frac{C_{out}^{imp}(t)}{A_n}, \quad A_n = \int_0^\infty C_{out}^{imp}(t')dt' \tag{2.20}$$

where A_n is the area under the C_{out}^{imp} versus time curve. A plot of $C(t)$ versus time is referred to as the **C** curve. However, it can be easily seen from Eqs. (2.7), (2.19), and (2.20) that $C = E$, and hence these two quantities will be used interchangeably. Fig. 2.3 shows some typical responses to step and impulse as well as sinusoidal inputs.

2.1.4 E *and* F *Functions for Ideal Reactors*

As an illustration of the application of the results of Section 2.1.3, we derive the E and F functions for the two ideal flow reactors introduced in Chapter 1, namely, the CSTR and the PFR.

2.1.4.1 Continuous Stirred Tank Reactor

We write the unsteady-state material balance (in − out = accumulation) for the tracer passing through the CSTR:

$$q_r C_{in} - q_r C_{out} = \frac{d}{dt}(V C_{out}) \tag{2.21}$$

where V is the volume of the CSTR and we have made use of the fact that the concentration of tracer within the CSTR equals that at the exit, C_{out}. Dividing Eq. (2.21) by q_r and rearranging, we obtain

$$\tau \frac{dC_{out}(t)}{dt} + C_{out}(t) = C_{in}(t)$$

where $\tau = V/q_r$ is the space time of the CSTR. Taking the Laplace transform,[7] we get

$$\tau \left[s\widehat{C}_{out}(s) - C_{out}(0) \right] + \widehat{C}_{out}(s) = \widehat{C}_{in}(s).$$

Noting that $C_{out}(0) = 0$ and rearranging, we get

$$\widehat{C}_{out}(s) = \frac{1}{\tau s + 1}\widehat{C}_{in}(s). \tag{2.22}$$

Using the convolution formula associated with the product of the Laplace transforms of two functions, we may invert Eq. (2.22) to obtain

$$C_{out}(t) = L^{-1}\left[\widehat{C}_{in}(s) \cdot \frac{1}{\tau s + 1} \right] = \frac{1}{\tau} \int_0^t C_{in}(t')\exp\left(-\frac{t - t'}{\tau}\right)dt' \quad \text{[CSTR]} \tag{2.23}$$

where L^{-1} refers to the inverse Laplace transform. It is to be noted that Eqs. (2.22) and (2.23) apply to a CSTR for *any* type of tracer input. Eq. (2.22) gives the relationship between C_{out} and C_{in} in the Laplace domain, whereas Eq. (2.23) gives the relation in the time domain.

(**Note**: The inverse Laplace transform of a product of two functions may be expressed in terms of a convolution integral as[7]

$$L^{-1}[f(s) \cdot g(s)] = \int_0^t f(t')g(t-t')dt' = \int_0^t f(t-t')g(t')dt'.$$

In Eq. (2.23), $f(s) = \widehat{C}_{in}(s)$, $g(s) = 1/(\tau s + 1)$, and $L^{-1}[1/(\tau s + 1)] = e^{-t/\tau}/\tau$.)

For a *step* input (Eq. 2.15), Eq. (2.23) yields

$$C_{out}^{step}(t) = \frac{C_\infty \exp(-t/\tau)}{\tau} \int_0^t \exp(t'/\tau)dt'$$

from which

$$C_{out}^{step}(t) = C_\infty[1 - \exp(-t/\tau)]. \quad \text{[CSTR]} \tag{2.24}$$

From Eqs. (2.16) and (2.24), we obtain

$$F^{CSTR}(t) = 1 - \exp(-t/\tau). \tag{2.25}$$

Differentiating Eq. (2.25) with respect to t and using Eq. (2.4), we obtain

$$E^{CSTR}(t) = \frac{1}{\tau} \exp(-t/\tau). \tag{2.26}$$

2.1.4.2 Plug Flow Reactor

We assume a tubular reactor of constant cross-sectional area A_C. Under the PFR assumption 1.85, the continuity Eq. (1.14) reduces (for no reaction) to

$$\frac{\partial C}{\partial t} + v\frac{\partial C}{\partial z} = 0 \tag{2.27}$$

where $C = C(z,t)$ is the concentration of the tracer and v is the fluid velocity. It is to be noted that, by the plug flow assumption, the velocity profile is flat at any location z along the length of the reactor. The assumption of constant volumetric flow rate along the length of the reactor means that the velocity is the same at all points in the reactor. The general solution of Eq. (2.27) can be verified to be of the form[8]

$$C(z,t) = \phi(t - z/v) \tag{2.28}$$

where ϕ is an arbitrary function. Since $C_{in}(t) = C(0,t)$, we obtain, from Eq. (2.28),

$$\phi(t) = C_{in}(t). \tag{2.29}$$

Thus

$$C(z,t) = C_{in}(t - z/v). \tag{2.30}$$

Given that $C_{out}(t) = C(L,t)$, where L is the length of the reactor, we get, from Eq. (2.30),

$$C_{out}(t) = C_{in}(t - L/v). \tag{2.31}$$

From the definition of space time, we observe

$$\tau = \frac{V}{q_r} = \frac{A_C L}{q_r} = \frac{L}{v}. \tag{2.32}$$

From Eqs. (2.31) and (2.32), we obtain

$$C_{out}(t) = C_{in}(t - \tau). \quad \text{[PFR]} \tag{2.33}$$

Eq. (2.33) is the relation between the inlet and outlet tracer concentration for a tubular reactor much like Eq. (2.23) is for the CSTR. It represents the commonly assumed and intuitively inferable fact that, for a PFR, the outlet tracer concentration is just the inlet concentration shifted by the space time τ. Here, we have shown it to be a rigorous consequence of the continuity equation for the PFR. For a step input (Eq. 2.15), Eq. (2.33) yields

$$C_{out}^{step}(t) = C_\infty u(t - \tau) \quad \text{[PFR]}$$

and hence from Eq. (2.16), we obtain

$$F^{PFR}(t) = u(t - \tau). \tag{2.34}$$

For an impulse input (Eq. 2.17), Eq. (2.33) implies that

$$C_{out}^{imp}(t) = \frac{n_0}{q_r} \delta(t - \tau). \tag{2.35}$$

Hence, from Eq. (2.19), we get

$$E^{PFR}(t) = \delta(t - \tau). \tag{2.36}$$

Eq. (2.36) can also be obtained from Eq. (2.34) by using Eq. (2.4) and using the fact that the Dirac delta distribution may be regarded as the derivative of the step function in a generalized sense.

2.1.5 Statistics of Residence Time Distribution

We will deal with two measures of the distribution, the first moment or the mean of the RTD, $\bar{t}_E$, and the second moment or variance, σ_E^2, which is the square of the standard deviation σ_E. These are defined as follows:

$$\bar{t}_E = \int_0^\infty tE(t)dt; \quad \sigma_E^2 = \int_0^\infty (t - \bar{t}_E)^2 E(t)dt = \int_0^\infty t^2 E(t)dt - \bar{t}_E^2. \tag{2.37}$$

The second expression in the equation for σ_E^2 follows from the fact that

$$\sigma_E^2 = \int_0^\infty (t - \bar{t}_E)^2 E(t)dt = \int_0^\infty t^2 E(t)dt - 2\bar{t}_E \int_0^\infty tE(t)dt + \bar{t}_E^2 \int_0^\infty E(t)dt$$

$$= \int_0^\infty t^2 E(t)dt - 2\bar{t}_E^2 + \bar{t}_E^2 = \int_0^\infty t^2 E(t)dt - \bar{t}_E^2$$

where we have made use of the definition of $\bar{t}_E$ (Eq. 2.37) and the normalization condition (Eq. 2.7). From Eq. (2.26), we obtain the following values for the CSTR:

$$\bar{t}_E = \tau, \quad \sigma_E^2 = \tau^2. \quad \text{(CSTR)} \tag{2.38}$$

Similarly, for the PFR, using Eq. (2.36), we obtain

$$\bar{t}_E = \tau, \quad \sigma_E^2 = 0. \quad \text{(PFR)} \tag{2.39}$$

Eqs. (2.38) and (2.39) reveal that the mean residence time equals the space time for both the ideal reactors, a result more generally valid for any "closed" vessel, i.e., one without dispersion.[9] The result of zero variance for the PFR is consistent with Eq. (2.36) in that the E function for a PFR is the impulse function (the same form as the input in Fig. 2.3B), only shifted by the residence time τ.

2.1.6 Application of Tracer Information to Reactors

For the ideal flow reactors, PFR and CSTR, the assumptions regarding flow pattern and mixing resulted in design equations from which conversion can be calculated with only kinetics as the input. If the flow pattern in a reactor does not conform to that of either of the ideal flow reactors, then the RTD of the reactor is certainly one of the inputs that enables one to calculate reactor conversions. The RTD is a measure of the times spent by different fluid elements in the reactor and may be regarded as representing the state of "macromixing."[9] In addition, we also require a model for how fluid elements of different ages interact or mix with each other. This is often referred to as "micromixing."[9] Levenspiel[2] points out that, apart from kinetics, three factors influence reactor behavior, namely, the RTD, the "state of aggregation" of the fluid, and the "earliness and lateness of mixing."

It is possible for two reactor configurations to have identical RTDs but yet yield different conversions on account of the nature of mixing being different in the two cases. A classic example[10] is the case of an ideal CSTR followed by an ideal PFR versus an ideal PFR followed by an ideal CSTR (Fig. 2.4). It can be easily shown that the RTD in both cases is the same and given by

$$E(t) = 0, \quad t < \tau_1$$

$$= \frac{e^{-(t-\tau_1)/\tau_2}}{\tau_2}, \quad t \geq \tau_1 \tag{2.40}$$

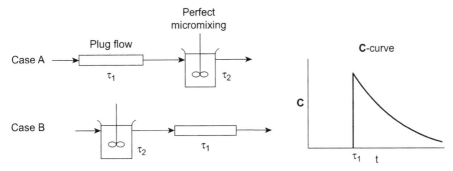

Figure 2.4

Kramers' example of a plug flow tubular reactor (residence time τ_1) and an ideal stirred tank (residence time τ_2) in series. Note that the **C**-curve is the same for both configurations.

where τ_1 and τ_2 are the space times of the PFR and CSTR, respectively. However, the exit conversion of the two reactor configurations depends on the kinetics of the reaction taking place in the reactors. For a first-order ($-r_A = k_1 C_A$) reaction, the exit concentration of the reactant A is the same for the two configurations and given by

$$C_{A,out} = C_{A,in}\frac{e^{-k_1\tau_1}}{1 + k_1\tau_2} \quad \text{(First order reaction)} \tag{2.41}$$

However, for a second-order reaction, the configuration of PFR followed by a CSTR can be shown to yield a higher conversion than that of the CSTR followed by a PFR.[9] This is attributed to the fact that while mixing is provided early on in the CSTR–PFR configuration, it is delayed in the PFR–CSTR configuration. We may construct a similar example for a half-order reaction with the following input data:

$$-r_A = k_{1/2}C_A^{1/2}, \quad C_{A,in} = 1.0 \text{ mol/L}, \quad \tau_1 = \tau_2 = 1 \text{ min}, \quad k_{1/2} = 1.0 \text{ (mol/L)}^{1/2}/\text{min}.$$

A straightforward calculation shows that in the case of the PFR followed by CSTR, $C_{A,out} = 0.04$ mol/L, whereas for the CSTR followed by PFR, $C_{A,out} = 0.014$ mol/L. These results indicate that for a half-order reaction, the CSTR followed by PFR yields a higher conversion.

The zero-order reaction ($-r_A = k_0$) provides an anomalous case. Here, both configurations lead to identical exit concentrations and conversions (even though order is less than one) and the results are given below:

$$\begin{aligned} C_{A,out} &= C_{A,in} - k_0(\tau_1 + \tau_2), \quad (\tau_1 + \tau_2) \leq C_{A,in}/k_0 \\ &= 0, \quad (\tau_1 + \tau_2) > C_{A,in}/k_0. \end{aligned} \quad \text{(Zero} - \text{order reaction)} \tag{2.42}$$

The above results indicate that delayed mixing provides a favorable condition for reaction orders greater than one, whereas early mixing favors higher conversion for orders less than one but not equal to zero. The nature of mixing appears to have no effect on the conversion level for first-order or zero-order reactions. In the case of first-order reactions, which are also referred to sometimes as linear processes, the independence of conversion on the nature of mixing is attributed to the conversion being independent of the initial or inlet concentration levels in the ideal reactors.[9]

But it is clear from the above example that, apart from the RTD, the nature of micromixing[2] can also influence reactor conversions depending on the kinetics. Just as knowledge about micromixing can affect predictions about the rates of reactions, the reverse is true; measurements of rates of reaction can give valuable information about the extent to which micromixing is occurring in a complex fluid flow field, such as that in a stirred tank.[11]

We can classify models of mixing under three categories:

1. zero-parameter models;
2. one-parameter models; and
3. two-parameter models.

In this chapter, we focus on (1) and (2). Under (1), we discuss the segregated and maximum mixedness models in Section 2.2. Under (2), we consider the tanks-in-series and dispersion models in Sections 2.3 and 2.4, respectively. A brief discussion of two-parameter models is contained in Section 2.5.

2.2 Zero-Parameter Models—Complete Segregation and Maximum Mixedness Models

In the complete segregation model, molecules with different residence times do not mix and are assumed to behave like batch reactors each undergoing conversion in accord with the time they spend in the reactor. Thus mixing is delayed to the maximum extent possible. A fluid exhibiting such a behavior is referred to as a "macrofluid."[2,9] Given that $E(t)$ represents the probability distribution function of the residence times, we may obtain the exit concentration of reactant A (undergoing a single reaction) as an average in the following manner:

$$C_{A,out}^{seg} = \int_0^\infty C_A^{batch}(t')E(t')dt'. \tag{2.43}$$

A good example where the segregated model applies is the conversion obtained when solid particles undergo a reaction in a rotary kiln (Fig. 2.5).

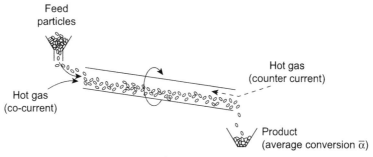

Figure 2.5
Average conversion from residence time distribution—example of rotary kiln.

We have seen how the CSTR represents a case of complete, early mixing with the result that as soon as the reactants enter the reactor, they assume the concentrations within the reactor. Mixing ensures uniform concentrations within the reactor, which are equal to that at the exit of the reactor. This assumption of complete mixedness leads to the RTD given by Eq. (2.26). For any other reactor, while this level of mixing on a molecular scale is not possible, we may consider a state of "maximum mixedness *compatible with a given residence time distribution*."[5] The equation corresponding to this condition may be shown to be[5,9]

$$\frac{dC_A}{d\lambda} = -r_A + \frac{E(\lambda)}{1 - F(\lambda)} \left(C_A - C_{A,in}\right) \tag{2.44}$$

where λ is a parameter that represents the time it takes for a fluid element to travel from a point in the reactor to the reactor exit. Eq. (2.44) must be solved along with the condition

$$\lim_{\lambda \to \infty} C_A = C_{A,in}. \tag{2.45}$$

Furthermore, we note that the exit concentration of A as per the maximum mixedness model is given by

$$C_A(\lambda = 0) = C_{A,out}^{mm} \tag{2.46}$$

where *mm* refers to "maximum mixedness."

Thus, the segregated and maximum mixedness models represent two extremes of micromixing. Hence for a reactor with a given RTD, these models can be used to obtain bounds on conversion achievable for a given reaction. Neither of these models contain a parameter that needs to be determined from the RTD data. Hence they are referred to as zero-parameter models. In the subsequent sections, we apply Eqs. (2.43) and (2.44) to get precise results and additional insights into the nature and degree of mixing and its effect on conversion for reactions exhibiting different types of kinetics.

2.2.1 Special Case of First-Order Reactions: Equivalence of the Segregated and Maximum Mixedness Models

In Section 2.1.6, we saw that the nature of mixing has no effect on reactor conversion for a first-order reaction. In this section, we will reinforce this idea within the context of zero-parameter models.

For a first-order reaction, $-r_A = k_1 C_A$ and hence Eq. (2.44) reduces to

$$\frac{dC_A}{d\lambda} + \left[\frac{E(\lambda)}{F(\lambda) - 1} - k_1\right] C_A = \frac{E(\lambda)}{F(\lambda) - 1} C_{A,in} \tag{2.47}$$

Eq. (2.47) is a first-order differential equation of the form

$$\frac{dy}{dx} + P(x)y = Q(x)$$

and can be solved by the integrating factor method in which the integrating factor is $e^{\int P(x)dx}$. The solution is of the form

$$y e^{\int P(x)dx} = \int Q(x) e^{\int P(x)dx} dx + c$$

where c is an integration constant. The integrating factor for Eq. (2.47) is

$$\exp\left(\int \left[\frac{E(\lambda)}{F(\lambda) - 1} - k\right] d\lambda\right) = e^{-k\lambda}(F(\lambda) - 1)$$

on noting that $E(\lambda) = F'(\lambda)$ and $\int \frac{F'(\lambda)}{F(\lambda)-1} d\lambda = \ln(F(\lambda) - 1)$ Given the condition 2.45, the solution to Eq. (2.47) may be shown to be

$$C_A(\lambda) = \frac{e^{k_1\lambda}}{1 - F(\lambda)} C_{A,in} \int_\lambda^\infty E(\lambda') e^{-k_1\lambda'} d\lambda'. \tag{2.48}$$

The exit concentration may be obtained from Eqs. (2.46) and (2.48) as

$$C_{A,out}^{mm} = C_{A,in} \int_0^\infty E(\lambda') e^{-k_1\lambda'} d\lambda' \tag{2.49}$$

given that $F(0) = 0$(Eq. 2.6).

For the segregated model, we need to use Eq. (2.43) and hence we require $C_A^{batch}(t')$. For a first-order reaction, this is given by

$$C_A^{batch}(t') = C_{A0}\, e^{-k_1 t'}$$

where $C_{A0} = C_A(t = 0)$. Substituting into Eq. (2.43) with $C_{A0} = C_{A,in}$, we obtain

$$C_{A,out}^{seg} = C_{A,in} \int_0^\infty E(t')e^{-k_1 t'} dt'. \tag{2.50}$$

The integral in Eq. (2.50) is identical to that in Eq. (2.49); only the (dummy) variable of integration is different. Hence,

$$C_{A,out}^{mm} = C_{A,out}^{seg} \quad \text{I order reaction} \tag{2.51}$$

It must be noted that the result 2.51 has been obtained for an arbitrary RTD. Thus if a first-order reaction is carried out in a reactor with an RTD $E(t)$, then one may simply use Eq. (2.50) to calculate the exit concentration and hence the conversion. Furthermore, because the segregated model and maximum mixedness models represent two extremes of mixing, we may conclude that the degree of mixing does not have any effect on conversion for a first-order reaction.

Given that the PFR and CSTR also represent two extremes levels of mixing, we now examine the relationship between the ideal reactors and the zero-parameter models.

2.2.2 PFR and Zero-Parameter Models

For a PFR, $E(t)$ is given by Eq. (2.36). Substituting into Eq. (2.43), we get

$$C_{A,out}^{seg,PFR} = \int_0^\infty C_A^{batch}(t')\delta(t' - \tau)dt' = C_A^{batch}(\tau). \tag{2.52}$$

Recall that the design equations for batch reactor can be carried over to that of the PFR for constant density mixtures provided the time in the batch reactor equations is replaced by the space time of the PFR (Section 1.7.1 of Chapter 1, Eq. 1.99). Thus, we infer from Eq. (2.52) that

$$C_{A,out}^{seg,PFR} = C_{A,out}^{PFR(ideal)}. \tag{2.53}$$

That is, when the $E(t)$ of the PFR is substituted into the segregated model equation, the exit concentration or conversion is the same as that given by the ideal PFR. As pointed out by Zwietering,[5] the RTD of the PFR "leaves no freedom for different degrees of segregation. Complete segregation is the only possibility." Viewed from another angle, since in an ideal PFR all fluid elements have the same residence time, the question of mixing of fluid elements of different residence times does not arise at all. Thus complete segregation is inherent in the PFR assumptions. Such is not the case for the RTD of the CSTR, which we now discuss.

2.2.3 Residence Time Distribution of the CSTR and the Zero-Parameter Models

We have seen that the assumption of complete mixedness leads to the RTD of Eq. (2.26) for the CSTR. However, it is entirely possible for a reactor, which is not a CSTR to exhibit an identical RTD. All results of this and the next section correspond to such a reactor.

From Eqs. (2.43) and (2.26), the expression for the exit concentration is, for the segregated model,

$$C_{A,out}^{seg} = \frac{1}{\tau} \int_0^\infty C_A^{batch}(t') e^{-t'/\tau} dt'. \tag{2.54}$$

In the case of the maximum mixedness model, from Eqs. (2.25) and (2.26), we note that

$$1 - F(\lambda) = exp(-\lambda/\tau), \quad E(\lambda) = \frac{1}{\tau} exp(-\lambda/\tau), \quad \frac{E(\lambda)}{1 - F(\lambda)} = \frac{1}{\tau}.$$

Substituting into Eq. (2.44), we get

$$\frac{dC_A}{d\lambda} = -r_A + \frac{1}{\tau}\left(C_A - C_{A,in}\right) \tag{2.55}$$

which has to be solved along with the condition 2.45. Assuming

$$-r_A + \frac{1}{\tau}\left(C_A - C_{A,in}\right) \neq 0 \tag{2.56}$$

in an interval $\left(\overline{C}_A, C_{A,in}\right)$, we may integrate Eq. (2.55) over that interval to obtain

$$\int_{\overline{C}_A}^{C_{A,in}} \frac{dC_A}{-r_A + \frac{1}{\tau}\left(C_A - C_{A,in}\right)} = \lim_{\lambda \to \infty} \int_{\overline{\lambda}}^{\lambda} d\lambda'$$

where $\overline{C}_A = C_A(\overline{\lambda})$. Clearly, the right-hand side of the above equation goes to infinity but the left-hand side integral is finite because the limits are finite and the integrand is finite on account of condition 2.56. This only means that the assumption 2.56 is not correct. Hence we conclude that

$$-r_A + \frac{1}{\tau}\left(C_A - C_{A,in}\right) = 0. \tag{2.57}$$

From Eq. (2.55), we note that $dC_A/d\lambda = 0$ for finite λ and hence $C_A(\lambda) = C_A(\lambda = 0) = C_{A,out}$. Hence, Eq. (2.57) may be rewritten as

$$C_{A,out}^{mm} = C_{A,in} + r_A \tau.$$

The above equation can be easily seen to be a rearrangement of the ideal CSTR design equation (Eq. 1.118 for a constant density system). We thus conclude that

$$C_{A,out}^{mm} = C_{A,in} + r_A \tau = C_{A,out}^{CSTR(ideal)}. \tag{2.58}$$

where r_A is to be evaluated at the reactor concentrations, which are equal to the exit concentrations. What we have shown is that if the RTD of the CSTR is substituted into the maximum mixedness model, then we obtain the ideal CSTR design equation. Thus the ideal CSTR equations may be regarded as corresponding to the state of maximum mixedness consistent with the RTD represented by Eq. (2.26). However, as is evident from Eq. (2.54), complete segregation is also consistent with the RTD, Eq. (2.26), of the CSTR. Thus the whole range of segregation (or mixing) is possible with the RTD of the CSTR. We now utilize this feature to analyze the predictions of the two zero-parameter models for various kinetics.

2.2.4 Bounds on Conversion: Some General Rules

2.2.4.1 Zero-Order Kinetics

For a zero-order reaction,

$$-r_A = k_0 \tag{2.59}$$

and hence Eqs. (2.57) and (2.58) result in

$$\begin{aligned} C_{A,out}^{mm} &= C_{A,in} - k_0 \tau, \quad \tau \le C_{A,in}/k_0 \\ &= 0, \quad \tau > C_{A,in}/k_0. \end{aligned} \tag{2.60}$$

For the segregated model, we note that

$$\begin{aligned} C_A^{batch}(t') &= C_{A0} - k_0 t', \quad t' \le C_{A0}/k_0 \\ &= 0, \quad t' > C_{A0}/k_0. \end{aligned} \tag{2.61}$$

where $C_{A0} = C_A(t = 0)$. Substituting into Eq. (2.43), with $C_{A0} = C_{A,in}$, $E(t)$ given by Eq. (2.26) and integrating, we obtain

$$C_{A,out}^{seg} = C_{A,in} - k_0 \tau + k_0 \tau \, e^{-C_{A,in}/k_0\tau}. \tag{2.62}$$

It must be noted that even though the upper limit in Eq. (2.43) is ∞, the effective upper limit is only $C_{A,in}/k_0$ on account of Eq. (2.61).

Comparing Eqs. (2.60) and (2.62), we first note a qualitative difference. The maximum mixedness model gives complete conversion for a finite space time, whereas the segregated model predicts complete conversion only in the asymptotic limit $\tau \to \infty$ similar to the case of a first-order reaction for the ideal reactors. Furthermore, the exit concentration predicted by the segregated model for a given space—time is always greater than that of the maximum mixedness model, or equivalently the conversion predicted by the segregated model is always lower than that of the maximum mixedness model for a zero-order reaction.

2.2.4.2 First-Order Kinetics

We have shown earlier (Eq. 2.51) that, for first-order reactions, the segregated and maximum mixedness models yield the same conversion. This result was derived for a reactor with an arbitrary RTD. Certainly, we should be able to verify it for the RTD of the CSTR. In fact, substituting for $E(t')$ from Eq. (2.26) in Eq. (2.50) (valid for a first-order reaction) and integrating, we obtain

$$C_{A,out}^{seg} = \frac{C_{A,in}}{\tau} \int_0^\infty e^{-t'/\tau} e^{-k_1 t'} dt' = \frac{C_{A,in}}{1 + k_1\tau}. \quad \text{(I order reaction)} \tag{2.63}$$

Clearly, this is the result we get if we substitute

$$-r_A = k_1 C_A = k_1 C_{A,out}$$

in Eq. (2.57) or Eq. (2.58) for the maximum mixedness model, which for the RTD given by Eq. (2.26) results in the ideal CSTR equations.

2.2.4.3 Second-Order Kinetics

Here $-r_A = k_2 C_A^2$. For the maximum mixedness model, we get, from Eq. (2.58),

$$C_{A,out}^{mm} = C_{A,in} \left(\frac{-1 + \sqrt{1 + 4k_2 C_{A,in}\tau}}{2k_2 C_{A,in}\tau} \right). \quad \text{(II order reaction)} \tag{2.64}$$

For the segregated model, we note, for a second-order reaction,

$$C_A^{batch}(t') = \frac{C_{A0}}{1 + k_2 C_{A0} t'}. \tag{2.65}$$

Substituting into Eq. (2.43) with $E(t)$ given by Eq. (2.28) and with $C_{A0} = C_{A,in}$, we obtain

$$C_{A,out}^{seg} = \frac{C_{A,in}}{\tau} \int_0^\infty \frac{e^{-t'/\tau}}{1 + k_2 C_{A,in} t'} dt'. \quad \text{(II order reaction)} \tag{2.66}$$

Using the transformation

$$x = \frac{1 + k_2 C_{A,in} t'}{k_2 C_{A,in}\tau},$$

Eq. (2.66) may be written as

$$C_{A,out}^{seg} = \frac{e^{1/k_2 C_{A,in}\tau}}{k_2\tau} \int_{1/k_2 C_{A,in}\tau}^\infty \frac{e^{-x}}{x} dx.$$

Table 2.1: Comparison of predictions of segregated and maximum mixedness models for second-order kinetics.

$k_2 C_{A,in}\tau$	$\alpha = 1/(k_2 C_{A,in}\tau)$	$E_1(\alpha)$	$C_{A,out}^{seg}/C_{A,in}$ (2.67)	$C_{A,out}^{mm}/C_{A,in}$ (2.64)
0.1	10	4.157×10^{-6}	0.9156	0.916
0.2	5	1.148×10^{-3}	0.852	0.854
1.0	1	0.2194	0.596	0.618
10	0.1	1.8229	0.201	0.270

Results apply for the RTD of the CSTR, Eq. (2.26).

The ratio $C_{A,out}^{seg}/C_{A,in}$ may be written in terms of the exponential integral E_1 as

$$\frac{C_{A,out}^{seg}}{C_{A,in}} = \alpha e^{\alpha} E_1(\alpha), \quad \alpha = 1/(k_2 C_{A,in}\tau), \quad E_1(\alpha) = \int_{\alpha}^{\infty} \frac{e^{-x}}{x}\,dx. \tag{2.67}$$

From the table of exponential integrals,[12] Table 2.1 can be constructed based on Eqs. (2.64) and (2.67).

The table shows that the value of $C_{A,out}^{seg}/C_{A,in}$ as given by Eq. (2.67) is less than that of $C_{A,out}^{mm}/C_{A,in}$ as given by Eq. (2.64) for a given value of the dimensionless quantity $k_2 C_{A,in}\tau$. The difference is negligible for small values of $k_2 C_{A,in}\tau$ but progressively increases. Thus the segregated model predicts a higher conversion compared to the maximum mixedness model for a reactor exhibiting the RTD of Eq. (2.26) and in which a second-order reaction is carried out.

We now summarize the results of this section. For a reactor with RTD that of the CSTR, the maximum mixedness model predicts a higher conversion for a zero-order reaction, whereas the segregated model predicts a higher conversion for a second-order reaction. For a first-order reaction, the conversions predicted by both models are the same. Given that the maximum mixedness and completely segregated models represent two extremes of micromixing for a given state of macromixing, we may regard the conversion predictions of these two models as representing the two extremes of conversion possible. Thus we may infer the following rules, for a reactor with a given RTD:

1. For reaction orders p such that $0 \le p < 1$, the segregation model gives a lower bound to the conversion, whereas the maximum mixedness model gives an upper bound. For orders such that $p > 1$, the converse is true.
2. For first-order reactions ($p = 1$), both models give identical conversions. Thus the degree of mixing does not have any effect on the conversion.

(**Note**: The results of this section leading to rules (1) and (2) above are consistent with the conclusions obtained in Section 2.1.6 that delayed mixing favors reactions of orders

Table 2.2: Comparison of predictions of segregated and maximum mixedness models for half-order kinetics.

$\beta = k_{1/2}C_{A,in}^{-1/2}\tau$	$C_{A,out}^{seg}/C_{A,in}$ (2.69)	$C_{A,out}^{mm}/C_{A,in}$ (2.68)
0.1	0.905	0.9049
1.0	0.432	0.382
10	0.06	0.01
100	0.007	0.0001

Results apply for the RTD of the CSTR, Eq. (2.26).

greater than one, whereas early mixing favors those with orders less than one. But there is one caveat. In Section 2.1.6, we saw that the zero-order reaction presented an anomalous instance for reaction orders less than one. Here, the zero-order reaction follows the same trend as any other positive order less than one. In fact, it can be shown that, for a half-order reaction,

$$\frac{C_{A,out}^{mm}}{C_{A,in}} = 1 - \beta\left(1 + 0.25\beta^2\right)^{1/2} + \frac{1}{2}\beta^2, \tag{2.68}$$

$$\frac{C_{A,out}^{seg}}{C_{A,in}} = 1 - \beta + \frac{1}{2}\beta^2\left(1 - e^{-2/\beta}\right) \tag{2.69}$$

where $\beta = k_{1/2}C_{A,in}^{-1/2}\tau$. Table 2.2 clearly shows that the exit concentration predicted by the maximum mixedness model is always less than that predicted by the segregated model or the conversion predicted by the maximum mixedness model is higher, which is in agreement with the general rule outlined above. It can also be seen that in both cases, the exit concentration asymptotically approaches zero as the space time increases for a fixed inlet concentration and temperature. Given that the maximum mixedness model for the RTD of Eq. (2.26) is equivalent to the ideal CSTR, this asymptotic trend toward complete conversion for a fractional order reaction is to be contrasted with the achievability of complete conversion in a finite time or space time for both batch and PFRs. For the zero-order reaction, nevertheless, complete conversion is achievable for a finite time or space time for all three ideal reactors.)

2.3 Tanks-in-Series Model

This is the first of the single parameter models that we discuss. The single parameter in the tanks-in-series model is the number N of tanks (CSTRs) in series that would be required to match the experimentally obtained RTD for a reactor in a sense that would be explained in Section 2.3.1. First, we obtain expressions for the distribution functions E and F for N CSTRs in series and discuss their important features and statistical properties.

Eq. (2.22) relates the inlet and outlet tracer concentration for a single CSTR in the Laplace domain. If we consider a series of N tanks each of space time τ_i, $i = 1,\ldots, N$, then we may apply Eq. (2.22) to the nth tank in the series as follows:

$$\widehat{C}_n(s) = \frac{1}{\tau_n s + 1}\widehat{C}_{n-1}(s) \tag{2.70}$$

Applying Eq. (2.70) first to the Nth CSTR, we obtain

$$\widehat{C}_N(s) = \widehat{C}_{out}(s) = \frac{1}{\tau_N s + 1}\widehat{C}_{N-1}(s) \tag{2.71}$$

Applying Eq. (2.70) now with $n = N - 1$, we may express $\widehat{C}_{N-1}(s)$ in terms of $\widehat{C}_{N-2}(s)$ and using Eq. (2.71), we may then express $\widehat{C}_{out}(s)$ in terms of $\widehat{C}_{N-2}(s)$ as follows:

$$\widehat{C}_{out}(s) = \frac{1}{\tau_N s + 1}\cdot\frac{1}{\tau_{N-1} s + 1}\widehat{C}_{N-2}(s).$$

Proceeding in this manner, we can obtain

$$\widehat{C}_{out}(s) = \frac{1}{\tau_N s + 1}\cdot\frac{1}{\tau_{N-1} s + 1}\cdots\frac{1}{\tau_1 s + 1}\widehat{C}_{in}(s). \tag{2.72}$$

Eq. (2.72) may be compactly written as

$$\widehat{C}_{out}(s) = \left(\prod_{j=1}^{N}\frac{1}{\tau_j s + 1}\right)\widehat{C}_{in}(s) \tag{2.73}$$

where $\prod$ represents the product of the N factors. If the space times of each of the CSTRs is the same and equal to τ_i, then Eq. (2.73) simplifies to

$$\widehat{C}_{out}(s) = \left(\frac{1}{\tau_i s + 1}\right)^{N}\widehat{C}_{in}(s) \tag{2.74}$$

For an impulse input, Eq. (2.17) applies and correspondingly

$$\widehat{C}_{in}^{imp}(s) = \frac{n_0}{q_r}.$$

Substituting into Eq. (2.74), we get

$$\widehat{C}_{out}^{imp}(s) = \frac{n_0}{q_r}\left(\frac{1}{\tau_i s + 1}\right)^{N}.$$

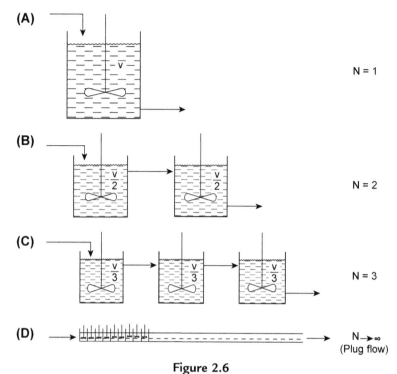

Figure 2.6
Sets of equal stirred tanks in series, each set having the same total volume V.

Taking the inverse Laplace transform,[7] we obtain

$$C_{out}^{imp}(t) = \frac{n_0}{q_r} \frac{1}{(N-1)!} \frac{1}{\tau_i^N} t^{N-1} e^{-t/\tau_i}.$$

From Eq. (2.19), we get

$$E_N^{\tau_i}(t) = \frac{1}{(N-1)!} \frac{1}{\tau_i^N} t^{N-1} e^{-t/\tau_i} \quad N \text{ CSTRs in series (fixed } \tau_i) \tag{2.75}$$

It must be noted that Eq. (2.75) represents a case where the space time of each CSTR is fixed at τ_i. Hence the total space time of the N CSTRs in series is $N\tau_i$ and increases with N. It would be interesting to modify Eq. (2.75) for the case in which the total space time of the CSTR configuration is fixed at, say, τ. This means that as N increases, the space time of each CSTR ($=\tau/N$) decreases (Fig. 2.6), and we expect the RTD of such a configuration to approach that of the PFR as $N \to \infty$. For this case, the RTD can be obtained by substituting for τ_i in Eq. (2.75) with τ/N. Thus we get

$$E_N^\tau(t) = \frac{N^N}{(N-1)!} \frac{1}{\tau^N} t^{N-1} e^{-Nt/\tau} \quad N \text{ CSTRs in series(fixed } \tau). \tag{2.76}$$

We examine the statistical properties of the two distributions. Considering Eq. (2.75), we note that the mean residence time from Eq. (2.37) reduces to

$$\bar{t}_{E_N^{\tau_i}} = \frac{1}{(N-1)!} \frac{1}{\tau_i^N} \int_0^\infty t^N e^{-t/\tau_i} dt. \tag{2.77}$$

Using the transformation $t' = t/\tau_i$, we may express Eq. (2.77) as

$$\bar{t}_{E_N^{\tau_i}} = \frac{\tau_i}{(N-1)!} \int_0^\infty (t')^N e^{-t'} dt'. \tag{2.78}$$

For integer N, the integral $\int_0^\infty (t')^N e^{-t'} dt'$ equals $N!$[7] Thus Eq. (2.78) yields

$$\bar{t}_{E_N^{\tau_i}} = N\tau_i. \tag{2.79}$$

A similar procedure using Eq. (2.76) leads to

$$\bar{t}_{E_N^\tau} = \tau. \tag{2.80}$$

Eqs. (2.79) and (2.80) are essentially equivalent in the sense that the mean residence time in both cases equals the total space time of the CSTR configuration. However, in the case of Eq. (2.79), the mean residence time increases with increase in the number of CSTRs, whereas in the case of Eq. (2.80), it remains constant. A similar analysis yields the following expressions for the variance:

$$\sigma_{E_N^{\tau_i}}^2 = N\tau_i^2, \quad \sigma_{E_N^\tau}^2 = \tau^2/N. \tag{2.81}$$

Again, we note the qualitative difference between the variance in the two cases. In the case where τ_i is fixed, the variance increases with N, whereas in the case where the total space time of the CSTR configuration is fixed at τ, the variance goes to zero as N is increased. The former case corresponds to progressive increase in the space time as more CSTRs are added, thereby increasing the spread of the RTD (Fig. 2.7). In the latter case, the total space time (τ) is fixed and hence as N increases, the space time is split among smaller and smaller CSTRs and the RTD progressively tends toward that of a PFR of space time τ (zero variance, Eq. 2.39).

We now examine certain other qualitative features of the RTDs given by Eqs. (2.75) and (2.76) that reinforce the above conclusions. We have seen that in the case of a single CSTR, $E(t)$ is a monotonically decreasing function of time (Eq. 2.28). It can be shown that

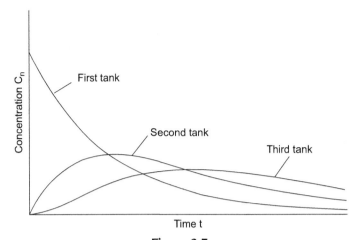

Figure 2.7
Progress of a pulse of tracer through a series of three stirred tanks.

for $N \geq 2$, $E(t)$ has a maximum (Fig. 2.7). The values of the time at which this occurs and the maximum value of the function for the two distributions 2.75 and 2.76 are given below:

$$t\left(E_N^{\tau_i}\big|_{\max}\right) = (N-1)\tau_i, \quad E_N^{\tau_i}\big|_{\max} = \frac{(N-1)^{N-1}}{(N-1)!} \frac{e^{-(N-1)}}{\tau_i}, \tag{2.82}$$

$$t\left(E_N^{\tau}\big|_{\max}\right) = (1 - 1/N)\tau, \quad E_N^{\tau}\big|_{\max} = \frac{(N-1)^{N-1}}{(N-1)!} \frac{N\,e^{-(N-1)}}{\tau}. \tag{2.83}$$

From Eq. (2.82), one can easily show that

$$t\left(E_N^{\tau_i}\big|_{\max}\right) \to \infty \ \text{ as } \ N \to \infty, \quad E_N^{\tau_i}\big|_{\max} \to 1/\tau_i \ \text{ as } \ N \to \infty. \tag{2.84}$$

On the other hand, from Eq. (2.83), we get

$$t\left(E_N^{\tau}\big|_{\max}\right) \to \tau \ \text{ as } \ N \to \infty, \quad E_N^{\tau}\big|_{\max} \to \infty \ \text{ as } \ N \to \infty. \tag{2.85}$$

Eq. (2.85) reinforces the earlier conclusion that the RTD of a sequence of CSTRs of *fixed total space time* tends toward that of a PFR as the number of CSTRs in series increases.

The trend toward a PFR in the limit of large N can also be represented in terms of scaled variables. Such a representation can be obtained from either Eq. (2.75) or Eq. (2.76). In the case of Eq. (2.75), we may define

$$\theta \equiv t/(N\tau_i), \quad E_N^\theta \equiv N\tau_i E_N^{\tau_i}(t)$$

or in the case of Eq. (2.76), we define

$$\theta \equiv t/\tau, \quad E_N^\theta \equiv \tau E_N^{\tau}(t) \tag{2.86}$$

In either case, we obtain

$$E_N^\theta(\theta) = \frac{N^N}{(N-1)!}\theta^{N-1} e^{-N\theta}. \tag{2.87}$$

The mean and standard deviation of the dimensionless distribution defined by Eq. (2.87) are defined as:

$$\bar{\theta}_E = \int_0^\infty \theta E(\theta) d\theta; \quad \sigma_\theta^2 = \int_0^\infty (\theta - \bar{\theta}_E)^2 E(\theta) d\theta = \int_0^\infty \theta^2 E(\theta) d\theta - \bar{\theta}_E^2. \tag{2.88}$$

Applying these definitions to E_N^θ as given by Eq. (2.87), we get

$$\bar{\theta}_{E_N^\theta} = 1, \quad \sigma_{E_N^\theta}^2 = 1/N. \tag{2.89}$$

Furthermore, we can show that

$$\theta\left(E_N^\theta\big|_{\max}\right) = (1 - 1/N), \quad E_N^\theta\big|_{\max} = N\frac{(N-1)^{N-1}}{(N-1)!} e^{-(N-1)}. \tag{2.90}$$

Thus

$$\theta\left(E_N^\theta\big|_{\max}\right) \to 1 \text{ as } N \to \infty, \quad E_N^\theta\big|_{\max} \to \infty \text{ as } N \to \infty. \tag{2.91}$$

On comparing Eqs. (2.89) and (2.91) with Eqs. (2.79)–(2.81) and Eqs. (2.84) and (2.85), we note that the limiting behavior of E_N^θ follows that of E_N^τ and is in contrast to that of $E_N^{\tau_i}$.

One may also obtain the F curve for N CSTRs in series.[13] We consider the case of fixed total space time τ so that the space time of each CSTR is given by $\tau_i = \tau/N$. Thus applying Eq. (2.74) for the step input (Eq. 2.15) for which $\widehat{C}_{in}(s) = C_\infty/s$, we get

$$C_{out}^{step}(t) = L^{-1}\left(\frac{C_\infty}{[(\tau/N)s + 1]^N}\frac{1}{s}\right).$$

From the tables,[14] we have

$$L^{-1}\left(\frac{1}{s[s+a]^n}\right) = a^{-n}\left(1 - e^{-at} e_{n-1}(at)\right), \quad e_{n-1}(z) = 1 + \frac{z}{1!} + \frac{z^2}{2!} + \cdots \frac{z^{n-1}}{(n-1)!}.$$

Applying the above formula, we obtain

$$C_{out}^{step}(t) = C_\infty\left\{1 - e^{-Nt/\tau}\left[1 + \frac{Nt}{\tau} + \frac{1}{2!}\left(\frac{Nt}{\tau}\right)^2 + \cdots \frac{1}{(N-1)!}\left(\frac{Nt}{\tau}\right)^{N-1}\right]\right\}.$$

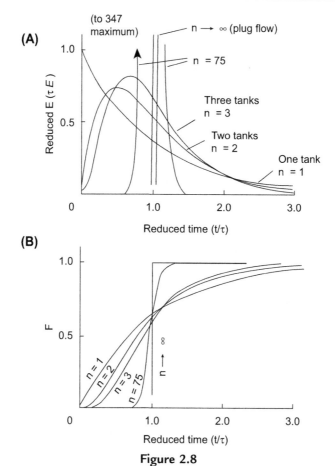

Figure 2.8

Curves for tracer leaving different sets of stirred tanks in series, each set having the same total space time τ. (A) **E**-curves: E^θ versus θ. (B) **F**-curves: F versus θ.

Hence, from Eq. (2.16), we obtain

$$F_N^\tau(t) = 1 - e^{-Nt/\tau}\left[1 + \frac{Nt}{\tau} + \frac{1}{2!}\left(\frac{Nt}{\tau}\right)^2 + \cdots \frac{1}{(N-1)!}\left(\frac{Nt}{\tau}\right)^{N-1}\right].$$

In terms of the dimensionless time θ, we obtain

$$F_N^\tau(\theta) = 1 - e^{-N\theta}\left[1 + N\theta + \frac{1}{2!}(N\theta)^2 + \cdots \frac{1}{(N-1)!}(N\theta)^{N-1}\right]. \qquad (2.92)$$

Fig. 2.8 shows a plot of E_N^θ, denoted as reduced E, and F_N^τ, denoted as F, versus the reduced time θ, given by Eq. (2.86). In both plots, we see the trend toward PFR as $N \to \infty$.

2.3.1 Predicting Reactor Conversion From Tanks-in-Series Model

From tracer data, one may construct the $E(t)$ curve and determine the mean and variance. Given that the space time of the reactor (τ) is specified, we may use Eq. (2.81) to determine N. Then the conversion expected from the reactor may be determined if the kinetics is known. For the first-order kinetics, the concentration of reactant A at the exit may be determined from

$$C_{A,out} = \frac{C_{A,in}}{(1 + k_1 \tau/N)^N} \tag{2.93}$$

which results from repeated application of Eq. (2.63) that applies for a first-order reaction in an ideal CSTR. There is no guarantee that the value of N obtained from Eq. (2.81) would turn out to be an integer, but Eq. (2.93) can still be used. For second-order reactions, the relation between the inlet and exit concentration of A for the rth reactor is given by (see Eq. 2.64)

$$C_{Ar} = \frac{-1 + \sqrt{(1 + 4k_2\tau_r C_{Ar-1})}}{2k_2\tau_r} \tag{2.94}$$

where τ_r is the space time of the rth reactor and is here independent of r, and it is given by τ/N for all r. Successive application of Eq. (2.94) can give the conversion expected from N reactors in series. Again, if N turns out to be a noninteger, then an interpolation between the values obtained for the nearest integer less than and greater than N can be used to get an estimate of the exit concentration or conversion.

2.4 Dispersed Plug Flow Model

2.4.1 Axial Dispersion and Model Development

The dispersed plug flow model can be regarded as the first stage of development from the simple idea of plug flow along a pipe. The fluid velocity and the concentrations of any dissolved species are assumed to be uniform across any section of the pipe, but here mixing or "dispersion" in the direction of flow (i.e., in the axial z-direction) is taken into account (Fig. 2.9). The axial mixing is described by a flux, which is related to the axial concentration gradient through a parameter D_L in exactly the same way as in molecular diffusion. Thus the total molar flux of any species i in the z-direction is assumed to be given by

$$N_i = C_i v_i = -D_L \frac{\partial C_i}{\partial z} + C_i u \tag{2.95}$$

where C_i is the concentration of species i, v_i is the velocity of species i, u is the mean fluid velocity, and D_L is the *dispersion coefficient* in the longitudinal direction. Although

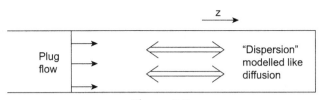

Figure 2.9

The dispersed plug flow model with mixing (dispersion) in axial z-direction; dispersion coefficient D_L.

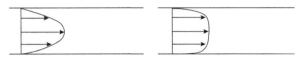

Figure 2.10

Actual velocity distribution in a pipe: (A) laminar flow and (B) turbulent flow.

Eq. (2.95) resembles the equation for molecular diffusion, the physical processes involved in flow dispersion are distinctly different.

Investigations into the underlying flow mechanisms that actually cause axial mixing in a pipe have shown that, in both laminar and turbulent flow, the nonuniform velocity profiles (see Fig. 2.10) are primarily responsible. Fluid near the center of the pipe travels more quickly than that near the wall, the overall result being mixing in the axial direction. This effect is most easily visualized for laminar flow in a pipe, although in reactor design the turbulent regime is generally the more important in pipes, vessels, and packed beds.

Consider a liquid, initially colorless, passing in laminar flow along a pipe, then imagine the incoming stream to be suddenly switched to a red liquid of the same viscosity, etc. The boundary between the two liquids is initially a plane disc at the inlet of the pipe (Fig. 2.11A). Because the liquid velocity is a maximum at the center of the pipe and zero at the walls, as time proceeds this boundary becomes stretched out into a conelike shape with a parabolic profile (Fig. 2.11B). After the tip of this cone has passed a particular point such as **A** (Fig. 2.11C), a section **AA′** across the pipe will show a central red core, whereas the surrounding liquid will still be colorless. The fraction of red liquid averaged across the section will increase with time of flow, giving the impression of mixing in the axial direction. In laminar flow in tubes of small length-to-diameter ratio, the sharpness of the boundary between the red and colorless liquid is only slightly blurred by the occurrence of molecular diffusion because, in liquids, diffusion is slow. With gases, however, the effect of diffusion will be greater (see, for example, Section 2.4.6, Fig. 2.19 for the effect of Schmidt Number on dispersion coefficient). In turbulent flow, the effect of radial eddy diffusion is to make this boundary so diffuse that concentrations become effectively uniform across the pipe, as required by the *dispersed plug flow* model. Even in laminar flow, for long small-diameter tubes, there exists a regime known as

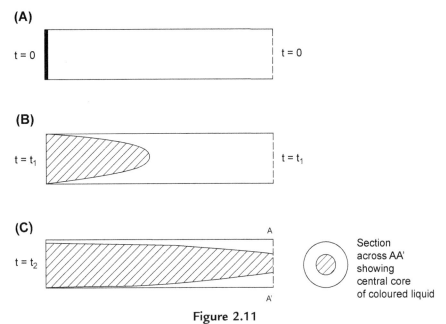

Figure 2.11
Sudden switch from clear to colored liquid in laminar flow in a pipe.

Taylor—Aris[15–17] dispersion in which radial diffusion is sufficient to produce an effectively uniform concentration over any cross section of the tube. The general conclusion that, even for turbulent flow, axial dispersion in a pipe is caused primarily by differences in velocity at different radial positions rather than by turbulent eddy diffusion in the axial direction may seem strange at first sight to the reader.

If the radial diffusion or radial eddy transport mechanisms considered above are insufficient to smear out any radial concentration differences, then the simple *dispersed plug flow* model becomes inadequate to describe the system. It is then necessary to develop a mathematical model for simultaneous radial and axial dispersion incorporating both radial and axial dispersion coefficients. This is especially important for fixed bed catalytic reactors and packed beds generally.

The model developed below is for dispersion in the axial direction only. Because the underlying mechanisms producing axial dispersion are complex as the discussion above shows, Eq. (2.95) is best regarded as essentially a mathematical definition of the dispersion process and the dispersion coefficient D_L. Note that the dispersed plug flow model uses just D_L to describe axial mixing and for that reason is classed as a "one-parameter model."

2.4.2 Basic Differential Equation

Consider a fluid flowing steadily along a uniform pipe as depicted in Fig. 2.12; the fluid will be assumed to have a constant density so that the mean velocity u is constant

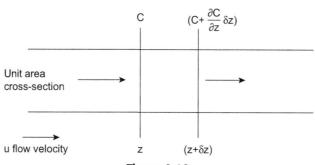

Figure 2.12
Material balance over an element of pipe.

(assuming one-dimensional flow). Let the fluid be carrying along the pipe a small amount of a tracer, which has been injected at some point upstream as a pulse distributed uniformly over the cross section; the concentration C of the tracer is sufficiently small not to affect the density. Because the system is not in a steady state with respect to the tracer distribution, the concentration will vary with both z the position in the pipe and, at any fixed position, with time; i.e., C is a function of both z and t but, at any given value of z and t, C is assumed to be uniform across that section of pipe. To obtain the differential equation governing $C(z, t)$, we specialize the continuity Eq. (1.14) to the one-dimensional case under consideration. Thus, we get

$$\frac{\partial C_i}{\partial t} + \frac{\partial}{\partial z}(C_i v_i) = r_i \tag{2.96}$$

for any species i. Applying the equation to the tracer with $C_i \equiv C$, substituting for $C_i v_i = N_i$ from Eq. (2.95) and with $r_i = 0$ (since tracer experiments typically do not involve chemical reactions), we obtain

$$\frac{\partial C}{\partial t} + \frac{\partial}{\partial z}\left(-D_L\frac{\partial C}{\partial z} + Cu\right) = 0$$

which, for constant D_L and u, reduces to

$$\frac{\partial C}{\partial t} = D_L\frac{\partial^2 C}{\partial z^2} - u\frac{\partial C}{\partial z}. \tag{2.97}$$

This is the basic differential equation governing the transport of a dilute tracer substance along a pipe. Being a partial differential equation, its solution, which gives the concentration C as a function of z and t, will be very much dependent on the boundary conditions that apply to any particular case. More insight may be obtained by introducing the dimensionless variables $z' = z/L$ and $\theta = ut/L$. This leads to

$$\frac{\partial C}{\partial \theta} = \frac{D_L}{uL}\frac{\partial^2 C}{\partial z'^2} - \frac{\partial C}{\partial z'}. \tag{2.98}$$

Eq. (2.98) reveals the dimensionless parameter D_L/uL, referred to as the *dispersion number*, as the defining parameter of the dispersion model.

(**Note**: Eq. (2.97) can also be derived by performing an elemental balance as indicated in Fig. 2.12 and as pointed out in Section 1.7.1 for the tubular reactor).

2.4.3 Response to an Ideal Pulse Input of Tracer

Fig. 2.13 shows, in principle, the result of injecting an instantaneous pulse of tracer into fluid in a pipe in which the differential equation above applies. Following the injection of the pulse at time $t = 0$, profiles of concentration versus position z in the pipe are shown at three successive "snapshots" in time, t_1, t_2, and t_3. The inset graph shows how the concentration of tracer would change with time for measurements taken at a fixed position $z = L$. Although this graph may look roughly similar in shape to the "snapshot" profiles in the pipe, it is distinctly different in principle. When the results of such measurements are "normalized" (see Section 2.1.3), the "**C**-curve" for the tracer response is obtained.

Mathematically, the solution to the partial differential Eq. (2.97) for a pulse input of n_0 moles of tracer into an *infinitely long* pipe of cross-sectional area A_C is[2]:

$$C(z,t) = \frac{n_0}{2A_C\sqrt{\pi D_L t}} \exp\left(-\frac{(z - ut)^2}{4D_L t}\right) \tag{2.99}$$

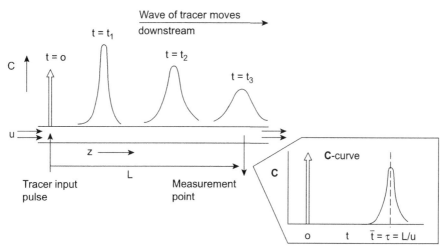

Figure 2.13

Response of the dispersed plug flow model to a pulse input of tracer. Three "snapshot" concentration profiles at different times t_1, t_2, and t_3 are shown. The inset shows the "**C**-curve" derived from measurements at a fixed position, $z = L$. If axial dispersion occurs to only a small extent (small D_L/uL), the **C**-curve is almost symmetrical.

That this equation is indeed a solution of Eq. (2.97) may be verified by partial differentiation with respect to t and z and substituting. It also satisfies the initial condition corresponding to an ideal pulse, i.e., when $t \to 0+$, $z = 0$, and $C \to \infty$; but, at $z \neq 0$, $C = 0$. In addition, at any time $t > 0$, the total moles of tracer anywhere in the pipe must be equal to n_0. That is,

$$n_0 = \int_{-\infty}^{\infty} C(z; t) A_C \, dz$$

At any particular time $t = t_1$, Eq. (2.99) becomes:

$$C(z, t_1) = \frac{n_0}{2A_C \sqrt{\pi D_L t_1}} \exp\left(-\frac{(z - ut_1)^2}{4D_L t_1} \right) \tag{2.100}$$

which has the same form as the general equation for a Gaussian distribution (Eq. 2.104). The "snapshot" concentration profiles of C versus z shown in Fig. 2.13 are therefore *symmetrical* Gaussian-shaped curves.

Now let us see what happens to the tracer concentration at a fixed position $z = L$; Eq. (2.99) then becomes:

$$C(L, t) = C_L(t) = \frac{n_0}{2A_C \sqrt{\pi D_L t}} \exp\left(-\frac{(L - ut)^2}{4D_L t} \right) \tag{2.101}$$

The fact that all the tracer introduced eventually passes through the location $z = L$ is represented by the condition that

$$n_0 = \int_0^{\infty} C_L(t) q_r dt$$

where q_r is the volumetric flow rate, assumed constant. The above condition may be easily verified by integrating Eq. 2.101 and noting that $q_r = uA_C$. As an equation for C_L as a function of t, Eq. 2.101, however, does not in general give a symmetrical Gaussian curve. The reason why it is not symmetrical can be seen from Fig. 2.14. If the concentration wave of tracer moving along the pipe changes its shape significantly while passing a fixed observation point, then the graph of the observed concentration C_L versus time will be skewed.

Mathematically, the skewness of the C_L versus t curve can be identified with the presence of t in Eq. (2.101) in the positions arrowed:

$$C_L(t) = \frac{n_0}{2A_C \sqrt{\pi D_L t}} e^{-\frac{(L - ut)^2}{4D_L t}} \tag{2.102}$$

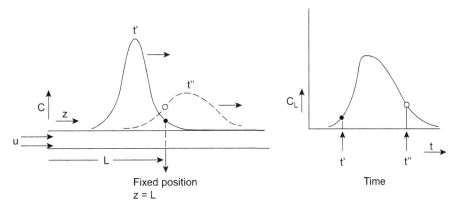

Figure 2.14

Dispersion of an ideal pulse of tracer injected into a pipe. Formation of an asymmetric curve of C versus t at a fixed point due to change in shape of a symmetrical concentration wave as it broadens while passing the point. D_L/uL value is large.

In many instances, however, the change in shape of the concentration wave in passing the observation point is negligibly small. Mathematically, this corresponds to a value $D_L/uL < 0.01$, approximately. In this case the arrowed t in the equation above may be replaced by $\bar{t} = L/u$, the mean residence time of the pulse in the section of pipe between $z = 0$ and $z = L$. (Because the shape of the concentration wave is now considered not to change in passing the observation point, this is also the time at which the peak of the wave passes $z = L$.)

Then the term $\dfrac{n_0}{2A_C\sqrt{\pi D_L t}}$ becomes $\dfrac{n_0}{2A_C\sqrt{\pi D_L \bar{t}}}$ or $\dfrac{n_0}{2A_C\sqrt{\dfrac{\pi D_L}{uL}uL\bar{t}}}$ or $\dfrac{n_0}{A_C\sqrt{2\pi}\sqrt{\dfrac{2D_L}{uL}u^2\bar{t}^2}}$ or

$\dfrac{n_0}{A_Cu\sqrt{2\pi}\sqrt{\dfrac{2D_L}{uL}\bar{t}^2}}$.

Also $e^{-\frac{(L-ut)^2}{4D_L\,t}}$ becomes $e^{-\frac{(L-ut)^2}{4D_L\,\bar{t}}}$ or $e^{-\frac{(L-ut)^2}{4D_L\,L/u}}$ or $e^{-\frac{(\bar{t}-t)^2}{(4D_L)(L/u^3)}}$ or $e^{-\frac{(\bar{t}-t)^2}{4(D_L/uL)(L/u)^2}}$ or $e^{-\frac{(t-\bar{t})^2}{2(2D_L/uL)\bar{t}^2}}$.

Thus, Eq. (2.102) can be written as

$$C(L,t) \simeq \frac{n_0}{A_C u}\,\frac{1}{\sqrt{2\pi}\sqrt{(2D_L/uL)\bar{t}^2}}\,\exp\left(-\frac{(t-\bar{t})^2}{2(2D_L/uL)\bar{t}^2}\right) \qquad (2.103)$$

We may compare this equation with the general form of equation for a Gaussian distribution:

$$f = \frac{1}{\sqrt{2\pi\sigma^2}}\,e^{-\frac{(x-\bar{x})^2}{2\sigma^2}} \qquad (2.104)$$

where f is the frequency function for an observation of magnitude x, $\bar{x}$ is the mean, and σ^2 is the variance. We see that they correspond, apart from the $n_0/A_C u$ factor, if:

$$\sigma^2 = \frac{2D_L}{uL}\bar{t}^2 \tag{2.105}$$

Eqs. (2.103) and (2.104) can be matched exactly if the C_L versus t curve is converted into a **C**-curve (Section 2.1.3) by means of a normalizing factor. That is,

$$\mathbf{C}_L = \frac{C_L}{\int_0^\infty C_L dt}$$

Here the normalizing quantity $\int_0^\infty C_L dt$ is:

$$\int_0^\infty C_L dt = \frac{n_0}{A_C u}\frac{1}{\sqrt{2\pi}\sigma}\int_0^\infty \exp\left(-\frac{(t-\bar{t})^2}{2\sigma^2}\right) dt = \frac{n_0}{A_C u}\frac{1}{2}\left(1 + erf\frac{\bar{t}}{\sqrt{2}\sigma}\right)$$

where[7] $erf(y) = \frac{2}{\sqrt{\pi}}\int_0^y \exp(-x^2) dx$. From Eq. (2.105),

$$\frac{\bar{t}}{\sqrt{2}\sigma} = \frac{1}{2}\sqrt{\frac{uL}{D_L}}.$$

Thus, for, $D_L/uL < 0.01$, $\bar{t}/(\sqrt{2}\sigma) > 5$ and hence $erf\left[\bar{t}/(\sqrt{2}\sigma)\right] \approx 1$. Thus, under the approximation for which Eq. (2.103) is valid, we obtain

$$\int_0^\infty C_L dt \approx \frac{n_0}{A_C u}.$$

We conclude therefore that, for small values of the dispersion number ($D/uL < 0.01$), the **C**-curve for a pulse input of tracer into a pipe is symmetrical around $t = \bar{t}$ and corresponds exactly to Eq. (2.104) for a Gaussian distribution function:

$$\mathbf{C}_L = \frac{1}{\sqrt{2\pi}\sqrt{(2D_L/uL)\bar{t}^2}}\exp\left(-\frac{(t-\bar{t})^2}{2(2D_L/uL)\bar{t}^2}\right) \tag{2.106}$$

The above equations can also be expressed in a fully dimensionless form. Using the dimensionless time θ introduced above in Eq. (2.98) and which may also be written as $\theta = t/\bar{t}$ since $\bar{t} = L/u$ (note that in dimensionless form the mean time $\bar{\theta} = 1$), then Eq. (2.106) becomes:

$$\mathbf{C}_L = \frac{1}{\bar{t}}\frac{1}{\sqrt{2\pi}\sqrt{(2D_L/uL)}}\exp\left(-\frac{(1-\theta)^2}{2(2D_L/uL)}\right) \tag{2.107}$$

Carrying out a further stage of normalization on this equation, i.e., introducing $\mathbf{C}_\theta = \bar{t}\,\mathbf{C}$ so that the area under the $\mathbf{C}_\theta$ versus θ curve is unity, then:

$$\mathbf{C}_\theta(L, \boldsymbol{\theta}) = \frac{1}{\sqrt{2\pi}\sqrt{(2D_L/uL)}} \exp\left(-\frac{(1-\boldsymbol{\theta})^2}{2(2D_L/uL)}\right) \tag{2.108}$$

which has a mean $\bar{\theta} = 1$ and a variance $\sigma^2 = 2D_L/uL$. Essentially this is the solution to Eq. (2.98) expressed in terms of the dimensionless time and distance and evaluated at $z' = 1$.

2.4.4 Experimental Determination of Dispersion Coefficient From a Pulse Input

Provided that the introduction of an ideal pulse input of tracer can be achieved experimentally, Eq. (2.105) allows the determination of dispersion coefficients from measurements of tracer concentration taken at a sampling point distance L downstream. Again, in an ideal situation, a large number of measurements of concentration would be taken so that a continuous graph of C versus t could be plotted (Fig. 2.15). The two most important parameters that characterize such a curve are (1) the mean time $\bar{t}$ that indicates when the wave of tracer passes the measuring point and (2) the variance σ^2 that indicates how much the tracer has spread out during the measurement time, i.e., the width of the curve. Thus, given a smooth continuous pulse response curve (in general, it could be symmetrical or not):

$$\text{Area under the curve} = \int_0^\infty C\,dt \tag{2.109}$$

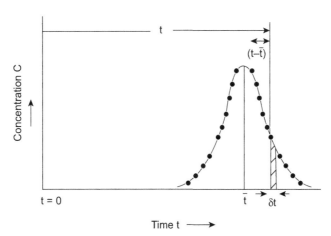

Figure 2.15

Experimental determination of dispersion coefficient—ideal pulse injection. Expected symmetrical distribution of concentration measurements. D_L/uL value is small.

$$\text{Mean (or first moment) } \bar{t} = \frac{\displaystyle\int_0^\infty Ct\,dt}{\displaystyle\int_0^\infty C\,dt} \tag{2.110}$$

$$\text{Variance (or second moment about the mean) } \sigma^2 = \frac{\displaystyle\int_0^\infty C(t-\bar{t})^2\,dt}{\displaystyle\int_0^\infty C\,dt}. \tag{2.111}$$

From Eq. (2.37), the variance can also be expressed as:

$$\sigma^2 = \frac{\displaystyle\int_0^\infty Ct^2\,dt}{\displaystyle\int_0^\infty C\,dt} - \bar{t}^2 \tag{2.112}$$

These relationships can be applied either to the C versus t curve directly or to the already normalized $\mathbf{C}$-curve, in which case the area under the curve $\int_0^\infty C\,dt$ is then unity. There are several different types of experimental data that might be obtained.

2.4.4.1 Many Equally Spaced Points

If we have a large number of data points equally spaced at a time interval Δt, the above relationships for mean and variance may be written in finite difference form:

$$\bar{t} = \frac{\sum C_i t_i (\Delta t)}{\sum C_i (\Delta t)} \qquad \sigma^2 = \frac{\sum C_i t_i^2 (\Delta t)}{\sum C_i (\Delta t)} - \bar{t}^2$$

That is,

$$\bar{t} = \frac{\sum C_i t_i}{\sum C_i} \tag{2.113}$$

$$\sigma^2 = \frac{\sum C_i t_i^2}{\sum C_i} - \bar{t}^2 \tag{2.114}$$

2.4.4.2 Relatively Few Data Points but Each Concentration C_i Measured Instantaneously at Time t_i (Fig. 2.16A)

Linear interpolation may be used; this is equivalent to joining the data points with straight lines. Thus, the area under the $C-t$ curve is approximated by the sum of trapezium-shaped increments.

$$\bar{t} = \frac{\displaystyle\sum_{i=1}^{n-1}\left\{\left(\frac{C_{i+1}+C_i}{2}\right)\left(\frac{t_{i+1}+t_i}{2}\right)(t_{i+1}-t_i)\right\}}{\displaystyle\sum_{i=1}^{n-1}\left\{\left(\frac{C_{i+1}+C_i}{2}\right)(t_{i+1}-t_i)\right\}} \tag{2.115}$$

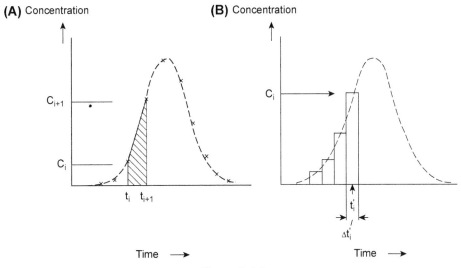

Figure 2.16

Experimental determination of dispersion coefficient: (A) treatment of data by linear interpolation; (B) treatment of mixing-cup data.

$$\sigma^2 = \frac{\sum\limits_{i=1}^{n-1}\left\{\left(\dfrac{C_{i+1}+C_i}{2}\right)\left(\dfrac{t_{i+1}+t_i}{2}\right)^2(t_{i+1}-t_i)\right\}}{\sum\limits_{i=1}^{n-1}\left\{\left(\dfrac{C_{i+1}+C_i}{2}\right)(t_{i+1}-t_i)\right\}} - \bar{t}^2 \tag{2.116}$$

Note that the terms $(t_{i+1} - t_i)$ in the numerator and denominator of these expressions will cancel out if the concentration data are for equal time intervals.

2.4.4.3 Data Collected by a "Mixing Cup"

Over a series of time intervals Δt_i, each sample is collected by allowing a small fraction of the fluid to flow steadily into a little cup; these samples are stirred to make them uniform and then analyzed. Each sample thus represents the concentration in some way averaged over the interval Δt_i. When these concentrations are plotted as the stepwise graph shown in Fig. 2.16B, the area under the graph is equal to the summation $\Sigma C_i \Delta t_i$ but, to approximate Eqs. (2.110) and (2.111), the appropriate times at which to take moments (first and second) are not immediately apparent because of the unknown nature of the concentration averaging process. Arbitrarily, we take the mid-increment value t_i' as shown.

$$\bar{t} = \frac{\sum\limits_{i} C_i t_i' \Delta t_i}{\sum\limits_{i} C_i \Delta t_i} \tag{2.117}$$

Similarly,

$$\sigma^2 = \frac{\sum_i C_i t_i'^2 \Delta t_i}{\sum_i C_i \Delta t_i} - \bar{t}^2 \qquad (2.118)$$

Here also the Δt_i in both numerator and denominator will cancel for equal time intervals.

Example 2.1

Numerical Calculation of Residence Time and Dispersion Number.

A pulse of tracer is introduced at the inlet of a vessel through which fluid flows at a steady rate. Samples, each taken virtually instantaneously from the outlet stream at time t_i after injection, are analyzed with the following results:

Time t_i(s)	0	20	40	60	80	100	120	140	160	180	200
Concentration C_i (kmol/m³) × 10³	0	0	0	0	0.4	5.5	16.2	11.1	1.7	0.1	0

Calculate the mean residence time of the fluid in the vessel and the dispersion number.

Solution

Although for a closed vessel the boundary conditions are different from the case of an open pipe (see Section 2.4.5), we will assume that the above methods of data treatment apply because, as will subsequently be verified, the dispersion number in this example is small.

Method A: Finite difference form—equal time intervals

This is the simplest method, although we cannot expect a very high degree of accuracy because the number of data points is relatively small.

Time t_i	0	20	40	60	80	100	120	140	160	180	200
$C_i(\times 10^3)$	0	0	0	0	0.4	5.5	16.2	11.1	1.7	0.1	0
$C_i t_i(\times 10^3)$	0	0	0	0	32	550	1,944	1,554	272	18	0
$C_i t_i^2(\times 10^3)$	0	0	0	0	2560	55,000	233,280	217,280	43,520	3240	0

Note that because evaluation of $\bar{t}$ and σ^2 involves ratios, multiplication of the concentrations by a common factor (here 10^3) to avoid inconveniently large or small numbers does not affect the result.

Thus from Eqs. (2.113) and (2.114):

$$\bar{t} = \frac{\sum C_i t_i}{\sum C_i} = \frac{4370}{35} = \underline{124.9 \text{ s}}$$

and

$$\sigma^2 = \frac{\sum C_i t_i^2}{\sum C_i} - \bar{t}_i^2 = \frac{555,160}{35} - \left(\frac{4370}{35}\right)^2 = 272.4 \text{ s}^2$$

therefore

$$\frac{\sigma^2}{\bar{t}^2} = \frac{272.4}{124.9^2} = 0.0175$$

Hence the dispersion number $\left(\frac{D_L}{uL}\right) = \frac{0.0175}{2} = \underline{0.0087}$.

Note that the dispersion number is thus confirmed as being "small," i.e., <0.01 and within the range for which the Gaussian approximation to the **C**-curve is acceptable.

Method B: Treatment of data by linear interpolation

	40	60	70	80	90	100	110	120	130	140	150	160	170	180	190	200
t_i	40	60		80		100		120		140		160		180		200
$C_i(10^3)$	0	0		0.4		5.5		16.2		11.1		1.7		0.1		0
$\frac{(t_{i+1}+t_i)}{2}$			70		90		110		130		150		170		190	
$\frac{(C_{i+1}+C_i)}{2}$			0.2		2.95		10.85		13.65		6.4		0.9		0.05	
$\left(\frac{C_{i+1}+C_i}{2}\right)\left(\frac{t_{i+1}+t_i}{2}\right)$			14		265.5		1,193.5		1,774.5		960		153		9.5	
$\left(\frac{C_{i+1}+C_i}{2}\right)\left(\frac{t_{i+1}+t_i}{2}\right)^2$			980		23,895		131,285		230,685		144,000		26,010		1,805	

$$\sum = 35$$
$$\sum = 4370$$
$$\sum = 558,660$$

Using Eqs. (2.115) and (2.116):

$$\bar{t} = \frac{4370}{35} = \underline{124.9 \text{ s}}$$

and

$$\sigma^2 = \frac{558,660}{35} - 124.9^2 = 15,961.7 - 15,598.3 = 372.4 \text{ s}^2$$

Hence

$$\frac{\sigma^2}{\bar{t}^2} = \frac{372.4}{124.9^2} = 0.0239$$

Hence

$$\text{Dispersion Number } \left(\frac{D_L}{uL}\right) = \frac{0.0239}{2} = \underline{0.0119}$$

The above example demonstrates that treatment of the basic data by different numerical methods can produce distinctly different results. The discrepancy between the results in this case is, in part, due to the inadequacy of the data provided; the data points are too few in number and their precision is poor. A lesson to be drawn from this example is that tracer experiments set up with the intention of measuring dispersion coefficients accurately need to be very carefully designed. As an alternative to the pulse injection method considered here, it is possible to introduce the tracer as a continuous sinusoidal concentration wave (Fig. 2.3C), the amplitude and frequency of which can be adjusted. Also there is a variety of different ways of numerically treating the data from either pulse or sinusoidal injection so that more weight is given to the most accurate and reliable of the data points. There has been extensive research to determine the best experimental method to adopt in particular circumstances.[13,18]

2.4.5 Further Development of Tracer Injection Theory

Some interesting and useful conclusions can be drawn from further examination of the response of a system to an input of tracer although the mathematical derivation of these conclusions[13] is too involved to be considered here.

2.4.5.1 Significance of the Boundary Conditions

In the early days of the development of the dispersed plug flow model, there was considerable controversy over the precise formulation of the boundary conditions that should apply to the basic partial differential Eq. (2.97) in various situations.[1,19] For a simple straight length of pipe as depicted in Fig. 2.13 the flow is undisturbed as it passes the injection and observation points, and the dispersion coefficient has the same value both upstream and downstream of each of these boundary locations. Under these circumstances the boundary is described as "open." For a reaction vessel, however, it may be more realistic to assume plug flow in the inlet pipe and dispersed flow in the vessel itself. In this case the inlet boundary is said to be "closed," i.e., there is no possibility of dispersion of the tracer back upstream into the plug flow region. For an example of boundary conditions applied to the case of a first-order reaction, see Section 2.4.7.

2.4.5.2 Dispersion Coefficients From Nonideal Pulse Data

The exact formulation of the inlet and outlet boundary conditions becomes important only if the dispersion number (D_L/uL) is large (>0.01). Fortunately, when D_L/uL is small (<0.01) and the **C**-curve approximates to a normal Gaussian distribution, differences in behavior between open and closed types of boundary condition are not significant. Also, for small dispersion number D_L/uL, it has been shown rather surprisingly that we *do not need to have ideal pulse injection* to obtain dispersion coefficients from **C**-curves. A tracer pulse of any arbitrary shape is introduced at any convenient point upstream and the

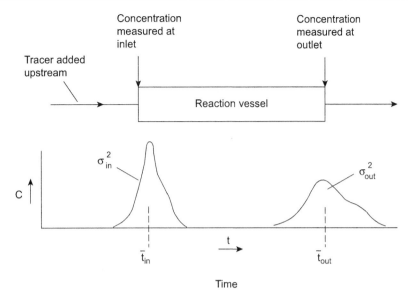

Figure 2.17

Determination of dispersion coefficient from measurements of **C**-curves at both inlet and outlet of reaction vessel. Tracer added at any convenient point upstream as nonideal pulse. D_L/uL value is small.

concentration measured over a period of time at both inlet and outlet of a reaction vessel whose dispersion characteristics are to be determined, as in Fig. 2.17. The means $\bar{t}_{in}$ and $\bar{t}_{out}$ and the variances σ^2_{in} and σ^2_{out} for each of the **C**-curves are found.

Then if:

$$\Delta\bar{t} = \bar{t}_{out} - \bar{t}_{in} \quad and \quad \Delta\sigma^2 = \sigma^2_{out} - \sigma^2_{in}$$
$$\Delta L/u = \Delta\bar{t} \quad and \quad 2(D_L/uL) = \Delta\sigma^2 \Big/ (\Delta\bar{t})^2$$

where ΔL is the length of the reaction vessel. The mean residence time $\Delta\bar{t}$ may alternatively be expressed in terms of the volume V and the volumetric flow rate, i.e., $\Delta\bar{t} = V/q_r$.

Example 2.2
Tracer measurements with two sampling points

To investigate the dispersion characteristics of a reaction vessel, a nonreactive tracer is injected at a convenient location some way upstream from the inlet to the vessel. Subsequently, samples (taken over very short periods of time) are collected from two positions, one being the inlet pipe to the vessel and the other at the vessel outlet, with the following results:

First sampling point:

Time t_i (min)	≤1.5	2.0	2.5	3.0	3.5	4.0	4.5	5.0	5.5	≤6
Concentration C_i (kmol/m^3) $\times$ 10^3	<0.1	0.2	4.7	14.6	23.5	18.5	5.7	1.6	0.4	<0.1

Second sampling point:

Time t_i (min)	≤5.5	6.0	6.5	7.0	7.5	8.0	8.5	9.0	9.5	10.0	10.5	≥11
Concentration C_i (kmol/m^3) $\times$ 10^3	<0.1	0.2	1.4	5.2	10.9	14.1	12.3	7.7	3.7	1.4	0.4	<0.1

Estimate (1) the mean residence time in the vessel and (2) the value of the dispersion number for the vessel.

Solution

Proceeding in the same way as for method A in the preceding example, and for convenience multiplying each concentration by 10^3:

First sampling point:

$$\sum (C_i)_1 = 69.2; \quad \sum (C_i t_i)_1 = 248.05; \quad \sum (C_i t_i^2)_1 = 912.975$$

$$\bar{t}_1 = \frac{248.05}{69.2} = 3.585 \text{ min}$$

$$\sigma_1^2 = \frac{912.975}{69.2} - 3.585^2 = 0.341 \text{ min}^2$$

Second sampling point:

$$\sum (C_i)_2 = 57.3; \quad \sum (C_i t_i)_2 = 468.45; \quad \sum (C_i t_i^2)_2 = 3867.075$$

$$\bar{t}_1 = \frac{468.45}{57.3} = 8.175 \text{ min}$$

$$\sigma_2^2 = \frac{3867.075}{57.3} - 8.175^2 = 0.658 \text{ min}^2$$

then

$$\Delta \bar{t} = \bar{t}_2 - \bar{t}_1 = 8.175 - 3.585 = \underline{4.59 \text{ min}}$$

Thus, 4.59 min is the mean residence time in the vessel.

$$\Delta \sigma^2 = \sigma_2^2 - \sigma_1^2 = 0.658 - 0.341 = 0.317 \text{ min}^2$$

$$\frac{\Delta \sigma^2}{(\Delta \bar{t})^2} = \frac{0.31}{4.59^2} = 0.0150 = 2\left(\frac{D_L}{u\Delta L}\right)$$

Therefore,

$$\text{Dispersion number } \left(\frac{D_L}{u\Delta L}\right) = \frac{0.0150}{2} = 0.0075$$

This value thus satisfies the condition that the dispersion number must be small (<0.01) for the above method to apply.

Note that, because the same total amount of tracer must pass the two observation points (assuming that no tracer is lost by adsorption on the vessel walls—which can happen), then we should expect $\int_0^\infty C_i dt$ to be the same at both positions; i.e., because the time interval between samples is the same (0.5 min) for each set of data, we would expect $\sum (C_i)_1 = \sum (C_i)_2$. The fact that they are distinctly different demonstrates the inaccuracies that may enter this type of calculation. Concentrations below 0.1×10^{-3} kmol/m^3 are not taken into account; this means that possibly a long drawn-out "tail" of concentrations below this value is ignored even though the area under this part of the **C**-curve may be significant in total, especially for the more widely spread curve from the second sampling point.

2.4.5.3 Pulse of Tracer Moving Through a Series of Vessels

The principles of additivity of residence times and additivity of variances can be extended to several reaction vessels in series (Fig. 2.18). Thus, for vessels **A**, **B**, and **C**, if $\bar{t}_A$, $\bar{t}_B$, and $\bar{t}_C$ are the residence times and σ_A^2, σ_B^2, and σ_C^2 are the variances that would result from a pulse of tracer passing through each of the vessels in turn, then the overall residence time will be $\bar{t}_A + \bar{t}_B + \bar{t}_C$ and the overall increase in the variance of a pulse will be $\sigma_A^2 + \sigma_B^2 + \sigma_C^2$. The limitation is again that the dispersion number for the vessel must be small.

2.4.6 Values of Dispersion Coefficients From Theory and Experiment

As might be expected, the dispersion coefficient for flow in a circular pipe is determined mainly by the Reynolds number Re. Fig. 2.19 shows the dispersion coefficient plotted in the dimensionless form (D_L/ud) versus the Reynolds number $Re = \rho u d/\mu$.[2,20] In the turbulent region, the dispersion coefficient is affected also by the wall roughness, whereas in the laminar region, where molecular diffusion plays a part, particularly in the radial direction, the dispersion coefficient is dependent on the Schmidt number $Sc(\mu/\rho D)$, where

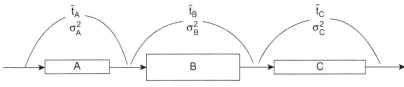

Figure 2.18
Additivity of mean residence times $\bar{t}_A$, $\bar{t}_B$, and $\bar{t}_C$ and variances σ_A^2, σ_B^2, and σ_C^2 for reaction vessels A, B, and C (connected in series by plug flow sections). D_L/uL values are small for each vessel.

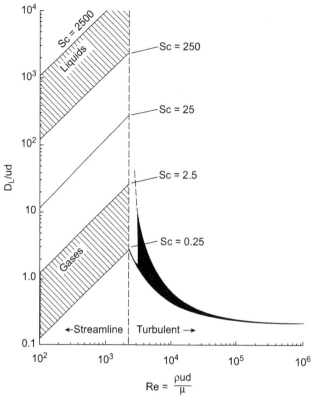

Figure 2.19
Dimensionless axial dispersion coefficients for fluids flowing in circular pipes. In the turbulent region, graph shows upper and lower limits of a band of experimentally determined values. In the laminar region the lines are based on the theoretical Eq. 2.120.

D is the molecular diffusion coefficient. For the laminar flow region where the Taylor—Aris theory[15–17] (Section 2.4.1) applies:

$$D_L = D + \frac{1}{192} \frac{u^2 d^2}{D} \tag{2.119}$$

where the first term represents the small additional effect on the dispersion coefficient of molecular diffusion in the axial direction. Eq. (2.119) may also be written:

$$\frac{D_L}{ud} = \frac{1}{Re\ Sc} + \frac{1}{192} Re\ Sc \tag{2.120}$$

This relation is plotted in Fig. 2.19, the first term having only a small effect in the Reynolds number range shown, except at low Schmidt numbers.

Dispersion coefficients for packed beds are usually plotted in the form of a similar dimensionless group, the Peclet number ud_p/D_L, which uses the diameter of the particles of the bed d_p as the characteristic length rather than the bed diameter.

Another interesting application of the data in Fig. 2.19 for dispersion coefficients in turbulent flow is in calculating the mixing that occurs in long pipelines. Many refined petroleum products are distributed by pipelines, which may extend over hundreds of kilometers. The same pipeline is used to convey several different products, each being pumped through the line for several hours before a sudden switch is made to the next product. At the receiving end of the line, some mixing will be found to have occurred at the interface between the two products, and a slug of the off-specification mixture needs to be diverted for reprocessing. Knowing the dispersion coefficient and the product specification required, the volume of mixture to be discarded may be calculated. An example of this calculation can be found in the Omnibook of Levenspiel.[21]

2.4.7 Dispersed Plug Flow Model With First-Order Chemical Reaction

We will consider a dispersed PFR in which a homogeneous irreversible first-order reaction takes place, the rate equation being $-r_A = k_1 C_A$ where A is a reactant. The reaction is assumed to be confined to the reaction vessel itself, i.e., it does not occur in the feed and outlet pipes. The temperature, pressure, and density of the reaction mixture will be considered uniform throughout. We will also assume that the flow is steady and that sufficient time has elapsed for conditions in the reactor to have reached a steady state. To obtain the governing differential equation, we consider Eq. (2.96). At steady state, $\partial C_i / \partial t = 0$. Substituting for $C_i v_i$ from Eq. (2.95) with $i \to A$, we get, for constant D_L and u,

$$-D_L \frac{d^2 C_A}{dz^2} + u \frac{dC_A}{dz} = r_A,$$

which, for a first-order reaction, reduces to

$$D_L \frac{d^2 C_A}{dz^2} - u \frac{dC_A}{dz} - k_1 C_A = 0. \tag{2.121}$$

The solution to this equation (unlike the partial differential Eq. 2.97) has been shown not to depend on the precise formulation of the inlet and outlet conditions, i.e., whether they are "open" or "closed."[22] In the following derivation, however, the reaction vessel is considered to be "closed," i.e., it is connected at the inlet and outlet by piping in which plug flow occurs and, in general, there is a flow discontinuity at both inlet and outlet. The boundary conditions to be used will be those that properly apply to a closed vessel. (See Section 2.4.5 regarding the significance of the boundary conditions for open and closed systems.)

To understand these boundary conditions, let us consider that the inlet pipe in which ideal plug flow occurs has the same diameter (shown by broken lines) as the reactor itself (Fig. 2.20). Inside the reactor, across any section perpendicular to the z-direction, the flux

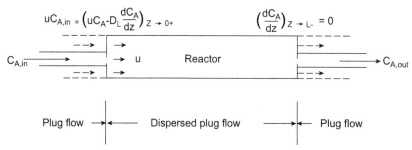

Figure 2.20
Dispersed plug flow with first-order homogeneous reaction $-r_A = k_1 C_A$; steady state, illustrating boundary conditions for a "closed" vessel.

of the reactant, i.e., the rate of transfer, is made up of two contributions, the convective flow uC_A and the diffusion-like dispersive flow $-D_L \frac{dC_A}{dz}$ (Eq. 2.95). That is,

$$\text{Flux within the reactor} = \left(uC_A - D_L \frac{dC_A}{dz} \right)$$

Because there is no dispersion in the inlet pipe, upstream of the reactor, there is only the convective contribution. That is,

$$\text{Flux into the reactor} = uC_{A,in}$$

Across the plane at the inlet to the reactor, *these two fluxes must be equal*; otherwise there would be a net accumulation at this plane. That is,

$$uC_{A,in} = \left(uC_A - D_L \frac{dC_A}{dz} \right)_{z \to 0+} \tag{2.122}$$

$z \to 0+$ meaning that z tends to zero from the positive direction. However, we do not expect the concentrations on either side of this inlet plane to be the same. On entering, the flow passes from an unmixed plug flow region to a mixed dispersed flow region. There will be a fall in concentration as the fluid enters the reactor in just the same way as there is a fall in concentration when the feed enters a stirred tank reactor (Section 2.3).

Now consider the outlet pipe from the reactor. The same argument as was used for the inlet plane (Eq. 2.122) applies to the flux across the outlet plane because again there can be no accumulation. That is,

$$\left(uC_A - D_L \frac{dC_A}{dz} \right)_{z \to L-} = uC_{A,out} \tag{2.123}$$

However, the fluid is now passing from a mixed region to an unmixed plug flow region; again as in a stirred tank reactor, the fluid leaving the reactor must have the same concentration of reactant as the fluid just inside the outlet plane. Thus, if $C_{A,L}$ is the concentration at $z = L-$ just inside the reactor and $C_{A,out}$ is the concentration in the fluid leaving, then $C_{A,L} = C_{A,out}$. But if $C_{A,L} = C_{A,out}$, then from the equality of flux (Eq. 2.123) we must have:

$$-D_L \left(\frac{dC_A}{dz} \right)_{z \to L-} = 0$$

or

$$\left(\frac{dC_A}{dz} \right)_{z \to L-} = 0 \tag{2.124}$$

To solve the differential Eq. (2.121) subject to the boundary conditions 2.122 and 2.124, we adopt the standard method of substituting the trial solution $C_A = Ae^{mz}$ and obtain the auxiliary equation:

$$D_L m^2 - um - k_1 = 0$$

which has the solutions:

$$m = u(1 + a)/2D_L \quad \text{and} \quad m = u(1 - a)/2D_L$$

where $a = (1 + 4k_1 D_L/u^2)^{1/2}$, i.e.,

$$C_A = B_1 e^{\frac{uz}{2D_L}(1+a)} + B_2 e^{\frac{uz}{2D_L}(1-a)} \tag{2.125}$$

where B_1 and B_2 are arbitrary constants to be evaluated from the boundary conditions.

Differentiating Eq. (2.125):

$$\frac{dC_A}{dz} = \frac{u}{2D_L} \left\{ B_1(1 + a)e^{\frac{uz}{2D_L}(1+a)} + B_2(1 - a)e^{\frac{uz}{2D_L}(1-a)} \right\} \tag{2.126}$$

At $z \to 0+$, from Eq. (2.125): $C_{A,0+} = B_1 + B_2$ and from Eq. (2.126): $\left(\frac{dC_A}{dz} \right)_{0+} = \frac{u}{2D_L}(B_1(1 + a) + B_2(1 - a))$ and using the inlet boundary condition Eq. (2.122):

$$u(C_{A,in} - C_{A,0+}) = -D_L \left(\frac{dC_A}{dz} \right)_{0+}$$

$$u(C_{A,in} - (B_1 + B_2)) = -D_L \frac{u}{2D_L}(B_1(1 + a) + B_2(1 - a))$$

or

$$\left(C_{A,in} - (B_1 + B_2)\right) = -\frac{1}{2}\left(B_1(1+a) + B_2(1-a)\right) \tag{2.127}$$

At $z \to L-$, from Eq. (2.126) and the boundary condition (Eq. 2.124):

$$\left(\frac{dC_A}{dz}\right)_L = 0 = B_1(1+a)e^{\frac{uL}{2D_L}(1+a)} + B_2(1-a)e^{\frac{uL}{2D_L}(1-a)} \tag{2.128}$$

Solving the simultaneous Eqs. (2.127) and (2.128) for B_1 and B_2 and substituting in Eq. (2.125) gives the concentration C_A at any position z in the reactor:

$$\frac{C_A}{C_{A,in}} = \frac{2e^{\frac{uz}{2D_L}}\left[(a+1)e^{\frac{au}{2D_L}(L-z)} + (a-1)e^{-\frac{au}{2D_L}(L-z)}\right]}{\left[(a+1)^2 e^{\frac{auL}{2D_L}} - (a-1)^2 e^{-\frac{auL}{2D_L}}\right]} \tag{2.129}$$

When $z = L$, $C_{A,L} = C_{A,out}$, the concentration of reactant **A** at the outlet of the reactor, so that:

$$\frac{C_{A,out}}{C_{A,in}} = \frac{4ae^{\frac{uL}{2D_L}}}{(a+1)^2 e^{\frac{auL}{2D_L}} - (a-1)^2 e^{\frac{auL}{2D_L}}} \tag{2.130}$$

2.4.7.1 Case of Small D_L/uL

Simplification of Eq. (2.130) for the case of small deviations from ideal plug flow (small D_L/uL) gives some interesting and useful results. From the definition of a that appears in Eqs. (2.125) and (2.130):

$$a = \left[1 + 4k_1 D_L/u^2\right]^{1/2} = \left[1 + 4k_1\tau\left(\frac{D_L}{uL}\right)\right]^{1/2}$$

where τ is the residence time in the reactor ($\tau = L/u$). This expression is then expanded by the binomial series and, neglecting cubic and higher order terms:

$$a = 1 + 2k_1\tau\left(\frac{D_L}{uL}\right) - 2(k_1\tau)^2\left(\frac{D_L}{uL}\right)^2 \tag{2.131}$$

Dividing numerator and denominator of Eq. (2.130) by $e^{\frac{uL}{2D_L}}$:

$$\frac{C_{A,out}}{C_{A,in}} = \frac{4a}{(a+1)^2 e^{\frac{uL}{2D_L}(a-1)} - (a-1)^2 e^{-\frac{uL}{2D_L}(a+1)}}$$

The second term in the denominator will be small compared with the first term because:

1. a is close to 1 and the factor $(a-1)^2$ will be small,
2. the exponent on **e** is large and negative, because u_L/D_L is the reciprocal of D_L/uL, which is small.

Thus, neglecting this second term:

$$\frac{C_{A,out}}{C_{A,in}} = \frac{4a}{(a+1)^2 e^{\frac{uL}{2D_L}(a-1)}} \tag{2.132}$$

Substituting from Eq. (2.131) for a in the exponential term and noting that when a is close to 1, $4a/(a+1)^2$ can be set approximately equal to 1:

$$\frac{C_{A,out}}{C_{A,in}} = \exp\left[-k_1\tau + (k_1\tau)^2\left(\frac{D_L}{uL}\right) \right] \tag{2.133}$$

2.4.7.2 Comparison With a Simple Plug Flow Reactor

The case of a simple PFR follows very easily from the above result for small values of D_L/uL. When there is no dispersion, D_L will be zero and Eq. (2.133) becomes:

$$\frac{C_{A,out}}{C_{A,in}} = \exp(-k_1\tau) \tag{2.134}$$

as expected from the basic treatment of PFRs in Chapter 1.

It is interesting to compare a reactor with dispersion (Eq. 2.130) with a simple PFR (Eq. 2.134). The effect of dispersion at a given desired fractional conversion for the reactant (i.e., at a given value of $C_{A,out}/C_{A,in}$) is to increase the volume required for reactor V_{td} compared with the volume V_t needed in the case of simple plug flow. This effect is shown quantitatively in Fig. 2.21, according to Levenspiel,[2] where V_{td}/V_t is plotted against $C_{A,out}/C_{A,in}$ for different values of the parameter $k_1\tau$. It can be seen that as D_L/uL increases so does the size of the dispersed PFR. The limit is reached when $D_L/uL \rightarrow \infty$, which corresponds to complete mixing similar to what would be obtained in a single stirred tank reactor. A similar chart for a second-order reaction may be found in Levenspiel.[2]

2.4.8 Applications and Limitations of the Dispersed Plug Flow Model

For reactors and process vessels where the departure from plug flow is not entirely negligible but not particularly large, the dispersed plug flow model is widely used. Because of the practical difficulties of introducing tracer substances into process streams feeding

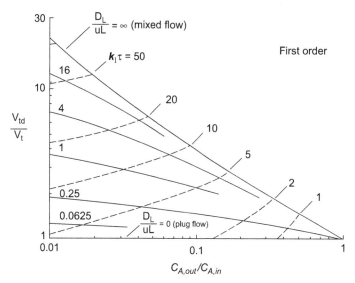

Figure 2.21

Comparison between plug flow reactors with and without axial dispersion. V_{td} is the volume with dispersion and V_t is the volume of the simple plug flow reactor required for the same conversion, i.e., the same value of the ratio $C_{A,out}/C_{A,in}$.

large-scale production units, most of the dispersion coefficients reported in the literature are for small-scale laboratory or pilot plant equipment. The use of the plug flow dispersion model is not necessarily restricted to homogeneous reactors, and the model may be applied to one or more of the streams in gas–solid, gas–liquid, and multiphase reactors, as discussed later under the heading of the types of reactor concerned.

However, if the flow deviates markedly from plug flow (i.e., if D_L/uL is large), it is most likely that the real situation does not conform to the physical processes implicit in the model, namely that the mixing is produced by a large number of independent random fluctuations. It is then questionable whether the model should be used. Some forms of tracer response curves following an ideal pulse type of injection give clear indication that the dispersed plug flow model is likely to be inadequate. Some examples of these are shown in Fig. 2.22. The occurrence of a double maximum as in Fig. 2.22A suggests that some bypassing is taking place. The very skewed curve in Fig. 2.22B possibly indicates that a well-mixed region is preceded or followed by a nearly plug flow region similar to the idealized configuration shown in Fig. 2.4. The third curve in Fig. 2.22C shows a long tail, suggesting the presence of a dead water zone with slow transfer of tracer back into the main stream.

When D_L/uL is found to be large and the tracer response curve is skewed, as in Fig. 2.22B, but without a significant delay, a continuous stirred tanks-in-series model

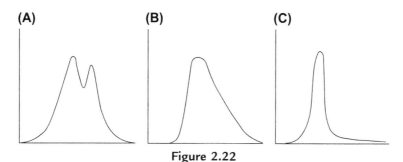

Figure 2.22

Types of tracer response curves showing indications that the dispersed plug flow model is unlikely to be applicable: (A) two maxima; (B) very skewed; and (C) long tail.

(Section 2.3) may be found to be more appropriate. The tracer response curve will then resemble one of those in Fig. 2.7 or Fig. 2.8. The number of tanks N may be found as explained in Section 2.3 through Eq. (2.81). Calculations of the mean and variance of an experimental curve can be used to determine either a dispersion coefficient D_L or a number of tanks N. Thus each of the models can be described as a "one-parameter model," the parameter being D_L in one case and N in the other. It should be noted that the value of N calculated in this way will not necessarily be integral, but this can be accommodated in the more mathematically general form of the tanks-in-series model as described by Nauman and Buffham.[13]

It can thus be seen that in some ways the dispersed plug flow model and the tanks-in-series model are complementary to each other. As $D_L/uL \rightarrow \infty$, the flow corresponds to a single ideally mixed stirred tank. For a series of tanks, as $N \rightarrow \infty$ for a fixed total space time, the flow corresponds to ideal plug flow (see Section 2.3). Generally, small departures from plug flow are better modeled by means of a dispersion coefficient D_L; large deviations are probably better treated by a tanks-in-series model thereby matching more closely the real physical processes. In more complicated situations such as those in Fig. 2.22, mixed modes involving more than just a single parameter need to be used as described in the next section.

2.5 Models Involving Combinations of the Basic Flow Elements

As well as regions corresponding to ideal plug flow, dispersed plug flow, an ideal stirred tank, and a series of N stirred tanks, as mentioned above a further type of region is a stagnant or dead water zone (Fig. 2.1A). In the construction of mixed models, we may also need to introduce bypass flow, internal recirculation, and cross-flow (which involves interchange but no net flow between regions). Of course, the more complicated the model, the larger is the number of parameters, which have to be determined by matching **C**- or **F**-curves. A wide variety of possibilities exist; for example, over 20 somewhat different models have been proposed for representing a gas–solid fluidized bed.[4]

In conclusion the following example shows how a real stirred tank might be modeled.[23]

Example 2.3

As a model of a certain poorly agitated continuous stirred tank reactor of total volume V, it is supposed that only a fraction w is well mixed, the remainder being a dead water zone. Furthermore, a fraction f of the total volume flow rate q_r fed to the tank bypasses the well-mixed zone completely (Fig. 2.23A).

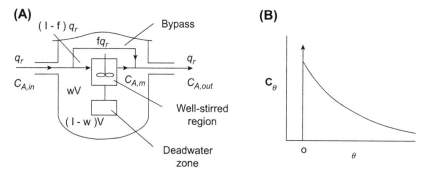

Figure 2.23

Model of a poorly agitated continuous stirred-tank reactor. (A) Flow model: fraction f of flow q_r bypasses; only a fraction w of tank volume V is well stirred. (B) Equivalent $\mathbf{C}_\theta$ curves for pulse input.

1. If the model is correct, sketch the **C**-curve expected if a pulse input of tracer is applied to the reactor.
2. On the basis of the model, determine the fractional conversion $\alpha_{A,out}$ that will be obtained under steady conditions if the feed contains a reactant, which undergoes a homogeneous first-order reaction with rate constant k_1.

Solution

1. Because of the bypass that is assumed to have no holding capacity, part of the input pulse appears instantly in the outlet stream. If the input pulse is normalized to unity, the magnitude of this immediate output pulse is f. The remainder of the $\mathbf{C}_\theta$-curve (Fig. 2.23B) is a simple exponential decay similar to the curve for one tank shown in Fig. 2.8A. However, in the present case, because only a fraction $(1 - f)$ of the unit pulse enters the well-mixed region of volume wV, the equation of the curve with $\theta = q_r t/V$ is:

$$\mathbf{C}_\theta = (1 - f)^2 \frac{1}{w} \exp\left\{ -\frac{(1 - f)}{w} \theta \right\}$$

as may be shown by an analysis similar to that leading to Eq. (2.75). Note that values for f and w could be determined by comparing experimentally obtained data with this expression.

2. Consider a material balance over the well-mixed region only, in the steady state. Let $C_{A,in}$ be the reactant inlet concentration and $C_{A,m}$ be the intermediate concentration at the outlet of this region before mixing with the bypass stream.

$$\underset{\text{Inflow}}{q_r(1-f)C_{A,in}} \quad - \quad \underset{\text{Outflow}}{q_r(1-f)C_{A,m}} \quad - \quad \underset{\text{Reaction}}{k_1 C_{A,m}wV} \quad = 0$$

Hence

$$C_{A,m} = \frac{C_{A,in}}{1 + \dfrac{k_1 wV}{q_r(1-f)}}$$

Next consider mixing of this stream with the bypass giving a final exit concentration $C_{A,out}$.

$$q_r(1-f)C_{A,m} + fq_r C_{A,in} = q_r C_{A,out}$$

therefore

$$\frac{C_{A,out}}{C_{A,in}} = (1-f)\frac{C_{A,m}}{C_{A,in}} + f$$

The fractional conversion at the exit:

$$\alpha_{A,out} = 1 - \frac{C_{A,out}}{C_{A,in}}$$

Substituting for $C_{A,out}$ and eliminating $C_{A,m}$, we find after some manipulation:

$$\alpha_{A,out} = \frac{1}{\dfrac{q_r}{k_1 wV} + \dfrac{1}{(1-f)}}$$

Note that if $f = 0$, $w = 1$, and writing, $V/q_r = \tau$, this expression reduces to:

$$\alpha_{A,out} = \frac{k_1 \tau}{1 + k_1 \tau}$$

as expected for a single ideally mixed stirred tank reactor.

Nomenclature

		Units in SI System	Dimensions N, L, T
A	Strength of impulse	kmol s/m^3	$NL^{-3}T$
A_C	Cross-sectional area	m^2	L^2
A_n	Area under concentration–time curve	kmol s/m^3	$NL^{-3}T$
a	Parameter $(1 + 4k_1 D_L/u^2)^{1/2}$	–	–
B_1	Integration constant	kmol/m^3	NL^{-3}

Continued

		Units in SI System	Dimensions N, L, T
B_2	Integration constant	kmol/m^3	NL^{-3}
C	Tracer concentration	kmol/m^3	NL^{-3}
C_i	Tracer concentration in sample i	kmol/m^3	NL^{-3}
C_{in}	Tracer concentration in inlet stream	kmol/m^3	NL^{-3}
C_L	Tracer concentration at position L	kmol/m^3	NL^{-3}
C_{out}	Tracer concentration in outlet stream	kmol/m^3	NL^{-3}
C_A	Concentration of reactant	kmol/m^3	NL^{-3}
$C_{A,in}$	Concentration of reactant at inlet	kmol/m^3	NL^{-3}
$C_{A,out}$	Concentration of reactant at exit	kmol/m^3	NL^{-3}
$C_{A,m}$	Concentration of reactant at intermediate position in stirred tank model	kmol/m^3	NL^{-3}
C_∞	Magnitude of concentration for a step change	kmol/m^3	NL^{-3}
C	Outlet response of concentration divided by area under concentration–time curve	s^{-1}	T^{-1}
C$_\theta$	Nondimensionalized **C** $(= \bar{t}C)$	—	—
D	Molecular diffusivity	m^2/s	L^{2}T^{-1}
D_L	Dispersion coefficient	m^2/s	L^{2}T^{-1}
d	Diameter of pipe	m	L
d_p	Diameter of particle	m	L
E	Exit age or residence time distribution function	s^{-1}	T^{-1}
F	Cumulative distribution function	—	—
f	Fraction of feed bypassing reactor	—	—
k_0	Zero-order rate constant	kmol/m^3 s	NL^{-3}T^{-1}
k_1	First-order rate constant	s^{-1}	T^{-1}
k_2	Second-order rate constant	m^3/kmol s	N^{-1}L^{3}T^{-1}
L	Length of pipe or reactor	m	L
m	Parameter in trial solution	m^{-1}	L^{-1}
N	Number of tanks in a series	—	—

		Units in SI System	Dimensions N, L, T
N_i	Flux of species i due to dispersion	$kmol/m^2\,s$	$NL^{-2}T^{-1}$
n_0	Moles of tracer injected	$kmol$	N
q_r	Volumetric flow rate	m^3/s	L^3T^{-1}
r_A	Rate of generation of species A due to reaction	$kmol/m^3\,s$	$NL^{-3}T^{-1}$
t	Time	s	T
$\bar{t}$	Mean time	s	T
$\overline{t_E}$	Mean of residence time distribution	s	T
t_i	Time for sample i	s	T
t_i'	Value of t_i at midpoint of Δt_i	s	T
u	Velocity in pipe	m/s	LT^{-1}
$u(t)$	Unit step function	—	—
V	Volume of vessel; total volume if more than one tank	m^3	L^3
V_t	Volume of plug flow tubular reactor	m^3	L^3
V_{td}	Volume of dispersed plug flow tubular reactor	m^3	L^3
v	Velocity of fluid in plug flow reactor	m/s	LT^{-1}
z	Length coordinate in axial direction along pipe or tubular Reactor	m	L
$z\prime$	Dimensionless variable z/L	—	—
α	Dimensionless number, $1/(k_2 C_{A,in}\tau)$	—	—
α_A	Fractional conversion of A	—	—
$\overline{\alpha}_A$	Average conversion	—	—
$\alpha_{A,out}$	Fractional conversion at exit	—	—
β	Dimensionless number $k_{1/2} C_{A,in}^{-1/2}\tau$	—	—
$\delta(t)$	Dirac delta distribution	s^{-1}	T^{-1}
θ	Dimensionless time $t/\bar{t}$ or t/τ	—	—
$\overline{\theta}$	Mean value of θ	—	—
σ^2	Variance	s^2	T^2
$\sigma_A^2,\ \sigma_B^2$	Variance for vessel **A**, vessel **B**	s^2	T^2
σ_E^2	Variance for the residence time distribution function $E(t)$	s^2	T^2
τ	Residence time in a pipe or reactor L/u or V/q_r	s	T
Re	Reynolds number ($\rho u d/\mu$)	—	—
Sc	Schmidt number ($\mu/\rho D$)	—	—

References

1. Danckwerts PV. Continuous flow systems. Distribution of residence times. *Chem Eng Sci* 1953;**2**:1.
2. Levenspiel O. *Chemical reaction engineering*. 3rd ed. Wiley; 1999.
3. Levenspiel O, Bischoff KB. Patterns of flow in chemical process vessels. In: Drew TB, Hoopes JW, Vermeulen T, editors. *Advances in chemical engineering*, vol. 4. Academic Press; 1963.
4. Kunii D, Levenspiel O. *Fluidization engineering*. 2nd ed. Butterworth-Heinemann; 1991.
5. Zwietering TN. The degree of mixing in continuous flow systems. *Chem Eng Sci* 1959;**11**:1.
6. Jaynes ET. *Probability theory, the logic of science*. Cambridge University Press; 2003.
7. Mickley HS, Sherwood TS, Reed CE. *Applied mathematics in chemical engineering*. McGraw-Hill; 1957.
8. Rhee H-K, Aris R, Amundson NR. *First-order partial differential equations, volume 1, theory and application of single equations*. Prentice-Hall; 1986.
9. Fogler HS. *Elements of chemical reaction engineering*. 3rd ed. Prentice-Hall; 1999.
10. Kramers H. Physical factors in chemical reaction engineering. *Chem Eng Sci* 1958;**8**:45.
11. Bourne JR. Mixing in single phase chemical reactors. In: Harnby N, Edwards MF, Nienow AW, editors. *Mixing in the process industries*. Butterworth-Heinemann; 1992. Chapter 10.
12. Abramowitz M, Stegun IA. *Handbook of mathematical functions with formulas, graphs, and mathematical tables*. Dover; 1972.
13. Nauman EB, Buffham BA. *Mixing in continuous flow systems*. Wiley; 1983.
14. Polyanin AD, Manzhirov AV. *Handbook of mathematics for engineers and scientists*. Chapman and Hall; 2007.
15. Taylor GI. Dispersion of soluble matter in solvent flowing slowly through a tube. *Proc R Soc A* 1953;**219**:186.
16. Aris R. On the dispersion of a solute in a fluid flowing through a tube. *Proc R Soc A* 1956;**235**:67.
17. Probstein RF. *Physicochemical hydrodynamics*. Butterworth-Heinemann; 1989.
18. Abbi YP, Gunn DJ. Dispersion characteristics from pulse response. *Trans Inst Chem Eng* 1976;**54**:225.
19. Wehner JF, Wilhelm RH. Boundary conditions of a flow reactor. *Chem Eng Sci* 1956;**6**:89.
20. Wen CY, Fan LT. *Models for flow systems and chemical reactors*. Dekker; 1975.
21. Levenspiel O. *Chemical reactor omnibook*. Corvallis, Oregon: OSU Book Stores; 1989.
22. Froment GF, Bischoff KB. *Chemical reactor analysis and design*. 2nd ed. Wiley; 1989. p. 535.
23. Cholette A, Cloutier L. Mixing efficiency determinations for continuous flow systems. *Can J Chem Eng* 1959;**37**:105.

CHAPTER 3

Gas—Solid Reactions and Reactors

Ravikrishnan Vinu

Indian Institute of Technology Madras, Chennai, India

Learning Outcomes

- Calculate effective diffusivity of gases through the catalyst pores by accounting for molecular diffusivity, Knudsen diffusivity, porosity, and tortuosity of the catalyst pellet.

- Develop expressions for concentration profile of species undergoing various reactions, viz., first-order, nth-order, parallel, and series reactions, in a porous catalyst pellet of different geometries.

- Evaluate the importance of Thiele modulus and effectiveness factor in determining the rate-limiting regime—pore diffusion versus reaction.

- Evaluate the effect of nonisothermal conditions on maximum temperature rise and effectiveness factor for a catalyst pellet for exothermic reactions.

- Derive the expression for overall rate of a catalytic reaction with both external mass transfer and internal pore diffusion.

- Formulate Langmuir—Hinshelwood—Hougen—Watson-type rate expressions for surface-catalyzed reactions involving elementary steps such as adsorption, reaction, and desorption using quasiequilibrium approach, and understand the mechanism of catalyst deactivation.

- Design packed bed catalytic reactors operating under isothermal, nonisothermal (exothermic), and nonideal conditions using mass and energy balance equations, and graphical approach.

- Write the balance equations relating solid conversion with time for gas—solid noncatalytic reactions under the influence of gas film, reaction, and solid product layer resistances.

3.1 Introduction

The design of heterogeneous chemical reactors falls into a special category because an additional complexity enters into the problem. We must now concern ourselves with the transfer of matter between phases, as well as considering the fluid dynamics and chemistry of the system. Thus, in addition to an equation describing the rate at which the chemical reaction proceeds, one must also provide a relationship or algorithm to account for the various physical processes that occur. For this purpose it is convenient to classify the reactions as gas–solid, gas–liquid, and gas–liquid–solid processes. This chapter will be concerned with gas–solid reactions, especially those for which the solid is a catalyst for the reaction.

The very great importance of heterogeneous catalysis in the chemical and petroleum industries can be seen from the examples shown in Table 3.1. Catalytic reactions such as the synthesis of ammonia, which has been practiced now for nearly a century, and the cracking of hydrocarbons provide us with fertilizers, fuels, and many of the basic intermediates of the chemical industry. As understood from the classic definition of catalysis in chemistry, the key function of a catalyst is to speed up the rate of a desired reaction. In practice this enables the desired reaction to take place under conditions of temperature and pressure such that the rates of competing reactions giving undesirable by-products are very low. Thus the degree of *selectivity* of a catalyst in promoting the desired reaction, while appearing to suppress unwanted reactions, is a very important consideration in developing or choosing a catalyst.

Some of the reactions listed in Table 3.1, particularly hydrogenations, require the simultaneous contacting of a gas, a liquid, and a solid catalyst, and reactors for this type of system are considered in Chapter 4. Otherwise, gas–solid catalytic reactors involve passing a reactant gas through a bed of catalyst particles (an exception being the multichannel "monolith" reactors now widely used for pollution control of automobile engine exhausts). In the most common *fixed bed* type of reactor, the particles are stationary and form a packed bed through which the gas passes downward (to avoid lifting the particles from the support plate), or in some designs radially (especially when a low pressure drop is required). For fixed bed reactors, the catalyst particles are typically of a size 1–20 mm equivalent diameter and can be of a variety of shapes, as shown in Fig. 3.1. The shape chosen depends partly on how easy that shape of catalyst particle is to manufacture and partly on the need to make the catalyst material readily accessible to the reactant being transferred from the gas phase. Thus the hollow cylinder in Fig. 3.1 has the advantage of making the interior of the particle more accessible as well as of providing a low pressure drop for the flow of the reactant mixture through a packed bed of these particles. The other principal type of reactor is a *fluidized bed*, in which the particles are mobile and generally much smaller in size, being typically 50–100 μm in diameter.

In a gas–solid catalytic reactor, the reaction itself takes place on the surface of the solid. For this reason, unless the reaction is inherently very fast, it is an advantage to have the particles of catalyst in a porous form giving a large internal surface area available for the reaction. In a reactor using porous catalyst particles, the reactant must first be transported through the gas phase to the surface of the particle and then diffuse along the honeycomb of pores toward the interior. The reactant will then adsorb at all of the accessible active surface in the interior and be chemically transformed into a product, which subsequently

Table 3.1: Selected Heterogeneous Catalysts of Industrial Importance.[a]

Reaction	Catalyst and Reactor Type (Continuous Operation Unless Otherwise Noted)
Dehydrogenation	
C_4H_{10} (butane) $\rightarrow$ butenes	$Cr_2O_3 \cdot Al_2O_3$ (fixed bed, cyclic)
Butenes $\rightarrow C_4H_6$ (butadiene)	Fe_2O_3 promoted with Cr_2O_3 and K_2CO_3
$C_6H_5C_2H_5 \rightarrow C_6H_5CH=CH_2$ (ethyl benzene $\rightarrow$ styrene)	Fe_2O_3 promoted with Cr_2O_3 and K_2CO_3 (fixed bed, in presence of steam)
CH_4 or other hydrocarbons $+ H_2O \rightarrow CO + H_2$ (steam reforming)	Supported Ni (fixed bed)
$(CH_3)_2CHOH \rightarrow CH_3COCH_3 + H_2$ (isopropanol $\rightarrow$ acetone + hydrogen)	ZnO
$CH_3CH(OH)C_2H_5 \rightarrow CH_3COC_2H_5 + H_2$	ZnO
Hydrogenation	
Edible fats and oils	Ni on a support (slurry reactor, batch)
Various hydrogenations of fine organic chemicals	Pd or Pt on carbon (slurry reactor, usually batch)
$C_6H_6 + 3H_2 \rightarrow C_6H_{12}$	Ni or noble metal on support (fixed bed or slurry reactor)
$N_2 + 3H_2 \rightarrow 2NH_3$	Fe promoted with Al_2O_3, K_2O, CaO, and MgO (adiabatic fixed beds)
$C_2H_2 \rightarrow C_2H_6$ (selective hydrogenation of C_2H_2 impurity in C_2H_4 from thermal-cracking plant)	Pd on Al_2O_3 or sulfided Ni on support (adiabatic fixed bed)
Oxidation	
$SO_2 + \frac{1}{2}O_2$ (air) $\rightarrow SO_3$	V_2O_5 plus K_2SO_4 on silica (adiabatic, fixed beds)
$2NH_3 + \frac{5}{2}O_2$ (air) $\rightarrow 2NO + 3H_2O$	90% Pt-10% Rh wire gauze, oxidizing conditions
$NH_3 + CH_4 +$ air $\rightarrow$ HCN (Andrussow process)	90% Pt-10% Rh wire gauze, under net reducing conditions
$C_{10}H_8$ or 1, $2\text{-}C_6H_4(CH_3)_2 + O_2 \rightarrow C_6H_4(CO)_2O$ (naphthalene or *o*-xylene + air $\rightarrow$ phthalic anhydride)	V_2O_5 on titania (multitube fixed bed)
$n\text{-}C_4H_{10} + O_2 \rightarrow C_4H_2O_3$ (butane + air $\rightarrow$ maleic anhydride)	Vanadia-phosphate (multitube fixed bed or fluidized bed)
$C_2H_4 + \frac{1}{2}O_2 \rightarrow (CH_2)_2O$ (ethylene oxide)	Ag on $\alpha\text{-}Al_2O_3$, promoted with Cl and Cs (multitube fixed bed)
$CH_3OH + O_2 \rightarrow CH_2O + H_2$ and/or H_2O	Ag (adiabatic reactor) or $Fe_2(MoO_4)_3$ (multitube fixed bed)
$C_3H_6 + O_2 \rightarrow CH_2\!=\!CHCHO$ (acrolein) and/or $CH_2\!=\!CHCOOH$ (acrylic acid)	Bismuth molybdate plus other components
$C_3H_6 + NH_3 + \frac{3}{2}O_2 \rightarrow CH_2\!=\!CHCN + 3H_2O$	Complex metal molybdates (fluidized bed)
Complete oxidation of CO and hydrocarbons, for pollution control	Pt or Pd, or both, on monolith support
Simultaneous control of CO, hydrocarbons, and NO_x in engine exhaust	Same, plus Rh, with careful control of oxidizing/reducing conditions
Simultaneous control of NO_x and SO_x in flue gases	Vanadia on titania with addition of NH_3
$C_2H_4 + \frac{1}{2}O_2 + CH_3COOH \rightarrow CH_3COOCH=CH_2$ (vinyl acetate)	Pd on acid-resistant support (vapor phase, multitube fixed bed)

Continued

Table 3.1: Selected Heterogeneous Catalysts of Industrial Importance.[a]**—cont'd**

Reaction	Catalyst and Reactor Type (Continuous Operation Unless Otherwise Noted)
Dehydrogenation	
$C_4H_8 + \frac{1}{2}O_2 \rightarrow C_4H_6 + H_2O$	Promoted ferrite spinels
Acid-Catalyzed Reactions	
Catalytic cracking	Zeolite in $SiO_2 \cdot Al_2O_3$ matrix plus other ingredients (transport reactor)
Hydrocracking	Pd on zeolite in an amorphous matrix; NiMo on silica–alumina; various other dual-function catalysts (adiabatic fixed beds)
Paraffin isomerization	Pt on H-mordenite zeolite in alumina matrix
Catalytic reforming	Pt, Pt-Re, or Pt-Sn on acidified Al_2O_3 or on zeolite in matrix (adiabatic, fixed beds, or moving bed, with interstage heating)
Polymerization	H_3PO_4 on clay (fixed bed)
Hydration, e.g., propylene to isopropyl alcohol	Mineral acid or acid-type ion-exchange resin (fixed bed)
$CH_3OH + isoC_4H_8 \rightarrow$ methyl tertiary butyl ether (MTBE)	Acid-type ion-exchange resin
Reactions of Synthesis Gas	
$CO + 2H_2 \rightarrow CH_3OH$	Cu^I-ZnO promoted with Al_2O_3 (adiabatic, fixed beds with interstage cooling, or multitube fixed bed)
$CO + 3H_2 \rightarrow CH_4 + H_2O$ (methanation)	Supported Ni (fixed bed)
$CO + H_2 \rightarrow$ paraffins, etc. (Fischer–Tropsch synthesis)	Fe or Co with promoters (multitube fixed bed or transport reactor)
Other	
Oxychlorination (e.g., $C_2H_4 + 2HCl + \frac{1}{2}O_2 \rightarrow C_2H_4Cl_2 + H_2O$)	$CuCl_2/Al_2O_3$ with KCl promoter
Hydrodesulfurisation, hydrodenitrogenation, hydrotreating	$CoMo/Al_2O_3$ or $NiMo/Al_2O_3$, sulfided (adiabatic, fixed beds with interstage cooling)
$SO_2 + 2H_2S \rightarrow 3S + 2H_2O$ (Claus process)	Al_2O_3 (fixed beds)
$H_2O + CO \rightarrow CO_2 + H_2$ (water–gas shift)	Fe_3O_4 promoted with Cr_2O_3 (adiabatic fixed bed); for a second, lower temperature stage, Cu-ZnO on Al_2O_3; CoMo on support

[a]Reproduced by kind permission from Satterfield.[1]

desorbs, diffuses to the exterior of the particle, and is then transported to the bulk fluid phase. For such reactions the designer usually has to assess the amount of solid material to pack into a reactor to achieve a specified conversion. The rate of transport of gas to and from the surface of the solid and within the pore structure of the particle will affect both the dimensions of the reactor unit required and the selection of particle size. Compared with a reaction uninfluenced by transport processes, the complexity of the design problem

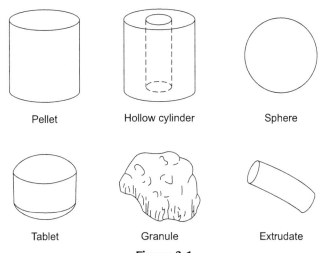

Figure 3.1
Various forms of catalyst particles.

is exacerbated because, in the steady state, the rate of each of the transport processes is coupled with the rates of adsorption, of chemical reaction, and of desorption.

Whatever the nature of the reaction and whether the vessel chosen for the operation be a packed tubular reactor or a fluidized bed, the essence of the design problem is to estimate the size of reactor required. This is achieved by solving the transport and chemical rate equations appropriate to the system. Prior to this, however, the operating conditions, such as initial temperature, pressure, and reactant concentrations, must be chosen and a decision must be made, concerning the type of reactor to be used. For example, one might have selected a packed tubular reactor operated adiabatically, or perhaps a fluidized bed reactor fitted with an internal cooling arrangement. Such operating variables constitute the design conditions and must be chosen before a detailed mathematical model is constructed.

3.2 Mass Transfer Within Porous Solids

For gas—solid heterogeneous reactions, particle size and average pore diameter will influence the reaction rate per unit mass of solid when internal diffusion is a significant factor in determining the rate. The actual mode of transport within the porous structure will depend largely on the pore diameter and the pressure within the reactor. Before developing equations that will enable us to predict reaction rates in porous solids, a brief consideration of transport in pores is pertinent.

3.2.1 The Effective Diffusivity

The diffusion of gases through the tortuous narrow channels of a porous solid generally occurs by one or more of three mechanisms. When the mean free path of the gas molecules is considerably greater than the pore diameter, collisions between molecules in the gas are much less numerous than those between molecules and pore walls. Under these conditions the mode of transport is Knudsen diffusion. When the mean free path of the gas molecules is much smaller than the pore diameter, gaseous collisions will be more frequent than collisions of molecules with pore walls, and under these circumstances, molecular diffusion occurs. A third mechanism of transport, which is possible when a gas is adsorbed on the inner surface of a porous solid, is surface diffusion. Transport occurs by the movement of molecules over the surface in the direction of decreasing surface concentration. Although there is not much evidence on this point, it is unlikely that surface diffusion is of any importance in catalysis at elevated temperatures. Nevertheless, surface diffusion may contribute to the overall transport process in low-temperature reactions of some vapors. Finally, it should be borne in mind that when a total pressure difference is maintained across a pore, as for some catalytic cracking reactions, forced flow in pores is likely to occur, transport being due to a total concentration or pressure gradient.

Both Knudsen and molecular diffusion can be described adequately for homogeneous media. However, a porous mass of solid usually contains pores of nonuniform cross section, which pursue a very tortuous path through the particle and which may intersect with many other pores. Thus the flux predicted by an equation for normal bulk diffusion (or for Knudsen diffusion) should be multiplied by a geometric factor, which takes into account the tortuosity and the fact that the flow will be impeded by that fraction of the total pellet volume, which is solid. It is therefore expedient to define an effective diffusivity D_e in such a way that the flux of material may be thought of as flowing through an equivalent homogeneous medium. We may then write

$$D_e = D\frac{\varepsilon}{\tau'} \tag{3.1}$$

where D is the molecular diffusion coefficient, ε is the porosity of the particles, and τ' is a tortuosity factor.

We thus imply that the effective diffusion coefficient is calculated on the basis of a flux resulting from a concentration gradient in a homogeneous medium, which has been made equivalent to the heterogeneous porous mass by invoking the geometric factor ε/τ'. Experimental techniques for estimating the effective diffusivity include diffusion and flow through pelletized particles. A common procedure is to expose the two faces of a porous disc of the material to the pure components at the same total pressure. An interesting method proposed by Barrer and Grove[2] relies on the measurement of the time lag required to reach a steady pressure gradient, whereas a technique employed by Gunn and Pryce[3] depends on measuring the dispersion of a sinusoidal pulse of tracer in a bed packed with the porous material. Gas chromatographic methods for evaluating dispersion coefficients have also been employed.[4]

Just as one considers two regions of flow for homogeneous media, one may have molecular or Knudsen transport for heterogeneous media.

3.2.1.1 The Molecular Flow Region

Hirschfelder, Curtiss, and Bird[5] obtained a theoretical expression for the molecular diffusion coefficient for two interdiffusing gases, modifying the kinetic theory of gases by taking into account the nature of attractive and repulsive forces between gas molecules. Their expression for the diffusion coefficient has been successfully applied to many gaseous binary mixtures and represents one of the best methods for estimating unknown values. On the other hand, Maxwell's formula, modified by Gilliland[6] and discussed in Volume 1, Chapter 10, also gives satisfactory results. Experimental methods for estimating diffusion coefficients rely on the measurement of flux per unit concentration gradient. An extensive tabulation of experimental diffusion coefficients for binary gas mixtures is to be found in a report by Eerkens and Grossman.[7]

To calculate the effective diffusivity in the region of molecular flow, the estimated value of D must be multiplied by the geometric factor ε/τ', which is descriptive of the heterogeneous nature of the porous medium through which diffusion occurs.

The porosity ε of the porous mass is included in the geometric factor to account for the fact that the flux per unit total cross section is ε times the flux, which would occur if there were no solid present. The porosity may conveniently be measured by finding the particle density ρ_p in a pyknometer using an inert nonpenetrating liquid. The true density ρ_s of the solid should also be found by observing the pressure of a gas (which is not adsorbed) before and after expansion into a vessel containing a known weight of the material. The ratio ρ_p/ρ_s then gives the fraction of solid present in the particles, and $(1 - \rho_p/\rho_s)$ is the porosity.

The tortuosity is also included in the geometric factor to account for the tortuous nature of the pores. It is the ratio of the path length, which must be traversed by molecules in diffusing between two points within a pellet to the direct linear separation between those points. Theoretical predictions of τ' rely on somewhat inadequate models of the porous structure, but experimental values may be obtained from measurements of D_e, D, and ε.

3.2.1.2 The Knudsen Flow Region

The region of flow where collisions of molecules with the container walls are more frequent than intermolecular gaseous collisions was the subject of detailed study by Knudsen[8] early in the 20th century. From geometrical considerations it may be shown[9] that, for the case of a capillary of circular cross section and radius r, the proportionality factor is $8\pi r^3/3$. This results in a Knudsen diffusion coefficient:

$$D_K = \frac{8r}{3}\sqrt{\frac{RT}{2\pi M}} \tag{3.2}$$

This equation, however, cannot be directly applied to the majority of porous solids because they are not well represented by a collection of straight cylindrical capillaries. Everett[10] showed that pore radius is related to the specific surface area S_g per unit mass and to the specific pore volume V_g per unit mass by the following equation:

$$r = \frac{2}{\alpha}\frac{V_g}{S_g} \tag{3.3}$$

where α is a factor characteristic of the particular pore geometry. Values of α depend on the pore structure, and for uniform nonintersecting cylindrical capillaries, $\alpha = 1$. Although an estimation of precise values of V_g and S_g from experimental data is obtained by the somewhat arbitrary selection of points on an adsorption isotherm representing complete pore filling and the completion of a monolayer, some significance may be given to an average pore dimension derived from pore volume and surface area measurements. Thus if, for the purposes of calculating a Knudsen diffusion coefficient, the pore model adopted consists of nonintersecting cylindrical capillaries and the radius computed from Eq. (3.3) is a radius r_e equivalent to the radius of a cylinder having the same surface to volume ratio as the pore, then Eq. (3.2) may be applied. In terms of the porosity ε, specific surface area S_g, and particle density ρ_p (mass per unit total particle volume, including the volume occupied by pore space),

$$D_K = \frac{16}{3}\frac{\varepsilon}{\rho_p S_g}\sqrt{\frac{RT}{2\pi M}} \tag{3.4}$$

In the region of Knudsen flow the effective diffusivity D_{eK} for the porous solid may be computed in a similar way to the effective diffusivity in the region of molecular flow, i.e., D_K is simply multiplied by the geometric factor.

3.2.1.3 The Transition Region

For conditions where Knudsen or molecular diffusion does not predominate, Scott and Dullien[11] obtained a relation for the effective diffusivity. The formula they obtained for a binary mixture of gases is

$$D_e = \frac{1}{\dfrac{1}{D_{eM}} + \dfrac{1}{D_{eK}} - \dfrac{x_A\left(1 + N'_B/N'_A\right)}{D_{eM}}} \tag{3.5}$$

where D_{eM} and D_{eK} are the effective diffusivities in the molecular and Knudsen regions, respectively; N'_A and N'_B are the molar fluxes of species **A** and **B**, respectively; and x_A is the mole fraction of **A**.

If equimolecular counterdiffusion takes place $N'_A = -N'_B$ (see Volume 1, Chapter 10) and the total pressure is constant, we obtain from Eq. (3.5) an expression for the effective self-diffusion coefficient in the transition region:

$$\frac{1}{D_e} = \frac{1}{D_{eM}} + \frac{1}{D_{eK}} \tag{3.6}$$

This result has also been obtained independently by other workers.[12,13]

3.2.1.4 Forced Flow in Pores

Many heterogeneous reactions give rise to an increase or decrease in the total number of moles present in the porous solid due to the reaction stoichiometry. In such cases there will be a pressure difference between the interior and exterior of the particle, and forced flow occurs. When the mean free path of the reacting molecules is large compared with the pore diameter, forced flow is indistinguishable from Knudsen flow and is not affected by pressure differentials. When, however, the mean free path is small compared with the pore diameter and a pressure difference exists across the pore, forced flow (Poiseuille flow: see Volume 1, Chapter 3) resulting from this pressure difference will be superimposed on molecular flow. The diffusion coefficient D_p for forced flow depends on the square of the pore radius and on the total pressure difference ΔP:

$$D_P = \frac{-\Delta P r^2}{8\mu} \tag{3.7}$$

The viscosity of most gases at atmospheric pressure is of the order of 10^{-7} Ns/m^2, so for pores of about 1 μm radius, D_P is approximately 10^{-5} m^2/s. Molecular diffusion coefficients are of similar magnitude so that in small pores forced flow will compete with molecular diffusion. For fast reactions accompanied by an increase in the number of moles, an excess pressure is developed in the interior recesses of the porous particle, which results in the forced flow of excess product and reactant molecules to the particle exterior. Conversely, for pores greater than about 100 μm radius, D_P is as high as 10^{-3} m^2/s, and the coefficient of diffusion that will determine the rate of intraparticle transport will be the coefficient of molecular diffusion.

Except in the case of reactions at high pressure, the pressure drop that must be maintained to cause flow through a packed bed of particles is usually insufficient to produce forced flow in the capillaries of the solid, and the gas flow is diverted around the exterior periphery of the pellets. Reactants then reach the interior of the porous solid by Knudsen or molecular diffusion.

3.3 Chemical Reaction in Porous Catalyst Pellets

A porous solid catalyst, whose behavior is usually specific to a particular system, enhances the approach to equilibrium of a gas-phase chemical reaction. Employment of such a material therefore enables thermodynamic equilibrium to be achieved at moderate temperatures in a comparatively short time interval.

When designing a heterogeneous catalytic reactor, it is important to know, for the purposes of calculating throughputs, the rate of formation of desired product on the basis of unit reactor volume and under the hydrodynamic conditions obtaining in the reactor. Whether the volume is defined with respect to the total reactor volume, including that occupied by solid, or with respect to void volume is really a matter of convenience. What is important, however, is either that rate data for the reaction be known for physical conditions identical to those prevailing within the reactor (this usually means obtaining rate data in situ) or, alternatively, that rate data obtained in the absence of mass transport effects be available. If pilot plant experiments are performed, conditions can usually be arranged to match those within a larger reactor, and then the rate data obtained can be immediately applied to a reactor design problem because the measurements will have taken into account mass transfer effects. If rate data from laboratory experiments are utilized, it is essential to ensure that they are obtained in the absence of mass and heat transfer effects. This being so, the rate should be multiplied by two factors to transpose the experimental rate to the basis of reaction rate per unit volume of a reactor packed with catalyst particles. If we wish to calculate throughput on the basis of total reactor volume, the bed voidage e should be taken into account and the rate should be multiplied by $(1 - e)$, which is the fraction of reactor volume occupied by solid. Account must also be taken of mass transfer effects and so the rate is multiplied by an *effectiveness factor*, η, which is defined as the ratio of the rate of reaction in a pellet to the rate at which reaction would occur if the concentration and temperature within the pellet were the same as the respective values external to the pellet. The factor η therefore accounts for the influence, which concentration and temperature gradients (which exist within the porous solid and result in mass and heat transfer effects) have on the chemical reaction rate. The reaction rate per unit volume of reactor is thus written:

$$r_v = (1 - e)k\mathrm{f}(C_A)\eta \qquad (3.8)$$

where $\mathrm{f}(C_A)$ is that function of concentration of the reactants, which describes the specific rate per unit volume in the absence of any mass and heat transfer effects within the particle and k is the specific rate constant per unit volume of reactor. An experimental determination of η merely involves comparing the observed rate of reaction with the rate of reaction on catalyst pellets sufficiently small for diffusion effects to be negligible.

3.3.1 Isothermal Reactions in Porous Catalyst Pellets

Thiele,[14] who predicted how in-pore diffusion would influence chemical reaction rates, employed a geometric model with isotropic properties. Both the effective diffusivity and the effective thermal conductivity are independent of position for such a model. Although idealized geometric shapes are used to depict the situation within a particle, such models, as we shall see later, are quite good approximations to practical catalyst pellets.

The simplest case we shall consider is that of a first-order chemical reaction occurring within a rectangular slab of porous catalyst, the edges of which are sealed so that diffusion occurs in one dimension only. Fig. 3.2 illustrates the geometry of the slab. Consider that the first-order irreversible reaction,

$$\mathbf{A} \rightarrow \mathbf{B}$$

occurs within the volume of the particle and suppose its specific velocity constant on the basis of unit surface area is k_a. For heterogeneous reactions uninfluenced by mass transfer effects, experimental values for rate constants are usually based on unit surface area. The corresponding value in terms of unit total volume of particle would be $\rho_p S_g k_a$ where ρ_p is the apparent density of the catalyst pellet and S_g is the specific surface area per unit mass of the solid including the internal pore surface area. We shall designate the specific rate

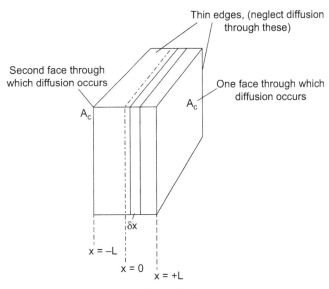

Figure 3.2
Geometry of slab model for catalyst pellets.

constant based on unit volume of particle as **k**. A component material balance for **A** across the element δx gives

$$-A_c D_e \left(\frac{dC_A}{dx} \right)_x = -A_c D_e \left(\frac{dC_A}{dx} \right)_{x+\delta x} + k C_A A_c \delta x \qquad (3.9)$$

because, in the steady state, the flux of **A** into the element at x must be balanced by the flux out of the element at $(x + \delta x)$ plus the amount lost by reaction within the volume element $A_c \delta x$. If the concentration gradient term at the point $(x + \delta x)$ in Eq. (3.9) is expanded in a Taylor series about the point x and differential coefficients of order greater than two are ignored, the equation simplifies to

$$\frac{d^2 C_A}{dx^2} - \frac{k C_A}{D_e} = 0 \qquad (3.10)$$

An analogous equation may be written for component **B**. By reference to Fig. 3.2, it will be seen that, because the product **B** diffuses outward, its flux is positive. Reaction produces **B** within the slab of material and hence makes the term depicting the rate of formation of **B** in the material balance equation positive, resulting in an equation similar in form to Eq. (3.10).

The boundary conditions for the problem may be written by referring to Fig. 3.2. At the exterior surface of the slab, the concentration will be that corresponding to the conditions in the bulk gas phase, provided there is no resistance to mass transfer in the gas phase. Hence

$$C_A = C_{A\infty} \quad \text{at } x = \pm L \qquad (3.11)$$

At the center of the slab, considerations of symmetry demand that

$$\frac{dC_A}{dx} = 0 \quad \text{at } x = 0 \qquad (3.12)$$

so that the net flux through the plane at $x = 0$ is zero, diffusion across this boundary being just as likely in the direction of increasing x as in the direction of decreasing x. The solution of Eq. (3.9) with the boundary conditions given by Eqs. (3.11) and (3.12) is

$$C_A = C_{A\infty} \frac{\cosh \lambda x}{\cosh \lambda L} \qquad (3.13)$$

where λ denotes the quantity $\sqrt{(k/D_e)}$. Eq. (3.13) describes the concentration profile of **A** within the catalyst slab. In the steady state the total rate of consumption of **A** must be equal to the total flux of **A** at the external surfaces. From symmetry, the total transfer across the external surfaces will be twice that at $x = L$. Thus

$$-A_c k \int_{-L}^{+L} C_A(x)dx = -2A_c D_e \left(\frac{dC_A}{dx} \right)_{x = \pm L}$$

If the whole of the catalyst surface area were available to the exterior concentration $C_{A\infty}$, there would be no diffusional resistance and the rate would then be $-2A_cLk\ C_{A\infty}$. The ratio of these two rates is the effectiveness factor:

$$\eta = \frac{-A_ck \int_{-L}^{+L} C_A(x)dx}{-2A_cLkC_{A\infty}} = \frac{-2A_cD_e\left(\dfrac{dC_A}{dx}\right)_{x=L}}{-2A_cLkC_{A\infty}} \tag{3.14}$$

By evaluating either the integral or the differential in the numerator of Eq. (3.14), the effectiveness factor may be calculated. In either case, by substitution from Eq. (3.13),

$$\eta = \frac{\tanh\lambda L}{\lambda L} = \frac{\tanh\phi}{\phi} \tag{3.15}$$

where

$$\phi = \lambda L = L\sqrt{\frac{k}{D_e}} \tag{3.16}$$

where ϕ is a dimensionless quantity known as the Thiele modulus. It is defined as the ratio of timescale of diffusion into the pore to that of the reaction inside the pore. The reaction rate per unit volume of the reactor for a first-order reaction can be written as $\eta(1-e)$ $kC_{A\infty}$. If the function η is plotted from Eq. (3.15), corresponding to the case of a first-order irreversible reaction in a slab with sealed edges, it may be seen from Fig. 3.3 that when $\phi < 0.2$, η is close to unity. Under these conditions there would be no diffusional resistance, for the rate of chemical reaction is not limited by diffusion. On the other hand when $\phi > 5.0$, $\eta = 1/\phi$ is a good approximation, and for such conditions, internal diffusion

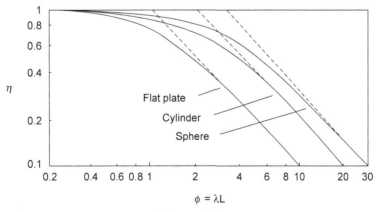

Figure 3.3
Effectiveness factors for flat plate, cylinder, and sphere.

is the rate-determining process. Between these two limiting values of ϕ, the effectiveness factor is calculated from Eq. (3.15) and the rate process is in a region where neither in-pore mass transport nor chemical reaction is overwhelmingly rate determining.

Only a very limited number of manufactured catalysts could be approximately described by the slab model, but there appear to be many which conform to the shape of a cylinder or sphere. Utilizing the same principles as for the slab, it may be shown (see the next example) that for a cylinder of radius r_0 sealed at the flat ends, the effectiveness factor is

$$\eta = \frac{2}{\lambda r} \frac{\mathbf{I}_1(\lambda r_0)}{\mathbf{I}_0(\lambda r_0)} \qquad (3.17)$$

where $\mathbf{I}_0$ and $\mathbf{I}_1$ denote zero- and first-order modified Bessel functions of the first kind.[15] For a sphere of radius r_0 (see Example 3.1),

$$\eta = \frac{3}{\lambda r_0} \left\{ \coth(\lambda r_0) - \frac{1}{\lambda r_0} \right\} \qquad (3.18)$$

Gunn[16] discusses the case of the hollow cylindrical catalyst particle. Such catalyst particles reduce the difficulties caused by excessive pressure drops.

Example 3.1
Derive an expression for the effectiveness factor of a cylindrical catalyst pellet, sealed at both ends, in which a first-order chemical reaction occurs.

Solution
The pellet has cylindrical symmetry about its central axis. Construct an annulus with radii $(r + \delta r)$ and r, and consider the diffusive flux of material into and out of the cylindrical annulus, length L.

A material balance for the reactant **A** gives (see Fig. 3.4)

Diffusive flux in at r − Diffusive flux out at $(r + \delta r)$ = Amount reacted in volume $2\pi L r \delta r$.

That is,

$$\left\{ -2\pi D_e L \left(r \frac{dC_A}{dr} \right)_r \right\} - \left\{ -2\pi D_e L \left(r \frac{dC_A}{dr} \right)_{r+\delta r} \right\} = 2\pi L r \delta r k C_A$$

Expanding the first term and ignoring terms higher than $(\delta r)^2$,

$$\frac{d^2 C_A}{dr^2} + \frac{1}{r} \frac{dC_A}{dr} - \lambda^2 C_A = 0 \quad \text{where } \lambda = \sqrt{\frac{k}{D_e}}$$

This is a standard modified Bessel equation of zero order whose solution is[15]

$$C_A = A\mathbf{I}_0(\lambda r) + B\mathbf{K}_0(\lambda r)$$

where $\mathbf{I}_0$ and $\mathbf{K}_0$ represent zero-order modified Bessel functions of the first and second kind, respectively.

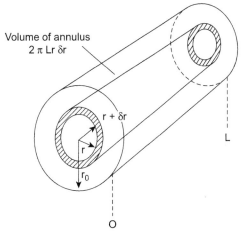

Figure 3.4
Geometry of cylindrical catalyst pellet.

The boundary conditions for the problem are $r = r_0$, $C_A = C_{A0}$; $r = 0$, C_A is finite. Because C_A remains finite at $r = 0$ and $\mathbf{K}_0(0) = \infty$, then we must put $B = 0$ to satisfy the physical conditions. Substituting the boundary conditions therefore gives the solution:

For the cylinder,

$$\frac{C_A}{C_{A0}} = \frac{\mathbf{I}_0(\lambda r)}{\mathbf{I}_0(\lambda r_0)}$$

$$\eta = \frac{2\pi r D_e L (dC_A/dr)_{r_0}}{\pi r_0^2 L k C_{A0}}$$

From the relation between C_A and r, $\left(\frac{dC_A}{dr}\right)_{r_0} = \lambda C_{A0} \frac{\mathbf{I}_1(\lambda r)}{\mathbf{I}_0(\lambda r)}$

Since

$$\frac{d}{dr}\{\mathbf{I}_0(\lambda r)\} = \mathbf{I}_1(\lambda r)$$

So

$$\eta = \frac{2}{\lambda r_0} \frac{\mathbf{I}_1(\lambda r_0)}{\mathbf{I}_0(\lambda r_0)}$$

Example 3.2
Derive an expression for the effectiveness factor of a spherical catalyst pellet in which a first-order isothermal reaction occurs.

Solution
Take the origin of coordinates at the center of the pellet, radius r_0, and construct an infinitesimally thin shell of radii $(r + \delta r)$ and r (see Fig. 3.5). A material balance for the reactant **A** across the shell gives

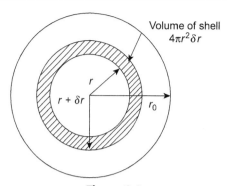

Figure 3.5
Geometry of spherical catalyst pellet.

Diffusive flux in at r — Diffusive flux out at $(r + \delta r)$ = Amount reacted in volume $4\pi r^2 \delta r$.

That is,

$$\left\{ -4\pi D_e \left(r^2 \frac{dC_A}{dr} \right)_r \right\} - \left\{ -4\pi D_e \left(r^2 \frac{dC_A}{dr} \right)_{r+\delta r} \right\} = 4\pi r^2 \delta r k C_A$$

Expanding the first term and ignoring terms higher than $(\delta r)^2$,

$$\frac{d^2 C_A}{dr^2} + \frac{2}{r} \frac{dC_A}{dr} = \frac{k}{D_e} C_A$$

or

$$\frac{1}{r^2} \frac{d}{dr} \left\{ r^2 \frac{dC_A}{dr} \right\} - \lambda^2 C_A = 0 \text{ where } \lambda = \sqrt{\frac{k}{D_e}}$$

Substituting $C_A = f(r)/r$,

$$f''(r) - \lambda^2 f(r) = 0$$

Therefore

$$f(r) = A e^{\lambda r} + B e^{-\lambda r}$$

The boundary conditions for the problem are $r = r_0$, $C_A = C_{A0}$; $r = 0$, $\frac{dC_A}{dr} = 0$. At r_0 we have $f(r_0) = C_{A0} r_0$. Substituting these boundary conditions,

$$f(r) = C_A r = \frac{C_{A0} r_0 \sinh(\lambda r)}{\sinh(\lambda r_0)}$$

Now for a sphere, $\eta = \frac{4\pi r_0^2 D_e (dC_A/dr)_{r_0}}{4/3 \pi r_0^3 k C_{A0}}$ From the relation between C_A and r,

$$\left(\frac{dC_A}{dr} \right)_{r_0} = \frac{C_{A0}}{r_0} \{ \lambda r_0 \coth(\lambda r_0) - 1 \}$$

Hence

$$\eta = \frac{3}{\lambda r_0} \left\{ \coth(\lambda r_0) - \frac{1}{\lambda r_0} \right\}$$

The Thiele moduli for the cylinder and sphere differ from those for the slab. In the case of the slab we recall that $\phi = \lambda L$, whereas for the cylinder it is conveniently defined as $\phi = \lambda r_0/2$ and for the sphere as $\phi = \lambda r_0/3$. In each case the reciprocal of this corresponds to the respective asymptote for the curve representing the slab, cylinder, or sphere. We may note here that the ratio of the geometric volume V_p of each of the models to the external surface area S_x is L for the slab, $r_0/2$ for the cylinder, and $r_0/3$ for the sphere. Thus, if the Thiele modulus is defined as

$$\phi = \lambda \frac{V_p}{S_x} = \frac{V_p}{S_x}\sqrt{\frac{k}{D_e}}, \tag{3.19}$$

the asymptotes become coincident. The asymptotes for large ϕ correspond to $\eta = 1/\phi$ for any shape of particle because, as Aris[17] points out, diffusion is rate determining under these conditions and reaction occurs, therefore, in only a very thin region of the particle adjacent to the exterior surface. The curvature of the surface is thus unimportant.

The effectiveness factor for the slab model may also be calculated for reactions other than first order. It turns out that when the Thiele modulus is large, the asymptotic value of η for all reactions is inversely proportional to the Thiele modulus, and when the latter approaches zero, the effectiveness factor tends to unity. However, just as we found that the asymptotes for a first-order reaction in particles of different geometry do not coincide unless we choose a definition for the Thiele modulus, which forces them to become superimposed, we find that the asymptotes for reaction orders $n = 0$, 1, and 2 do not coincide unless we define a generalized Thiele modulus:

$$\overline{\phi} = \frac{V_p}{S_x}\sqrt{\frac{n+1}{2}\frac{kC_{A\infty}^{(n-1)}}{D_e}} \tag{3.20}$$

The modulus $\overline{\phi}$ defined by Eq. (3.20) has the advantage that the asymptotes to η are approximately coincident for all particle shapes and for all reaction orders except $n = 0$; for this latter case[18] $\eta = 1$ for $\overline{\phi} < 2$ and $\eta = 1/\overline{\phi}$ for $\overline{\phi} > 2$. Thus η may be calculated from the simple slab model, using Eq. (3.20) to define the Thiele modulus. The curve of η as a function of $\overline{\phi}$ is therefore quite general for practical catalyst pellets. For $\overline{\phi} > 3$ it is found that $\eta = 1/\overline{\phi}$ to an accuracy within 0.5%, whereas the approximation is within 3.5% for $\overline{\phi} > 2$. It is best to use this generalized curve (i.e., η as a function of $\overline{\phi}$) because the asymptotes for different cases can then be made almost to coincide. The errors involved in using the generalized curve are probably no greater than errors perpetrated by estimating values of parameters in the Thiele modulus.

3.3.2 Effect of Intraparticle Diffusion on Experimental Parameters

When intraparticle diffusion occurs, the kinetic behavior of the system is different from that which prevails when chemical reaction is rate determining. For conditions of diffusion control, ϕ will be large, and then the effectiveness factor $\eta(= 1/\phi \tanh \phi$, from Eq. 3.15) becomes $\overline{\phi}^{-1}$. From Eq. (3.19), it is seen therefore that η is proportional to $k^{-1/2}$. The chemical reaction rate on the other hand is directly proportional to k so that, from Eq. (3.8) at the beginning of this section, the overall reaction rate is proportional to $k^{1/2}$.

Because the specific rate constant is directly proportional to $e^{-E/RT}$, where E is the activation energy for the chemical reaction in the absence of diffusion effects, we are led to the important result that for a diffusion-limited reaction the rate is proportional to $e^{-E/2RT}$. Hence the apparent activation energy E_D, measured when reaction occurs in the diffusion controlled region, is only half the true value:

$$E_D = E/2 \tag{3.21}$$

A further important result that arises because of the functional form of $\overline{\phi}$ is that the apparent order of reaction in the diffusion-controlled region differs from that which is observed when chemical reaction is rate determining. Recalling that the reaction order is defined as the exponent n to which the concentration $C_{A\infty}$ is raised in the equation for the chemical reaction rate, we replace $f(C_A)$ in Eq. (3.8) by $C_{A\infty}^n$. Hence the overall reaction rate per unit volume is $(1-e)\eta k C_{A\infty}^n$. When diffusion is rate determining, η is (as already mentioned) equal to ϕ^{-1}; from Eq. (3.20) it is therefore proportional to $C_{A\infty}^{-(n-1)/2}$. Thus the overall reaction rate depends on $C_{A\infty}^n C_{A\infty}^{-(n-1)/2} = C_{A\infty}^{(n+1)/2}$. The apparent order of reaction n_D as measured when reaction is dominated by intraparticle diffusion effects is thus related to the true order as follows:

$$n_D = (n+1)/2 \tag{3.22}$$

A zero-order reaction thus becomes a half-order reaction, a first-order reaction remains first order, whereas a second-order reaction has an apparent order of 3/2 when strongly influenced by diffusional effects. Because k and n are modified in the diffusion-controlled region then, if the rate of the overall process is estimated by multiplying the chemical reaction rate by the effectiveness factor (as in Eq. 3.8), it is imperative to know the true rate of chemical reaction uninfluenced by diffusion effects.

Oftentimes there is a need to choose the right catalyst pellet size to reduce or eliminate the intraparticle diffusion in packed bed reactors. The relative rates of a catalytic first-order reaction with two different pellet sizes, L_1 and L_2, can be written as $(r_{v1}/r_{v2}) = (\eta_1 k C_{A\infty})/(\eta_2 k C_{A\infty})$. Under diffusion-free or reaction-limited regime, the rates are equal owing to $\eta_1 = \eta_2 = 1$. However, under diffusion-limited regime, $(r_{v1}/r_{v2}) = L_2/L_1$, as $\eta = 1/\overline{\phi}$. This relationship can be used to assess the rate-limiting step for catalytic reactions when the rate data are available at two or more catalyst particle sizes.

The functional dependence of other parameters on the reaction rate also becomes modified when diffusion determines the overall rate. If we write the rate of reaction for an nth-order reaction in terms of Eq. (3.8) and substitute the general expression obtained for the effectiveness factor at high values of $\overline{\phi}$, where η is proportional to $1/\overline{\phi}$ and $\overline{\phi}$ is defined by Eq. (3.20), we obtain

$$r_v = (1-e)kC_{A\infty}^n \eta = (1-e)kC_{A\infty}^n \frac{S_x}{V_p} \sqrt{\frac{2D_e}{(n+1)kC_{A\infty}^{n-1}}} \tag{3.23}$$

The way in which experimental parameters are affected when intraparticle diffusion is important may be deduced by inspection of Eq. (3.23). Referring the specific rate constant to unit surface area, rather than unit reactor volume, the term $(1 - e)\,k$ is equivalent to $\rho_b S_g k_a$ where ρ_b is the bulk density of the catalyst and S_g is its surface area per unit mass. On the other hand, the rate constant k appearing in the denominator under the square root sign in Eq. (3.23) is based on unit particle volume and is therefore equal to $\rho_p S_g k_a$ where ρ_p is the particle density. Thus, if bulk diffusion controls the reaction, the rate becomes dependent on the square root of the specific surface area, rather than being directly proportional to surface area, in the absence of transport effects. We do not include the external surface area S_x in this reckoning because the ratio V_p/S_x is, for a given particle shape, an independent parameter characteristic of the particle size. On the other hand, if Knudsen diffusion determines the rate, then because the effective diffusivity for Knudsen flow is inversely proportional to the specific surface area (Eq. 3.4), the reaction rate becomes independent of surface area.

The pore volume per unit mass V_g (a measure of the porosity) is also a parameter, which is important and is implicitly contained in Eq. (3.23). Because the product of the particle density ρ_p and specific pore volume V_g represents the porosity, then ρ_p is inversely proportional to V_g. Therefore, when the rate is controlled by bulk diffusion, it is proportional not simply to the square root of the specific surface area but also to the product of $\sqrt{S_g}$ and $\sqrt{V_g}$. If Knudsen diffusion controls the reaction, then the overall rate is directly proportional to V_g because the effective Knudsen diffusivity contained in the quantity $\sqrt{(D_e/\rho_p)}$ is, from Eq. (3.4), proportional to the ratio of the porosity ε and the particle density ρ_p.

Table 3.2 summarizes the effect that intraparticle mass transfer effects have on parameters involved explicitly or implicitly in the expression for the overall rate of reaction.

3.3.3 Nonisothermal Reactions in Porous Catalyst Pellets

So far the effect of temperature gradients within the particle has been ignored. Strongly, exothermic reactions generate a considerable amount of heat that, if conditions are to remain stable, must be transported through the particle to the exterior surface where it may

Table 3.2: Effect of Intraparticle Diffusion on Various Parameters.

Rate-Determining Step	Order	Activation Energy	Surface Area	Pore Volume
Chemical reaction	n	E	S_g	Independent
Bulk diffusion	$(n+1)/2$	$E/2$	$\sqrt{S_g}$	V_g
Knudsen diffusion	$(n+1)/2$	$E/2$	Independent	V_g

then be dissipated. Similarly an endothermic reaction requires a source of heat, and in this case the heat must permeate the particle from the exterior to the interior. In any event a temperature gradient within the particle is established, and the chemical reaction rate will vary with position.

We may consider the problem by writing a material and heat balance for the slab of catalyst depicted in Fig. 3.2. For an irreversible first-order exothermic reaction, the material balance on reactant **A** is

$$\frac{d^2 C_A}{d x^2} - \frac{k C_A}{D_e} = 0 \quad \text{(equation 3.10)}$$

A heat balance over the element δx gives

$$\frac{d^2 T}{dx^2} + \frac{(-\Delta H)k C_A}{k_e} = 0 \tag{3.24}$$

where ΔH is the enthalpy change resulting from reaction and k_e is the effective thermal conductivity of the particle defined by analogy with the discussion on effective diffusivity leading to Eq. (3.1). In writing these two equations, it should be remembered that the specific rate constant k is a function of temperature, usually of the Arrhenius form ($k = Ae^{-\mathbf{E}/\mathbf{R}T}$, where A is the frequency factor for reaction). These two simultaneous differential equations are to be solved together with the boundary conditions:

$$C_A = C_{A^\infty} \quad \text{at} \ x = \pm L \tag{3.25}$$

$$T = T_\infty \quad \text{at} \ x = \pm L \tag{3.26}$$

$$\frac{dC_A}{dx} = \frac{dT}{dx} = 0 \ \text{at} \ x = 0 \tag{3.27}$$

Because of the nonlinearity of the equations, the problem can only be solved in this form by numerical techniques.[18,19] However, an approximation may be made, which gives an asymptotically exact solution,[20] or, alternatively, the exponential function of temperature may be expanded to give equations, which can be solved analytically.[21,22] A convenient solution to the problem may be presented in the form of families of curves for the effectiveness factor as a function of the Thiele modulus. Fig. 3.6 shows these curves for the case of a first-order irreversible reaction occurring in spherical catalyst particles. Two additional independent dimensionless parameters are introduced into the problem, and these are defined as

$$\beta = \frac{(-\Delta H)D_e C_{A^\infty}}{k_e T_\infty} \tag{3.28}$$

$$\varepsilon' = \mathbf{E}/\mathbf{R}T \tag{3.29}$$

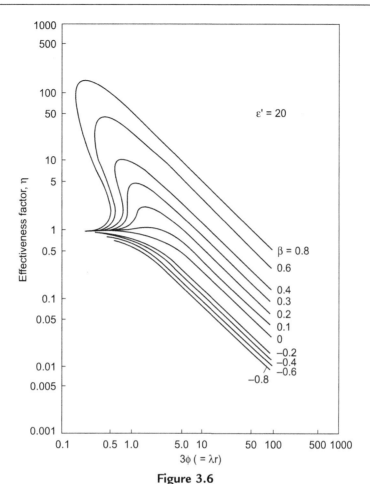

Figure 3.6

Effectiveness factor as a function of the Thiele modulus for nonisothermal conditions for a sphere.[21]

The parameter β represents the maximum temperature difference that could exist in the particle relative to the temperature at the exterior surface, for if we recognize that in the steady state, the heat flux within an elementary thickness of the particle is balanced by the heat generated by chemical reaction then

$$k_e \frac{dT}{dx} = \Delta H \, D_e \frac{dC_A}{dx} \tag{3.30}$$

If Eq. (3.30) is then integrated from the exterior surface where $T = T_\infty$ and $C_A = C_{A\infty}$ to the center of the particle where (say) $T = T_M$ and $C_A = C_{AM}$, we obtain

$$\frac{T_M - T_\infty}{T_\infty} = \frac{(-\Delta H)D}{k_e T_\infty}(C_{A\infty} - C_{AM}) \tag{3.31}$$

When the Thiele modulus is large, C_{AM} is effectively zero and the maximum difference in temperature between the center and exterior of the particle is $(-\Delta H)D_e C_{A\infty}/k_e$. Relative to the temperature outside the particle, this maximum temperature difference is therefore β. For exothermic reactions, β is positive, whereas for endothermic reactions, it is negative. The curve in Fig. 3.6 for $\beta = 0$ represents isothermal conditions within the pellet. It is interesting to note that for a reaction in which $-\Delta H = 10^5$ **kJ/kmol**, $k_e = 1$ **W/mK**, $D_e = 10^{-5}$ m^2/s, and $C_{A\infty} = 10^{-1}$ kmol/m^3, the value of $T_M - T_\infty$ is 100°C. In practice, much lower values than this are observed but it does serve to show that serious errors may be introduced into calculations if conditions within the pellet are arbitrarily assumed to be isothermal.

On the other hand, it has been argued that the resistance to heat transfer is effectively within a thin gas film enveloping the catalyst particle.[23] Thus, for the whole practical range of heat transfer coefficients and thermal conductivities, the catalyst particle may be considered to be at a uniform temperature. Any temperature increase arising from the exothermic nature of a reaction would therefore be across the fluid film rather than in the pellet interior.

Fig. 3.6 shows that, for exothermic reactions ($\beta > 0$), the effectiveness factor may exceed unity. This is because the increase in rate caused by the temperature rise inside the particle more than compensates for the decrease in rate caused by the negative concentration gradient, which effects a decrease in concentration toward the center of the particle. A further point of interest is that, for reactions that are highly exothermic and at low value of the Thiele modulus, the value of η is not uniquely defined by the Thiele modulus and the parameters β and ε'. The shape of the curves in this region indicates that the effectiveness factor may correspond to any one of three values for a given value of the Thiele modulus. In effect there are three different conditions for which the rate of heat generation within the particle is equal to the rate of heat removal. One condition represents a metastable state, and the remaining two conditions correspond to a region in which the rate is limited by chemical reaction (relatively low temperatures) and a region where there is diffusion limitation (relatively high temperatures). The region of multiple solutions in Fig. 3.6, however, corresponds to large values of β and ε' seldom encountered in practice.

McGreavy and Thornton[23] have developed an alternative approach to the problem of identifying such regions of unique and multiple solutions in packed bed reactors. Recognizing that the resistance to heat transfer is probably due to a thin gas film surrounding the particle, but that the resistance to mass transfer is within the porous solid, they solved the mass and heat balance equations for a pellet with modified boundary conditions. Thus the heat balance for the pellet represented by Eq. (3.24) was replaced by

$$h(T - T_\infty) = k_G(-\Delta H)(C_{A\infty} - C_A) \qquad (3.32)$$

and solved simultaneously with the mass balance represented by Eq. (3.10). Boundary conditions represented by Eqs. (3.25) and (3.26) were replaced by

$$D_e \frac{dC_A}{dx} = k_G(C_{A\infty} - C_A) \ \text{ at } \ x = L \tag{3.33}$$

and

$$k_e \frac{dT}{dx} = h(T - T_\infty) \ \text{ at } \ x = L, \tag{3.34}$$

respectively.

A modified Thiele modulus may be defined by rewriting $\phi = L\sqrt{k/D_e}$ (see Eq. 3.16) in the form:

$$\phi' = L\sqrt{\frac{A}{D_e}} \tag{3.35}$$

where A is the frequency factor in the classical Arrhenius equation. The numerical solution is then depicted in Fig. 3.7,[23] which resembles a plot of effectiveness factor as a

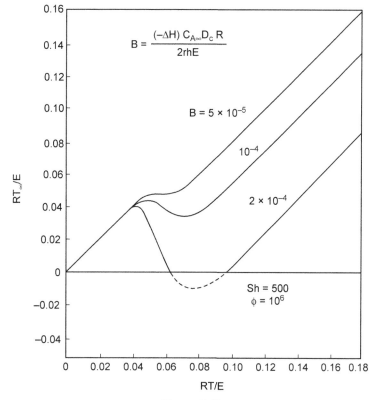

Figure 3.7
Multiple states for catalyst pellets in reactor.

function of Thiele modulus (cf. Fig. 3.6). An effectiveness factor chart describes the situation for a given single particle in a packed bed (and is therefore of limited value in reactor design), whereas Fig. 3.7 may be used to identify the region of multiple solutions for the whole reactor. If local extrema are calculated from Fig. 3.7 by finding conditions for which $dT_\infty/dT = 0$, the bounds of T_∞ may be located. It is then possible to predict the region of multiple solutions corresponding to unstable operating conditions. Fig. 3.8, for example, shows two reactor trajectories, which would intersect the region of metastable conditions within the reactor. Such a method can predict regions of instability for the packed reactor rather than for a single particle, because the use has been made of the modified Thiele modulus employing the kinetic Arrhenius factor A, which is independent of position along the bed.

3.3.4 Criteria for Diffusion Control

In assessing whether a reactor is influenced by intraparticle mass transfer effects, Weisz and Prater[24] developed a criterion for isothermal reactions based on the observation that the effectiveness factor approaches unity when the generalized Thiele modulus is of the order of unity. It has been shown[17] that the effectiveness factor for all catalyst geometries and reaction orders (except zero order) tends to unity when the generalized Thiele modulus falls below a value of one. Because η is about unity when $\phi < \sqrt{2}$ for zero-order reactions, a quite general criterion for diffusion control of simple isothermal reactions not

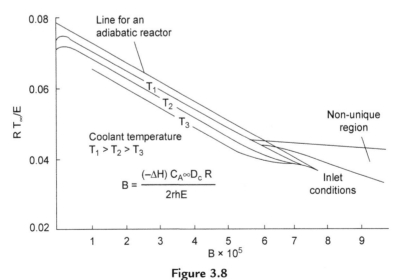

Figure 3.8
Reactor trajectories in adiabatic and cooled catalyst beds.

affected by product inhibition is $\phi < 1$. Because the Thiele modulus (see Eq. 3.19) contains the specific rate constant for chemical reaction, which is often unknown, a more useful criterion is obtained by substituting r'_v/C_{A^∞} (for a first-order reaction) for k to give

$$\left\{\frac{V_p}{S_x}\right\}^2 \frac{r'_v}{D_e C_{A^\infty}} < 1 \tag{3.36}$$

where r'_v is the measured rate of reaction per unit volume of catalyst particle.

Petersen[25] points out that this criterion is invalid for more complex chemical reactions whose rate is retarded by products. In such cases the observed kinetic rate expression should be substituted into the material balance equation for the particular geometry of particle concerned. An asymptotic solution to the material balance equation then gives the correct form of the effectiveness factor. The results indicate that the inequality 3.36 is applicable only at high partial pressures of product. For low partial pressures of product (often the condition in an experimental differential tubular reactor), the criterion will depend on the magnitude of the constants in the kinetic rate equation.

The usual experimental criterion for diffusion control involves an evaluation of the rate of reaction as a function of particle size. At a sufficiently small particle size the measured rate of reaction will become independent of particle size, and the rate of reaction can be safely assumed to be independent of intraparticle mass transfer effects. At the other extreme, if the observed rate is inversely proportional to particle size, the reaction is strongly influenced by intraparticle diffusion. For a reaction whose rate is inhibited by the presence of products, there is an attendant danger of misinterpreting experimental results obtained for different particle sizes when a differential reactor is used; under these conditions, the effectiveness factor is sensitive to changes in the partial pressure of product.

Weisz and Hicks[21] showed that when reaction conditions within the particle are nonisothermal, a suitable criterion defining conditions under which a reaction is not controlled by mass and heat transfer effects in the solid is

$$\left\{\frac{V_p}{S_x}\right\}^2 \frac{r'_v}{D_e C_{A^\infty}} \exp\left\{\frac{\varepsilon'\beta}{1+\beta}\right\} < 1 \tag{3.37}$$

3.3.5 Selectivity in Catalytic Reactions Influenced by Mass and Heat Transfer Effects

It is rare that a catalyst can be chosen for a reaction such that it is entirely specific or unique in its behavior. More often than not products additional to the main desired product are generated concomitantly. The ratio of the specific chemical rate constant of a desired reaction to that for an undesired reaction is termed the kinetic selectivity factor (which we shall designate by S) and is of central importance in catalysis. Its magnitude is determined by the relative rates at which adsorption, surface reaction, and desorption occur in the overall process and, for consecutive reactions, whether or not the intermediate product forms a localized or mobile adsorbed complex with the surface. In the case of two parallel competing catalytic reactions a second factor, the thermodynamic factor, is also of importance. This latter factor depends exponentially on the difference in free energy changes associated with the adsorption–desorption equilibria of the two competing reactants. The thermodynamic factor also influences the course of a consecutive reaction where it is enhanced by the ability of the intermediate product to desorb rapidly and also the reluctance of the catalyst to readsorb the intermediate product after it has vacated the surface.

The kinetic and thermodynamic selectivity factors are quantities, which are functions of the chemistry of the system. When an active catalyst has been selected for a particular reaction (often by a judicious combination of theory and experiment), we ensure that the kinetic and thermodynamic factors are such that they favor the formation of desired product. Many commercial processes, however, employ porous catalysts because this is the best means of increasing the extent of surface at which the reaction occurs. Chemical engineers are therefore interested in the effect, which the porous nature of the catalyst has on the selectivity of the chemical process.

Wheeler[26] considered the problem of chemical selectivity in porous catalysts. Although he employed a cylindrical pore model and restricted his conclusions to the effect of pore size on selectivity, the following discussion will be based on the simple geometrical model of the catalyst pellet introduced earlier (see Fig. 3.2 and Section 3.3.1).

3.3.5.1 Isothermal Conditions

Sometimes it may be necessary to convert only one component in a mixture into a desired product. For example, it may be required to dehydrogenate a six-membered cycloparaffin in the presence of a five-membered cycloparaffin without affecting the latter. In this case it is desirable to select a catalyst, which favors the reaction:

$$A \xrightarrow{k_1} B$$

when it might be possible for the reaction:

$$X \xrightarrow{k_1} Y$$

to occur simultaneously.

Suppose the desired product is **B** and also suppose that the reactions occur in a packed tubular reactor in which we may neglect both longitudinal and radical dispersion effects. If both reactions are first order, the ratio of the rates of the respective reactions is

$$\frac{r'_{vA}}{r'_{vX}} = \frac{k_1 C_A}{k_2 C_X} \tag{3.38}$$

If the reactions were not influenced by in-pore diffusion effects, the intrinsic kinetic selectivity would be $k_1/k_2 (= S)$. When mass transfer is important, the rate of reaction of both **A** and **X** must be calculated with this in mind. From Eq. (3.9), the rate of reaction for the slab model is

$$-A_c D_e = \left(\frac{dC_A}{dx}\right)_{x=\pm L}$$

The concentration profile of **A** through the slab is given by

$$C_A = C_{A^\infty} \frac{\cosh(\lambda x)}{\cosh(\lambda L)} \quad \text{(Eq. 3.13)}$$

so that by differentiation of C_A we obtain the rate of decomposition of **A**:

$$\dot{r}_{vA} = \frac{-A_c D_e}{V_p} \left(\frac{dC_A}{dx}\right)_{x=L} = \frac{-A_c D_e C_{A^\infty}}{L V_p} \phi_1 \tanh \phi_1 \tag{3.39}$$

where $\phi_1 = L\sqrt{k_1/D_e}$ is the Thiele modulus pertaining to the decomposition of **A**. A similar equation may be written for the decomposition of **X**:

$$\dot{r}_{vX} = \frac{-A_c D_e C_{X^\infty}}{L V_p} \phi_2 \tanh \phi_2 \tag{3.40}$$

where ϕ_2 corresponds to the Thiele modulus for the decomposition of **X**. If we were dealing with a general type of catalyst pellet, then because of the properties of the general Thiele parameter $\overline{\phi}$, we need only replace ϕ_1 and ϕ_2 by $\overline{\phi}_1$ and $\overline{\phi}_2$, respectively, and substitute (V_p/S_x) for the characteristic dimension L. The ratio of the rates of decomposition of **A** and **X** then becomes

$$\frac{\dot{r}_{vA}}{\dot{r}_{vX}} = \frac{C_{A^\infty}}{C_{X^\infty}} \frac{\overline{\phi}_1 \tanh \overline{\phi}_1}{\overline{\phi}_2 \tanh \overline{\phi}_2} \tag{3.41}$$

Although Eq. (3.41) is only applicable to competing first-order reactions in catalyst particles, at large values of $\overline{\phi}$ where diffusion is rate controlling, the equation is equivalent at the asymptotes to equations obtained for reaction orders other than one.

Because $\overline{\phi}$ is proportional to $\sqrt{k}$, we may conclude that for competing simultaneous reactions strongly influenced by diffusion effects (where $\overline{\phi}$ is large and $\tanh \overline{\phi} \approx 1$), the selectivity depends on $\sqrt{S}$ (where $S = k_1/k_2$), the square root of the ratio of the respective rate constants. The corollary is that, for such reactions, maximum selectivity is displayed by small sized particles and, in the limit, if the particle size is sufficiently small (small $\overline{\phi}$ so that $\tanh\overline{\phi} \simeq \overline{\phi}$), the selectivity is the same as for a nonporous particle, i.e., S itself.

We should add a note of caution here, however, for in the Knudsen flow region, D_e is proportional to the pore radius. When the pores are sufficiently small for Knudsen diffusion to occur, then the selectivity will also be influenced by pore size. Maximum selectivity would be obtained for small particles, which contain large diameter pores.

When two simultaneous reaction paths are involved in a process, the routes may be represented by:

A classical example of this type of competitive reaction is the conversion of ethanol by a copper catalyst at about 300°C. The principal product is acetaldehyde, but ethylene is also evolved in smaller quantities. If, however, an alumina catalyst is used, ethylene is the preferred product. If **B** is the desired product in the above reaction scheme, then the selectivity may be found by comparing the respective rates of formation of **B** and **C**. Adopting the slab model for simplicity and remembering that the rates of formation of **B** and **C** must be equal to the flux of **B** and **C** at the exterior surface of the particle in the steady state, assuming that the effective diffusivities of **B** and **C** are equal,

$$\frac{r'_{vB}}{r'_{vC}} = \left(\frac{dC_B}{dx}\right)_{x=L} \bigg/ \left(\frac{dC_C}{dx}\right)_{x=L} \tag{3.42}$$

The respective fluxes may be evaluated by writing the material balance equations for each component and solving the resulting simultaneous equations. If the two reactions are of the same kinetic order, then it is obvious from the form of Eq. (3.42) that the selectivity is unaffected by mass transfer in pores. If, however, the kinetic orders of the reactions differ, then, as the example below shows, the reaction of the lower kinetic order is favored. The rate of formation of **B** with respect to **C** would therefore be impeded and the selectivity would be reduced if the order of the reaction producing **B** were less than that for the reaction producing **C**. In such cases the highest selectivity would be obtained by the use of small diameter particles or particles in which the effective diffusivity is high.

Example 3.3

Two gas-phase concurrent irreversible reactions occur isothermally in a flat slab-shaped porous catalyst pellet. The desirable product **B** is formed by a first-order chemical reaction, and the wasteful product **C** is formed by a zero-order reaction. Deduce an expression for the catalyst selectivity.

Solution

Taking a material balance across the element δx in Fig. 3.2, the flux in at $(x + \delta x)$ minus the flux out at x is equal to the amount reacted in volume $2A_c\delta x$ where A_c represents the area of each of the faces. If the slab is thin, diffusion through the edges may be neglected. For the three components, the material balance equations therefore become

$$\frac{d^2C_A}{dx^2} - f^2 C_A = g^2$$

$$\frac{d^2C_B}{dx^2} + f^2 C_A = 0$$

$$\frac{d^2C_C}{dx^2} = -g^2$$

where $f^2 = k_1/D_e$ and $g^2 = k_2/D_e$. The boundary conditions are

$$x = \pm L, \quad C_A = C_{A0}, \quad C_B = C_C = 0$$

and

$$x = 0, \quad \frac{dC_A}{dx} = \frac{dC_B}{dx} = \frac{dC_C}{dx} = 0$$

The solution satisfying the above set of differential equations is $C_A = Ae^{fx} - g^2/f^2$ where the term g^2/f^2 represents the particular integral. On inserting the boundary conditions, the complete solution is

$$C_A = \left(C_{A0} + \frac{g^2}{f^2}\right) \frac{\cosh(fx)}{\cosh(fL)} - \frac{g^2}{f^2}$$

The concentration C_B may now be found, but because the selectivity will be given by

$$S = \frac{(dC_B/dx)_{x=L}}{(dC_C/dx)_{x=L}}$$

the material balance equation need only be integrated once. We obtain

$$\left(\frac{dC_B}{dx}\right)_{x=L} = -f^2 \int_0^L \left\{\left(C_{A0} + \frac{g^2}{f^2}\right) \frac{\cosh(fx)}{\cosh(fL)} - \frac{g^2}{f^2}\right\} dx = -f\left(C_{A0} + \frac{g^2}{f^2}\right)\tanh(fL) + g^2 L$$

and

$$\left(\frac{dC_C}{dx}\right)_{x=L} = -g^2 L$$

Then writing $\phi = L\sqrt{\frac{k_1}{D_e}}$

$$S = \left(\frac{k_1}{k_2}C_{A0} + 1\right) \frac{\tanh \phi}{\phi} - 1$$

Another important class of reactions, which is common in petroleum reforming reactions, may be represented by the following scheme:

$$A \xrightarrow{k_1} B \xrightarrow{k_2} C$$

and exemplified by the dehydrogenation of six-membered cycloparaffins to aromatics (e.g., cyclohexane converted to cyclohex-1-ene and ultimately benzene) catalyzed by transition metals and metal oxides. Again we suppose **B** to be the desired product while **C** is a waste product. (If **C** were the desired product, we would require a low selectivity for the formation of **B**). On the basis of first-order kinetics and using the flat plate model, the material balance equation for component **B** is

$$D_e \frac{d^2 C_B}{dx^2} = k_2 C_B - k_1 C_A \tag{3.43}$$

For component **A** the material balance equation is, as previously obtained,

$$D_e \frac{d^2 C_A}{dx^2} = k_1 C_A \quad \text{(equation 3.10)}$$

The boundary conditions for the problem are

$$C_A = C_{A\infty} \text{ and } C_B = C_{B\infty} \quad \text{at} \quad x = L \tag{3.44}$$

$$dC_A/dx = dC_B/dx = 0 \quad \text{at} \quad x = 0 \tag{3.45}$$

Solving the two simultaneous linear differential equations with the above boundary conditions leads to

$$C_A = C_{A\infty} \frac{\cosh(\lambda_1 x)}{\cosh(\lambda_1 L)} \quad \text{(equation 3.13)}$$

and

$$C_B = C_{A\infty} \left\{ \frac{k_1}{k_1 - k_2} \right\} \left\{ \frac{\cosh(\lambda_2 x)}{\cosh(\lambda_2 L)} - \frac{\cosh(\lambda_1 x)}{\cosh(\lambda_1 L)} \right\} + C_{B\infty} \frac{\cosh(\lambda_2 x)}{\cosh(\lambda_2 L)} \tag{3.46}$$

where λ_1 and λ_2 are $\sqrt{k_1/D_e}$ and $\sqrt{k_2/D_e}$, respectively. The selectivity of the reaction will be the rate of formation of **B** with respect to **A** which, in the steady state, will be equal to the ratio of the fluxes of **B** and **A** at the exterior surface of the particle. Thus

$$-\frac{(dC_B/dx)_{x=L}}{(dC_A/dx)_{x=L}} = \left\{ \frac{k_1}{k_1 - k_2} \right\} \left\{ 1 - \frac{\phi_2 \tanh \phi_2}{\phi_1 \tanh \phi_1} \right\} - \frac{C_{B\infty}}{C_{A\infty}} \frac{\phi_2 \tanh \phi_2}{\phi_1 \tanh \phi_1} \tag{3.47}$$

Although the ratio of the fluxes of **B** and **A** at the exterior surface of the particle is really a point value, we may conveniently regard it as representing the rate of change of the concentration of **B** with respect to **A** at a position in the reactor corresponding to the location of the particle. The left side of Eq. (3.47) may thus be replaced by $-dC_B/dC_A$ where C_B and C_A are now gas-phase concentrations. With this substitution and for large values of ϕ_1 and ϕ_2, one of the limiting forms of Eq. (3.47) is obtained:

$$-\frac{dC_B}{dC_A} = \frac{\sqrt{S}}{1 + \sqrt{S}} - \frac{C_B}{C_A} \frac{1}{\sqrt{S}} \tag{3.48}$$

where S is the kinetic selectivity ($=k_1/k_2$). Integrating Eq. (3.48) from the reactor inlet (where the concentration of **A** is, say, C_{A0} and that of **B** is taken as zero) to any point along the reactor,

$$\frac{C_B}{C_A} = \left\{\frac{S}{S-1}\right\}\left\{\left(\frac{C_A}{C_{A0}}\right)^{\frac{1-\sqrt{S}}{\sqrt{S}}} - 1\right\} \tag{3.49}$$

The other limiting form of Eq. (3.47) is obtained when ϕ_1 and ϕ_2 are small:

$$-\frac{dC_B}{dC_A} = 1 - \frac{1}{S}\frac{C_B}{C_A} \tag{3.50}$$

which, on integration, gives

$$\frac{C_B}{C_A} = \left\{\frac{S}{S-1}\right\}\left\{\left(\frac{C_A}{C_{A0}}\right)^{\frac{1-S}{S}} - 1\right\} \tag{3.51}$$

Comparison of Eqs. (3.48) and (3.50) shows that at low effectiveness factors (when ϕ_1 and ϕ_2 are large) the selectivity is less than it would be if the effectiveness factor for each reaction were near unity (small ϕ_1 and ϕ_2). The consequence of this is seen by comparing Eqs. (3.49) and (3.51), the respective integrated forms of Eqs. (3.48) and (3.50). The yield of **B** is comparatively low when in-pore diffusion is a rate-limiting process and, for a given fraction of **A** reacted, the conversion to **B** is impeded. A corollary to this is that it should be possible to increase the yield of a desired intermediate by using smaller catalyst particles or, alternatively, by altering the pore structure in such a way as to increase the effective diffusivity. If, however, the effectiveness factor is below about 0.3 (at which value the yield of **B** becomes independent of diffusion effects), a large reduction in pellet size (or large increase in D_e) is required to achieve any significant improvement in selectivity. Wheeler[26] suggests that when such a drastic reduction in pellet size is necessary, a fluidized bed reactor may be used to improve the yield of an intermediate.

3.3.5.2 Nonisothermal Conditions

The influence that the simultaneous transfer of heat and mass in porous catalysts has on the selectivity of first-order concurrent catalytic reactions was investigated by Østergaard.[27] As shown previously, selectivity is not affected by any limitations due to mass transfer when the process corresponds to two concurrent first-order reactions:

However, with heat transfer between the interior and exterior of the pellet (made possible by temperature gradients resulting from an exothermic diffusion-limited reaction), the selectivity may be substantially altered. For the flat plate model, the material and heat balance equations to solve are

$$\frac{d^2 C_A}{dx^2} - \left(\frac{k_1 + k_2}{D_e}\right) C_A = 0 \tag{3.52}$$

$$\frac{d^2 C_B}{dx^2} + \frac{k_1}{D_e} C_A = 0 \tag{3.53}$$

and

$$\frac{d^2 T}{dx^2} + \frac{(-\Delta H_1)k_1 + (-\Delta H_2)k_2}{k_e} C_A = 0 \tag{3.54}$$

where

$$k_i = A_i \exp(-E_i/RT), \quad i = 1, 2 \tag{3.55}$$

and ΔH_1 and ΔH_2 correspond to the respective enthalpy changes. The boundary conditions are as follows:

$$\frac{dC_A}{dx} = \frac{dC_B}{dx} = \frac{dT}{dx} = 0 \quad \text{at} \quad x = 0 \tag{3.56}$$

$$C_A = C_{A\infty}, \quad C_B = C_{B\infty}, \quad T = T_\infty \quad \text{at} \quad x = L \tag{3.57}$$

This is a two-point boundary value problem and, because of the nonlinearity of the equations, cannot be solved analytically. If, however, $E_1 = E_2$, the selectivity is the same as if there were no resistance to either heat or mass transfer (see Example 3.4). For the case where $\Delta H_1 = \Delta H_2$ but $E_1 \neq E_2$, the selectivity is determined by the effects of

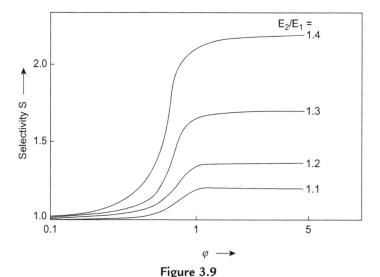

Figure 3.9
Selectivity as a function of the Thiele modulus for nonisothermal conditions.

simultaneous heat and mass transfer. Fig. 3.9 shows that if the activation energy of the desired reaction is lower than that of the reaction leading to the wasteful product ($E_2/E_1 > 1$), the best selectivity is obtained for high values of the Thiele modulus. When $E_2/E_1 < 1$, a decrease in selectivity results. In the former case the selectivity approaches an upper limit asymptotically, and it is not worth while increasing the Thiele modulus beyond a value where there would be a significant decrease in the efficiency of conversion.

Example 3.4

Show that the selectivity of two concurrent first-order reactions occurring in flat-shaped porous catalyst pellets is independent of the effect of either heat or mass transfer if the activation energies of both reactions are equal.

Solution

For the concurrent first-order reactions, the mass and heat transfer equations are

$$\frac{d^2 C_A}{dx^2} - \frac{k_1 + k_2}{D_e} C_A = 0$$

$$\frac{d^2 C_B}{dx^2} - \frac{k_1}{D_e} C_A = 0$$

$$\frac{d^2 T}{dx^2} + \frac{(-\Delta H_1)k_1 + (-\Delta H_2)k_2}{k_e} C_A = 0$$

where $k_i = A_i \exp(-E_i/RT)$, $i = 1, 2$. The boundary conditions are given by

$$\frac{dC_A}{dx} = \frac{dC_B}{dx} = \frac{dT}{dx} = 0 \quad \text{at} \quad x = 0$$

$$C_A = C_{A\infty}, \quad C_B = C_{B\infty}, \quad T = T_\infty \quad \text{at} \quad x = L$$

When $E_1 = E_2$ we see that $k_1/k_2 = A_1/A_2$—a ratio independent of temperature. Hence, if the mass transfer equations are divided, they may, for the case $E_1 = E_2$, be integrated directly and the gradients evaluated at the slab surface, $x = L$. Thus

$$\left(\frac{dC_A}{dx}\right)_{x=L} = \frac{\left(1 + \frac{A_2}{A_1}\right)}{\left(1 + \frac{(-\Delta H_2)A_2}{(-\Delta H_1)A_1}\right)} \frac{k_e}{D_e(-\Delta H_1)} \left(\frac{dT}{dx}\right)_{x=L}$$

$(dC_B/dx)_{x=L}$ may be found similarly. The selectivity is determined by the ratio of the reaction rates at the surface. In the steady state, this is equal to the ratio of the fluxes of **C** and **B** at the slab surface. Hence we obtain for the selectivity

$$\frac{\left(\frac{dC_C}{dx}\right)_{x=L}}{\left(\frac{dC_B}{dx}\right)_{x=L}} = \frac{\left(\frac{dC_A}{dx}\right)_{x=L}}{\left(\frac{dC_B}{dx}\right)_{x=L}} - 1 = 1 + \frac{A_2}{A_1} - 1 = \frac{A_2}{A_1}$$

and this is the same result as would have been obtained if there were no resistance to either mass or heat transfer within the pellets.

3.3.5.3 Selectivity of Bifunctional Catalysts

Certain heterogeneous catalytic processes require the presence of more than one catalyst to achieve a significant yield of desired product. The conversion of *n*-heptane to *i*-heptane, for example, requires the presence of a dehydrogenation catalyst, such as platinum, together with an isomerization catalyst such as silica−alumina. In this particular case the *n*-heptane would be dehydrogenated by the platinum catalyst and the product isomerized to *i*-heptene which, in turn, would be hydrogenated to give finally *i*-heptane. When the hydrogenation and the isomerization are carried out simultaneously, the catalyst is said to act as a bifunctional catalyst. The recent trend is toward the production of catalyst in which both functions are built into the individual catalyst particles. This has the great advantage that it is not necessary to ensure that the component particles are mixed together well.

Many organic reactions involving the upgrading of petroleum feedstocks are enhanced if a bifunctional catalyst is used. Some of the reactions that take place may be typified by one or more of the following:

$$1: \quad A \overset{X}{\to} B \overset{Y}{\to} C$$

$$2: \quad A \overset{X}{\rightleftharpoons} B \overset{Y}{\rightleftharpoons} C$$

$$3: \quad A \overset{X}{\rightleftharpoons} B \overset{Y}{\to} C; \quad B \overset{X}{\to} D$$

in which **A** represents the initial reactant, **B** and **D** represent unwanted products, and **C** represents the desired product. **X** and **Y** represent hydrogenation and isomerization catalysts, respectively. These reaction schemes implicitly assume the participation of hydrogen and pseudo−first-order reaction kinetics. To a first approximation, the assumption of first-order chemical kinetics is not unrealistic, for Sinfelt[28] has shown that under some conditions many reactions involving upgrading of petroleum may be represented by first-order kinetics. Gunn and Thomas[29] examined mass transfer effects accompanying such reactions occurring isothermally in spherical catalyst particles containing the catalyst components **X** and **Y**. They demonstrated that it is possible to choose the volume fraction of **X** in such a way that the formation of **C** may be maximized. Curves 1, 2, and 3 in Fig. 3.10 indicate that a tubular reactor may be packed with discrete spherical particles of **X** and **Y** in such a way that the throughput of **C** is maximized. For given values of the kinetic constants of each step, the effective diffusivity, and the particle size, the amount of **C** formed is maximized by choosing the ratio of **X** to **Y** correctly. Curve 1 corresponds to the irreversible reaction 1 and requires more of component **X** than either reaction 2 or 3 for the same value of chosen parameters. On the other hand, the mere fact that reaction 2 has a reversible step means that more of **Y** is required to produce

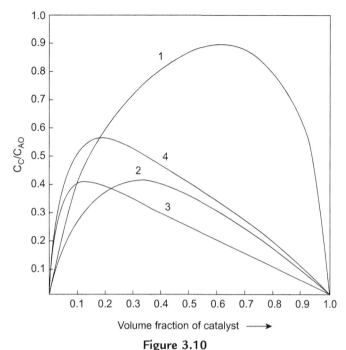

Figure 3.10
Optimum yields of desired product in bifunctional catalyst systems.

the maximum throughput of **C**. When a second end product **D** results, as in reaction 3, even more of catalyst component **Y** is needed if the formation of **D** is catalyzed by **X**.

Even better yields of **C** result if components **X** and **Y** are incorporated in the same catalyst particle, rather than if they exist as separate particles. In effect, the intermediate product **B** no longer has to be desorbed from particles of the **X**-type catalyst, transported through the gas phase and thence readsorbed on **Y**-type particles prior to reaction. Resistance to intraparticle mass transfer is therefore reduced or eliminated by bringing **X**-type catalyst sites into close proximity to **Y**-type catalyst sites. Curve 4 in Fig. 3.10 illustrates this point and shows that for such a composite catalyst, containing both **X** and **Y** in the same particle, the yield of **C** for reaction 3 is higher than it would have been had discrete particles of **X** and **Y** been used (curve 3).

The yields of **C** found for reactions 1, 2, and 3 were evaluated with the restriction that the catalyst composition should be uniform along the reactor length. However, there seems to be no point in **Y** being present at the reactor inlet where there is no **B** to convert. Similarly, there is little point in much **X** being present at the reactor outlet where there may be only small amounts of **A** remaining unconverted to **B**. Thomas and Wood[30] examined this question for reaction schemes represented by reactions 1 and 2. For an irreversible consecutive reaction, such as 1, they showed analytically that the optimal profile consists of two catalyst zones, one containing pure **X** and the other containing pure **Y**. When one of the steps is reversible, as in 2, numerical optimization techniques showed that a catalyst composition changing along the reactor is superior to either a constant catalyst composition or a two-zone reactor. However, it was also shown that, in most cases, a bifunctional catalyst of constant composition will give a yield, which is close to the optimum, thus obviating the practical difficulty of packing a reactor with a catalyst having a variable composition along the reactor length. The exception noted was that for reactions in which there may be a possibility of the desired product undergoing further unwanted side reactions, the holding time should be restricted and a two-zone reactor, containing zones of pure **X** and pure **Y**, is to be preferred. Similar results were obtained independently by Gunn[31] who calculated, using a numerical method, the maximum yield of desired product for discrete numbers of reactor stages containing different proportions of the two catalyst components. Jackson[32] has produced an analytical solution to the problem for the case of reaction 2. Jenkins and Thomas[33] have demonstrated that, for the reforming of methylcyclopentane at 500°C (773 K) and 1 bar pressure, theoretical predictions of optimum catalyst compositions based on a kinetic model are close to the results obtained employing an experimental hill-climbing procedure.

3.3.6 Catalyst Deactivation and Poisoning

Catalysts can become deactivated in a number of ways during the course of the operation. A common cause of deactivation is the sintering of the particles to form a continuous matter as a result of the development of local hot spots within the body of the reactor. Another is the deposition of carbon on the surface of the particles, as in the case of the fluidized cracking of hydrocarbons in the oil industry. In this case the spent catalyst is transported to a second fluidized bed in which the carbon coating is burned off and the regenerated catalyst is then recycled.

Catalysts may also become poisoned when the feed stream to a reactor contains impurities, which are deleterious to the activity of the catalyst. Particularly strong poisons are substances whose molecular structure contains lone electron pairs capable of forming covalent bonds with catalyst surfaces. Examples are ammonia, phosphine, arsine, hydrogen sulfide, sulfur dioxide, and carbon monoxide. Other poisons include hydrogen, oxygen, halogens, and mercury. The surface of a catalyst becomes poisoned by virtue of the foreign impurity adsorbed within the porous structure of the catalyst and covering a fraction of its surface, thus reducing the overall activity. The detailed reaction mechanism of catalyst poisoning is described in Section 3.5.6. The reactants participating in the desired reaction must now be transported to the unpoisoned part of the surface before any further reaction ensues, and so poisoning increases the average distance over which the reactants must diffuse prior to reaction at the surface. We may distinguish between two types of poisoning:

1. homogeneous poisoning, in which the impurity is distributed evenly over the active surface; and
2. selective poisoning in which an extremely active surface first becomes poisoned at the exterior surface of the particle and then progressively becomes poisoned toward the center of the particle.

When homogeneous poisoning occurs, because no reaction will be possible on the poisoned fraction (ζ, say, as shown in Fig. 3.11) of active surface, it is reasonable to suppose that the intrinsic activity of the catalyst is in proportion to the fraction of active surface remaining unpoisoned. To find the ratio of activity of the poisoned catalyst to the activity of an unpoisoned catalyst, one would compare the stationary flux of reactant to the particle surface in each case. For a first-order reaction occurring in a flat plate (slab) of catalyst, one finds (see Example 3.5) this ratio to be

$$F = \frac{\sqrt{1-\zeta}\,\tanh\{\phi\sqrt{1-\zeta}\,\}}{\tanh \phi} \tag{3.58}$$

where ϕ is the Thiele modulus for a first-order reaction occurring in a flat plate of catalyst and ζ is the fraction of active surface poisoned. The two limiting cases of Eq. (3.58) correspond to extreme values of ϕ. When ϕ is small, F becomes equal to $(1 - \zeta)$ and the activity decreases linearly with the amount of poison added, as shown by curve 1 in Fig. 3.12. On the other hand, when ϕ is large $F = \sqrt{1 - \zeta}$ and the activity decreases less rapidly than linearly due to the reactants penetrating the interior of the particle to a lesser extent for large values of the Thiele modulus (curve 2, Fig. 3.12).

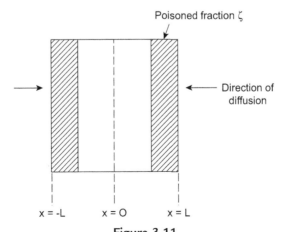

Figure 3.11
Geometry of partially poisoned (selective) slab.

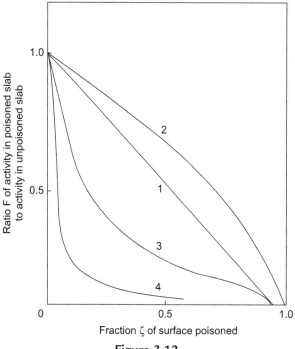

Figure 3.12
Catalyst poisoning.

Example 3.5

A fraction ζ of the active surface of some porous slab-shaped catalyst pellets becomes poisoned. The pellets are used to catalyze a first-order isothermal chemical reaction. Find an expression for the ratio of the activity of the poisoned catalyst to the original activity of the unpoisoned catalyst when (1) homogeneous poisoning occurs and (2) selective poisoning occurs.

Solution

(1) If homogeneous poisoning occurs, the activity decreases in proportion to the fraction $(1 - \zeta)$ of surface remaining unpoisoned. In the steady state the rate of reaction is equal to the flux of reactant to the surface. The ratio of activity F of the poisoned slab to the unpoisoned slab will be equal to the ratio of the reactant fluxes under the respective conditions. Hence

$$F = \left(\frac{dC_A}{dx}\right)'_{x=L} \bigg/ \left(\frac{dC_A}{dx}\right)_{x=L}$$

where the prime denotes conditions in the poisoned slab.

Now from Eq. (3.13) the concentration of reactant is a function of the distance the reactant has penetrated the slab. Thus

$$C_{AL} = C_{A\infty}\frac{\cosh(\lambda x)}{\cosh(\lambda L)} \quad \text{where } \lambda = \sqrt{\frac{k}{D_e}}.$$

If the slab were poisoned, the activity would be $k(1 - \zeta)$ rather than k and then

$$C_{AL} = C_{A\infty}\frac{\cosh(\lambda' x)}{\cosh(\lambda' L)} \quad \text{where } \lambda' = \sqrt{\frac{k(1-\zeta)}{D_e}}$$

Evaluating the respective fluxes at $x = L$,

$$F = \frac{\sqrt{1 - \zeta}\tanh\{\phi\sqrt{(1 - \zeta)}\,\}}{\tanh \varphi} \quad \text{where } \phi = L\sqrt{\frac{k}{D_e}}$$

(2) When selective poisoning occurs, the exterior surface of the porous pellet becomes poisoned initially and the reactants must then be transported to the unaffected interior of the catalyst before reaction may ensue. When the reaction rate in the unpoisoned portion is chemically controlled, the activity merely falls off in proportion to the fraction of surface poisoned. However, if the reaction is diffusion limited, in the steady state the flux of reactant past the boundary between poisoned and unpoisoned surfaces is equal to the chemical reaction rate (see Fig. 3.11). Thus

Flux of reactant at the boundary between poisoned and unpoisoned portion of slab $= D_e\dfrac{C_{A\infty} - C_{AL}}{\zeta L}$

Reaction rate in unpoisoned length $(1-\zeta)L = D_e\left(\dfrac{dC_A}{dx}\right)_{x=(1-\zeta)L}$

The concentration profile in the unpoisoned length is, by analogy with Eq. (3.13),

$$C_A = C_{AL}\frac{\cosh(\lambda x)}{\cosh\{\lambda(1 - \zeta)L\}}$$

Therefore

$$D_e\left(\frac{dC_A}{dx}\right)_{x=(1-\zeta)L} = \frac{D_e}{L}C_{AL}\phi \tanh\{\phi(1 - \zeta)\}$$

where $\phi = \lambda L = L\sqrt{\frac{k}{D_e}}.$
In the steady state then

$$\frac{D_e(C_{A\infty} - C_{AL})}{\zeta L} = \frac{D_e}{L}C_{AL}\phi \tanh\{\phi(1 - \zeta)\}$$

Solving the above equation explicitly for C_{AL},

$$C_{AL} = \frac{C_{A\infty}}{1 + \phi\zeta \tanh\{\varphi(1 - \zeta)\}}$$

Hence the rate of reaction in the partially poisoned slab is

$$\frac{D_e(C_{A\infty} - C_{AL})}{L} = \frac{C_{A\infty}D_e}{L}\frac{\phi\tanh\{\phi(1-\zeta)\}}{1+\phi\zeta\tanh\{\phi(1-\zeta)\}}$$

In an unpoisoned slab, the reaction rate is $\left(\frac{C_{A\infty}D_e}{L}\right)\phi\tanh\phi$ and so

$$F = \frac{\tanh\{\phi(1-\zeta)\}}{1+\phi\zeta\tanh\{\phi(1-\zeta)\}}\frac{1}{\tanh\phi}$$

For an active catalyst, $\tanh\{\phi(1-\zeta)\} \to 1$ and then $F \to (1+\phi\zeta)^{-1}$.
Selective poisoning occurs with very active catalysts. Initially, the exterior surface is poisoned and then, as more poison is added, an increasing depth of the interior surface becomes poisoned and inaccessible to reactant. If the reaction rate in the unpoisoned portion of catalyst happens to be chemically controlled, the reaction rate will fall off directly in proportion to the fraction of surface poisoned and the activity decreases linearly with the amount of poison added (curve 1, Fig. 3.12). When the reaction is influenced by diffusion, in the steady state the flux of reactant past the boundary between the poisoned and unpoisoned parts of the surface will equal the reaction rate in the unpoisoned portion. For the slab model, as the foregoing example shows, the ratio of activity in a poisoned catalyst to that in an unpoisoned catalyst is

$$F = \frac{\tanh\{\varphi(1-\zeta)\}}{1+\varphi\zeta\tanh\{\varphi(1-\zeta)\}}\frac{1}{\tanh\varphi} \tag{3.59}$$

For large values of the Thiele modulus, the fraction $\phi(1-\zeta)$ will usually be sufficiently large that $F = (1+\phi\zeta)^{-1}$. Curve 3 in Fig. 3.12 depicts selective poisoning of active catalysts near the particle exterior and is the function represented by Eq. (3.59). Curve 4 describes the effect of selective poisoning for large values of the Thiele modulus. For the latter case the activity decreases drastically, after only a small amount of poison has been added.
It is apparent from the above discussion that ζ is time dependent, for an increasing amount of poison is being added quite involuntarily as the impure feed continually flows to the catalyst contained in the reactor. The general problem of obtaining $\zeta(t)$ is complex but, in principle, can be treated by solving the unsteady-state conservation equation for the poison. When the poison is strongly and rapidly adsorbed, the fraction of surface poisoned is dependent on the square root of the time for which the feed has been flowing. In general, if $\zeta(t)$ is known, a judgment can be made concerning the optimum time for which the catalyst may be used before its activity falls to a value, which produces an uneconomical throughput or yield of product.[34-37]

3.4 Mass Transfer From a Fluid Stream to a Solid Surface

Under some circumstances there will be a resistance to the transport of material from the bulk fluid stream to the exterior surface of a catalyst particle. When such a resistance to mass transfer exists, the concentration C_A of a reactant in the bulk fluid will differ from its concentration C_{Ai} at the solid–gas interface. Because C_{Ai} is usually unknown, it is necessary to eliminate it from the rate equation describing the external mass transfer process. Because, in the steady state, the rates of all of the steps in the process are equal, it is possible to obtain an overall rate expression in which C_{Ai} does not appear explicitly.

Based on such analyses, which of course do imply a film model in which the resistance to mass transfer is supposed to be confined to a film of finite thickness (see Volume 1, Chapter 10), it is possible to estimate the effect that mass transport external to the solid surface has on the overall reaction rate. For equimolar counterdiffusion of a component **A** in the gas phase, the rate of transfer of **A** from the bulk gas to the interface can be expressed as

$$N_A = k_G(C_{AG} - C_{Ai}) \tag{3.60}$$

where k_G is the gas film mass transfer coefficient per unit external surface area, and C_{AG} and C_{Ai} are the molar concentrations of the component **A** in the bulk gas and at the interface, respectively. The right-hand side of Eq. (3.60) contains the mass transfer coefficient k_G, which is used if the driving force is expressed in terms of gas concentrations. Because of the stoichiometric demands imposed by chemical reaction, equimolar counterdiffusion of components may not necessarily occur and the effects of bulk flow must be taken into account.

The rate r_v' of chemical reaction per unit volume of particle for a first-order reaction is given by

$$r_v' = kC_{Ai}\eta \tag{3.61}$$

where η is the efficiency factor, defined by Eq. (3.8), which allows for mass transfer effects in the pores of the catalyst articles. The rate is proportional to the concentration C_{Ai} of reactant at the interface. In the steady state, the flux of reactant will balance the rate of chemical reaction. From Eqs. (3.60) and (3.61) we find the unknown concentration, because

$$N_A a = r_v' \tag{3.62}$$

where a is the interfacial area per unit volume. Hence

$$C_{Ai} = \frac{k_G a C_{AG}}{k_G a + k\eta} \tag{3.63}$$

The rate at which the overall process of mass transfer and chemical reaction occurs may be found by substituting for C_{Ai} in Eq. (3.61) to give

$$r_v' = \left\{ \frac{1}{(k\eta)^{-1} + (k_G a)^{-1}} \right\} C_{AG} \tag{3.64}$$

If, under the operating conditions, $k_Ga \gg k\eta$, the overall rate approaches $k\eta C_{AG}$. In this case the chemical reaction is said to be rate determining. If, on the other hand, $k_Ga \ll k$, the rate approaches $k_Ga C_{AG}$ and the transport process is rate determining.

The transfer coefficient can be correlated in the form of a dimensionless Sherwood number $Sh(=k_G d_p/D)$. The particle diameter d_p is often taken to be the diameter of the sphere having the same area as the (irregular shaped) pellet. Thaller and Thodos[38] correlated the mass transfer coefficient in terms of the gas velocity u and the Schmidt number $Sc(=\mu/\rho D)$:

$$j_d = \frac{Sh}{ReSc^{1/3}} = \frac{k_G}{u} Sc^{2/3} \tag{3.65}$$

where j_d is the mass transfer factor (see Volume 1, Chapter 12, and Volume 2, Chapter 4) and Re is the Reynolds number ($u\rho d_p/\mu$) based on the particle diameter.

The mass transfer factor has also been correlated as a function of the Reynolds number only and thus taking into account only of hydrodynamic conditions. If e is the voidage of the packed bed and the total volume occupied by all of the catalyst pellets is V_p, then the total reactor volume is $V_p/(1-e)$. Hence the rate of mass transfer of component **A** per unit volume of reactor is $N_A S_x(1-e)/V_p$. If we now consider a case in which only external mass transfer controls the overall reaction rate, we have

$$r_v = \frac{N_A S_x(1-e)}{V_p} = (1-e)\frac{S_x}{V_p}k_G(C_A - C_{Ai}) \tag{3.66}$$

Alternatively, Eq. (3.66) may be written in terms of the j-factor. The unknown interface concentration C_{Ai} can now be eliminated in the usual way, by equating the rate of mass transfer to the rate of chemical reaction.

We see that, in principle, the overall reaction rate can be expressed in terms of coefficients such as the reaction rate constant and the mass transfer coefficient. To be of any use for design purposes, however, we must have knowledge of these parameters. By measuring the kinetic constant in the absence of mass transfer effects and using correlations to estimate the mass transfer coefficient, we are really implying that these estimated parameters are independent of one another. This would only be true if each element of external surface behaved kinetically as all other surface elements. Such conditions are only fulfilled if the surface is uniformly accessible. It is fortuitous, however, that predictions of overall rates based on such assumptions are often within the accuracy of the kinetic information, and for this reason, values of k and k_G obtained independently are frequently employed for substitution into overall rate expressions.

3.5 Chemical Kinetics of Heterogeneous Catalytic Reactions

To complete the discussion of factors involved in the design of gas—solid heterogeneous catalytic reactors, we will examine several aspects of the kinetics of chemical reactions occurring in the presence of a catalyst surface. We consider the following equilibrium reaction for heuristic purposes:

$$A + B \rightleftharpoons P$$

The elementary steps involved in this reaction occurring at a vacant active site **V** of the catalyst can be written as follows:

$$\text{Adsorption of } \mathbf{A}: \quad \mathbf{A} + \mathbf{V} \underset{\mathbf{k}_{Ad}}{\overset{\mathbf{k}_{Aa}}{\rightleftharpoons}} \mathbf{A} - \mathbf{V}$$

$$\text{Adsorption of } \mathbf{B}: \quad \mathbf{B} + \mathbf{V} \underset{\mathbf{k}_{Bd}}{\overset{\mathbf{k}_{Ba}}{\rightleftharpoons}} \mathbf{B} - \mathbf{V}$$

$$\text{Surface reaction}: \quad \mathbf{A} - \mathbf{V} + \mathbf{B} - \mathbf{V} \underset{\mathbf{k}'_S}{\overset{\mathbf{k}_S}{\rightleftharpoons}} \mathbf{P} - \mathbf{V} + \mathbf{V}$$

$$\text{Desorption of } \mathbf{P}: \quad \mathbf{P} - \mathbf{V} \underset{\mathbf{k}_{Pa}}{\overset{\mathbf{k}_{Pd}}{\rightleftharpoons}} \mathbf{P} + \mathbf{V}$$

We will now develop rate expressions for each of the above elementary steps in the surface catalytic reaction. The rate of adsorption of species **A** can be expressed in terms of the product of the prevailing partial pressure of adsorbate and the extent of free surface, whereas the rate of desorption is proportional to the fraction of surface covered. The net rate of adsorption of **A** in molecules per unit area will be

$$r''_{Aa} = \mathbf{k}_{Aa} P_A \left(1 - \sum_j \theta_j \right) - \mathbf{k}_{Ad} \theta_A \tag{3.67}$$

where θ_A represents the fraction of active sites covered by **A**, $\sum_j \theta_j$ is the total fraction of sites occupied by all species (**A**, **B**, **P**), and $\left(1 - \sum_j \theta_j \right)$ is the fraction of active sites left vacant. The surface concentration of **A**, C_A, can be defined as the number of molecules of A (n_A) adsorbed per surface area of the catalyst (S'). Similarly, the concentration of vacant (C_V) and total active sites (C_S) can be defined as n_V/S' and n_S/S', respectively, where n_V and n_S are the number of vacant sites and the total available active sites. The fraction θ_A and $\left(1 - \sum_j \theta_j \right)$ can be written in terms of concentrations as C_A/C_S and

C_V/C_S, respectively. C_V is $\left(C_S - \sum_j C_j\right)$. In terms of surface concentrations, Eq. (3.67)

can be written as[1]

$$r''_{Aa} = k_{Aa}P_A\frac{C_V}{C_S} - k_{Ad}\frac{C_A}{C_S} \tag{3.68}$$

Similar expressions for adsorption of the other reactant B and desorption of the product P can be written as follows:

$$r''_{Ba} = k_{Ba}P_B\frac{C_V}{C_S} - k_{Bd}\frac{C_B}{C_S} \tag{3.69}$$

$$r''_{Pd} = k_{Pd}\frac{C_P}{C_S} - k_{Pa}P_P\frac{C_V}{C_S} \tag{3.70}$$

The net rate of the bimolecular surface chemical reaction will be the difference between the rates of the forward and reverse reactions. The forward rate is determined by the simultaneous presence of chemisorbed **A** and **B**. Therefore, it will be proportional to the number of pairs of adjacent catalyst sites occupied by **A** and **B**, which is given by the product of the number n_A of adsorbed reactant molecules per surface area of the catalyst (S') and the ratio of occupied to unoccupied sites. This latter quantity is expressed as $\theta_B/(1 - \theta_A - \theta_B)$. The reverse rate, on the other hand, is proportional to the product of the number n_P of adsorbed product molecules per surface area of the catalyst (S') and the

fraction of active sites left vacant $\left(1 - \sum_j \theta_j\right)$. The net rate of the surface reaction per

unit surface is thus

$$r''_S = k_S\frac{n_A}{S'}\left(\frac{\theta_B}{1 - \theta_A - \theta_B}\right) - k'_S\frac{n_P}{S'}\left(1 - \sum_j \theta_j\right) \tag{3.71}$$

Based on the preceding discussion, the surface reaction rate can be written in terms of concentrations as

$$r''_S = k_S\frac{C_AC_B}{C_S} - k'_S\frac{C_PC_V}{C_S} \tag{3.72}$$

provided the surface coverage is sufficiently small that $1 >> (\theta_A + \theta_B)$, which condition is usually fulfilled during catalysis.It should be noted that the partial pressures in Eqs. (3.68)–(3.70) are those corresponding to the values at the interface between solid and gas. If

---

[1]   The net rates of adsorption, desorption, and reaction of species are expressed in terms of number of molecules per unit area per unit time, $L^{-2}T^{-1}$. It must be transposed to units of moles per unit time per unit catalyst mass for reactor design purposes. This is accomplished by multiplying r'' by S_g/N, where **N** is Avogadro's number.

there were no resistance to transport through the gas phase, then, as discussed in an earlier section, the partial pressures will correspond to those in the bulk gas.

The overall chemical rate at steady state may be written in terms of the partial pressures of **A**, **B**, and **P** by equating the rates r''_{Aa}, r''_{Pd}, and r''_S and eliminating the surface concentrations C_A, C_B and C_P and C_V from Eqs. (3.68), (3.70), and (3.72). The final equation so obtained is cumbersome and unwieldy and contains several constants which cannot be determined independently for practical reasons. This approach to obtain the overall rate of the surface-catalyzed reaction is known as the pseudo–steady-state approach. For relatively simpler reaction mechanisms involving isomerization of alkanes such as *n*-pentane to iso-pentane over Pt on alumina catalyst, the pseudo–steady-state approach yields manageable rate expression.[39,40] However, for most of the other reaction mechanisms, it is convenient to adopt the pseudoequilibrium approach in which one parameter, adsorption, surface reaction, or desorption, is rate determining.

3.5.1 Adsorption of a Reactant as the Rate-Determining Step

If the adsorption of **A** is the rate-determining step in the sequence of adsorption, surface reaction, and desorption processes, then Eq. (3.68) will be the appropriate equation to use for expressing the overall chemical rate. To be of use, however, it is first necessary to express C_A, C_V, and C_S in terms of the partial pressures of reactants and products. To do this an approximation is made; it is assumed that all processes except the adsorption of **A** are at equilibrium. Thus the processes involving **B** and **P** are in a state of pseudoequilibrium, i.e., $r''_{Ba} = r''_{Pd} = r''_S = 0$. The surface concentration of **B** and **P** can therefore be expressed in terms of equilibrium constants, K_B and K_P from Eqs. (3.69) and (3.70).

$$K_B = \frac{k_{Ba}}{k_{Bd}} = \frac{C_B}{P_B C_V} \tag{3.73}$$

$$K_P = \frac{k_{Pa}}{k_{Pd}} = \frac{C_P}{P_S C_V} \tag{3.74}$$

Eq. (3.72), the equation for the surface reaction between **A** and **B**, can be written in terms of the equilibrium constant for the surface reaction:

$$K_S = \frac{k_S}{k'_S} = \frac{C_P C_V}{C_A C_B} \tag{3.75}$$

Substituting Eqs. (3.73)–(3.75) into Eq. (3.68),

$$r''_{Aa} = \frac{k_{Aa} C_V}{C_S} \left\{ P_A - \frac{K_P}{K_A K_B K_C} \frac{P_P}{P_B} \right\} \tag{3.76}$$

where K_A is the ratio k_{Aa}/k_{Pd}. The ratio C_V/C_S can be written in terms of partial pressures and equilibrium constants because the concentration of vacant sites is the difference between the total concentration of sites and the sum of the surface concentrations of **A, B,** and **P**:

$$C_V = C_S - (C_A + C_B + C_P) \tag{3.77}$$

and so, from Eqs. (3.73)−(3.75),

$$\frac{C_V}{C_S} = \frac{1}{1 + K_B P_B + K_P P_P + \dfrac{K_P}{K_B K_S} \dfrac{P_P}{P_B}} \tag{3.78}$$

If we define the equilibrium constant for the adsorption of **A** as

$$K_A = \frac{C_A}{P_A C_V} \tag{3.79}$$

and recognize that the thermodynamic equilibrium constant for the overall equilibrium is

$$K = \frac{P_P}{P_A P_B} \tag{3.80}$$

then Eq. (3.76) gives, for the rate of chemical reaction,

$$r''\big|_{\text{Adsorption of A rate determining}} = \frac{k_{Aa}\left\{P_A - \dfrac{1}{K}\dfrac{P_P}{P_B}\right\}}{1 + K_B P_B + K_P P_P + \dfrac{K_A}{K}\dfrac{P_P}{P_B}} \tag{3.81}$$

It should be noted that Eq. (3.81) contains a driving force term in the numerator. This is the driving force tending to drive the chemical reaction toward the equilibrium state. The collection of terms in the denominator is usually referred to as the adsorption term, because terms such as $K_B P_B$ represent the retarding effect of the adsorption of species **B** on the rate of disappearance of **A**. New experimental techniques enable the constants K_B, etc. to be determined separately during the course of a chemical reaction,[41] and hence, if it were found that the adsorption of **A** controls the overall rate of conversion, Eq. (3.81) could be used directly as the rate equation for design purposes. If, however, external mass transfer were important, the partial pressures in Eq. (3.81) would be values at the interface and an equation (such as Eq. 3.66) for each component would be required to express interfacial partial pressures in terms of bulk partial pressures. If internal diffusion were also important, the overall rate equation would be multiplied by an effectiveness factor either estimated experimentally or alternatively obtained by theoretical considerations similar to those discussed earlier.

3.5.2 Surface Reaction as the Rate-Determining Step

If, on the other hand, surface reaction determined the overall chemical rate, Eq. (3.72) would represent the rate. If it is assumed that a pseudoequilibrium state is reached for each of the adsorption–desorption processes, then, by a similar method to that already discussed for reactions where adsorption is rate determining, it can be shown that the rate of chemical reaction is (for a Langmuir–Hinshelwood (LH) mechanism)

$$r''|_{\text{Surface reaction rate determining}} = \frac{C_S k_S K_A K_B \left\{ P_A P_B - \frac{1}{K} P_P \right\}}{(1 + K_A P_A + K_B P_B + K_P P_P)^2} \tag{3.82}$$

This equation also contains a driving force term and an adsorption term. A similar equation may be derived for the case of an Eley–Rideal mechanism, in which the gaseous or physically adsorbed **B** interacts with chemisorbed **A**. In this case, there is no elementary step for the adsorption of **B**, and the rate of surface reaction can be written as

$$r''_S = k_S C_A P_B - k'_S \frac{C_P C_V}{C_S} \tag{3.83}$$

The final rate expression can be derived as

$$r''|_{\text{Surface reaction rate determining}} = \frac{k_S K_A \left\{ P_A P_B - \frac{K_P}{K} P_P \right\}}{(1 + K_A P_A + K_P P_P)} \tag{3.84}$$

Comparing Eqs. (3.82) and (3.84), it is evident that the adsorption group in the denominator in the latter does not contain the term $K_B P_B$ signifying that the species **B** is not chemisorbed on the catalyst surface. Moreover, the exponent of the adsorption group is 2 (dual sites) when both species **A** and **B** are adsorbed on the active sites, whereas it is unity when only **A** is adsorbed. The exponent of the adsorption group is generally related to the number of catalyst active sites involved in the elementary reaction step.

3.5.3 Desorption of a Product as the Rate-Determining Step

In this case pseudoequilibrium is assumed for the surface reaction and for the adsorption–desorption processes involving **A** and **B**. By similar methods to those employed in deriving Eq. (3.81), it can be shown that the rate of chemical reaction is

$$r''|_{\text{Desorption of P rate determining}} = \frac{k_{Pa} K \left\{ P_A P_B - \frac{1}{K} P_P \right\}}{1 + K_A P_A + K_B P_B + K K_P P_A P_B} \tag{3.85}$$

when the desorption of **P** controls the rate. It should be noted that this equation also contains a driving force term and an adsorption term.

The approach just described to determine the kinetics of heterogeneous catalytic reactions in presence of different rate-limiting steps is known as Langmuir–Hinshelwood (LH) or Langmuir–Hinshelwood–Hougen–Watson (LHHW) approach. It is based on the adsorption principles proposed by Langmuir, which was further developed by the other scientists. The inherent assumptions of this approach are (1) all catalytic sites are alike, (2) the adsorbed molecule occupies only one active site, (3) the surface reaction occurs in a single step, (4) adsorption kinetics and equilibria obey Langmuir isotherm, and (5) single adsorption, reaction, or desorption step is rate limiting.[39,42]

The mathematical complexity of the rate expressions derived under different limiting steps can be reduced to match the experimentally observed rate expression if some information is available on the nature of the reaction and that of the adsorbed species. If the elementary reaction step is irreversible, then the driving force group in Eqs. (3.82) and (3.84) contains the terms corresponding to only the forward reaction. If one or more species is known to be insignificantly adsorbed on the catalyst surface, then the terms $K_i P_i$ in the denominator corresponding to these species can be ignored relative to those species that are mostly adsorbed. The species with low coverage on the catalyst active sites are termed "low-abundance catalyst-surface species", whereas those that entirely cover the catalyst active sites are termed "most abundant catalyst-surface species".[42]

3.5.4 Rate-Determining Steps for Other Mechanisms

In principle it is possible to write down the rate equation for any rate-determining chemical step assuming any particular mechanism. To take a specific example, the overall rate may be controlled by the adsorption of **A** and the reaction may involve the dissociative adsorption of **A**, only half of which then reacts with adsorbed **B** by a LH mechanism. Generally, species such as O_2 and H_2 adsorb dissociatively on certain catalyst metal sites. It can be shown that the term corresponding to dissociative adsorption of species in the adsorption group is $\sqrt{K_{O_2} P_{O_2}}$. The basic rate equation that represents such a process can be transposed into an equivalent expression in terms of partial pressures and equilibrium constants by methods similar to those employed to obtain the rate Eqs. (3.81), (3.82), and (3.85). Table 3.3 contains a number of selected mechanisms for each of which the basic rate equation, the driving force term, and the adsorption term are given.

For reactions occurring in presence of bimetallic catalysts, it is possible that some species adsorb preferentially over one type of metal active site, whereas some others prefer to

Table 3.3: Structure of Reactor Design Equations.

Reaction		Mechanism	Driving Force	Adsorption Term
1. $A \rightleftharpoons P$ (Equilibrium constant K is dimensionless)	(i)	Adsorption of **A** controls rate	$P_A - \dfrac{P_P}{K}$	$1 + -\frac{K_A}{K}P_P + K_P P_P$
	(ii)	Surface reaction controls rate, single site mechanism	$P_A - \dfrac{P_P}{K}$	$1 + -K_A P_P + K_P P_P$
	(iii)	Surface reaction controls rate, adsorbed **A** reacts with adjacent vacant site	$P_A - \dfrac{P_P}{K}$	$(1 + -K_A P_P + K_P P_P)^2$
	(iv)	Desorption of **P** controls rate	$P_A - \dfrac{P_P}{K}$	$1 + K K_P P_P + K_P P_P$
	(v)	**A** dissociates when adsorbed and adsorption controls rate	$P_A - \dfrac{P_P}{K}$	$\left\{ 1 + \left(\frac{K_A}{K} - P_P\right)^{1/2} + K_P P_P \right\}^2$
	(vi)	**A** dissociates when adsorbed and surface reaction controls rate	$P_A - \dfrac{P_P}{K}$	$\left\{ 1 + (K_A P_A)^{1/2} + K_P P_P \right\}^2$
2. $A + B \rightleftharpoons P$ (Equilibrium constant K has dimensions $\mathbf{M^{-1}LT^2}$)		Langmuir— Hinshelwood mechanism (adsorbed **A** reacts with adsorbed **B**)		
	(i)	Adsorption of **A** controls rate	$P_A - \dfrac{P_P}{K P_B}$	$1 + \frac{K_A P_P}{K P_B} + K_B P_B + K_P P_P$
	(ii)	Surface reaction controls rate	$P_A P_B - \dfrac{P_P}{K}$	$\left\{ 1 + K_A P_A + K_B P_B + K_p P_P \right\}^2$
	(iii)	Desorption of **P** controls rate	$P_A P_B - \dfrac{P_P}{K}$	$1 + K_A P_A + K_B P_B + K K_p P_P P_B$

1. The expression for the rate of reaction in terms of partial pressures is proportional to the driving force divided by the adsorption term.

2. To derive the kinetic expression for irreversible reactions, simply omit the second term in the driving force.

3. If two products are formed (**P** and **Q**), the equations are modified by (1) multiplying the second term of the driving force by P_Q and (2) adding $K_Q P_Q$ within the bracket of the adsorption term.

4. If **A** or **B** dissociates during the process of chemisorption, both the driving force and the adsorption term should be modified (see for example cases 1(v) and 1(vi) above). For a full discussion of such situations see Hougen and Watson.[43]

adsorb on the other metal active site. Under such circumstances, the derivation of rate expression involves two separate catalyst site balance equations.

$$C_{S1} = C_{V1} + \sum_{i, i \neq j} C_i$$

$$C_{S2} = C_{V2} + \sum_{j, j \neq i} C_j$$

C_{S1} and C_{S2} denote the total concentration of active sites of type 1 and type 2, respectively, whereas C_{V1} and C_{V2} denote the concentration of vacant active sites of type 1 and type 2, respectively. The final rate expression thus derived for the case of molecular adsorption of species will contain the adsorption group in the denominator,

which is $\left(1 + \sum_{i,i\neq j} K_i C_i\right)\left(1 + \sum_{j,j\neq i} K_j C_j\right)$.

A useful empirical approach to the design of heterogeneous chemical reactors often consists of selecting a suitable equation, such as one in Table 3.3, which, with numerical values substituted for the kinetic and equilibrium constants, represents the chemical reaction in the absence of mass transfer effects. Graphical methods are often employed to aid the selection of an appropriate equation,[44] and the constants are determined by a least squares approach.[43] It is important to stress, however, that while the equation selected may well represent the experimental data, it does not necessarily represent the true mechanism of reaction; as is evident from Table 3.3, many of the equations are similar in form. Nevertheless the equation will adequately represent the behavior of the reaction for the conditions investigated and can be used for design purposes. Initial rate method is a useful technique to assess the rate-limiting step in a catalytic reaction. It involves plotting the initial rate of the reaction calculated experimentally using the reactant or product concentration variation with time in the initial time period with initial total pressure or concentration of the reactant. A fuller and more detailed account of a wide selection of mechanisms and the dependence of initial rates on initial total pressure for different reaction stoichiometries has been given by Hougen and Watson[43] and Helfferich.[42]

3.5.5 Examples of Rate Equations for Industrially Important Reactions

For most gas—solid catalytic reactions, usually a rate equation corresponding to one form or another of the Hougen and Watson type described above can be found to fit the experimental data by a suitable choice of the constants that appear in the adsorption and driving force terms. The following examples have been chosen to illustrate this type of rate equation. However, there are some industrially important reactions for which rate equations of other forms have been found to be more appropriate, of particular importance being ammonia synthesis and sulfur dioxide oxidation.[40]

(1) The hydration of ethylene to give ethanol was studied by Mace and Bonilla[45] using a catalyst of tungsten trioxide WO_3 on silica gel at a total pressure of 136 bar:

$$C_2H_4 + H_2O \rightleftharpoons C_2H_5OH$$
$$\mathbf{A} \qquad \mathbf{B} \qquad \mathbf{M}$$

They concluded that the applicable rate equation was

$$r'' = \frac{kK_AK_B(P_AP_B - P_M/K_P)}{(1 + K_AP_A + K_BP_B)^2} \tag{3.86}$$

Eq. (3.86) was consistent with a mechanism whereby the reaction between ethylene and water adsorbed on the surface was rate controlling but without strong adsorption of the product ethanol. Commercially at the present time, phosphoric acid on kieselguhr is the preferred catalyst.

(2) In the process outlined in Chapter 1 (Example 1.1), for the dehydrogenation of ethylbenzene to produce styrene in the presence of steam, the reaction

$$C_6H_5 \cdot CH_2 \cdot CH_3 = C_6H_5 \cdot CH : CH_3 + H_2$$
$$\textbf{E} \qquad\qquad \textbf{S} \qquad\quad \textbf{H}$$

is catalyzed by iron oxide. Potassium carbonate also is present in the catalyst to promote the reaction of carbon with steam and so to keep the catalyst free from carbon deposits. Present-day catalysts contain in addition other promoters such as molybdenum and cerium oxides, Mo_2O_3 and Ce_2O_3.[46] Kinetic studies using an older type of catalyst resulted in the rate equation[47]:

$$r'' = k\left[P_E - \frac{P_SP_H}{K_P}\right] \tag{3.87}$$

This equation can be interpreted as indicating a surface reaction being rate controlling (although the controlling step could be adsorption), but all the species involved being only weakly adsorbed; i.e., the adsorption terms K_EP_E, K_SP_S, K_HP_H, which would otherwise appear in the denominator, are all much less than unity.

Rate equations for the promoted catalysts currently used have the following form[46]:

$$r'' = \frac{kK_E\left(P_E - \dfrac{P_HP_S}{K_P}\right)}{(1 + K_EP_E + K_SP_S)} \tag{3.88}$$

This indicates that, for these catalysts, the adsorption terms K_EP_E and K_SP_S are significant compared with unity.

3.5.6 Mechanism of Catalyst Poisoning

Catalyst deactivation by poisoning can occur when some reactants in the feed or by-products or poisoning molecules irreversibly get chemisorbed on the active sites, thereby reducing the availability of active sites for the reaction of interest. Let us consider the following heuristic reaction in which

$$\textbf{A} + \textbf{B} \rightarrow \textbf{P(poison)} + \textbf{Q(desired)}$$

The elementary steps involved in this reaction occurring at a vacant active site **V** of the catalyst can be written as follows:

$$\text{Adsorption of A}: \quad A + V \underset{k_{Ad}}{\overset{k_{Aa}}{\rightleftharpoons}} A - V$$

$$\text{Adsorption of B}: \quad B + V \underset{k_{Bd}}{\overset{k_{Ba}}{\rightleftharpoons}} B - V$$

$$\text{Surface reaction}: \quad A - V + B - V \overset{k_S}{\rightarrow} P - V + Q - V$$

$$\text{Desorption of Q}: \quad Q - V \underset{k_{Qa}}{\overset{k_{Qd}}{\rightleftharpoons}} Q + V$$

Considering the surface reaction to be rate limiting, the following expression can be written for the rate of the reaction:

$$r''_S = k_S \frac{C_A C_B}{C_S} \tag{3.89}$$

Following the steps described in the preceding sections, C_A and C_B can be written as $K_A P_A C_V$ and $K_B P_B C_V$, respectively. The concentration of available active sites on the catalyst can be written as follows:

$$C_S = C_V + C_A + C_B + C_P + C_Q \tag{3.90}$$

As the catalyst poison adsorbs irreversibly on the catalyst active site, its surface concentration cannot be described using an equilibrium relationship, while C_Q can be written as $K_Q P_Q C_V$. To express C_V in terms of the other measurable quantities, the following expression can be written:

$$C_V = \frac{C_S - C_P}{1 + K_A P_A + K_B P_B + K_Q P_Q} \tag{3.91}$$

The rate expression thus becomes

$$r''_S = \frac{C_A C_B}{C_S} C_V = \frac{k_S K_A K_B P_A P_B C_S}{(1 + K_A P_A + K_B P_B + K_Q P_Q)^2} \left(1 - \frac{C_P}{C_S}\right)^2 \tag{3.92}$$

The term C_P/C_S denotes the fraction of sites that are unavailable for adsorption of reactants and products or the fraction of sites poisoned by by-product **P**. The fraction of sites that are available for adsorption, which varies with time, can be defined as the activity function, $a_c(t)$, such that[39]

$$r''_S\Big]_{\text{with catalyst deactivation}} = r''_S\Big]_{\text{fresh catalyst}} \times a_c(t) \tag{3.93}$$

By this way, the rate of reaction with the fresh catalyst is separated from the nondimensional catalytic activity, $a_c(t)$. This allows the independent determination of the rate parameters corresponding to the fresh catalyst via bench-scale experiments. The function $a_c(t)$ is usually determined by empirical fitting. Many functional forms of $a_c(t)$, including zeroth-order decay, first-order decay, second-order decay, and *n*th-order decay laws, are reported for a variety of catalytic reactions.[39] For example, the first-order catalyst decay can be expressed in the form of a rate equation:

$$r_d = -\frac{da_c}{dt} = k_d a_c \tag{3.94}$$

Integrating the above equation in the limits of $a_c = 1$ to a_c in the time range of $t = 0$ to t yields $a_c = \exp(-k_d t)$. The deactivation rate constant, $k_d(T)$, is dependent on temperature, and can be expressed in the form of Arrhenius equation with preexponential factor and activation energy of deactivation. Several other mechanisms of catalyst deactivation are described in the review by Bartholomew.[48]

3.6 Design Calculations

The problems that a chemical engineer has to solve when contemplating the design of a chemical reactor packed with a catalyst or reacting solid are, in principle, similar to those encountered during the design of an empty reactor, except that the presence of the solid somewhat complicates the material and heat balance equations. The situation is further exacerbated by the designer having to predict and avoid those conditions, which might lead to instability within the reactor.

We shall consider, in turn, the various problems, which have to be faced when designing isothermal, adiabatic, and other nonisothermal tubular reactors, and we shall also briefly discuss fluidized bed reactors. Problems of instability arise when inappropriate operating conditions are chosen and when reactors are started up. A detailed discussion of this latter topic is outside the scope of this chapter, but, because reactor instability is undesirable, we shall briefly inspect the problems involved.

3.6.1 Packed Tubular Reactors

In a substantial majority of the cases where packed tubular reactors are employed, the flow conditions can be regarded as those of plug flow. However, dispersion (already discussed in Chapter 2 in relation to homogeneous reactors) may result in lower conversions than those obtained under truly plug flow conditions.

In the first instance, the effects of dispersion will be disregarded. Then the effects of dispersion in the axial (longitudinal) direction only will be taken into account, and this discussion will then be followed by considerations of the combined contributions of axial and radial (transverse) dispersion.

3.6.1.1 Behavior of Reactors in the Absence of Dispersion
3.6.1.1.1 Isothermal Conditions

The design equation for the isothermal fixed bed tubular reactor with no dispersion effects represents the simplest form of reactor to analyze. No net exchange of mass or energy

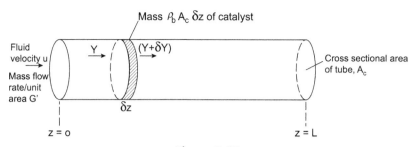

Figure 3.13
One-dimensional tubular reactor.

occurs in the radial direction, so transverse dispersion effects can also be neglected. If we also suppose that the ratio of the tube length to particle size is large (>ca.50), then we can safely ignore longitudinal dispersion effects compared with the effect of bulk flow. Hence, in writing the conservation equation over an element δz of the length of the reactor (Fig. 3.13), we may consider that the fluid velocity u is independent of radial position; this implies a flat velocity profile (plug flow conditions) and ignores dispersion effects in the direction of flow.

Suppose that a mass of catalyst, whose bulk density is ρ_b, is contained within an elementary length δz of a reactor of uniform cross section A_c. The mass of catalyst occupying the elementary volume will therefore be $\rho_b A_c \delta z$. Let G' be the steady-state mass flow rate per unit area to the reactor. If the number of moles of product per unit mass of fluid entering the elementary section is Y and the amount emerging is $(Y + \delta Y)$, the net difference in mass flow per unit area across the element will be $G' \delta U$. In the steady state this will be balanced by the amount of product formed by chemical reaction within the element so

$$G' \delta Y = r_Y^M \rho_b \delta z \tag{3.95}$$

where the isothermal reaction rate r_Y^M is expressed in units of moles per unit time per unit mass of catalyst.[2] To calculate the length of reactor required to achieve a conversion corresponding to an exit concentration Y_L, Eq. (3.87) is integrated from Y_0 (say) to Y_L, where the subscripts 0 and L refer to inlet and exit conditions, respectively. Thus

$$L = \frac{G'}{\rho_b} \int_{Y_0}^{Y_L} \frac{dY}{r_Y^M} \tag{3.96}$$

---

[2] If intraparticle diffusion effects are important and an effectiveness factor η is employed (as in Eq. 3.8) to correct the chemical kinetics observed in the absence of transport effects, then it is necessary to adopt a stepwise procedure for solution. First the pellet equations (such as 3.10) are solved to calculate η for the entrance to the reactor and then the reactor Eq. (3.95) may be solved in finite difference form, thus providing a new value of Y at the next increment along the reactor. The whole procedure may then be repeated at successive increments along the reactor.

If the reaction rate is a function of total pressure as well as concentration, and there is a pressure drop along the reactor due to the solid packing, the conversion within the reactor will be affected by the drop in total pressure along the tube. In most cases it is perfectly reasonable to neglect the change in kinetic energy due to the expansion of the fluid in comparison with the pressure force and the component of the drag force in the direction of flow created by the surface of the solid particles. The pressure drop dP along an elementary length dz of the packed bed may thus be written in terms of R_1, the component of the drag force per unit surface area of particles in the direction of flow:

$$-\frac{dP}{dz} = \frac{R_1}{d'_m} = \left(\frac{R_1}{\rho u_1^2}\right) \frac{\rho u_1^2}{d'_m} \tag{3.97}$$

in which u_1 is the mean velocity in the voids ($= u/e$ where u is the average velocity as measured over the whole cross-sectional area of the bed and e is the bed voidage) and d'_m ($=$ volume of voids/total surface area) is one quarter of the hydraulic mean diameter. Ergun[49] has correlated the friction factor $\left(R_1/\rho u_1^2\right)$ in terms of the modified Reynolds number. In principle, therefore, a relation between P and z may be obtained by integration of Eq. (3.97) and it is therefore possible to allow for the effect of total pressure on the reaction rate by substituting $P = \mathrm{f}(z)$ in the expression for the reaction rate. Eq. (3.88) may then be integrated directly.

When R_Y is not a known function of Y, an experimental program to determine the chemical reaction rate is necessary. If the experimental reactor is operated in such a way that the conversion is sufficiently small—small enough that the flow of product can be considered to be an insignificant fraction of the total mass flow—the reactor is said to be operating differentially. Provided the small change in composition can be detected quantitatively, the reaction rate may be directly determined as a function of the mole fraction of reactants. On this basis a differential reactor provides only initial rate data. It is therefore important to carry out experiments in which products are added at the inlet, thereby determining the effect of any retardation by products. An investigation of this kind over a sufficiently wide range of conditions will yield the functional form of r_Y and, by substitution in Eq. (3.96), the reactor size and catalyst mass can be estimated. If it is not possible to operate the reactor differentially, conditions are chosen so that relatively high conversions are obtained and the reactor is now said to be an integral reactor. By operating the reactor in this way, Y may be found as a function of W/G' and r_Y may be determined, for any given Y, by evaluating the slope of the curve at various points.

If axial dispersion were an important effect, the reactor performance would tend to fall below that of a plug flow reactor. In this event an extra term must be added to Eq. (3.95) to account for the net dispersion occurring within the element of bed. This subject is considered later.

Example 3.6

Carbon disulfide is produced in an isothermally operated packed tubular reactor at a pressure of 1 bar by the catalytic reaction between methane and sulfur.

$$CH_4 + 2S_2 = CS_2 + 2H_2S$$

At 600°C (=873 K) the reaction is chemically controlled, the rate being given by

$$r_Y^M = \frac{8 \times 10^{-2} \exp(-115,000/RT) P_{CH_4} P_{S_2}}{1 + 0.6 P_{S_2} + 2.0 P_{CS_2} + 1.7 P_{H_2S}} \quad \text{kmol/s(kg catalyst)}$$

where the partial pressures are expressed in bar, the temperature in K, and **R** as kJ/kmol K. Estimate the mass of catalyst required to produce 1 tonne per day of CS_2 at a conversion level of 90%.

Solution

Assuming plug flow conditions within the reactor, the following equation applies:

$$L = \frac{G'}{\rho_b} \int_0^{Y_L} \frac{dY}{r_Y^M}$$

where Y is the concentration of CS_2 in moles per unit mass, G' is the mass flow per unit area, and ρ_b is the catalyst bulk density. In terms of catalyst mass, this becomes

$$W = G \int_0^{Y_L} \frac{dY}{r_Y^M}$$

where G is now the molar flow rate through the tube.

The rate expression is given in terms of partial pressures, and this is now rewritten in terms of the number of moles χ of methane converted to CS_2. Consider 1 mol of gas entering the reactor, then at any cross section (distance z along the tube from the inlet) the number of moles of reactant and product may be written in terms of χ as follows:

	Moles at Inlet	Moles at Position z
CH_4	1/3	$1/3 - \chi$
S_2	2/3	$2/3 - 2\chi$
CS_2	—	χ
H_2S	—	2χ
Total	1.00	1.00

For a pressure of 1 bar, it follows that the partial pressures in the rate expression become

$$P_{CH_4} = (1/3 - \chi); \quad P_{S_2} = (2/3 - 2\chi); \quad P_{CS_2} = \chi \text{ and } P_{H_2S} = 2\chi$$

Now if M is the mean molecular weight of the gas at the reactor inlet, then χ kmol of CS_2 is produced for every M kg of gas entering the reactor. Hence

$$Y = \frac{\chi}{M}$$

and the required catalyst mass is now

$$W = \frac{G}{M} \int_0^{\chi_L} \frac{d\chi}{r_Y^M}$$

For an expected conversion level of 0.9, the upper limit of the integral will be given by $\frac{\chi_L}{\frac{1}{3.} - \chi_L}$

$= 0.9$, i.e., by $\chi_L = 0.158$. Thus 1 tonne per day ($= 1.48$ kmol/s CS_2) is produced by using an inlet flow of $G = \frac{1.48 \times 10^{-4} \times M}{0.158}$ kg/s.

Hence

$$W = \frac{533}{0.158} \int_0^{0.158} \frac{d\chi}{r_\chi^M}$$

Substituting the expressions for partial pressures into the rate equation, the mass of catalyst required may be determined by direct numerical (or graphical) integration. Thus

$$\underline{W = 23 \text{ kg}}$$

3.6.1.1.2 Adiabatic Conditions

Adiabatic reactors are more frequently encountered in practice than isothermal reactors. Because there is no exchange of heat with the surroundings, radial temperature gradients are absent. All of the heat generated or absorbed by the chemical reaction manifests itself by a change in enthalpy of the fluid stream. It is therefore necessary to write a heat balance equation for the reactor in addition to the material balance Eq. (3.95). Generally, heat transfer between solid and fluid is sufficiently rapid for it to be justifiable to assume that all the heat generated or absorbed at any point in the reactor is transmitted instantaneously to or from the solid. It is therefore only necessary to take a heat balance for the fluid entering and emerging from an elementary section δz. Referring to Fig. 3.13 and neglecting the effect of longitudinal heat conduction,

$$G' \bar{c}_p \delta T = \rho_b (-\Delta H) r_{YT}^M \delta z \tag{3.98}$$

where $\bar{c}_p$ is the mean heat capacity of the fluid, ΔH is the enthalpy change on reaction, and r_{YT}^M is the reaction rate (moles per unit time and unit mass of catalyst)—now a function of Y and T.

Simultaneous solution of the mass balance Eq. (3.95) and the heat balance Eq. (3.98) with the appropriate boundary conditions give z as a function of Y.

A simplified procedure is to assume $-\Delta H/\bar{c}_p$ as constant. If Eq. (3.98) (the heat balance equation) is divided by Eq. (3.95) (the mass balance equation) and integrated, we immediately obtain

$$T = T_0 + \frac{(-\Delta H)Y}{\bar{c}_p} \tag{3.99}$$

where T_0 is the inlet temperature. This relation implies that the adiabatic reaction path is linear. If Eq. (3.99) is substituted into the mass balance Eq. (3.95),

$$\frac{dL}{dz} = \frac{\rho_b}{G'} r^M_{YT_0} \tag{3.100}$$

where the reaction rate $r^M_{YT_0}$ along the adiabatic reaction path is now expressed as a function of Y only. Integration then gives z as a function of Y directly.

Because the adiabatic reaction path is linear, a graphical solution, also applicable to multibed reactors, is particularly apposite. (See Example 3.7 as an illustration.) If the design data are available in the form of rate data $r^M_{YT_0}$ for various temperatures and conversions, they may be displayed as contours of equal reaction rate in the (T, Y) plane. Fig. 3.14 shows such contours on which an adiabatic reaction path of slope $\bar{c}_p/-\Delta H$ and intercept T_0 on the abscissa is superimposed. The reactor size may be evaluated by computing

$$\frac{\rho_b L}{G'} = \int_{Y_0}^{Y_L} \frac{dY}{r^M_{YT_0}} \tag{3.101}$$

from a plot of $1/r^M_{YT_0}$ as a function of Y. The various values of $r^M_{YT_0}$ correspond to those points at which the adiabatic reaction path intersects the contours.

It is often necessary to employ more than one adiabatic reactor to achieve a desired conversion. In the first place, chemical equilibrium may have been established in the first reactor and it is then necessary to cool and/or remove the product before entering the second reactor. This, of course, is one good reason for choosing a catalyst that will function at the lowest possible temperature. Secondly, for an exothermic reaction, the

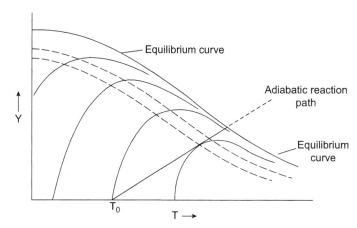

Figure 3.14
Graphical solution for an adiabatic reactor.

temperature may rise to a point at which it is deleterious to the catalyst activity. At this point the products from the first reactor are cooled before entering a second adiabatic reactor. To design such a system, it is only necessary to superimpose on the rate contours the adiabatic temperature paths for each of the reactors. The volume requirements for each reactor can then be computed from the rate contours in the same way as for a single reactor. It is necessary, however, to consider carefully how many reactors in series are required to operate them economically.

Should we wish to minimize the size of the system, it would be important to ensure that, for all conversions along the reactor length, the rate is at its maximum.[50,51] Because the rate is a function of conversion and temperature, setting the partial differential $\partial r_{YT}/\partial T$ equal to zero will yield, for an exothermic reaction, a relation $T_{MX}(Y_{MX})$, which is the locus of temperatures at which the reaction rate is a maximum for a given conversion. The locus T_{MX} of these points passes through the maxima of curves of Y as a function of T as shown in Fig. 3.15 as contours of constant rate. Thus, to operate a series of adiabatic reactors along an optimum temperature path, hence minimizing the reactor size, the feed is heated to some point A (Fig. 3.15) and the reaction is allowed to continue along an adiabatic reaction path until a point such as **B**, in the vicinity of the optimum temperature curve, is reached. The products are then cooled to C before entering a second adiabatic

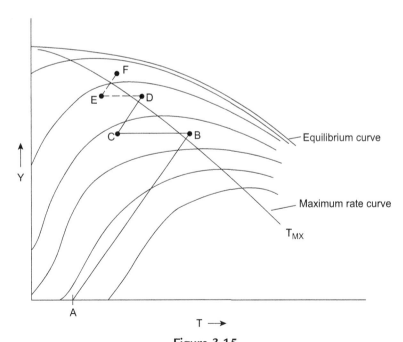

Figure 3.15
Optimum design of a two-stage and three-stage adiabatic reactor.

reactor in which reaction proceeds to an extent indicated by point D, again in the vicinity of the curve T_{MX}. The greater the number of adiabatic reactors in the series the closer the optimum path is followed. For a given number of reactors, say, three, there will be six design decisions to be made corresponding to the points A to F inclusive. These six decisions may be made in such a way as to minimize the capital and running costs of the system of reactors and heat exchangers. However, such an optimization problem is outside the scope of this chapter, and the interested reader is referred to Aris.[17] It should be pointed out, nevertheless, that the high cost of installing and operating heat transfer and control equipment so as to maintain the optimum temperature profile militates against its use. If the reaction is not highly exothermic, an optimal isothermal reactor system may be a more economic proposition and its size may not be much larger than the adiabatic system of reactors. Each case has to be examined on its own merits and compared with other alternatives.

A single-stage adiabatic reactor is illustrated in Fig 3.16, which shows the graded spherical packing used as a catalyst support at the bottom of the bed, and as a mechanical stabilizer for the bed at the top. For conditions where the gas flow rate is high and where the pressure drop allowable across the bed is low, much higher diameter:height ratios will be used.

Fig. 3.17 shows a radial-flow reactor with the inlet gas flowing downward at the walls of the vessel and then flowing radially inward through the bed before passing through a distributor into the central outlet. Uniform flow along the length of the reactor is ensured by the use of a distributor with appropriate resistance to the flow of gas.

Example 3.7

SO_3 is produced by the catalytic oxidation of SO_2 in two packed adiabatic tubular reactors arranged in series with intercooling between stages. The molar composition of the mixture entering the first reactor is 7% SO_2, 11% O_2, and 82% N_2. The inlet temperature of the reactor is controlled at 688 K and the inlet flow is 0.17 kmol/s. Calculate the mass of catalyst required for each stage so that 78% of the SO_2 is converted in the first stage and a further 20% in the second stage.

Thermodynamic data for the system are as follows:

Component	Mean specific heat in range 415–600°C (688–873K) (kJ/kmol K)
SO_2	51.0
SO_3	75.5
O_2	33.0
N_2	30.5

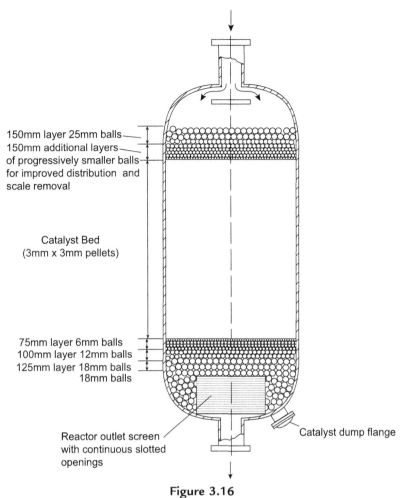

150mm layer 25mm balls
150mm additional layers
of progressively smaller balls
for improved distribution and
scale removal

Catalyst Bed
(3mm x 3mm pellets)

75mm layer 6mm balls
100mm layer 12mm balls
125mm layer 18mm balls
18mm balls

Reactor outlet screen
with continuous slotted
openings

Catalyst dump flange

Figure 3.16
Single-stage adiabatic reactor.

For the reaction

$$SO_2 + \tfrac{1}{2} O_2 \rightleftharpoons SO_3$$

the standard enthalpy change at 415°C (= 688 K) is $\Delta H^0_{600} = -97,500$ kJ/kmol.

Solution
The chemical reaction that ensues is

$$SO_2 + \tfrac{1}{2} O_2 \rightleftharpoons SO_3$$

and if 0.17×0.07 kmol of SO_2 is converted to χ kmol of SO_3, a material balance for the first reactor may be drawn up.

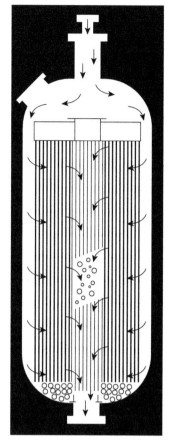

Figure 3.17
Radial-flow reactor.

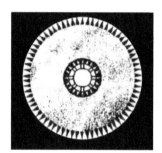

Component	Kilomoles at inlet	Kilomoles at outlet
SO_2	0.17×0.07	$0.17 \times 0.07K - \chi$
O_2	0.17×0.11	$0.17 \times 0.11 - \chi/2$
SO_3	—	χ
N_2	0.17×0.82	0.17×0.82

From the last column of the above table and the specific heat data, the difference in sensible heat between products and reactants is

$$H_p - H_R = (T - 688)\{[(0.17 \times 0.07) - \chi] \times 51.0 + [(0.17 \times 0.11) - \chi/2] \times 33.0 + [75.5\chi] + [0.17 \times 0.82 \times 30.5]\}$$ where T (**K**) is the temperature at the exit from the first reactor. The conversion x may be written as

$$x = \frac{\chi}{0.17 \times 0.07}$$

and on substitution and simplification,

$H_P - H_R = (T - 688) \, (5.476 + 0.095x)$A heat balance (in kJ) over the first reactor gives

$$H_P - H_R = \Delta H^0_{688} = 97,500\chi = 1160x$$

where $\Delta H^0_{688}(=-97,500 \text{ kJ/kmol})$ is the standard enthalpy of reaction at 688K.
From the two expressions obtained,

$$T = 688 + \frac{1160x}{5.476 + 0.095x} \approx 688 + 212x$$

This provides a linear relation between temperature and conversion in the first reactor.
Fig. 3.18 is a conversion chart for the reactant mixture and shows rate curves in the conversion—temperature plane. The line AA' is plotted on the chart from the linear T—x relation.
The rate of reaction at any conversion level within the reactor may therefore be obtained by

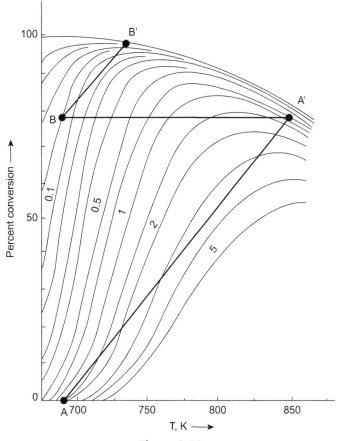

Figure 3.18

Graphical solution for design of SO_2 reactor. Figures on curves represent the reaction rate in units of (kmol s^{-1} kg^{-1} $\times$ 10^6).

reading the rate corresponding to the intersection of the line with various conversion ordinates. Now for the first reactor,

$$W_1 = 0.17 \times 0.07 \int_0^{0.78} \frac{dx}{r^M}$$

and this is easily solved by graphical integration as depicted in Fig. 3.19, by plotting $1/r^M$ versus conversion x and finding the area underneath the curve corresponding to the first reactor. The area corresponds to 0.336×10^6 kg s (kmol)$^{-1}$ and $W_1 = 4000$ kg = 4 tonnes. A similar calculation is now undertaken for the second reactor but, in this particular case, a useful approximation may be made. Because the reactant mixture is highly diluted with N_2, the same heat conservation equation may be used for the second reactor as for the first reactor. The line BB' starting at the point (688, 0.78) and parallel to the first line AA' may therefore be drawn. For the second reactor,

$$W_2 = 0.17 \times 0.07 \int_0^{0.2} \frac{dx}{r^M}$$

and this is also solved by the same procedure as before. $1/r^M$ is plotted as a function of x by noting the rate corresponding to the intersection of the line BB' with the selected conversion ordinate. The area under the curve $1/r^M$ versus x for the second reactor is 4.9×10^6 kg s/kmol. Hence $W_2 = 6 \times 10^4$ kg = 60 tonnes.

It will be noted that the catalyst requirement in the second reactor is very much greater because, being near to equilibrium, the reaction rate is between one and two orders of magnitude lower.

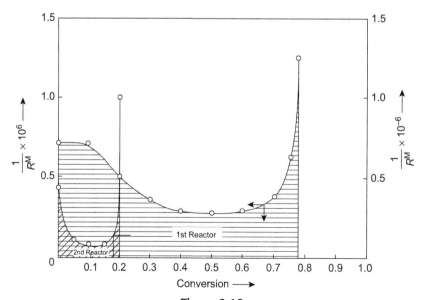

Figure 3.19
Graphical integration to determine catalyst masses.

3.6.1.1.3 Nonisothermal and Nonadiabatic Conditions

A useful approach to the preliminary design of a nonisothermal fixed bed reactor is to assume that all the resistance to heat transfer is in a thin layer near the tube wall. This is a fair approximation because radial temperature profiles in packed beds are parabolic with most of the resistance to heat transfer near the tube wall. With this assumption a one-dimensional model, which becomes quite accurate for small diameter tubes, is satisfactory for the approximate design of reactors. Neglecting diffusion and conduction in the direction of flow, the mass and energy balances for a single component of the reacting mixture are as follows:

$$G' \frac{dY}{dz} = \rho_b r_{YT}^M \quad \text{(equation 3.95)}$$

$$G' \bar{c}_p \frac{dT}{dz} = \rho_b (-\Delta H) r_{YT}^M - \frac{h}{a'} (T - T_W) \quad (3.102)$$

where h is the heat transfer coefficient (dimensions $\mathbf{MT^{-3}\theta^{-1}}$) expressing the resistance to the transfer of heat between the reactor wall and the reactor contents, a' is the surface area for heat transfer per unit hydraulic radius (equal to the area of cross section divided by the perimeter), and T_w is the wall temperature.

Inspection of the above equations shows that, if the wall temperature is constant, for a given inlet temperature, a maximum temperature is attained somewhere along the reactor length if the reaction is exothermic. It is desirable that this should not exceed the temperature at which the catalyst activity declines. In Fig. 3.20 the curve ABC shows a nonisothermal reaction path for an inlet temperature T_0 corresponding to A.

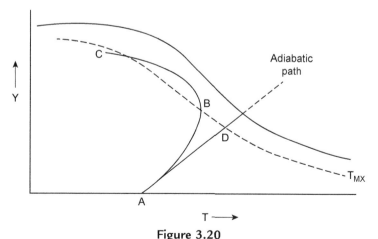

Figure 3.20
Reaction path of a cooled tubular reactor.

Provided $T_0 > T_W$, it is obvious that $\frac{dT}{dY} < \frac{-\Delta H}{\bar{c}_p}$ and the rate of temperature increase will be less than in the adiabatic case. The point B, in fact, corresponds to the temperature at which the reaction rate is at a maximum, and the locus of such points is the curve T_{MX} described previously. The maximum temperature attained from any given inlet temperature may be calculated by solving, using an iterative method,[52] the pair of simultaneous Eqs. (3.95) and (3.102) and finding the temperature at which $\frac{dr_{YT}^M}{dT}$ or, equivalently, $\frac{dT}{dY} = 0$. We will see later that a packed tubular reactor is very sensitive to change in wall temperature. It is therefore important to estimate the maximum attainable temperature for a given inlet temperature, from the point of view of maintaining both catalyst activity and reactor stability.

An important class of reactors is that for which the wall temperature is not constant but varies along the reactor length. Such is the case when the cooling tubes and reactor tubes form an integral part of a composite heat exchanger. Fig. 3.21A and B shows, respectively, cocurrent and countercurrent flow of coolant and reactant mixture, the coolant fluid being entirely independent and separate from the reactants and products. However, the reactant feed itself may be used as coolant before entering the reactor tubes and again may flow cocurrent or countercurrent to the reactant mixture (Fig. 3.21C and D, respectively). In each case, heat is exchanged between the reactant mixture and the cooling fluid. A heat balance for a component of the reactant mixture leads to

$$G'\bar{c}_p \frac{dT}{dz} = \rho_b(-\Delta H)r_{YT}^M - \frac{U}{a'}(T - T_C) \tag{3.103}$$

an equation analogous to Eq. 3.102 but in which T_C is a function of z and U is an overall heat transfer coefficient for the transfer of heat between the fluid streams. The variation in T_C may be described by taking a heat balance for an infinitesimal section of the cooling tube:

$$G'_C \bar{c}_{pC} \frac{dT_C}{dz} \pm \frac{U}{a'}(T - T_C) = 0 \tag{3.104}$$

where G'_C is the mass flow rate of coolant per unit cross section of the reactor tube and $\bar{c}_{pC}$ is the mean heat capacity of the coolant. If flow is cocurrent, the lower sign is used; if flow is countercurrent, the upper sign is used. Because the mass flow rate of the cooling fluid is based on the cross-sectional area of the reactor tube, the ratio $G'\bar{c}_p/G'_C\bar{c}_{pC}(= \Gamma)$ is a measure of the capacities of the two streams to exchange heat. In terms of the limitations imposed by the one-dimensional model, the system is fully described by Eqs. (3.103) and (3.104) together with the mass balance equation:

$$G' \frac{dY}{dz} = \rho_b r_{YT}^M \quad \text{(Eq. 3.95)}$$

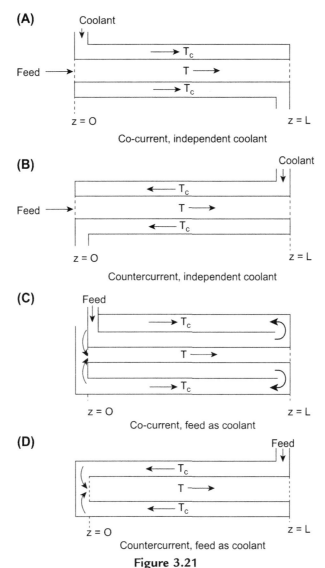

Figure 3.21

Cocurrent and countercurrent cooled tubular reactors: (A and B) independent coolant; (C and D) feed as coolant

The boundary conditions will depend on whether the flow is cocurrent or countercurrent, and whether or not the reactant mixture is used in part of the cooling process.

The reaction path in the T, Y plane could be plotted by solving the above set of equations with the appropriate boundary conditions. A reaction path similar to the curve ABC in Fig. 3.20 would be obtained. The size of reactor necessary to achieve a specified conversion could be assessed by tabulating points at which the reaction path crosses the constant rate contours, hence giving values of R_{YT} that could be used to integrate the mass balance Eq. (3.95). The reaction path would be suitable, provided the maximum temperature attained was not deleterious to the catalyst activity.

In the case of a reactor cooled by incoming feed flowing countercurrent to the reaction mixture (Fig. 3.21D), and typified by the ammonia converter sketched in Fig. 3.22, the boundary conditions complicate the integration of the equations. At the point $z = L$ where the feed enters the cooling tubes, $T_C = T_{CL}$, whereas at the reactor entrance ($z = 0$) the feed temperature is, of course, by definition the reactor inlet temperature, so here $T = T_0 = T_C$ and $Y = Y_0$. van Heerden[53] solved this two-point boundary value problem by assuming various values for T_0 and then integrating the equations to find the reactor exit temperature T_L and the incoming feed temperature T_{CL} at $z = L$. The quantity $(T - T_{CL})$ is a measure of the heat transferred between reactor and cooling tubes so a plot of $(T - T_{CL})$

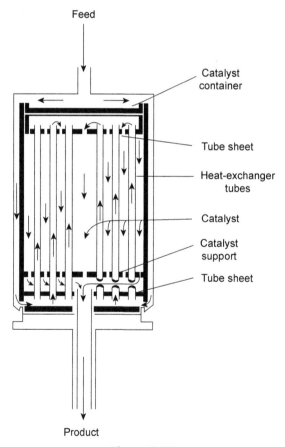

Figure 3.22
Ammonia converter.

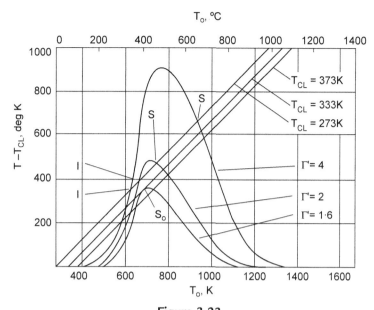

Figure 3.23
Heat transfer characteristics of countercurrent cooled tubular reactor.

versus T_0 will describe the heat exchanging capacity of the system for various reactor inlet conditions. Such curves are displayed in Fig. 3.23 with $\Gamma'\left(= UaL/G'c\bar{c}_{pC}\right)$ as a parameter. Superimposed on the diagram are parallel straight lines of unit slope, each line corresponding to the value of T_{CL}, the boundary condition at $z = L$. The solution to the three simultaneous equations and boundary conditions will be given by the intersections of the line T_{CL} with the heat transfer curves. In general an unstable solution I and a stable solution S are obtained. That S corresponds to a stable condition may be seen by the fact that if the system were disturbed from the semistable condition at I by a sudden small increase in T_0, more heat would be transferred between reactor and cooling tubes, because $(T - T_{CL})$ would now be larger and the heat generated by reaction would be sufficiently great to cause the inlet temperature to rise further until the point S was reached, whereupon the heat exchanging capacity would again match the heat generated by reaction. The region in which it is best to operate is in the vicinity of S_0, the point at which I and S are coincident. As the catalyst activity declines, the curves of $(T - T_{CL})$ versus T_0 are displaced downward and so the system will immediately become quenched. For continuing operation, therefore, more heat exchange capacity must be added to the system. This may be achieved by decreasing the mass flow rate. The temperature level along the reactor length therefore increases and the system now operates along a reaction path (such as ABC in Fig 3.20), displaced to regions of higher temperature but lower conversion. Production is thus maintained but at the cost of reduced conversion concomitant with the decaying catalyst activity.

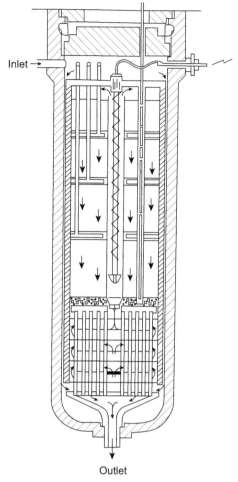

Inlet →

Outlet

Figure 3.24
Multiple-bed reactor for ammonia synthesis.

In industrial practice, a multiple-bed reactor (Fig. 3.24) is normally used for the synthesis of ammonia, rather than the single-stage reactor illustrated in Fig. 3.22. Because the reaction takes place at high pressures, the whole series of reactions is contained within a single pressure vessel, the diameter of which is minimized for reasons of mechanical design.

3.6.1.2 Dispersion in Packed Bed Reactors

The importance of dispersion and its influence on flow pattern and conversion in homogeneous reactors has already been studied in Chapter 2. The role of dispersion, both axial and radial, in packed bed reactors will now be considered. A general account of the nature of dispersion in packed beds, together with details of experimental results and their correlation, has already been given in Volume 2, Chapter 4. Those features that have a significant effect on the behavior of packed bed reactors will now be summarized. The equation for the material balance in a reactor will then be obtained for the case where plug flow conditions are modified by the effects of axial dispersion. Following this, the effect of simultaneous axial and radial dispersion on the nonisothermal operation of a packed bed reactor will be discussed.

3.6.1.2.1 The Nature of Dispersion

The effect that the solid packing has on the flow pattern within a tubular reactor can sometimes be of sufficient magnitude to cause significant departures from plug flow conditions. The presence of solid particles in a tube causes elements of flowing gas to become displaced randomly and therefore produces a mixing effect. An eddy diffusion coefficient can be ascribed to this mixing effect and becomes superimposed on the transport processes, which normally occur in unpacked tubes—either a molecular diffusion process at fairly low Reynolds numbers or eddy motion due to turbulence at high Reynolds numbers. Both transverse and longitudinal components of the flux attributed to this dispersion effect are of importance but operate in opposite ways. Transverse dispersion tends to bring the performance of the reactor closer to that which would be predicted by the simple design equation based on plug flow. On the other hand, longitudinal dispersion is inclined to invalidate the plug flow assumptions so that the conversion would be less than that would be expected if plug flow conditions obtained. The reason for this is that transverse mixing of the fluid elements helps to smooth out the parabolic velocity profile, which normally develops in laminar flow in an unpacked tube, whereas longitudinal dispersion in the direction of flow causes some fluid elements to spend less time in the reactor than they would if this additional component of flux due to eddy motion were not superimposed.

The magnitude of the dispersion effect due to transverse or radial mixing can be assessed by relying on theoretical predictions[54] and experimental observations,[55] which confirm that the value of the Peclet number $Pe(=ud_p/D$, where d_p is the particle diameter) for transverse dispersion in packed tubes is approximately 10. At bed Reynolds numbers of around 100, the diffusion coefficient to be ascribed to radial dispersion effects is about four times greater than the value for molecular diffusion. At higher Reynolds numbers the radial dispersion effect is correspondingly larger.

Longitudinal dispersion in packed reactors is thought to be caused by interstices between particles acting as mixing chambers. Theoretical analysis of a model[56] based on this assumption shows that the Peclet number Pe_l for longitudinal dispersion is about 2, and this has been confirmed by experiment.[55,57] Thus the diffusion coefficient for longitudinal dispersion is approximately five times that for transverse dispersion for the same flow conditions. The flux that results from the longitudinal dispersion effect is, however, usually much smaller than the flux resulting from transverse dispersion because axial concentration gradients are very much less steep than radial concentration gradients if the ratio of the tube length to diameter is large. Whether or not longitudinal dispersion is important depends on the ratio of the reactor length to particle size. If the ratio is less than about 100, then the flux resulting from longitudinal dispersion should be considered in addition to the flux due to bulk flow when designing the reactor.

3.6.1.2.2 Axial Dispersion

An axial (longitudinal) dispersion coefficient may be defined by analogy with Boussinesq's concept of eddy viscosity.[58] Thus both molecular diffusion and eddy diffusion due to local turbulence contribute to the overall dispersion coefficient or effective diffusivity in the direction of flow for the bed of solid. The moles of fluid per unit area and unit time an element of length δz entering by longitudinal diffusion will be $-D'_L(dY/dz)_z$, where D'_L is now the dispersion coefficient in the axial direction and has units $\mathbf{ML^{-1}T^{-1}}$ (because the concentration gradient has units $\mathbf{NM^{-1}L^{-1}}$). The amount leaving the element will be $-D'_L(dY/dz)_{z+\delta z}$. The material balance equation will therefore be

$$D'_L\left(\frac{dY}{dz}\right)_z + G'Y_z = -D'_L\left(\frac{dY}{dz}\right)_{z+\delta z} + G'Y_{z+\delta z} + \rho_b r_Y^M \delta z \qquad (3.105)$$

which on expansion becomes

$$D'_L\frac{d^2Y}{dz^2} - G'\frac{dY}{dz} = \rho_b r_Y^M \qquad (3.106)$$

The boundary condition at $z = 0$ may be written down by invoking the conservation of mass for an element bounded by the plane $z = 0$ and any plane upstream of the inlet where $Y = Y_0$, say,

$$G'Y_0 = G'(Y)_{z=0} - D'_L\left(\frac{dY}{dz}\right)_{z=0} \qquad (3.107)$$

from which we obtain the following condition:

$$z = 0, G'Y = G'Y_0 + D'_L\frac{dY}{dz} \qquad (3.108)$$

Similarly, at the reactor exit we obtain

$$z = L, \quad \frac{dY}{dz} = 0 \qquad (3.109)$$

because there is no further possibility of Y changing from Y_L after the reactor exit. Eq. (3.105) may be solved analytically for zero- or first-order reactions, but for other cases, resort to numerical methods is generally necessary. In either event, Y is obtained as a function of z and so the length L of catalyst bed required to achieve a conversion corresponding to an exit concentration Y_L may be calculated. The mass of catalyst corresponding to this length L is $\rho_b A_c L$.

Having pointed out the modifications to be made to a design based on the plug flow approach, it is salutary to note that axial dispersion is seldom of importance in fixed bed tubular reactors. This point is illustrated in Example 3.8.

Example 3.8

Show that the effect of axial dispersion on the conversion obtained in a typical packed bed gas—solid catalytic reactor is small. As the starting point, consider the following relationship (see Chapter 2, Eq. 2.133):

$$\frac{C_A}{C_{Ain}} = \exp\left[-k\tau + (k\tau)^2 \frac{D_L}{uL}\right] \tag{A}$$

which applies to a first-order reaction, rate constant k, carried out in a *homogeneous* tubular reactor in which the dispersion number D_L/uL is small. In Eq. (A), C_A is the concentration at a distance L along the reactor, C_{Ain} is the concentration at the inlet, D_L is the longitudinal dispersion coefficient for an empty tube homogeneous system, u is the linear velocity, and τ is the residence time (L/u). For a typical packed bed, take the value of the Peclet number $\left(u d_p / e D_L^*\right)$ as 2. D_L^* is the dispersion coefficient in a packed tube. Then show that for a fractional conversion of 0.99 and a ratio of d_p/L of 0.02, the length L of the reactor with axial dispersion exceeds the length of the simple ideal plug flow reactor by only 4.6% where d_p is the diameter of the particles.

Solution

Let L_p be the length of the simple ideal plug flow homogeneous reactor with no dispersion for the same value of C_A/C_{Ain}, and τ_p be the required residence time in the plug flow reactor, i.e., in Eq. (A) $D_L = 0$, so

$$\frac{C_A}{C_{Ain}} = \exp(-k\tau_p) \tag{B}$$

Therefore

$$\exp(-k\tau_p) = \exp\left[-k\tau + (k\tau)^2 \frac{D_L}{uL}\right]$$

i.e., dividing by $k\tau_p$,

$$\frac{\tau}{\tau_p} = 1 + k\frac{\tau}{\tau_p}\frac{D_L}{uL}$$

The second term on the right-hand side may be regarded as a "correction" term, which will be small compared to 1. Therefore, setting $\tau = \tau_p$ in this term,

$$L/L_p = \frac{\tau}{\tau_p} = 1 + k\tau_p\frac{D_L}{uL} = 1 + k\tau_p\frac{D_L}{u}\frac{1}{L}$$

For a packed bed, Peclet number $Pe = \frac{ud_p}{eD_L} = 2$ or $\frac{eD_L^*}{u} = \frac{d_p}{2}$ Furthermore, D_L for the empty tube may be replaced by eD_L^*. That is,

$$L/L_p = 1 + k\tau_p\frac{d_p}{2}\frac{1}{L}$$

From Eq. (B),

$$k\tau_p = \ln\frac{C_{Ain}}{C_A}$$

$$L/L_p = 1 + \left(\ln\frac{C_{Ain}}{C_A}\right)\frac{1}{2}\frac{d_p}{L}$$

Thus, for 99% conversion $\frac{C_{Ain}}{C_A} = \frac{1}{0.01}$ and $\frac{d_p}{L} = 0.02$.

Therefore

$$L/L_p = 1 + \ln\left(100 \times \frac{1}{2} \times 0.02\right) = 1 + 0.046$$

That is, effect of axial dispersion would be to increase length of plug flow reactor by 4.6%.

3.6.1.2.3 Axial and Radial dispersion—Nonisothermal Conditions

When the reactor exchanges heat with the surroundings, radial temperature gradients exist and this promotes transverse diffusion of the reactant. For an exothermic reaction, the reaction rate will be highest along the tube axis because the temperature there will be greater than at any other radial position. Reactants, therefore, will be rapidly consumed at the tube center resulting in a steep transverse concentration gradient causing an inward flux of reactant and a corresponding outward flux of products. The existence of radial temperature and concentration gradients, of course, renders the simple plug flow approach to design inadequate.

It is now essential to write the mass and energy balance equations for the two dimensions z and r. For the sake of completeness, we will include the effect of longitudinal dispersion and heat conduction and deduce the material and energy balances for one component in an elementary annulus of radius δr and length δz. We assume that equimolar counterdiffusion occurs, and by reference to Fig. 3.25, write down, in turn, the components of mass, which are entering the element in unit time longitudinally and radially:

Moles entering by longitudinal bulk flow $= 2\pi r \delta r G'(Y)_z$
Moles entering by transverse diffusion $= D'_r 2\pi r \delta z \left(\frac{\partial Y}{\partial r}\right)_r$
Moles entering by longitudinal diffusion $= D'_r 2\pi r \delta z \left(\frac{\partial Y}{\partial z}\right)_z$

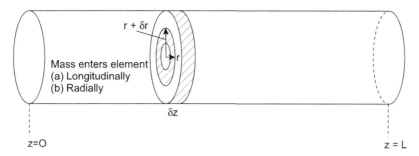

Figure 3.25
Two-dimensional tubular reactor.

In general, the longitudinal and radial dispersion coefficients D'_r and D'_L will differ. The moles leaving the element in unit time can be written similarly as a series of components:

Moles leaving by longitudinal bulk flow $= 2\pi r\delta r G'(Y)_{z+\delta z}$

Moles leaving by transverse diffusion $= -D'_r 2\pi(r + \delta r)\delta z\left(\frac{\partial Y}{\partial r}\right)_{r+\delta r}$

Moles leaving by longitudinal diffusion $= -D'_L 2\pi r\delta r\left(\frac{\partial Y}{\partial z}\right)_{z+\delta z}$

Moles of component produced by chemical reaction $= 2\pi r\delta r\delta z\rho_b r_{YT}^M$

In the steady state the algebraic sum of the moles entering and leaving the element will be zero. By expanding terms evaluated at $(z + \delta z)$ and $(r + \delta r)$ in a Taylor series about the points z and r, respectively, and neglecting second-order differences, the material balance equation becomes

$$D'_r\left\{\frac{\partial^2 Y}{\partial r^2} + \frac{1}{r}\frac{\partial Y}{\partial r}\right\} + D'_L\frac{\partial^2 Y}{\partial z^2} - G'\frac{\partial Y}{\partial z} = \rho_b r_{YT}^M \tag{3.110}$$

A heat balance equation may be deduced analogously:

$$k_r\left\{\frac{\partial^2 T}{\partial r^2} + \frac{1}{r}\frac{\partial T}{\partial r}\right\} + k_l\frac{\partial^2 T}{\partial z^2} - G'\bar{c}_p\frac{\partial T}{\partial z} = \rho_b(-\Delta H)r_{YT}^M \tag{3.111}$$

where k_r and k_l are the thermal conductivities in the radial and longitudinal directions, respectively, $\bar{c}_p$ is the mean heat capacity of the fluid, and $(-\Delta H)$ is the heat evolved per mole due to chemical reaction. When writing the boundary conditions for the above pair of simultaneous equations, the heat transferred to the surroundings from the reactor may be accounted for by ensuring that the tube wall temperature correctly reflects the total heat flux through the reactor wall. If the reaction rate is a function of pressure, then the momentum balance equation must also be invoked, but if the rate is insensitive or independent of total pressure, then it may be neglected.

It is useful at this stage to note the assumptions, which are implicit in the derivation of Eqs. (3.110) and (3.111). These are as follows: dispersion coefficients (both axial and radial) and thermal conductivity all constant, instantaneous heat transfer between the solid catalyst and the reacting ideal gas mixture, and changes in potential energy negligible. It should also be noted that the two simultaneous equations are coupled and highly nonlinear because of the effect of temperature on the reaction rate. Numerical methods of solution are therefore generally adopted and those employed are based on the use of finite differences. The neglect of longitudinal diffusion and conduction simplifies the equations considerably. If we also suppose that the system is isotropic, we can write a single effective diffusivity D'_e for the bed in place of D'_r and D'_L and a single effective thermal conductivity k_e.

The results of calculations typical of an exothermic catalytic reaction are shown in Figs. 3.26 and 3.27. It is clear that the conversion is higher along the tube axis than at other radial positions and that the temperature first increases to a maximum before decreasing at points further along the reactor length. The exothermic nature of the reaction, of course, leads to an initial increase in temperature but, because in the later stages of reaction the radial heat transfer to the wall and surroundings becomes larger than the heat evolved by chemical reaction, the temperature steadily decreases. The mean temperature and conversion over the tube radius for any given position along the reactor length may then be computed from such results and consequently the bed depth required for a given specified conversion readily found.

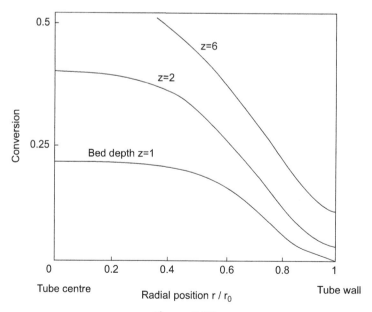

Figure 3.26
Conversion profiles as a function of tube length and radius.

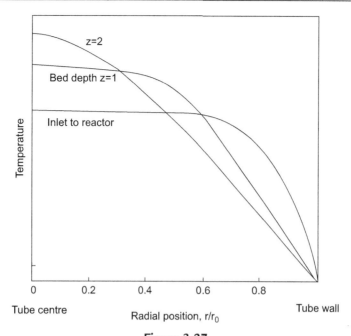

Figure 3.27
Temperature profiles as a function of tube length and radius.

Further advancements in the theory of fixed bed reactor design have been made,[59,60] but it is unusual for experimental data to be of sufficient precision and extent to justify the application of sophisticated methods of calculation. Uncertainties in the knowledge of effective thermal conductivities and heat transfer between gas and solid make the calculation of temperature distribution in the bed susceptible to inaccuracies, particularly in view of the pronounced effect of temperature on the reaction rate.

3.6.2 Thermal Characteristics of Packed Reactors

There are several aspects of thermal sensitivity and instability, which are important to consider in relation to reactor design. When an exothermic catalytic reaction occurs in a nonisothermal reactor, for example, a small change in coolant temperature may, under certain circumstances, produce undesirable hot spots or regions of high temperature within the reactor. Similarly, it is of central importance to determine whether or not there is likely to be any set of operating conditions, which may cause thermal instability in the sense that the reaction may either become extinguished or continue at a higher temperature level as a result of fluctuations in the feed condition. We will briefly examine these problems.

3.6.2.1 Sensitivity of Countercurrent Cooled Reactors

To illustrate the problem of thermal sensitivity, we will analyze the simple one-dimensional model of the countercurrent cooled packed tubular reactor described earlier and illustrated in Fig. 3.25. We have already seen that the mass and heat balance equations for the system may be written as

$$G'\frac{dY}{dz} = \rho_b r^M \quad \text{(Eq. 3.95)}$$

$$G'\bar{c}_p\frac{dT}{dz} = -\Delta H \rho_b r^M - \frac{h}{a'}(T - T_W) \quad \text{(Eq. 3.102)}$$

where T_W represents a constant wall temperature. Solution of these simultaneous equations with initial conditions $z = 0$, $T = T_0$, and $Y = Y_0$ enables us to plot the reaction path in the (T, Y) plane. The curve ABC, portrayed in Fig. 3.20, is typical of the reaction path that might be obtained. There will be one such curve for any given value T_0 of the reactor inlet temperature. By locating the points at which $dT/dY = 0$, the loci of maximum temperature may be plotted, each locus corresponding to a chosen value T_W for the constant coolant temperature along the reactor length. Thus, a family of curves such as that sketched in Fig. 3.28 would be obtained, representing loci of maximum temperatures for constant wall temperatures T_W, increasing sequentially from curve 1 to curve 5. Because the maximum temperature along the nonisothermal reaction path (point **B** in Fig. 3.20) must be less than the temperature given by the intersection of the adiabatic line with the locus of possible maxima (point **D** in Fig. 3.20), the highest temperature achieved in the reactor must be bounded by

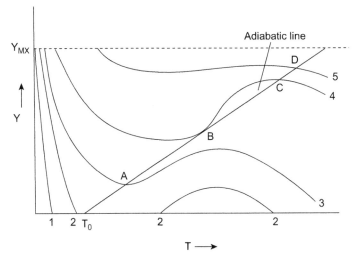

Figure 3.28
Loci of maximum wall temperatures.

this latter point. If the adiabatic line—and therefore the nonisothermal reaction path—lies entirely below the locus of possible maxima, then $dT/dY > 0$ and the temperature will increase along the reactor length because $dT/dz > 0$ for all points along the reactor; if, however, the adiabatic line lies above the locus, $dT/dY < 0$ and the temperature will therefore decrease along the reactor length. For a particular T_W corresponding to the loci 1 or 2 in Fig. 3.28, T_0, the inlet temperature to the reactor, must be the maximum temperature in the reactor because the adiabatic line lines entirely above these two curves making dT/dz always negative. If T_W corresponds to curve 3, however, we might expect the temperature to increase from T_0 to a maximum and thence to decrease along the remaining reactor length because the adiabatic line now intersects this particular locus at point **A**. In the case of curve 4 the adiabatic line is tangent to the maximum temperature locus at **B** and then intersects it at point **C** where the temperature would be much higher. The adiabatic line intersects curve 5 at point **D**. From this analysis, therefore, we are led to expect that, if the steady wall temperature of the reactor is increased through a sequence corresponding to curves 1 to 4, the maximum temperature in the reactor gradually increases from T_0 to **B** as T_W increases from T_{W1} to T_{W4}. If the wall temperature were to be increased to T_{W5} (the maximum temperature locus corresponding to curve 5), there would be a discontinuity and the maximum temperature in the reactor would suddenly jump to point **D**. This type of sensitivity was predicted by Bilous and Amundson[61] who calculated temperature maxima within a nonisothermal reactor for various constant wall temperatures. The results of their computations are shown in Fig. 3.29, and it is seen that, if the wall temperature increases from 300K to 335K, a temperature maximum appears in the reactor. Further increase in wall temperature causes the maximum to increase sharply to higher temperatures. Bilous and Amundson also found that sensitivity to heat transfer can produce temperature maxima in much the same way as sensitivity to change in wall temperature. In view of the nature of Eqs. (3.95) and (3.102), this is to be expected.

3.6.2.2 The Autothermal Region

Clearly it is desirable to utilize economically the heat generated by an exothermic reaction. If the heat dissipated by reaction can be used in such a way that the cold incoming gases are heated to a temperature sufficient to initiate a fast reaction, then by judicious choice of operating conditions the process may be rendered thermally self-sustaining. This may be accomplished by transferring heat from the hot exit gases to the incoming feed. The differential equations already discussed for the countercurrent self-cooled reactor are applicable in this instance. Combining the three simultaneous Eqs. (3.95), (3.103), and (3.104) (utilizing the upper sign in Eq. (3.104) for the

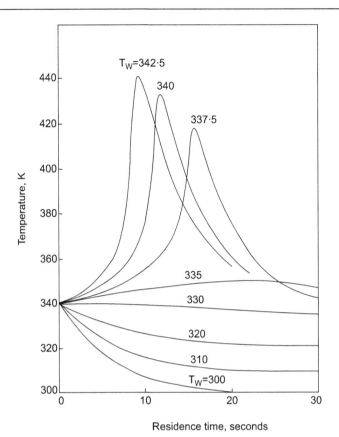

Figure 3.29
Sensitivity of reactor to change in wall temperature.

countercurrent case and remembering that for a self-cooled—or in this particular case self-heated—reactor $G'_c \bar{c}_{pc} = G' \bar{c}_p$), we obtain

$$G' \frac{d}{dz} \left\{ \frac{(-\Delta H)}{\bar{c}_p} Y - T + T_C \right\} = 0 \tag{3.112}$$

Integration of this equation yields

$$T_C = T - \frac{(-\Delta H)}{\bar{c}_p} (Y - Y_0) \tag{3.113}$$

Substituting Eq. (3.113) into the heat balance Eq. (3.103) for the reactor,

$$G' \frac{dT}{dz} = \frac{(-\Delta H)}{\bar{c}_p} \rho_b r^M - \frac{G'(-\Delta H)}{\bar{c}_p} \frac{\Gamma'}{L} (Y - Y_0) \tag{3.114}$$

where the heat exchanging capacity of the system has been written in terms of $\Gamma'(=UaL/G'\bar{c}_{pc})$. The mass balance equation

$$G'\frac{dY}{dz} = \rho_b r^M \quad \text{(Eq. 3.95)}$$

must be integrated simultaneously with Eq. (3.114) from $z = 0$ to $z = L$ with the initial conditions (at z = 0), $Y = Y_0$ and $T = T_0$ to obtain T and Y as functions of z. In this event the exit temperature T_L and concentration Y_L can, in principle, be computed. Numerical integration of the equations is generally necessary. To estimate whether the feed stream has been warmed sufficiently for the reaction to give a high conversion and be thermally self-supporting, we seek a condition for which the heat generated by reaction equals the heat gained by the cold feed stream. The heat generated per unit mass of feed by chemical reaction within the reactor may be written in terms of the concentration of product:

$$Q_1^M = (-\Delta H)(Y_L - Y_0) \tag{3.115}$$

whereas the net heat gained by the feed would be

$$Q_2^M = \bar{c}_p(T_L - T_{CL}) \tag{3.116}$$

The system is thermally self-supporting if $Q_1^M = Q_2^M$, i.e., if

$$T_C - T_{CL} = \frac{(-\Delta H)}{\bar{c}_p}(Y_L - Y_0) \tag{3.117}$$

The temperature of the nonreacting heating fluid at the reactor inlet is T_0 and at the exit T_{CL} so the difference, say ΔT, is

$$\Delta T = T_0 - T_{CL} = T_0 - T_L + \frac{(-\Delta H)}{\bar{c}_p}(Y_L - Y_0) \tag{3.118}$$

Because the numerical values of T_L and Y_L are dependent on the heat exchanging capacity (as shown by Eq. 3.114), the quantity on the right-hand side of Eq. (3.118) may be displayed as a function of the inlet temperature to the bed, T_0, with Γ' as a parameter. The three bell-shaped curves in Fig. 3.30 are for different values of Γ', and each represents the locus of values given by the right-hand side of this equation. The left-hand side of the equation may be represented by a straight line of unit slope through the point $(T_{CL}, 0)$. The points at which the line intersects the curve represent solutions to Eq. (3.118). However, we seek only a stable solution that coincides with a high yield. Such a solution would be represented by the point A in Fig. 3.30, for the straight line intersects the curve only once and at a bed temperature corresponding to a high yield of product. The condition for autothermal reaction would therefore be represented by the operating condition (T_{CL}, Γ'_3), i.e., the cold feed temperature should be $T_0 = T_{CL}$ and the heat exchange capacity (which determines the ratio L/G, the length of solid packing divided by the mass flow rate) should correspond to Γ'_3. The corresponding temperature profiles for the reactor and exchanger are sketched in Fig. 3.31.

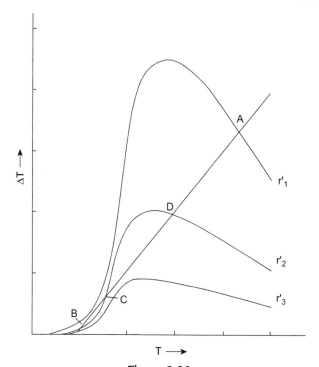

Figure 3.30
Temperature increase of feed as function of inlet temperature.

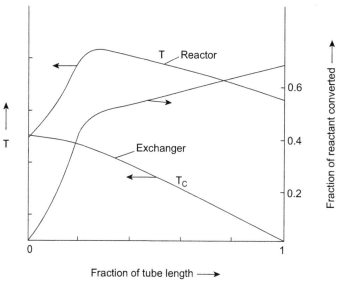

Figure 3.31
Countercurrent self-sustained reactor.

3.6.2.3 Stability of Packed Bed Tubular Reactors

Referring to Fig. 3.30, which in effect is a diagram to solve a heat balance equation, it may be seen that the line representing the heat gained by the feed stream intersects two of the heat generation curves at a single point and one of them at three points. As already indicated, intersection at one point corresponds to a solution which is stable, for if the feed state is perturbed there is no other state of thermal equilibrium at which the system could continue to operate. The intersection at point A corresponds to a state of thermal equilibrium at a high reaction temperature and therefore high conversion. On the other hand, although the system is in a state of thermal equilibrium at point **B**, the temperature and yield are low and the reaction is almost extinguished—certainly not a condition one would choose in practice. A condition of instability is represented by the intersection at C. Here any small but sudden decrease in feed temperature would cause the system to become quenched, for the equilibrium state would revert to **B**. A sudden increase in feed temperature would displace the system to D. Although D represents a relatively stable condition, such a choice should be avoided, for a large perturbation might result in the system restabilizing at B where the reaction is quenched. A reactor should therefore be designed such that thermal equilibrium can be established at only one point (such as A in Fig. 3.30) where the conversion is high.

It is interesting to note from Fig. 3.32 that if the feed to an independently cooled reactor (in which an exothermic reaction is occurring) is gradually varied through a sequence of increasing temperatures such as 1 to 5, the system will have undergone some fairly violent changes in thermal equilibrium before finally settling at point I. Corresponding to a

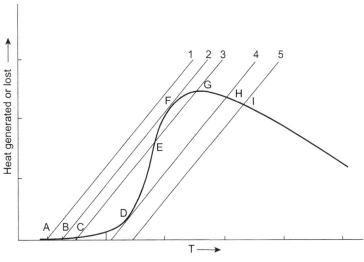

Figure 3.32
Stability states in reactors.

sequential increase in feed temperature from 1 to 4, the reactor will pass through states A, B, and C to D, the last point at which any slight increase in feed temperature causes the thermal equilibrium condition to jump to a point near H. If the feed temperature is then raised to 5, the system gradually changes along the smooth curve from H to I. On reducing the feed temperature from 5 to 2 on the other hand, the sequence of equilibrium states will alter along the path IHGF. Any further small decrease in feed temperature would displace the system to a point near **B** and subsequently along the curve **BA** as the feed temperature gradually changes from 2 to 1. Thus, there is a hysteresis effect when the system is taken through such a thermal cycle, and this is clearly illustrated in Fig. 3.33, which traces the path of equilibrium states as the feed temperature is first increased from 1 to 5 and subsequently returned to 1. These effects are important to consider when start-up or shutdown times are estimated.

Another type of stability problem arises in reactors containing reactive solid or catalyst particles. During chemical reaction the particles themselves pass through various states of thermal equilibrium, and regions of instability will exist along the reactor bed. Consider, for example, a first-order catalytic reaction in an adiabatic tubular reactor and further suppose that the reactor operates in a region where there is no diffusion limitation within the particles. The steady-state condition for reaction in the particle may then be expressed by equating the rate of chemical reaction to the rate of mass transfer. The rate of chemical reaction per unit reactor volume will be $(1 - e)kC_{Ai}$ because the effectiveness factor η is

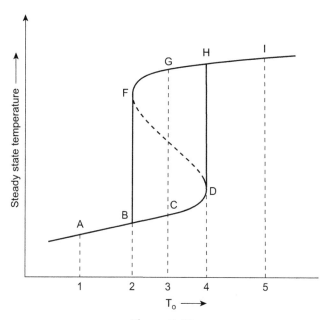

Figure 3.33
Hysteresis effect in packed bed reactor.

considered to be unity. From Eq. (3.66) the rate of mass transfer per unit volume is $(1 - e)$ $(S_x/V_p)k_G(C_{AG} - C_{Ai})$ so the steady-state condition is

$$kC_{Ai} = k_G \frac{S_x}{V_p}(C_{AG} - C_{Ai}) \tag{3.119}$$

Similarly, equating the heat generated by reaction to the heat flux from the particle,

$$(-\Delta H)kC_{Ai} = h\frac{S_x}{V_p}(T_i - T) \tag{3.120}$$

where h is the heat transfer coefficient between the particle and the gas, and T_i is the temperature at the particle surface.

Eliminating C_{Ai} from Eqs. (3.119) and (3.120),

$$T_i + T = (-\Delta H)\frac{V_p}{S_x}\frac{1}{h}\frac{kC_{AG}}{1 + \dfrac{V_p k}{S_x k_G}} \tag{3.121}$$

Now the maximum temperature T_{MX} achieved for a given conversion in an adiabatic reactor may be adduced from the heat balance:

$$T_{MX} = T + \frac{(-\Delta H)C_{AG}}{\rho \bar{c}_p} \tag{3.122}$$

which is analogous to Eq. (3.91), except that concentration has been expressed in terms of moles per unit volume, C_{AG} $(=\rho Y)$, with the gas density ρ regarded as virtually constant. Combining Eqs. (3.121) and (3.122), the temperature difference between the particle surface and the bulk gas may be expressed in dimensionless form:

$$\frac{T_i - T}{T_{MX} - T} = \frac{ak}{1 + bk} \tag{3.123}$$

where k is the usual Arrhenius function of temperature, a is the constant $\dfrac{\rho \bar{c}_p V_p}{S_x h}$, and b is the constant $\dfrac{V_p}{S_x k_G}$.

The condition of thermal equilibrium for a particle at some point along the reactor is thus established by Eq. (3.123). The function on the right-hand side of this equation is represented by a sigmoidal curve as shown in Fig. 3.34, which is a function of bed temperature $T(z)$ for some position z along the reactor. The left-hand side of Eq. (3.123) (as a function of T) is a family of straight lines all of which pass through the common pole point $(T_{MX}, 1)$ designated **P**. It is apparent from Fig. 3.34 that as the temperature increases along the bed, it is possible to have a sequence of states corresponding to lines such as PA, PB, PC, PD, and PE. As soon

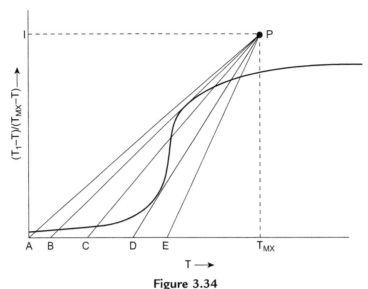

Figure 3.34
Stability states of particles in adiabatic bed.

as a position has been reached along the reactor where the temperature is B, there is the possibility of two stable temperatures for a particle. Such a situation would persist until a position along the reactor has been reached where the temperature is D, after which only one stable state is possible. Thus, there will be an infinite family of steady-state profiles within this region of the bed. Which profile actually obtains depends on start-up and bed entrance conditions. However, for steady-state operating conditions, the work of McGreavy and Thornton[23] offers some hope to designers in that it may be possible to predict and thus avoid regions of instability in cooled packed bed catalytic reactors.

3.6.3 Fluidized Bed Reactors

The general properties of fluidized beds are discussed in Volume 2, Chapter 6. Their application to gas—solid and catalytic reactions has certain advantages over the use of tubular type reactors. High wall-to-bed heat transfer coefficients enable heat to be abstracted from, or absorbed by, the reactor with considerable efficiency. A mechanical advantage is also gained by the relative ease with which solids may be conveyed and, because of solids mixing, the whole of the gas in the reactor is at substantially the same temperature. Extremely valuable is the large external surface area exposed by the solid to the gas; because of this, reactions limited by intraparticle diffusion will give a higher conversion in a fluidized bed than in a packed bed tubular reactor.

May[62] considered a model for describing reactions in a fluidized bed. He wrote the mass conservation equations for the bubble phase (identified with rising bubbles containing little or no solid particles) and the dense phase (in which the solid particles are thoroughly mixed). He tacitly assumed that mass transferred by cross-flow between the two phases was not a function of bed height and took an average concentration gradient for the whole height of the bed. Conversions within the bed were calculated by solving the conservation equations for particular cross-flow ratios. May compared the predicted conversion in beds in which there was no mixing of solids with beds in which there was complete back-mixing. It was found that conversion was smaller for complete back-mixing and relatively low cross-flow ratios than for no mixing and high cross-flow ratios. Thus, if back-mixing is appreciable, a larger amount of catalyst has to be used in the fluidized bed to achieve a given conversion than if a plug flow reactor, in which there is no back-mixing, is used. It was also shown that, for two concurrent first-order reactions, the extent of gas—solid contact affects the selectivity. Although the same total conversion may be realized, more catalyst is required when the contact efficiency is poor than when the contact efficiency is good, and proportionately more of the less reactive component in the feed would be converted at the expense of the more reactive component.

The difficulty with an approach in which flow and diffusive parameters are assigned to a model is that the assumptions do not conform strictly to the pattern of behavior in the bed. Furthermore, it is doubtful whether either solid dispersion or gas mixing can be looked on as a diffusive flux. Rowe[63] calculated, from a knowledge of the fluid dynamics in the bed, a mean residence time for particles at the surface of which a slow first-order gas—solid reaction occurred. Unless the particle size of the solid material was chosen correctly, most of the reactant gas passed through the bed as bubbles and had insufficient time to react. The most effective way of increasing the contact efficiency was to increase the particle size, for this caused more of the reactant gas to pass through the dense phase. Doubling the particle size almost halved the contact time required for the reactants to be completely converted. Provided that the gas flow is sufficiently fast to cause bubble formation in the bed, the heat transfer characteristics are good.

3.7 Gas—Solid Noncatalytic Reactors

There is a large class of industrially important heterogeneous reactions in which a gas or a liquid is brought into contact with a solid and reacts with the solid transforming it into a product. Among the most important are the reduction of iron oxide to metallic iron in a blast furnace; the combustion of coal particles in a pulverized fuel boiler or grate type furnaces; and the incineration of solid wastes. These examples also happen to be some of the most complex chemically. Further simple examples are the roasting of sulfide ores such as zinc blende:

$$ZnS + \frac{3}{2}O_2 \rightarrow ZnO + SO_2$$

and two reactions which are used in the carbonyl route for the extraction and purification of nickel:

$$NiO + H_2 \rightarrow Ni + H_2O$$
$$Ni + 4CO \rightarrow Ni(CO)_4$$

In the first of these reactions the product, impure metallic nickel, is in the form of a porous solid, whereas in the second, the nickel carbonyl is volatile and only the impurities remain as a solid residue. It can be seen from even these few examples that a variety of circumstances can exist. As an initial approach to the subject, however, it is useful to distinguish two extreme ways in which reaction within a particle can develop.

(1) *Shrinking core reaction mode*. If the reactant solid is nonporous, the reaction begins on the outside of the particle and the reaction zone moves toward the center. The core of unreacted material shrinks until it is entirely consumed. The product formed may be a coherent but porous solid like the zinc oxide from the oxidation of zinc sulfide above, in which case the size of the particle may be virtually unchanged (Fig. 3.35A(i)). On the other hand, if the product is a gas or if the solid product formed is a friable ash, as in the combustion of some types of coal, the particle decreases in size during the reaction until it

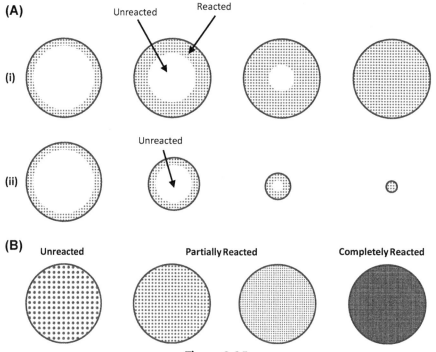

Figure 3.35
Stages in the reaction of a single particle by (A) shrinking core reaction mode with the formation of either (i) a coherent porous solid or (ii) friable ash or gaseous product and (B) progressive conversion reaction mode.

disappears (Fig. 3.35A(ii)). When the reactant solid is porous, a fast reaction may nevertheless still proceed via a shrinking core reaction mode because diffusion of the gaseous reactant into the interior of the solid will be a relatively slow process.

(2) *Progressive conversion reaction mode.* In a porous reactant solid, an inherently slower reaction can, however, proceed differently. If the rate of diffusion of the gaseous reactant into the interior of the particle can effectively keep pace with the reaction, the whole particle may be progressively and uniformly converted into the product (Fig. 3.35B). If the product is itself a coherent solid, the particle size may be virtually unchanged, but if the product is volatile or forms only a weak solid structure, the particle will collapse or disintegrate toward the end of the reaction.

3.7.1 Modeling and Design of Gas–Solid Reactors

The basic approach is to consider the problem in two parts. Firstly, the reaction of a single particle with a plentiful excess of the gaseous reactant is studied. A common technique is to suspend the particle from the arm of a thermobalance in a stream of gas at a carefully controlled temperature; the course of the reaction is followed through the change in weight with time. From the results a suitable kinetic model may be developed for the progress of the reaction within a single particle. Included in this model will be a description of any mass transfer resistances associated with the reaction and of how the reaction is affected by concentration of the reactant present in the gas phase.

The second part of the problem is concerned with the contacting pattern between gas and solid in the equipment to be designed. Material and thermal balances have to be considered, together with the effects of mixing in the solid and gas phases. These will influence the local conditions of temperature, gas composition, and fractional conversion applicable to any particular particle. The ultimate aim, which is difficult to achieve because of the complexity of the problem, is to estimate the overall conversion, which will be obtained for any particular set of operating conditions.

3.7.2 Single Particle Unreacted Core Models

Most of the gas–solid reactions that have been studied appear to proceed by the shrinking core reaction mode. In the simplest type of unreacted core model, it is assumed that there is a nonporous unreacted solid with the reaction taking place in an infinitely thin zone separating the core from a completely reacted product as shown in Fig. 3.36 for a spherical particle. Considering a reaction between a gaseous reactant **A** and a solid **B** and assuming that a coherent porous solid product is formed, five consecutive steps may be distinguished in the overall process:

1. Mass transfer of the gaseous reactant **A** through the gas film surrounding the particle to the particle surface.

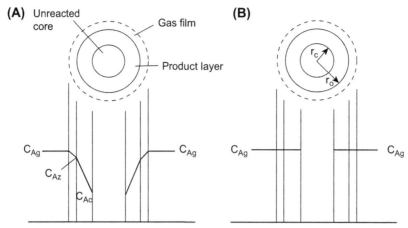

Figure 3.36
Unreacted core model, impermeable solid, showing gas-phase reactant concentration with
(A) significant gas film and product layer resistances and (B) negligible gas film and product
layer resistances.

2. Penetration of **A** by diffusion through pores and cracks in the layer of product to the surface of the unreacted core.
3. Chemical reaction at the surface of the core.
4. Diffusion of any gaseous products back through the product layer.
5. Mass transfer of gaseous products through the gas film.

If no gaseous products are formed, steps (4) and (5) cannot contribute to the resistance of the overall process. Similarly, when gaseous products are formed, their counterdiffusion away from the reaction zone may influence the effective diffusivity of the reactants, but gaseous products will not otherwise affect the course of the reaction unless it is a reversible one. Of the three remaining steps (1) to (3), if the resistance associated with one of these steps is much greater than the other resistances then, as with catalytic reactors, that step becomes rate determining for the overall reaction process.

The shrinking core reaction mode is not necessarily limited to nonporous unreacted solids. With a fast reaction in a porous solid, diffusion into the core and chemical reaction occur in parallel. The mechanism of the process is very similar to the mechanism of the catalyzed gas—solid reactions where the Thiele modulus is large, the effectiveness factor is small, and the reaction is confined to a thin zone (Section 3.3.1). This combination of reaction with core diffusion gives rise to a reaction zone which, although not infinitely thin but diffuse, nevertheless advances into the core at a steady rate.

Before a simple mathematical analysis is possible, a further restriction needs to be applied; it is assumed that the rate of advance of the reaction zone is small compared with the diffusional velocity of **A** through the product layer, i.e., that a pseudo-steady state exists.

Although general models in which external film mass transfer, diffusion through the product layer, diffusion into the core, and chemical reaction are all taken into account have been developed,[64] it is convenient to consider the special cases that arise when one of these stages is rate determining. Some of these correspond to the shrinking core mode of reaction, whereas others lead to the progressive conversion mode. In further sections, the shrinking code model is discussed for various rate-determining steps including chemical reaction, gas film mass transfer, and diffusion through solid product layer.

3.7.2.1 Unreacted Core Model—Chemical Reaction Control

Consider a reaction of stoichiometry,

$$\mathbf{A}(g) + b\mathbf{B}(s) = \text{Products}$$

which is first order with respect to the gaseous reactant **A**. If neither the gas film nor the solid product layer presents any significant resistance to mass transfer, the concentration of **A** at the reaction surface at the radial position r_c will be C_{AG}, the same as in the bulk of the gas (Fig. 3.36B).

If the reacting core is impermeable, reaction will take place at the surface of the core, whereas if the core has some degree of porosity, the combination of chemical reaction and limited core diffusivity will give rise to a more extended reaction zone. In either case, the overall rate of reaction will be proportional to the area of the reaction front.

Taking therefore unit area of the core surface as the basis for the reaction rate, and writing the first-order rate constant as k_s, then the rate at which moles of **A** are consumed in the reaction is given by

$$\frac{dn_A}{dt} = 4\pi r_c^2 k_s C_{AG} \tag{3.124}$$

If the core is porous, k_s is not a simple rate constant but incorporates the core diffusivity as well.

From the stoichiometry of the reaction,

$$\frac{1}{\nu_A}\frac{dn_A}{dt} = \frac{1}{\nu_B}\frac{dn_B}{dt}$$

where $\nu_A = -1$, $\nu_B = -b$, n_B is the moles of **B** in the core. $n_B = C_B \frac{4}{3}\pi r_c^3$, C_B being the molar density of the solid **B**. Thus

$$-\frac{1}{b}\frac{dn_B}{dt} = -\frac{1}{b}C_B 4\pi r_c^2 \frac{dr_c}{dt} \tag{3.125}$$

Hence, from Eqs. (3.124) and (3.125),

$$-C_B\frac{dr_C}{dt} = bk_s C_{AG}$$

Integrating,

$$-C_B \int_{r_0}^{r_c} dr_c = bk_s C_{AG} \int_0^t dt$$

$$t = \frac{C_B}{bk_s C_{AG}} (r_0 - r_c) \tag{3.126}$$

The time t_f for complete conversion corresponds to $r_c = 0$, i.e.,

$$t_f = \frac{C_B r_0}{bk_s C_{AG}} \tag{3.127}$$

When the core has a radius r_c and the fractional conversion for the particle as a whole is α_B, the ratio t/t_f is given by

$$t/t_f = (1 - r_c/r_0) = 1 - (1 - \alpha_B)^{1/3} \tag{3.128}$$

That is,

3.7.2.2 Unreacted Core Model—Gas Film Control

For the same reaction shown in the previous section, we now consider that the gas film offers significant resistance, whereas the resistance due to reaction and solid product layer is negligible. Generally, the concentration driving force for the mass transfer of reactant **A** through the gas film is given by the difference between the concentration in the bulk gas phase and that at the surface of the particle, $r = r_0$. However, if one assumes that gas film is the major resistance, then the concentration of reactant **A** at the particle surface is zero for an irreversible reaction. The rate at which moles of **A** are consumed in the process is given by

$$\frac{dn_A}{dt} = 4\pi r_0^2 k_G C_{AG} \tag{3.129}$$

k_G denotes the gas film mass transfer coefficient. From Eqs. 3.125 and 3.129,

$$-C_B \frac{r_c^2}{r_0^2} \frac{dr_C}{dt} = bk_G C_{AG} \tag{3.130}$$

Integrating Eq. (3.130) in the limits r_0 to r_c in the time range 0 to t yields

$$t = \frac{C_B r_0}{3bk_G C_{AG}} \left(1 - \frac{r_c^3}{r_0^3}\right) \tag{3.131}$$

The time t_f for complete conversion corresponds to $r_c = 0$, i.e.,

$$t_f = \frac{C_B r_0}{3bk_G C_{AG}} \tag{3.132}$$

When the core has a radius r_c and the fractional conversion for the particle as a whole is α_B, the ratio t/t_f is given by

$$t/t_f = \left(1 - \frac{r_c^3}{r_0^3}\right) = \alpha_B \tag{3.133}$$

3.7.2.3 Unreacted Core Model—Solid Product Layer Control

In combustion and gasification of solid fuels such as coal, biomass, and solid wastes, one frequently encounters the formation of a solid product layer or ash surrounding the unreacted core. Mild agitation or vibration is generally provided in traveling grate furnaces to separate the ash to achieve complete combustion of the carbon present in the core through the supply of secondary air. When the ash layer presents significant diffusional resistance for the transport of the gas-phase species **A** to react with the core, the following steady-state relationship can be written for the rate of transport of **A**:

$$\frac{dn_A}{dt} = 4\pi r_0^2 D_A \left(\frac{dC_A}{dr}\right)_{r_0} = 4\pi r_c^2 D_A \left(\frac{dC_A}{dr}\right)_{r_c} = 4\pi r^2 D_A \left(\frac{dC_A}{dr}\right)_r = \text{constant}$$

This is justified by the facts that the regression of the unreacted core surface is much slower than the transport of **A**, and the system has a spherical symmetry. For any radius r, the rate of consumption of **A** is given by

$$-\frac{dn_A}{dt} = -\frac{1}{b}\frac{dn_B}{dt} = 4\pi r^2 D_A \left(\frac{dC_A}{dr}\right)_r = \text{constant} \tag{3.134}$$

Integrating Eq. (3.134) in the limits of $r = r_0$ to r_c and $C_A = C_{AG}$ to 0, with the assumption that the concentration profile of A within the solid layer is at steady state (dn_B/dt is constant), yields

$$\left(-\frac{dn_B}{dt}\right)\int_{r_0}^{r_c}\frac{dr}{r^2} = 4\pi b D_A \int_{C_{AG}}^{0} dC_A \tag{3.135}$$

That is,

$$\left(-\frac{dn_B}{dt}\right)\left(\frac{1}{r_0} - \frac{1}{r_c}\right) = -4\pi b D_A C_{AG} \tag{3.136}$$

It is now possible to let the unreacted core to shrink with time and evaluate the rate of regression of the surface. Substituting Eq. (3.125) in 3.136 and integrating in the limits of $r_c = r_0$ to r_c in the time range 0 to t yields

$$C_B \int_{r_0}^{r_c}\left(\frac{1}{r_0} - \frac{1}{r_c}\right)r_c^2 dr_c = -b D_A C_{AG} \int_0^t dt \tag{3.137}$$

That is,

$$t = \frac{C_B r_0^2}{6bD_A C_{AG}} \left(1 - 3\frac{r_c^2}{r_0^2} + 2\frac{r_c^3}{r_0^3} \right) \tag{3.138}$$

The time t_f for complete conversion corresponds to $r_c = 0$, i.e.,

$$t_f = \frac{C_B r_0^2}{6bD_A C_{AG}} \tag{3.139}$$

When the core has a radius r_c and the fractional conversion for the particle as a whole is α_B, the ratio t/t_f is given by

$$t/t_f = \left(1 - 3\frac{r_c^2}{r_0^2} + 2\frac{r_c^3}{r_0^3} \right) = 1 - 3(1 - \alpha_B)^{2/3} + 2(1 - \alpha_B) \tag{3.140}$$

A more extended treatment and similar conversion time expressions for particles of different shapes may be found in the book by Szekely et al.[65] and Levenspiel.[66]

Example 3.9

Spherical particles of a sulfide ore 2 mm in diameter are roasted in an air stream at a steady temperature. Periodically small samples of the ore are removed, crushed, and analyzed with the following results:

Time (min)	15	30	60
Fractional conversion	0.334	0.584	0.880

Are these measurements consistent with a shrinking core and chemical reaction rate proportional to the area of the reaction zone? If so, estimate the time for complete reaction of the 2-mm particles, and the time for complete reaction of similar 0.5-mm particles.

Solution

Using the above data, time t can be plotted against $[1 - (1 - \alpha_B)^{1/3}]$ according to Eq. (3.120); a straight line confirms the assumed model and the slope gives t_f, the time to complete conversion.

Alternatively, calculate $t_f = t/[1 - (1 - \alpha_B)^{1/3}]$ from each data point:

Time (min)	15	30	60
Fractional conversion	0.334	0.584	0.880
Calculated t_f (min)	118	118	118

The constancy of t_f confirms the model; estimated time to complete conversion of 2 mm particles is thus 118 min.

Because $t_f \propto r_0$, the estimated time to complete conversion of 0.5-mm-diameter particles is $118.4 \times (0.5/2) = 29.6$ min.

3.7.2.4 Limitations of Simple Models—Solids Structure

In practice, reaction in a particle may occur within a diffuse front instead of a thin reaction zone, indicating a mechanism intermediate between the shrinking core and progressive conversion modes of reaction. Often it is not realistic to single out a particular step as being rate determining because there may be several factors, each of which affects the reaction to a similar extent. The conversion time expressions, namely, Eqs. (3.126), (3.131), and (3.138), that were derived by assuming the influence of only one resistance in each case can be combined to give an overall time of conversion such that $t_{overall} = t_{reaction\ alone} + t_{gas\ film\ alone} + t_{solid\ layer\ alone}$. This is possible because the resistances are in series and vary linearly with concentration of **A**.

The rate of consumption of the solid B under the influence of all three resistances can be written as follows, after combining Eqs. (3.124), (3.129), and (3.136) and the stoichiometry of the reaction.[66]

$$-\frac{1}{S_x}\frac{dn_B}{dt} = \frac{bC_{AG}}{\dfrac{1}{k_G} + \dfrac{r_0(r_c - r_0)}{D_A r_c} + \dfrac{r_0^2}{r_c^2 k_s}} \tag{3.141}$$

Here, S_x denotes the external surface area of the spherical particle $(4\pi r_0^2)$. The above equation can be expressed in terms of the rate of shrinkage of the unreacted core using Eq. (3.125) and solved to obtain the variation of r_c with time under the influence of all three resistances.

As a further complication there may be significant temperature gradients around and within the particle with fast chemical reactions; for example, in the oxidation of zinc sulfide, temperature differences of up to 55 K between the reaction zone and the surrounding furnace atmosphere have been measured.[67] Other complex processes such as combustion in a realistic furnace will also involve radiative heat transfer from the inner wall surface in addition to conduction and convective modes of heat transfer. Ranzi and coworkers[68] developed a comprehensive model of biomass combustion in a traveling grate combustor of 12 kW capacity by integrating the detailed kinetics of biomass pyrolysis, char gasification, and reactor and particle scale mass and energy balances to understand the variation of temperature and product composition at various layers of the bed with time. As biomass combustion involves both endothermic pyrolysis reactions and exothermic secondary reactions such as oxidation of volatiles and char, temperature gradients as high as 200 K were observed between different layers of biomass in the bed. Recent progress in understanding and modeling uncatalyzed gas–solid reactions has been based on the now well-established theory of catalyzed gas–solid reactions. The importance of characterizing the structure of the solids, i.e., porosity, pore size, and shape, internal surface area, and adsorption behavior, is now recognized. The problem with uncatalyzed reactions is that structural changes necessarily take place during the course of the reaction; pores in the reacting solid are enlarged as reaction proceeds; if a solid product is formed, there is a process of nucleation of the second solid phase. Furthermore, at the reaction temperature, some sintering of the product or reactant phases may occur and, if a highly exothermic reaction takes place in a thin reaction zone, drastic alterations in structure may occur near the reaction front.

Some reactions are brought about by the action of heat alone, for example, the thermal decomposition of carbonates, and baking bread and other materials. These constitute a special class of solid reactions somewhat akin to the progressive conversion type of reaction models but with the rate limited by the rate of heat penetration from the exterior.

3.7.2.5 Shrinking Particles and Film Growth

Certain solid materials such as pure carbon or polymers that do not basically contain inorganic matter can be burnt completely without the formation of a solid product layer. Some other examples of shrinking particles include reactive etching of silicon by chlorine gas in a patterned film and preparation of silicone monomers from silicon powder by reaction with hydrogen chloride in presence of copper catalyst. These two reactions are given by[69]

$$Si + 2Cl_2 = SiCl_4$$
$$Si + 2HCl = SiH_2Cl_2$$

Under such circumstances, surface reaction and mass transfer through the gas film are the only resistance offering steps. For a shrinking spherical particle undergoing the first-order reaction with respect to gaseous reactant **A**, the analysis will be the same as that outlined in Eqs. (3.124)–(3.128) when reaction is the rate-determining step. The variation of particle size with time is given by Eq. (3.126). However, when gas film resistance is significant, then the gas film mass transfer coefficient is not a constant, and it varies as the particle shrinks in size. The variation of k_G with radius of the sphere is given by the following correlation that is valid for flow over a sphere[66]:

$$Sh = 2 + 0.4\left(Re^{1/2} + 0.06Re^{2/3}\right)Sc^{0.4} \tag{3.142}$$

The characteristic length in this correlation is the diameter of the spherical particle, $2r_0$. For slow flow of gas around a small sphere, Eq. (3.142) reduces to $Sh = 2$. Thus, $k_G = D_A/r$. The rate of consumption of the solid can now be written as

$$-\frac{1}{b}\frac{dn_B}{dt} = -\frac{C_B 4\pi r^2}{b}\frac{dr}{dt} = 4\pi r^2\left(\frac{D_A}{r}\right)C_{AG} \tag{3.143}$$

Integrating,

$$-\int_{r_0}^{r} r\,dr = \frac{bD_A}{C_B} C_{AG} \int_0^t dt \tag{3.144}$$

That is,

$$t = \frac{C_B r_0^2}{2bC_{AG}D}\left(1 - \frac{r^2}{r_0^2}\right) = 1 - (1 - \alpha_B)^{2/3} \tag{3.145}$$

Another category of gas–solid reactions involves the formation of a solid layer as a result of transformation of reactant gas. Chemical vapor deposition is a classic process to fabricate thin solid films in microelectronic chips, wear resistant coatings, and functional materials such as graphene from gaseous reactants. The reactions are given by[69]

$$SiH_4 \rightarrow Si + 2H_2$$
$$SiH_4 + 2O_2 \rightarrow SiO_2 + 2H_2O$$
$$TiCl_4 + CH_4 \rightarrow TiC + 4HCl$$

These reactions show the decomposition of silane to form silicon coating in microelectronic circuit fabrication, formation of a silica film, and decomposition of titanium tetrachloride to form a hard film of titanium carbide.

Similar analysis that was adopted for the consumption of solids can also be utilized for their growth. For the reaction of stoichiometry, $A(g) = B(s) + C(g)$, the following expressions can be derived for the variation of film thickness, l, with time on a planar substrate with initial film thickness l_0.[69]

Only chemical reaction resistance: $l(t) = l_0 + \dfrac{k_s C_{AG} t}{C_B}$

Only gas film mass transfer resistance: $l(t) = l_0 + \dfrac{k_G C_{AG} t}{C_B}$

3.7.3 Types of Equipment and Contacting Patterns

There is a wide choice of contacting methods and equipment for gas–solid reactions. As with other solids-handling problems, the solution finally adopted may depend very much on the physical condition of the reactants and products, i.e., particle size, any tendency of the particles to fuse together, ease of gas–solid separation, etc. One type of equipment, the rotary kiln, has already been mentioned (Chapter 2, Fig. 2.5), and some further types of equipments suitable for continuous operation are shown in Fig. 3.37. The concepts of macromixing in the solid phase and dispersion in the gas phase as discussed in the previous section will be involved in the quantitative treatment of such equipment.

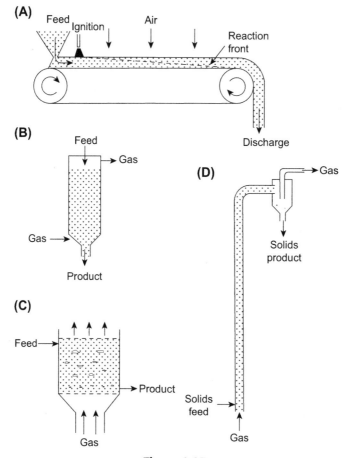

Figure 3.37
Types of reactor and contacting patterns for gas–solid reactions: (A) moving bed in cross-flow; (B) hopper type of reactor, particles moving downward in countercurrent flow; (C) fluidized bed reactor, particles well mixed; and (D) transfer line reactor, solids transported as a dilute phase in cocurrent flow.

The principle of the moving bed of Fig. 3.37A in cross-flow to the air supply is used for roasting zinc blende and for the combustion of large coal on chain grate stokers. Another kind of moving bed is found in hopper-type reactors in which particles are fed at the top and continuously move downward to be discharged at the bottom; an example is the decomposer used for nickel carbonyl where pure nickel is deposited on metallic nickel balls, which grow in size as they pass down the reactor (Fig. 3.37B).

Rotary kilns (Chapter 2, Fig. 2.5) have advantages where the solid particles tend to stick together as in cement manufacture and in the reduction and carbonylation steps in the purification of nickel. In rotary kilns, the flow of gas may be cocurrent or countercurrent to the solids.

If the solid particles can be maintained in the fluidized state without problems of agglomeration or attrition, the fluidized bed reactor (Fig. 3.37C) is likely to be preferred. For short contact times at high temperatures, the dilute phase transfer line reactor (Fig. 3.37D) has advantages.

Raked hearth reactors were once extensively used in the metals extraction industries but are now being superseded.

Semibatch reactors, where the gas passes through a fixed bed of solids that are charged and removed batchwise, are common for small-scale operations.

3.7.3.1 Fluidized Bed Reactor

As the following example shows, the fluidized bed reactor is one of the few types that can be analyzed in a relatively simple manner, providing (1) complete mixing is assumed in the solid phase and (2) a sufficient excess of gas is used so that the gas-phase composition may be taken as uniform throughout.

Example 3.10

Particles of a sulfide ore are to be roasted in a fluidized bed using excess air: the particles may be assumed spherical and uniform in size. Laboratory experiments indicate that the oxidation proceeds by the unreacted core mechanism with the reaction rate proportional to the core area, the time for complete reaction of a single particle being 16 min at the temperature at which the bed will operate. The particles will be fed and withdrawn continuously from the bed at a steady rate of 6 tonnes of product per hour (1.67 kg/s). The solids holdup in the bed at any time is 10 tonnes.

1. Estimate what proportion of the particles leaving the bed still contains some unreacted material in the core.
2. Calculate the corresponding average conversion for the product as a whole.
3. If the amount of unreacted material so determined is unacceptably large, what plant modifications would you suggest to reduce it?

Solution

The fluidized bed will be considered as a continuous stirred tank reactor in which ideal macromixing of the particles occurs. As shown in the section on mixing (Chapter 2, Section 2.1.3 and 2.1.4), in the steady state the required *exit age distribution* is the same as the **C**-curve obtained using a single shot of tracer. In fact the desired **C**-curve is identical with that derived in Chapter 2, Fig. 2.2, for a tank containing a liquid with ideal micromixing, but now the argument is applied to particles as follows:In Fig. 3.38 let W_F be the mass holdup of particles in the reactor, G_0 be the mass rate of outflow in the steady state (i.e., the holding time or residence time $\tau = W_F/G_0$), and c' be the number of tracer particles per unit mass of all

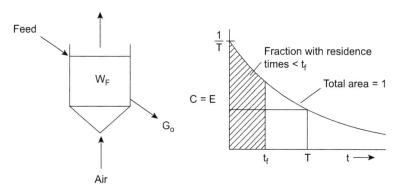

Figure 3.38
Flow of particles through a fluidized bed showing a normalized **C**-curve, which is identical with the exit age distribution function **E**.

particles. Consider a shot of n_{m0} tracer particles input at $t = 0$, giving an initial value $c'_0 = n_{m0}/W_F$ in the bed. Applying a material balance,

Inflow	−	Outflow	=	Accumulation
0	−	$c'G_0$	=	$W_F\dfrac{dc'}{dt}$

That is,

$$\frac{dc'}{dt} = -\frac{G_0}{W_F}c' = -\frac{c'}{\tau}; \quad \text{whence} \quad c' = c'_0 e^{-t/\tau}$$

For a normalized **C**-curve, we require

$$C = \frac{c'}{\int_0^\infty c'\,dt} = \frac{c'_0 e^{-t/\tau}}{c'_0 \int_0^\infty e^{-t/\tau}\,dt} = \frac{1}{\tau}e^{-t/\tau} = E$$

By definition the exit age distribution function **E** is such that the fraction of the exit stream with residence times between t and $t + \delta t$ is given by **E** δt.

(1) To determine the proportion of the particles that leave the bed still containing some unreacted material in the core, we calculate **F** that fraction of particles that leave the bed with residence times less than t_f, the time for complete reaction of a particle:

$$\mathbf{F} = \int_0^{t_f} \frac{1}{\tau}e^{-t/\tau}\,dt = 1 - e^{-t_f/\tau}$$

For the fluidized bed, $\tau = \frac{10}{6/60} = 100$ min. For a particle, $t_f = 16$ min.

Therefore, fraction with some unreacted material $= 1 - e^{-16/100} = \underline{0.148}$.

(2) The conversion α of reactant in a single particle depends on its length of stay in the bed. Working in terms of the fraction unconverted $(1 - \alpha)$ that, from Eq. (3.128), is given by

$$(1 - \alpha) = \left(1 - \frac{t}{t_f}\right)^3$$

the mean value of the fraction unconverted:

$$(1 - \bar{\alpha}) = \sum (1 - \alpha)\mathbf{E}\delta t$$

where $\mathbf{E}\,\delta t$ is the fraction of the exit stream, which has stayed in the reactor for times between t and $t + \delta t$ and in which $(1 - \alpha)$ is still unconverted, i.e.,

$$(1 - \bar{\alpha}) = \int_0^{t_f} (1 - \alpha)\frac{1}{\tau}e^{-t/\tau}dt$$

where t_f is taken as the upper limit rather than ∞ because even if the particle stays in the reactor longer than the time for complete reaction, the conversion cannot exceed 100% and for such particles $(1 - \alpha)$ is zero. Substituting for $(1 - \alpha)$,

$$(1 - \bar{\alpha}) = \int_0^{t_f} \left(1 - \frac{t}{t_f}\right)^3 \frac{1}{\tau}e^{-t/\tau}dt$$

Using repeated integration by parts,

$$(1 - \bar{\alpha}) = 1 - 3\left(\frac{\tau}{t_f}\right) + 6\left(\frac{\tau}{t_f}\right)^2 - 6\left(\frac{\tau}{t_f}\right)^3\left(1 - e^{-t_f/\tau}\right)$$

If t_f/τ is small, it is useful to expand the exponential when several of the above terms cancel,

$$(1 - \bar{\alpha}) = \frac{1}{4}\left(\frac{t_f}{\tau}\right) - \frac{1}{20}\left(\frac{t_f}{\tau}\right)^2 + \frac{1}{120}\left(\frac{t_f}{\tau}\right)^3 - \cdots$$

In the present problem $t_f/\tau = 16/100$. That is,

$$(1 - \bar{\alpha}) = 0.04 - 0.0013 + \cdots = 0.039$$

Hence the average conversion for the product $\bar{\alpha} = 0.961$.
(3) To reduce the amount of unreacted material, the following modifications could be suggested:

a. reduce the solids feed rate,
b. increase the solids holdup in the bed,
c. construct a second fluidized bed in series, and
d. partition the proposed fluidized bed.

Nomenclature

		Units in SI System	Dimension in M, N, L, T, θ
A	Frequency factor in Arrhenius equation	s^{-1}*	T^{-1}*
A_c	Cross-sectional area	m^2	L^2
a	Constant in Eq. (3.123)	—	—
a	Interfacial area per unit volume	m^{-1}	L^{-1}
a_B	Activity of component **B**	—	—
$\mathbf{a_c}$	Activity of a catalyst	—	—
a'	Surface area for heat transfer per unit hydraulic radius	m	L
B	$(=\frac{-\Delta H C_{A\infty} D_e \mathbf{R}}{2rh\mathbf{E}})$ Dimensionless parameter introduced by McGreavy and Thornton (see Fig. 3.7 and 3.8)	—	—
b	Constant in Eq. (3.123)	—	—
b	Stoichiometric ratio	—	—
C	Outlet response of concentration divided by area under $C-t$ curve	s^{-1}	T^{-1}
C	Total molar concentration	$kmol/m^3$	NL^{-3}
C_A	Molar concentration of component **A**	$kmol/m^3$	NL^{-3}
$\mathbf{C_A}$	Surface concentration of **A**		L^{-2}
C_{Ai}	Molar concentration at interface	$kmol/m^3$	NL^{-3}
$\mathbf{C_B}$	Surface concentration of **B**		L^{-2}
C_M	Molar concentration at center of pellet	$kmol/m^3$	NL^{-3}
$\mathbf{C_P}$	Surface concentration of **P**		L^{-2}
$\mathbf{C_Q}$	Surface concentration of **Q**		L^{-2}
$\mathbf{C_s}$	Concentration of available sites		L^{-2}
$\mathbf{C_V}$	Concentration of vacant sites		L^{-2}
$C_{A\infty}$	Molar concentration at infinity	$kmol/m^3$	NL^{-3}
$\bar{c}_p$	Mean heat capacity of fluid per unit mass	J/kgK	$L^2T^{-2}\theta^{-1}$
$\bar{c}_{pc}$	Mean heat capacity of coolant per unit mass	J/kgK	$L^2T^{-2}\theta^{-1}$
c'	Number of tracer particles per unit mass of all particles	kg^{-1}	M^{-1}
c'_0	Initial value of c'	kg^{-1}	M^{-1}
D	Molecular bulk diffusion coefficient	m^2/s	L^2T^{-1}
D_e	Effective diffusivity	m^2/s	L^2T^{-1}
D_{eK}	Effective diffusivity in Knudsen regime	m^2/s	L^2T^{-1}
D_{eM}	Effective diffusivity in molecular regime	m^2/s	L^2T^{-1}

		Units in SI System	Dimension in M, N, L, T, θ
D_K	Knudsen diffusion coefficient	m^2/s	L^2T^{-1}
D_p	Diffusion coefficient for forced flow	m^2/s	L^2T^{-1}
D'_e	Effective diffusivity based on concentration expressed as Y	kg/ms	$ML^{-1}T^{-1}$
D'_L	Dispersion coefficient in longitudinal direction based on concentration expressed as Y	kg/ms	$ML^{-1}T^{-1}$
D'_r	Radial dispersion coefficient based on concentration expressed as Y	kg/ms	$ML^{-1}T^{-1}$
d	Tube diameter	m	L
d_p	Particle diameter	m	L
d'_M	Volume of voids per total surface area	m	L
E	Exit age distribution	s^{-1}	T^{-1}
E	Activation energy per mole for chemical reaction	J/kmol	$MN^{-1}L^2T^{-2}$
E_D	Apparent activation energy per mole in diffusion controlled region	J/kmol	$MN^{-1}L^2T^{-2}$
e	Void fraction	—	—
F	C/C_∞	—	—
F	Ratio of activity in poisoned catalyst pellet to activity in unpoisoned pellet	—	—
G	Mass flow rate	kg/s	MT^{-1}
G_0	Mass outflow ratio in steady state	kg/s	MT^{-1}
G'	Mass flow per unit area	kg/m^2s	$ML^{-2}T^{-1}$
G'_C	Mass flow rate of coolant per unit cross section of reactor tube	kg/m^2s	$ML^{-2}T^{-1}$
ΔH	Difference of enthalpy between products and reactants (heat of reaction $= -\Delta H$)	J/kmol	$MN^{-1}L^2T^{-2}$
h	Heat transfer coefficient	W/m^2K	$MT^{-3}\theta^{-1}$
j	Mass transfer factor	—	—
K	Thermodynamic equilibrium constant—for $A + B \leftrightarrow P$	ms^2/kg	$M^{-1}LT^2$
	for $A \leftrightarrow P$	—	—
K_A	Adsorption—desorption equilibrium constant for **A**	ms^2/kg	$M^{-1}LT^2$
K_B	Adsorption—desorption equilibrium constant for **B**	ms^2/kg	$M^{-1}LT^2$
K_p	Adsorption—desorption equilibrium constant for **P**	ms^2/kg	$M^{-1}LT^2$

Continued

		Units in SI System	Dimension in M, N, L, T, θ
K_s	Equilibrium constant for surface reaction	—	—
k	Chemical rate constant for forward reaction	$s^{-1}*$	$T^{-1}*$
k'	Chemical rate constant for reverse reaction, per unit volume of particle or reactors	$s^{-1}*$	$T^{-1}*$
$\overline{k}$	Overall rate *or* reaction constant per unit volume	$s^{-1}*$	$T^{-1}*$
k	Chemical rate constant for forward surface reaction (molecules per unit area and unit time)	$m^{-2}s^{-1}$	$L^{-2}T^{-1}$
k'	Chemical rate constant for reverse surface reaction (molecules per unit area and unit time)	$m^{-2}s^{-1}$	$L^{-2}T^{-1}$
k_a	First-order rate constant of a reaction (expressed in terms of surface area of a catalyst)	$m\ s^{-1}$	$L\ T^{-1}$
k_{Aa}	Rate constant for adsorption of **A** (see Eq. 3.67)	$s/kg\ m$	$M^{-1}L^{-1}T$
k_{Ad}	Rate constant for desorption of **A** (see Eq. 3.67)	$m^{-2}s^{-1}$	$L^{-2}T^{-1}$
k_{Ba}	Rate constant for adsorption of **B** (see Eq. 3.69)	$s/kg\ m$	$M^{-1}L^{-1}T$
k_{Bd}	Rate constant for desorption of **B** (see Eq. 3.69)	$m^{-2}s^{-1}$	$L^{-2}T^{-1}$
k_d	Rate constant signifying catalyst deactivation	s^{-1}	T^{-1}
k_e	Effective thermal conductivity	W/mK	$MLT^{-3}\theta^{-1}$
k_G	Gas film mass transfer coefficient	m/s	$L\ T^{-1}$
k_l	Thermal conductivity in longitudinal direction	W/mK	$MLT^{-3}\theta^{-1}$
k_{pa}	Rate constant for adsorption of **P** (see Eq. 3.70) (molecules per unit area, unit time, and unit partial pressure of **P**)	s/kgm	$M^{-1}L^{-1}T$
k_{Pd}	Rate constant for desorption of **P** (see Eq. 3.70) (molecules per unit area and unit time)	$m^{-2}s^{-1}$	$L^{-2}T^{-1}$
k_r	Thermal conductivity in radial direction	W/mK	$MLT^{-3}\theta^{-1}$
k_s	Chemical rate constant for forward surface reaction (molecules per unit time)	s^{-1}	T^{-1}

		Units in SI System	Dimension in M, N, L, T, θ
k_s'	Chemical rate constant for reverse surface reaction (molecules per unit time)	s^{-1}	T^{-1}
k_s	Rate constant in Eq. (3.124)	m/s	LT^{-1}
k_e'	Rate constant for reverse surface reaction (molecules per unit time)	s^{-1}	T^{-1}
L	Length of reactor	m	L
M	Molecular weight	kg/kmol	MN^{-1}
N	Avogadro's number (6.023×10^{26} molecules per kmol or 6.023×10^{23} molecules per mole)	$kmol^{-1}$	N^{-1}
N'	Molar flux of material	$kmol/ms^2$	$NL^{-2}T^{-1}$
n	Order of chemical reaction	—	—
n_A	Number of moles of **A**	kmol	N
n_B	Number of moles of **B**	kmol	N
$\boldsymbol{n}_{M0}$	Number of tracer particles injected	—	—
$\boldsymbol{n}_A$	Number of adsorbed molecules of component **A**	—	—
$\boldsymbol{n}_D$	Apparent order of chemical reaction in diffusion controlled regime	—	—
$\boldsymbol{n}_P$	Number of adsorbed molecules of component **P**	—	—
$\boldsymbol{n}_S$	Number of available active sites	—	—
P	Pressure	N/m^2	$ML^{-1}T^{-2}$
P_A	Partial pressure of component **A**	N/m^2	$ML^{-1}T^{-2}$
P_{Ai}	Partial pressure of component **A** at interface	N/m^2	$ML^{-1}T^{-2}$
P_B	Partial pressure of component **B**	N/m^2	$ML^{-1}T^{-2}$
P_P	Partial pressure of component **P**	N/m^2	$ML^{-1}T^{-2}$
P_Q	Partial pressure of component **Q**	N/m^2	$ML^{-1}T^{-2}$
Q	Heat transferred per unit volume and unit time to surroundings	W/m^3	$ML^{-1}T^{-3}$
Q_1^M	Heat generated by chemical reaction per unit mass	J/kg	L^2T^{-2}
Q_2^M	Heat gain by feed to the reactor per unit mass	J/kg	L^2T^{-2}
R	Gas constant	J/kmolK	$MN^{-1}L^2T^{-2}\theta^{-1}$

Continued

		Units in SI System	Dimension in M, N, L, T, θ
R_1	Component of drag force per unit area in direction of flow	N/m^2	$\text{ML}^{-1}\text{T}^{-2}$
r^M	Reaction rate (moles per unit mass of catalyst and unit time)	kmol/kg s	$\text{NM}^{-1}\text{T}^{-1}$
r_{YT}^M	Reaction rate (moles per unit mass of catalyst and unit time) when a function of Y and T	kmol/kg s	$\text{NM}^{-1}\text{T}^{-1}$
r'_v	Reaction rate (moles per unit volume of reactor and unit time)	kmol/m^3 s	$\text{NL}^{-3}\text{T}^{-1}$
r_d	Rate of deactivation of catalyst	s^{-1}	T^{-1}
r'_v	Reaction rate (moles per unit volume of particle and unit time)	kmol/m^3 s	$\text{NL}^{-3}\text{T}^{-1}$
r''	Overall rate of surface catalyzed reaction (molecules per unit area and unit time)	m^{-2} s^{-1}	$\text{L}^{-2}\text{T}^{-1}$
r''_{Aa}	Rate of adsorption of **A** (molecules per unit area and unit time)	m^{-2}s^{-1}	$\text{L}^{-2}\text{T}^{-1}$
r''_{Ba}	Rate of adsorption of **B** (molecules per unit area and unit time)	m^{-2}s^{-1}	$\text{L}^{-2}\text{T}^{-1}$
r''_{Pd}	Rate of desorption of **P** (molecules per unit area and unit time)	m^{-2}s^{-1}	$\text{L}^{-2}\text{T}^{-1}$
r''_S	Rate of surface reaction (molecules per unit area and unit time)	m^{-2}s^{-1}	$\text{L}^{-2}\text{T}^{-1}$
r	Radius	m	L
r_c	Radial position in core at which reaction is taking place	m	L
r_0	Initial value of r_c or radius of sphere or cylinder	m	L
r_e	Radius of cylinder with same surface to volume ratio as pore	m	L
S	Catalyst selectivity	—	—
S_g	Specific surface area (per unit mass)	m^2/kg	M^{-1}L^2
S_x	External surface area	m^2	L^2
S'	Total catalyst surface area	m^2	L^2
T	Temperature	K	θ
T_C	Temperature of coolant stream	K	θ
t_{cL}	Temperature of coolant at reactor outlet	K	θ
T_{C0}	Temperature of coolant at reactor inlet	K	θ

		Units in SI System	Dimension in M, N, L, T, θ
T_i	Temperature at exterior surface of pellet	K	θ
T_L	Temperature at exit of reactor	K	θ
T_M	Temperature at particle center	K	θ
T_{MX}	Maximum temperature	K	θ
T_W	Wall temperature	K	θ
ΔT	$(= T_0 - T_{CL})$ Temperature difference of the nonreacting fluid over length of reactor	K	θ
t	Time	s	T
t_f	Time for complete conversion	s	T
U	Overall heat transfer coefficient	W/m²K	$MT^{-3}\theta^{-1}$
u	Gas velocity	m/s	LT^{-1}
u_1	Mean velocity in voids	m/s	LT^{-1}
V_g	Specific pore volume per unit mass	m³/kg	$M^{-1}L^3$
V_p	Particle volume	m³	L^3
v	Volumetric rate of flow	m³/s	L^3T^{-1}
W	Mass of catalyst	kg	M
W_F	Mass of particles in reactor (holdup)	kg	M
x	Distance in x-direction	m	L
x_A	Mole fraction of A	—	—
Y	Moles of reactant per unit mass of fluid	kmol/kg	NM^{-1}
Y_L	Moles per unit mass of reactant (or product) at reactor exit	kmol/kg	NM^{-1}
Y_{MX}	Moles per unit mass of reactant (or product) for which temperature is maximum	kmol/kg	NM^{-1}
Y_0	Moles per unit mass of reactant (or product) at reactor inlet	kmol/kg	NM^{-1}
z	Length variable for reactor tube	m	L
Pe	$(= ud_p/D)$ Peclet number	—	—
Re	$(= u\rho d_p/\mu)$ Reynolds number	—	—
Sc	$(= \mu/\rho D)$ Schmidt number	—	—
Sh	$(= k_G d_p/D)$ Sherwood number	—	—
α	Fractional conversion	—	—
$\bar{\alpha}$	Average value of α	—	—
β	Parameter representing maximum temperature difference in particle relative to external surface (Eq. 3.28)	—	—

Continued

		Units in SI System	Dimension in M, N, L, T, θ
Γ	$\left(= G'\bar{c}_p/G'_C\bar{c}_{pC}\right)$ Relative heat capacities of two fluid streams to exchange heat	—	—
Γ'	$\left(= UaL/G'_C\bar{c}_{pC}\right)$ Heat transfer factor	—	—
ε	Particle porosity	—	—
ε'	Dimensionless activation energy (Eq. 3.29)	—	—
ζ	Fraction of surface poisoned	—	—
η	Effectiveness factor		
θ_A	Fraction of surface occupied by component **A**	—	—
λ	Defined as $\sqrt{\frac{k}{D_e}}$ for first-order process (Eq. 3.15)	m^{-1}	L^{-1}
μ	Viscosity	Ns/m^2	$ML^{-1}T^{-1}$
ρ	Fluid density	kg/m^3	ML^{-3}
ρ_b	Bulk density	kg/m^3	ML^{-3}
ρ_p	Particle density	kg/m^3	ML^{-3}
ρ_s	True density of solid	kg/m^3	ML^{-3}
τ	Residence time	s	T
τ'	Tortuosity factor	—	—
ϕ	Thiele modulus (Eq. 3.19)	—	—
$\bar{\phi}$	Generalized Thiele modulus (Eq. 3.20)	—	—
ϕ'	Modified Thiele modulus (Eq. 3.35)	—	—

Subscripts

A	Refers to component **A**
B	Refers to component **B**
G	Refers to bulk value in gas phase
P	Refers to product **P**
X	Refers to component **X**
∞	Value as time approaches infinity
in	Inlet value

*For first-order reaction. Dimensions are a function of order of reaction.

References

1. Satterfield CN. *Heterogeneous catalysis in industrial practice*. 2nd ed. McGraw-Hill; 1991.
2. Barrer RM, Grove DM. Flow of gases and vapours in a porous medium and its bearing on adsorption problems: I. Steady state of flow, II. Transient flow. *Trans Faraday Soc* 1951;**47**:826–37.
3. Gunn DJ, Pryce C. Dispersion in packed beds. *Trans Inst Chem Eng* 1969;**47**:T341.
4. Macdonald WR, Habgood HW. Paper 45. Measurement of diffusivities in zeolites by gas chromatographic methods. In: *19th Canadian Chemical Engineering Conference, Edmonton*; 1969.

5. Hirshfelder JO, Curtiss CF, Bird RB. *Molecular theory of gases and liquids.* Wiley; 1954.
6. Gilliland ER. Diffusion coefficients in gaseous systems. *Ind Eng Chem* 1934;**26**:681.
7. Eerkens JW, Grossman LM. Evaluation of the diffusion equation and tabulation of experimental diffusion coefficients. In: *Tech. Report HE/150/150 (Univ. Calif. Inst. Eng. Res Dec. 5th)*; 1957.
8. Knudsen M. Die Gesetze der Molekularströmung und der inneren Reibungströmung der Gase durch Röhren. *Ann Phys* 1909;**28**:75 (Molekularströmung des Wasserstoffs durch Röhren und das Hitzdrahtmanometer. *Ann Phys* 1911;**35**:389).
9. Herzfeld K, Smallwood M. In: Taylor HS, editor. *Treatise on physical chemistry*, vol. 1. Princeton Univ. Press; 1931. p. 169.
10. Everett DH. The structure and properties of porous materials: some problems in the investigation of porosity by adsorption methods. *Colston Res Symp (Bristol)* 1958.
11. Scott DS, Dullien FAL. Diffusion of ideal gases in capillaries and porous solids. *AIChE J* 1962;**8**:113.
12. Bosanquet CH. The optimum pressure for a diffusion separation plant. In: *British TA report BR/507 (Sept. 27th)*; 1944.
13. Pollard WG, Present RD. Gaseous self-diffusion in long capillary tubes. *Phys Rev* 1948;**73**:762.
14. Thiele EW. Relation between catalytic activity and size of particle. *Ind Eng Chem* 1939;**31**:916.
15. Mickley HS, Sherwood TK, Reed CE. *Applied mathematics in chemical engineering.* 2nd ed. McGraw-Hill; 1957.
16. Gunn DJ. Diffusion and chemical reaction in catalysis. *Chem Eng Sci* 1967;**22**:1439.
17. Aris R. *Introduction to the analysis of chemical reactors*, vol. 236. Prentice-Hall; 1965.
18. Carberry JJ. The catalytic effectiveness factor under nonisothermal conditions. *AIChE J* 1961;**7**:350.
19. Tinkler JD, Metzner AB. Reaction rate in nonisothermal catalysts. *Ind Eng Chem* 1961;**53**:663.
20. Petersen EE. Nonisothermal chemical reaction in porous catalysts. *Chem Eng Sci* 1962;**17**:987.
21. Weisz PB, Hicks JS. The behaviour of porous catalyst particles in view of internal mass and heat diffusion effects. *Chem Eng Sci* 1962;**17**:265.
22. Gunn DJ. Nonisothermal reaction in catalyst particles. *Chem Eng Sci* 1966;**21**:383.
23. McGreavy C, Thornton JM. Generalized criteria for the stability of catalytic reactors. *Can J Chem Eng* 1970;**48**:187.
24. Weisz PB, Prater CD. Interpretation of measurements in experimental catalysis. *Adv Catal* 1954;**6**:143.
25. Petersen EE. A general criterion for diffusion influenced chemical reactions in porous solids. *Chem Eng Sci* 1965;**20**:587.
26. Wheeler A. Reaction rates and selectivity in catalyst pores. *Adv Catal* 1951;**3**:249.
27. Østergaard K. The influence of intraparticle heat and mass diffusion on the selectivity of parallel heterogeneous catalytic reactions. In: *Proc. 3rd Int. Cong. Catalysis, Amsterdam*, vol. 2; 1964. p. 1348.
28. Sinfelt JH. Bifunctional catalysis. In: Drew TB, Hoopes JW, Vermeulen T, editors. *Advances in chemical engineering*, vol. 5. Academic Press; 1964. p. 37.
29. Gunn DJ, Thomas WJ. Mass transport and chemical reaction in multifunctional catalyst systems. *Chem Eng Sci* 1965;**20**:89.
30. Thomas WJ, Wood RM. Use of maximum principle to calculate optimum catalyst composition profile for bifunctional catalyst systems contained in tubular reactors. *Chem Eng Sci* 1967;**22**:1607.
31. Gunn DJ. The optimisation of bifunctional catalyst systems. *Chem Eng Sci* 1967;**22**:963.
32. Jackson R. Optimal use of mixed catalysts. *J Optim Theor Appl* 1968;**2**:1.
33. Jenkins B, Thomas WJ. Optimum catalyst formulation for the aromatization of methylcyclopentane. *Can J Chem Eng* 1970;**48**:179.
34. Horn F. Optimum temperature regulation for continuous chemical processes. *Z Electrochem* 1961;**65**:209.
35. Chou A, Ray WH, Aris R. Simple control policies for reactors with catalyst decay. *Trans Inst Chem Eng* 1967;**45**:T153.
36. Jackson R. An approach to the numerical solution of time-dependent optimisation problems in two-phase contacting devices. *Trans Inst Chem Eng* 1967;**45**:T160.

37. Ogunye AF, Ray WH. Non-simple control policies for reactors with catalyst decay. *Trans Inst Chem Eng* 1968;**46**:T225.

38. Thaller L, Thodos G. The dual nature of a catalytic reaction: the dehydration of sec-butyl alcohol to methyl ethyl ketone at elevated pressures. *AIChE J* 1960;**6**:369.

39. Fogler HS. *Elements of chemical reaction engineering*. 3rd ed. Prentice-Hall; 1999.

40. Froment GF, Bischoff KB, de wilde J. *Chemical reactor analysis and design*. 3rd ed. John Wiley and Sons, Inc; 2011.

41. Tamaru K. Adsorption during the catalytic decomposition of formic acid on silver and nickel catalysts. *Trans Faraday Soc* 1959;**55**:824.

42. Helfferich F. Kinetics of multistep reactions. In: *Comprehensive Chemical Kinetics*, vol. 40. Elsevier Science; 2004.

43. Hougen OA, Watson KM. *Chemical process principles, Part 3*. Wiley; 1947. p. 938.

44. Thomas WJ, John B. Kinetics and catalysis of the reactions between sulphur and hydrocarbons. *Trans Inst Chem Eng* 1967;**45**:T119.

45. Mace CV, Bonilla CF. The conversion of ethylene to ethanol. *Chem Eng Prog* 1954;**50**:385.

46. Rase HF. *Fixed-bed reactor design and diagnostics*. Butterworths; 1990.

47. Wenner RR, Dybdal EC. Catalytic dehydrogenation of ethylbenzene. *Chem Eng Prog* 1948;**44**:275.

48. Bartholomew CH. Mechanisms of catalyst deactivation. *Appl Catal A Gen* 2001;**212**:17.

49. Ergun S. Fluid flow through packed columns. *Chem Eng Prog* February 1952;**48**(2):89.

50. Denbigh KG. Velocity and yield in continuous reaction systems. *Trans Faraday Soc* 1944;**40**:352.

51. Denbigh KG. Optimum temperature sequences in reactors. *Chem Eng Sci* 1958;**8**:125.

52. Petersen EE. *Chemical reaction analysis*, vol. 186. Prentice-Hall; 1965.

53. van Heerden C. Autothermic processes—properties and reactor design. *Ind Eng Chem* 1953;**45**:1242.

54. Wehner JF, Wilhelm RH. Boundary conditions of flow reactor. *Chem Eng Sci* 1956;**6**:89.

55. McHenry KW, Wilhelm RH. Axial mixing of binary gas mixtures flowing in a random bed of spheres. *AIChE J* 1957;**3**:83.

56. Aris R, Amundson NR. Longitudinal mixing or diffusion in fixed beds. *AIChE J* 1957;**3**:280.

57. Kramers H, Alberda G. Frequency-response analysis of continuous-flow systems. *Chem Eng Sci* 1953;**2**:173.

58. Boussinesq MJ. Essai sur la théorie des eaux courantes. *Mém Prés Div Sav Acad Sci Inst Fr* 1877;**23**:1−680 (see 31).

59. Deans HA, Lapidus L. A computational model for predicting and correlating the behavior of fixed-bed reactors: I. Derivation of model for nonreactive systems, II. Extension to chemically reactive systems. *AIChE J* 1960;**6**:656−63.

60. Beek J. Design of packed catalytic reactors. In: Drew TB, Hoopes JW, Vermeulen T, editors. *Advances in chemical engineering*, vol. 3. Academic Press; 1962. p. 203.

61. Bilous O, Amundson NR. Chemical reactor stability and sensitivity II. Effect of parameters on sensitivity of empty tubular reactors. *AIChE J* 1956;**2**:117.

62. May WG. Fluidized-bed reactor studies. *Chem Eng Prog* December 1959;**55**(12):49.

63. Rowe PN. Gas−solid reaction in a fluidized bed. *Chem Eng Prog* March 1964;**60**(3):75.

64. Wen CY. Noncatalytic heterogeneous solid fluid reaction models. *Ind Eng Chem* 1968;**60**(9):34.

65. Szekely J, Evans JW, Sohn HY. *Gas-solid reactions*. Academic Press; 1976.

66. Levenspiel O. *Chemical reaction engineering*. 3rd ed. John Wiley and Sons; 2003.

67. Denbigh KG, Beveridge GSG. The oxidation of zinc sulphide spheres in an air stream. *Trans Inst Chem Eng* 1962;**40**:23.

68. Ranzi E, Pierucci S, Aliprandi PC, Stringa S. Comprehensive and detailed kinetic model of a travelling grate combustor of biomass. *Energy fuels* 2011;**25**:4195.

69. Schmidt LD. *The engineering of chemical reactions*. 2nd ed. Oxford University Press; 2005.

Further Reading

Aris R. *Elementary chemical reactor analysis*. Prentice-Hall; 1969.

Aris R. *The mathematical theory of diffusion and reaction in permeable catalysts*, vols. I and II. Oxford University Press; 1975.

Bond JC. *Heterogeneous catalysis: principles and applications*. 2nd ed. Oxford Science Publications; 1987.

Campbell IM. *Catalysis at surfaces*. Chapman and Hall; 1988.

Carberry JJ. *Chemical and catalytic reaction engineering*. McGraw-Hill; 1976.

Doraiswamy LK, Sharma MM. Heterogeneous reactions. In: *Gas—solid and solid—solid reactions*, vol. 1. Wiley; 1984.

Gates BC. *Catalytic chemistry*. Wiley; 1992.

Lapidus L, Amundson NR. *Chemical reactor theory—a review*. Prentice-Hall; 1977.

Petersen EE. *Chemical reaction analysis*. Prentice-Hall; 1965.

Rase HF. *Chemical reactor design for process plants*, vols. 1 and 2. Wiley; 1977.

Richardson JT. *Principles of catalyst development*. Plenum Press; 1989.

Satterfield CN. *Mass transfer in heterogeneous catalysis*. M.I.T. Press; 1970.

Schmidt LD. *The engineering of chemical reactions*. 2nd ed. Oxford University Press; 2005.

Smith JM. *Chemical engineering kinetics*. 3rd ed. McGraw-Hill; 1981.

Szekely J, Evans JW, Sohn HY. *Gas-solid reactions*. Academic Press; 1976.

Thomas JM, Thomas WJ. *Introduction to the principles of heterogeneous catalysis*. Academic Press; 1967.

Twigg MV, editor. *Catalyst handbook*. 2nd ed. Wolfe Publishing; 1979.

CHAPTER 4

Gas–Liquid and Gas–Liquid–Solid Reactors

Ravikrishnan Vinu

Indian Institute of Technology Madras, Chennai, India

Learning Outcomes

- Develop expressions for rate and concentration profile for mass transfer with chemical reaction according to film theory

- Evaluate the importance of nondimensional numbers, viz., Hatta number and liquid enhancement factor, in assessing the intrinsic speed of reaction and reaction regime around the film

- Describe the salient aspects of and design principles involved in gas–liquid reactors such as packed column, agitated tank, and bubble column reactors with example cases

- Describe the construction of laboratory reactors such as laminar jet contactor, laminar film contactor, stirred cell reactor, and rapid thermal flow reactor to evaluate reaction kinetics

- Describe the design principles involved in modeling gas–liquid–solid reactors of suspended bed and fixed bed types by combining mass transfer and reaction steps

- Develop model to assess the effect of nonideality in flow pattern (e.g., axial dispersion) in trickle bed gas–liquid–solid reactor

---

4.1 Gas–Liquid Reactors

4.1.1 Gas–Liquid Reactions

In a number of important industrial processes, it is necessary to carry out a reaction between a gas and a liquid. Usually the objective is to make a particular product, for example, a chlorinated hydrocarbon such as chlorobenzene by the reaction of gaseous chlorine with liquid benzene. Sometimes the liquid is simply the reaction medium, perhaps containing a catalyst, and all the reactants and products are gaseous. In other cases the main aim is to separate a constituent such as CO_2 from a gas mixture; although pure water could be used to remove CO_2, a solution of caustic soda, potassium carbonate, or ethanolamines has the advantages of increasing both the absorption capacity of the liquid and the rate of absorption. Typical reactions involved in CO_2 absorption are given by

$$CO_2 + 2NaOH \rightarrow Na_2CO_3 + H_2O$$
$$CO_2 + 2RNH_2 \rightarrow RNHCOO^- + RNH_3^+$$

The subject of gas–liquid reactor design thus really includes absorption with chemical reaction, which is discussed in Volume 2, Chapter 12.

4.1.2 Types of Reactors

The types of equipment used for gas–liquid reactions are shown in Fig. 4.1. Fig. 4.1A shows a conventional packed column, which is often used when the purpose is to absorb a

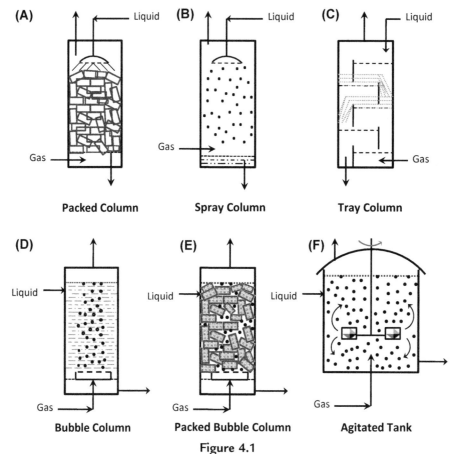

Figure 4.1
Equipment used for gas–liquid reactions: in (A), (B), and (C) liquid holdup is low; in (D), (E), and (F) liquid holdup is high.

constituent from a gas. The liquid is distributed in the form of films over the packing, and the gas forms the continuous phase. The pressure drop for the gas is relatively low, and the packed column is therefore very suitable for treating large volume flows of gas mixtures. Fig. 4.1B shows a spray column; as in the packed column, the liquid holdup is comparatively low, and the gas is the continuous phase. In the sieve tray column shown in Fig. 4.1C, however, and in bubble columns, gas is dispersed in the liquid passing over the tray. Because the tray is relatively shallow, the pressure drop in the gas phase is fairly low, and the liquid holdup, although a little larger than for a packed column, is still relatively small. The tray column is useful when stagewise operation is required and a relatively large volumetric flow rate of gas is to be treated.

When a high liquid holdup is required in the reactor, one of the types shown in Fig. 4.1D—F may be used. The bubble column (Fig. 4.1D) is simply a vessel filled with liquid, with a sparger ring at the base for dispersing the gas. In some cases a draught tube is used to direct recirculation of the liquid and to influence the bubble motion. One of the disadvantages of the simple bubble column is that coalescence of the bubbles tends to occur with the formation of large slugs whose upper surfaces are in the form of spherical caps. By packing the vessel with Raschig rings, for example, as in Fig. 4.1E, the formation of very large bubbles is avoided. The reactor thus becomes an ordinary packed column operated in a flooded condition and with a sparger to disperse the gas; naturally the maximum superficial gas velocity is much less than in an unflooded packed column. Finally in Fig. 4.1F an agitator is used to disperse the gas into the liquid contained in a tank. A vaned-disc type of impeller is normally used. An agitated tank provides small bubbles and thus a high interfacial area of contact between gas and liquid phases, but its greater mechanical complication compared with a simple bubble column is a disadvantage with corrosive materials and at high pressures and temperatures. If necessary, stagewise operation can be achieved by having several compartments in a vertical column, with impellers mounted on a common shaft (see Section 4.1.7).

4.1.3 Rate Equations for Mass Transfer With Chemical Reaction

In designing a gas—liquid reactor, there is the need not only to provide the required temperature and pressure for the reaction, but also to ensure adequate interfacial area of contact between the two phases. Although the reactor as such is classed as heterogeneous, the chemical reaction itself is really homogeneous, occurring either in the liquid phase, which is the most common, or in the gas phase, and only rarely in both. Many different varieties of gas—liquid reactions have been considered theoretically, and a large number of analytical or numerical results can be found in the literature.[1,2]

The absorption of a gas by a liquid with simultaneous reaction in the liquid phase is the most important case. There are several theories of mass transfer between two fluid phases (see Volume 1, Chapter 10; Volume 2, Chapter 12), but for the purpose of illustration the film theory will be used here. Results from the possibly more realistic penetration theory are similar numerically, although more complicated in their mathematical form.[3,4]

Consider a *second-order* reaction in the liquid phase between a substance **A**, which is transferred from the gas phase, and reactant **B**, which is in the liquid phase only. The gas will be taken as consisting of pure **A** so that complications arising from gas film resistance are avoided. The stoichiometry of the reaction is represented by

$$m\mathbf{A}_{(g \to l)} + n\mathbf{B}_{(l)} \to \text{Products}$$

with the rate equation

$$-r_A = k_2 C_A C_B \tag{4.1}$$

Note that

$$r_B = \frac{\nu_B}{\nu_A} r_A$$

where ν_A and ν_B are stoichiometric coefficients in the reaction and are given by $\nu_A = -m$, $\nu_B = -n$.

In the film theory, steady-state conditions are assumed in the film such that, in any volume element, the difference between the rate of mass transfer into and out of the element is just balanced by the rate of reaction within the element. Carrying out such a material balance on reactant **A**, the following differential equation results:

$$D_A \frac{d^2 C_A}{dx^2} - r_A = 0 \tag{4.2}$$

Similarly a material balance on **B** gives

$$D_B \frac{d^2 C_B}{dx^2} - r_B = 0 \tag{4.3}$$

Introducing the rate Eq. (4.1), the differential equations become

$$D_A \frac{d^2 C_A}{dx^2} - k_2 C_A C_B = 0 \tag{4.4}$$

$$D_B \frac{d^2 C_B}{dx^2} - k_2 C_A C_B \frac{\nu_B}{\nu_A} = 0 \tag{4.5}$$

Typical conditions within the film may be seen in Fig. 4.2A. Boundary conditions at the gas−liquid interface arefor **A**,

$$C_A = C_{Ai}, \quad x = 0 \tag{4.6}$$

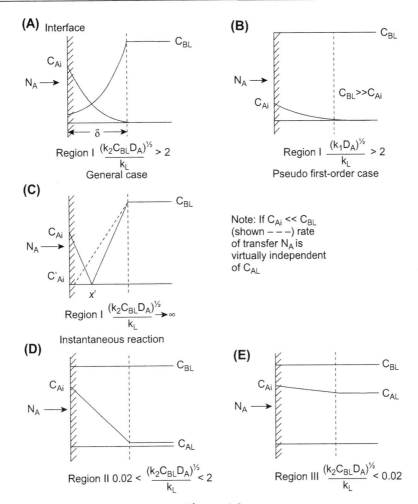

Figure 4.2

Liquid-phase concentration profiles for mass transfer with chemical reaction—film theory.

for **B**,

$$\frac{dC_B}{dx} = 0, \quad x = 0 \tag{4.7}$$

On the liquid side of the film where $x = \delta$, the boundary condition for **B** is simply

$$C_B = C_{BL}, \quad x = \delta \tag{4.8}$$

To obtain the boundary condition for **A**, we note that, except for the amount of **A** that reacts in the film, **A** is transferred across this boundary and reacts in the bulk of the liquid. For unit area of interface, the volume of this bulk liquid may be written as $[(\varepsilon_L/a) - \delta]$, where ε_L is the volume of liquid per unit volume of reactor space (i.e., the liquid holdup), a is the gas–liquid interfacial area per unit volume of reactor space (i.e., specific area), and δ is the thickness of the film. The boundary condition for **A** is thus:

$$-D_A\left(\frac{dC_A}{dx}\right)_{x=\delta} = k_2 C_{AL} C_{BL}\left(\frac{\varepsilon_L}{a} - \delta\right)$$

$$\underset{\text{across boundary}}{\underbrace{\text{Mass transfer}}} \qquad \underset{\text{bulk liquid}}{\underbrace{\text{Reaction in}}} \tag{4.9}$$

Sometimes, the net amount of A transported into the differential bulk volume by various mechanisms such as flow is also included in the right-hand side of the above boundary condition to account for the fact that the fluid element is not isolated from the surroundings.[5] Although a complete analytical solution of the set of Eqs. (4.4)–(4.9) is not possible, analytical solutions are obtainable for part of the range of variables[3,6] and a numerical solution has been obtained for the remainder.[7]

When the length scale is normalized to liquid-side film thickness, δ, the general solution to Eq. (4.4) can be written as follows:

$$C_A = K_1 \cosh\left(\beta\frac{x}{\delta}\right) + K_2 \sinh\left(\beta\frac{x}{\delta}\right) \tag{4.10}$$

β is given by

$$\beta = \sqrt{\frac{k_2 C_B \delta^2}{D_A}} = \sqrt{\frac{k_2 C_B D_A}{(D_A/\delta)^2}} = \frac{\sqrt{k_2 C_B D_A}}{k_L} \tag{4.11}$$

where k_L denotes the mass transfer coefficient for physical absorption of **A**, and the dimensionless parameter, β, called as Hatta number or Hatta modulus, is analogous to Thiele modulus used in reactions involving porous catalyst particles. The term $\beta^2 = \frac{k_2 C_B D_A}{k_L^2}$ signifies the ratio of timescale of diffusional transport to that of reaction in the liquid film.

The integration constants, K_1 and K_2, in Eq. (4.10) can be evaluated by using Eq. (4.6), and the simplified boundary condition that $C_A = C_{AL}$ at $x = \delta$. Importantly, the latter boundary condition needs to be used with caution as the concentration of **A** in the bulk liquid is usually determined using Eq. (4.9). This boundary condition will not be valid when the reaction is very fast and is completed within the liquid film (Refer Froment, Bischoff and De Wilde[5]).

The final solution is given by

$$C_A = \frac{C_{Ai} \sinh\left(\beta\left(1 - \frac{x}{\delta}\right)\right) + C_{AL} \sinh\left(\beta\frac{x}{\delta}\right)}{\sinh\beta} \tag{4.12}$$

The rate of the overall phenomena as seen from the interface or molar flux of A at the interface, $N_A|_{x=0} = -D_A \frac{dC_A}{dx}\big|_{x=0}$, is given by

$$N_A|_{x=0} = \frac{\beta}{\tanh \beta}\left(1 - \frac{C_{AL}}{C_{Ai}\cosh \beta}\right)k_L C_{Ai} = f_A k_L C_{Ai} \tag{4.13}$$

The term f_A signifies reaction factor or liquid enhancement factor. Comparison of the above equation with Eq. (4.14) that describes the molar flux of A when there is no reaction:

$$N_A = k_L(C_{Ai} - C_{AL}) \tag{4.14}$$

shows that when greater than 1, f_A represents the enhancement of the rate of transfer of **A** caused by the chemical reaction, as compared with pure physical absorption with a zero concentration of **A** in the bulk liquid.

The complete solution of the set of Eqs. (4.4)–(4.9) is shown in graphical form in Fig. 4.3, in which the reaction factor f_A is plotted against Hatta number. (The

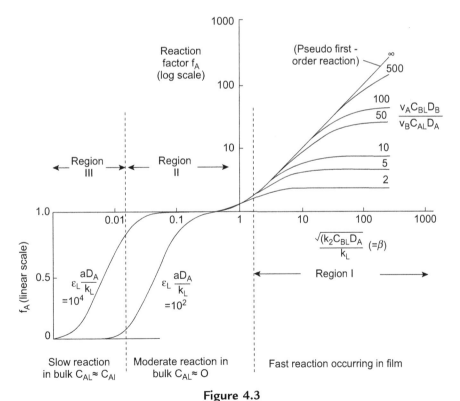

Figure 4.3

Reaction factor f_A for a second-order reaction (numerical solution) and pseudo–first-order reaction (analytical solution).

right-hand side of this diagram, region I, is the same as Fig. 12.12 of Volume 2.) The physical significance of the various regions covered by this diagram is important and is best appreciated by considering the corresponding concentration profiles of Fig. 4.2 as follows.

4.1.3.1 Rate of Transformation of A per Unit Volume of Reactor

The value of Hatta number, β, provides an important indication of whether a large specific interfacial area a or a large liquid holdup ε_L is required in a reactor to be designed for a particular reaction of rate constant k_2. Let the rate of transformation of **A** per unit volume of reactor be J_A. Three regions may be distinguished as shown in Fig. 4.3.

4.1.3.1.1 Region I: $\beta > 2$

The reaction is fast and occurs mainly in the film as **A** is being transported (Fig. 4.2A). C_{AL}, the concentration of **A** in the bulk of the liquid, is virtually zero. Thus

$$J_A = k_L C_{Ai} f_A a \tag{4.15}$$

Therefore, in this set of circumstances, J_A will be large if a is large; a large interfacial area a is required in the reactor, but the liquid holdup is not important. A packed column, for example, would be suitable.

If the concentration of **B** C_{BL} in the bulk liquid is much greater than C_{Ai}, a common case being where **A** reacts with a pure liquid **B**, a pseudo–first-order rate constant can be defined, $k_1 = k_2 C_{BL}$. Under these circumstances, the reaction factor becomes

$$f_A = \frac{\beta}{\tanh \beta}, \quad \text{where} \quad \beta = \frac{\sqrt{k_1 D_A}}{k_L} \tag{4.16}$$

The concentration profile is sketched in Fig. 4.2B. Note that for large β, $f_A \to \beta$ and hence

$$N_A = \sqrt{k_1 D_A} C_{Ai} \tag{4.17}$$

N_A is then independent of mass transfer coefficient k_L. Although C_{BL} can be considered as a constant across a horizontal slice, it needs not be the same over all heights of the equipment.

On the other hand, if the reaction between **A** and **B** is virtually instantaneous, as shown in Fig. 4.2C, the rate of transformation is determined mainly by the rate of mass transfer of **B** to the reaction zone. The reaction zone can be considered to be present at $x = x'$ in the vicinity of the interface. The concentration profiles of A and B are essentially linear, as their molar fluxes are constant, and are related by $\frac{N_A}{\nu_A} = -\frac{N_B}{\nu_B}$. The fluxes are given by

$$N_A = -\frac{D_A}{x'} C_{Ai} \tag{4.18}$$

$$-N_B = \frac{D_B}{\delta - x'} C_{BL} \tag{4.19}$$

By eliminating x', N_A can be written as

$$N_A = k_L C_{Ai} \left(1 + \frac{\nu_A D_B C_{BL}}{\nu_B D_A C_{Ai}} \right) \tag{4.20}$$

Using the above expression, the reaction factor can be expressed as follows:

$$f_A = \frac{k_L C_{Ai} \left(1 + \dfrac{\nu_A D_B C_{BL}}{\nu_B D_A C_{Ai}} \right)}{k_L (C_{Ai} - 0)} = 1 + \frac{\nu_A D_B C_{BL}}{\nu_B D_A C_{Ai}} \tag{4.21}$$

This shows that the reaction factor is independent of Hatta number, and this corresponds to a series of horizontal lines in region I in Fig. 4.3 for different values of $(\nu_A D_B C_{BL})/(\nu_B D_A C_{Ai})$. A classic example of an instantaneous reaction is the absorption of HCl in NaOH, which is essentially $H^+ + OH^- \rightarrow H_2O$. The curves connecting these horizontal lines with the straight line for pseudo–first-order reaction correspond to moderately fast second-order reactions.

4.1.3.1.2 Region II: $0.02 < \beta < 2$

This is an intermediate region in which the reaction is sufficiently fast to hold C_{AL} close to zero (Fig. 4.2D), but, although C_{AL} is small, nearly all the reaction occurs in the bulk of the liquid. The holdup ε_L is important because unless

$$\varepsilon_L \left/ \frac{aD_A}{k_L} \right. > 10^2$$

f_A will be substantially less than 1 in the part of the region (see Fig. 4.3) where when $f_A = 1$,

$$J_A = k_L C_{Ai} a \tag{4.22}$$

Thus, both interfacial area and holdup should be high. For example, an agitated tank will give a high value of J_A.

4.1.3.1.3 Region III: $\beta < 0.02$

The reaction is slow and occurs in the bulk of the liquid. Mass transfer serves to keep the bulk concentration C_{AL} of **A**, close to the saturation value C_{Ai} (Fig. 4.2E), and sufficient interfacial area should be provided for this purpose. However, a high liquid holdup is the more important requirement, and a bubble column is likely to be suitable. If $C_{AL} \approx C_{Ai}$, then

$$J_A = k_2 C_{Ai} C_{BL} \varepsilon_L \tag{4.23}$$

Near the boundary between the regions II and III, where virtually none of the reaction occurs in the film, the concentration C_{AL} of **A** in the bulk of the liquid is determined by the simple relation:

$$J_A \;=\; N_A a \;=\; \underset{\substack{\text{Mass transfer} \\ \text{through film}}}{k_L a(C_{Ai} - C_{AL})} \;=\; \underset{\substack{\text{Reaction in} \\ \text{bulk}}}{k_2 C_{AL} C_{BL} \varepsilon_L} \tag{4.24}$$

That is,

$$C_{AL} = C_{Ai}\left(1 + \frac{k_2 C_{BL}\varepsilon_L}{k_L a}\right)^{-1} \tag{4.25}$$

The above expressions contain the concentration of **A** at the interface C_{Ai}, which is experimentally difficult to determine. One way to eliminate C_{Ai} is to equate the fluxes of **A** through gas phase and liquid phase at steady state using Eq. (4.26). The partial pressure of **A** at the interface P_{Ai} can be related to C_{Ai} using Henry's law defined as $P_{Ai} = HC_{Ai}$ (see Volume 2, Chapter 12).

$$N_A = k_G(P_{AG} - P_{Ai}) = k_L(C_{Ai} - C_{AL}) \tag{4.26}$$

4.1.4 Choice of a Suitable Reactor

Choosing a suitable reactor for a gas–liquid reaction is a question of matching the characteristics of the reaction system, especially the reaction kinetics, with the characteristics of the reactors under consideration. As we have seen above, two of the most important characteristics of gas–liquid reactors are the specific interfacial area a and the liquid holdup ε_L. Table 4.1 shows some representative values of these quantities for various gas–liquid reactors.

Table 4.1: Comparison of specific interfacial area a and liquid holdup ε_L for various types of reactor.

Type of Contactor	Specific Area a m^2/m^3	Liquid Holdup ε_L (Fraction)
Spray column	60	0.05
Packed column (25 mm Raschig rings)	220	0.08
Plate column	150	0.15
Bubble contactor	200	0.85
Agitated tank	500	0.80

Example 4.1

You are asked to recommend, with reasons, the most suitable type of equipment for carrying out a gas–liquid reaction between a gas **A** and a solution of a reactant **B**. Particulars of the system are as follows:

Concentration of **B** in solution $= 5$ kmol/m^3
Diffusivity of **A** in the solution $= 1.5 \times 10^{-9}$ m^2/s
Rate constant of the reaction (**A** + **B** → Products; second order overall) $= 0.03$ m^3/kmol s.
For bubble dispersions (plate columns, bubble columns, agitated vessels), take k_L as having a range from 2×10^{-4} to 4×10^{-4} m/s. For a packed column, take k_L as having a range 0.5×10^{-4} to 1.0×10^{-4} m/s.

Solution

We first calculate the value of the Hatta number:

$$\beta = \frac{\sqrt{k_2 C_{BL} D_A}}{k_L}$$

From the data above,

$$\sqrt{k_2 C_{BL} D_A} = \sqrt{0.03 \times 5 \times 1.5 \times 10^{-9}} = 1.5 \times 10^{-5} \text{ m/s}$$

For the bubble dispersions,

$$\frac{\sqrt{k_2 C_{BL} D_A}}{k_L}$$

will therefore lie in the range

$$\frac{1.5 \times 10^{-5}}{2 \times 10^{-4}} = 0.075 \text{ to } \frac{1.5 \times 10^{-5}}{4 \times 10^{-4}} = 0.038$$

For a packed column, β will have the range

$$\frac{1.5 \times 10^{-5}}{0.5 \times 10^{-4}} = 0.03 \text{ to } \frac{1.5 \times 10^{-5}}{1.0 \times 10^{-4}} = 0.15$$

Referring to Fig. 4.3 these values lie in region II indicating that the reaction is only moderately fast and that a relatively high liquid holdup is required. A packed column would in any case therefore be unsuitable. We therefore conclude from the above considerations that an agitated tank, a simple bubble column, or a packed bubble column should be chosen. The final choice between these will depend on such factors as operating temperature and pressure, corrosiveness of the system, allowable pressure drop in the gas, and the possibility of fouling.

Note that to continue further with the design of the reactor by taking a value of f_A from Fig. 4.3, or using Eq. (4.22) or Eq. (4.24), requires a knowledge of ε_L and a for the bubble dispersion as well as a more exact value of k_L. These will depend on the type and configuration of the reactor chosen, the flow rates at which it is operated, and the physical properties of the chemical species involved.

4.1.5 Information Required for Gas–Liquid Reactor Design

Once the type of reactor has been chosen, designing the reactor from basic principles (as opposed to pilot plant construction and development) first of all requires some tentative decisions about the size, shape, and mechanical arrangement of the equipment in the light of knowledge of the flow rates to be used. These preliminary decisions will eventually be subject to confirmation or amendment as the design proceeds. Then there is a variety of other information that needs to be either estimated from the literature or measured experimentally if the opportunity exists. Altogether the kind of information required may be considered under three headings.

4.1.5.1 Kinetic Constants of the Reaction

The kinetics of the reaction needs to be known or measured, in particular, the rate constant and how it may be affected by temperature. Many gas–liquid reactions, such as chemical reactions generally, are accompanied by the evolution or absorption of heat. Even if there are arrangements within the reactor for the removal of heat (e.g., cooling coils in a stirred tank reactor), it is unlikely that the temperature will be maintained constant at all stages in the process. Experimental methods for measuring the kinetics of reactions are considered in a later section.

4.1.5.2 Physical Properties of the Gas and Liquid

The two most important physical properties of the system are the solubility of the gas and its diffusivity in the liquid.

The solubility of the gas is needed so that C_{Ai}, the concentration of the reactant **A** at the interface, can be calculated. The solubility is often expressed through the Henry's law constant H; this is defined by $P_{Ae} = HC_A$, where P_{Ae} is the partial pressure of the gas **A** at equilibrium with liquid in which the concentration of the dissolved gas is C_A. This immediately raises a problem for many systems: how can the solubility, which requires gas and liquid to be at equilibrium, be determined when the gas reacts with the liquid? The answer is by one of the several methods.

1. The Henry's law constant can be calculated, in addition to the reaction rate constant, from the results of experiments designed to investigate the kinetics of a gas–liquid reaction (see Section 4.1.9).
2. Where the second reactant **B** is dissolved in a solvent (as in the case of sodium hydroxide in water in the example below), the solubility of the reactant **A** (carbon dioxide in the example), in the solvent (water), can be used and a correction can be made for any secondary "salting out" effect of the reactant **B**.
3. The solubility can be estimated by semiempirical correlations based on the theory of solutions.

4. The solubility of an unreactive gas with a similar type of molecule (e.g., N_2O for CO_2) can be used with an adjustment based on relative solubilities in other solvents.

A similar problem exists in determining the diffusivity of a gas in a liquid with which it reacts. Diffusivities are not easy to measure accurately, even under the best experimental circumstances. As in the case of solubility, the diffusivity D_{AB} needed in the basic equations can be estimated from a semiempirical correlation, and usually this is the best method available.[8] As before, where the system involves a second reactant **B** dissolved in a solvent, it may be possible to measure experimentally the diffusivity in the solvent in the absence of **B**. A laminar jet or wetted-wall column (as described in Section 4.1.9 for measuring the kinetics of the reaction) would be suitable for this purpose.

Other physical properties required are viscosities, especially the viscosity of the liquid; densities of the liquid and gas; surface tension of the liquid, including the influence of surfactants (e.g., on bubble coalescence behavior); and, if the gas is a mixture, the gas-phase diffusivity of the reactant **A**. These physical properties are needed to evaluate the equipment characteristics as follows.

4.1.5.3 Equipment Characteristics

The remaining quantities that need to be known are the performance characteristics of the particular equipment that has been chosen. These are principally the liquid-phase mass transfer coefficient k_L, the interfacial area of contact per unit volume a, the liquid-phase volume fraction ε_L, and, if the gas is a mixture, the gas-phase mass transfer coefficient k_G or volume coefficient $k_G a$. In addition, if the gas or liquid flow in the reactor is to be modeled using the dispersed plug flow model, for example, then the dispersion coefficients will also need to be known. These performance characteristics are essentially physical quantities; they depend on the fluid mechanics of liquid films flowing over packing elements in packed columns or on the behavior of bubbles in bubble columns and agitated tanks. They will generally be the same for purely physical absorption of a gas as in the case where chemical reaction is occurring, unless there is strong local temperature or concentration gradients. For the design of a reactor, values of k_L, etc. can be estimated from correlations based on gas absorption experiments carried out on the same type of equipment. A number of such correlations for columns with various kinds of packing are discussed in Volume 2. Further correlations can be found for agitated tanks in the book by Tatterson[9] and in other literature sources, and for bubble columns in Deckwer's book.[10]

4.1.6 Examples of Gas—Liquid Reactors

In the following examples, the aim is to illustrate how the chemical reaction is coupled with the mass transfer processes. To this end the reactor operating characteristics such as k_L, a, ε_L, and the gas—liquid physical properties such as the Henry's law constant and liquid-phase diffusivity will be used, but the details of their calculation will not be considered.

4.1.6.1 Packed Column Reactors

The design of packed column reactors is very similar to the design of packed columns without reaction (Volume 2, Chapter 12). Usually, plug flow is assumed for both gas and liquid phases. Because packed columns are used for fast chemical reactions, often the gas-side mass transfer resistance is significant and needs to be taken into account. The calculation starts on the liquid side of the gas–liquid interface where the chemical reaction rate constant is compounded with the liquid-side mass transfer coefficient to give a reaction-enhanced liquid film mass transfer coefficient k'_L as in the following example. The specific interfacial area a for the irrigated packing is then introduced to give the volumetric coefficient $k'_L a$. This is then combined with the gas-side volumetric coefficient $k_G a$ to give an overall gas-phase coefficient $K_G a$ on which the performance of the column is finally based.

Example 4.2
A Packed Column Reactor

Carbon dioxide is to be removed from an air stream at 1 bar total pressure by absorption into a 0.5 M solution (0.5 kmol/m³) of NaOH at 20°C in a column of 1 m diameter packed with 25-mm ceramic rings. Air entering at a rate of 0.02 kmol/s will contain 0.1 mol% CO_2, which must be reduced to 0.005 mol% at the exit. (Such a very low CO_2 concentration might be required, for example, in the feed stream to an air liquefaction process.) The NaOH solution is supplied to the column at such a rate that its concentration is not appreciably changed in passing through the column. From the data below, calculate the height of packing required. Is a packed column the most suitable type of reactor for this purpose?
Data[3]: Second-order rate constant for the reaction

$$CO_2 + OH^- = HCO_3^- : k_2 = 9.5 \times 10^3 \text{ m}^3/\text{kmol s}$$

(In a solution of NaOH this reaction is followed instantaneously by

$$HCO_3^- + OH^- = CO_3^{2-} + H_2O$$

corresponding to an overall reaction $CO_2 + 2NaOH = Na_2CO_3 + H_2O$.)
For 0.5 M NaOH at 20°C:

Diffusivity of CO_2: 1.8×10^{-9} m²/s
Solubility of CO_2: Henry's law constant $H = 25$ bar m³/kmol (2.5×10^6 Nm/kmol)

Equipment performance characteristics (see Volume 2, Chapter 12):

Effective interfacial area of packing: 280 m²/m³
Film mass transfer coefficients:
   Liquid: $k_L = 1.2 \times 10^{-4}$ m/s
   Gas (volume coefficient): $k_G a = 0.056$ kmol/m³ s bar

Solution

Note first of all that the concentration of the OH^- ion, 0.5 M assuming complete dissociation, will be very much greater than the concentration of the carbon dioxide in the solution. The reaction can therefore be treated as a pseudo–first-order reaction with a rate constant

$$k_1 = k_2 \times C_{OH^-} = 9.5 \times 10^3 \times 0.5 = 4.75 \times 10^3 \text{ s}^{-1}$$

The value of the Hatta number β (Eq. 4.11) is thus

$$\beta = \sqrt{k_1 D_{CO_2}}/k_L = \sqrt{4.75 \times 10^3 \times 1.8 \times 10^{-9}}/1.2 \times 10^{-4} = 24.4$$

Referring to Fig. 4.3, it can be seen that with this value for β the system will lie in the fast reaction regime, region I, and that the packed column will be a suitable reactor. Also, β in Eq. (4.16) is sufficiently large for tanh β to be effectively 1, so that Eq. (4.17) applies:

$$N_{CO_2} = \sqrt{k_1 D_{CO_2}} C_{CO_2}$$

This equation can be written $N_{CO_2} = k'_L C_{CO_2}$, where $k'_L = \sqrt{k_1 D_{CO_2}}$ and may be regarded as a liquid film mass transfer coefficient enhanced by the fast chemical reaction. This is very convenient because it allows us to use the expression in Volume 2, Chapter 12 for combining liquid film and gas film coefficients to give an overall gas film coefficient; i.e.,

$$\frac{1}{K_G a} = \frac{1}{k_G a} + \frac{H}{k'_L a}$$

where k'_L has been substituted for k_L, which would apply in the absence of any chemical reaction. In this way we can take into account that, in this problem, the gas is not pure carbon dioxide but a dilute mixture in air, which will give rise to a gas-phase resistance. Substituting the numerical values, $k'_L = \sqrt{4.75 \times 10^3 \times 1.8 \times 10^{-9}} = 2.9 \times 10^{-3}$ m/s . and

$$\frac{1}{K_G a} = \frac{1}{0.056} + \frac{25}{2.9 \times 10^{-3} \times 280} = \frac{1}{0.056} + \frac{1}{0.032}$$

Therefore

$$K_G a = 0.021 \text{ kmol/m}^3 \text{ s bar}$$

This value of $K_G a$ will apply to any position in the column.

4.1.6.1.1 Height of Packing

We can now proceed to the second part of the calculation and find the height of packing required. Plug flow for the gas phase will be assumed; because the composition of the liquid is assumed not to change, the flow pattern in the liquid does not enter into the problem. Note that using $K_G a$ requires that the driving force for mass transfer be expressed in terms of gas-phase partial pressures.

Referring to Fig. 4.4, let flow rate of the carrier gas (air) per unit cross-sectional area be G kmol/m^2 s, and the mole fraction of CO_2 be y (because the gas is very dilute, the mole

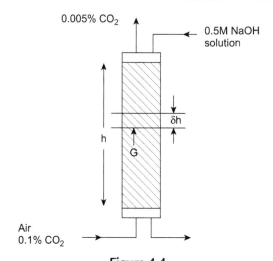

Figure 4.4
Absorption of CO_2 in 0.5 M NaOH solution.

fraction is virtually the same as the mole ratio, which should appear in the following material balance). Taking a balance across the element shown in Fig. 4.4,

$$-G\delta y = K_G a(P_A - P_{Ae})\delta h$$

where P_{Ae} is the partial pressure that would be in equilibrium with the bulk liquid; here, because the liquid is a concentrated solution of NaOH, the partial pressure of CO_2 P_{Ae} in equilibrium with it is virtually zero. Also, $P_A = yP$ where P is the total pressure.

Therefore

$$-G\delta y = K_G a y P \delta h$$

Therefore

$$h = \frac{G}{K_G a P} \int_{y_{out}}^{y_{in}} \frac{dy}{y} = \frac{G}{K_G a P} \ln \frac{y_{in}}{y_{out}}$$

and

$$G = \frac{0.02}{\pi \times 1^2/4} = 0.0255 \ \text{kmol/m}^2 \ \text{s}$$

Therefore

$$\text{Height of packing required } h = \frac{0.0255}{0.021 \times 1} \ \ln \frac{0.001}{0.00005} = \underline{3.64 \ \text{m}}$$

4.1.6.1.2 Confirmation of Pseudo–First-Order Behavior

When pseudo–first-order behavior is assumed for a reaction of this type, there is always the possibility that the second reactant **B**, here the NaOH, will not diffuse into the reaction zone sufficiently fast to maintain its concentration close to the value in the bulk liquid. A criterion for the pseudo–first-order assumption to be applicable has been shown by Danckwerts[3] to be

$$\beta < \frac{1}{2}\left(1 + \frac{v_A C_{BL} D_B}{v_B C_{Ai} D_A}\right)$$

For the present problem, $v_A/v_B = 0.5$, $D_B/D_A = 1.7$, and $C_{BL} = 0.5$.

The largest value that C_{Ai} could possibly have is that at the base of the column if there were no gas-phase resistance. Then

$$C_{Ai} = y_{in}P/H = (0.001 \times 1)/25 = 0.00004 \, \text{kmol/m}^3$$

The right-hand side of the above inequality is then 1.1×10^4, which is much greater than the value for Hatta number β, which has already been calculated as 24.4.

4.1.6.1.3 Further Comments

It should be appreciated that this example is one of the few very simple practical cases of a packed column reactor. The removal of carbon dioxide by reaction is a step in many important industrial processes, e.g., the purification of natural gas and of hydrogen in the manufacture of ammonia. Most of these processes[11,12] use as the liquid absorbent a solution that can be regenerated, for example, solutions of amines or potassium carbonate,[13] and the design of these columns is distinctly more complicated.

The value of the superficial gas velocity used in the packed column of the above example may be calculated as follows:

Gas mass flow rate per unit area $G = 0.0255 \, \text{kmol/m}^2$ s at a pressure of 1 bar and 20°C. That is,

$$\text{Volume of gas passing up the column in 1 s} = \frac{0.0255 \times 8314 \times 293}{1 \times 10^5}$$

$$= 0.621 \, \text{m}^3 \quad \text{over} \quad 1 \, \text{m}^2 \quad \text{cross} - \text{sectional area}$$

Thus, the superficial gas velocity is 0.621 m/s. Note that this is a much higher value than would typically be used in a bubble column.

4.1.6.2 Agitated Tank Reactors: Flow Patterns of Gas and Liquid

The most important assumptions in the design of agitated tank reactors are that *both* the liquid phase and the gas phase may be regarded as *well mixed*. That the liquid phase should be well mixed is the more obvious of these assumptions because one of the main functions of the impeller is to induce a rapid circulatory motion in the liquid. If the liquid circulation velocities are sufficiently intense, gas bubbles are drawn down with the liquid and recirculated to the impeller zone. Behind the blades of most types of gas dispersing impellers, gas-filled "cavities" are formed.[9,14] It has been found that both fresh gas entering the vessel from the sparger and recirculated bubbles merge into these cavities and are then redispersed in the highly turbulent vortices shed from the blade tips. As bubbles of the mixed gas rise toward the surface, some will separate into the headspace above the liquid, and the gas will leave the vessel. The assumption of ideal mixing of the gas phase therefore means that the composition of the gas in the bubbles is the same as that of the gas *leaving* the vessel. (Note that one of the problems in the design or scale-up of agitated tank gas–liquid reactors is that the flow pattern of the gas may tend more toward plug flow as the size of the tank increases.) A further assumption, borne out by many experimental investigations of mass transfer from gas bubbles, is that any gas film resistance is negligible compared with the liquid film resistance.

Example 4.3

An Agitated Tank Reactor

o-Xylene is to be oxidized in the liquid phase to *o*-methylbenzoic acid by means of air dispersed in an agitated vessel.[5,15]

The composition of the reaction mixture in the vessel will be maintained at a constant low conversion by continuous withdrawal of the product stream and a steady inflow of fresh *o*-xylene. Under these conditions the reaction rate is approximately independent of the *o*-xylene concentration and is pseudo-first order with respect to the dissolved oxygen concentration C_O.

That is,

$$-r_o = k_1 C_o$$

where r_o is the rate of consumption of oxygen, i.e., kmol O_2/s (m^3 liquid), and C_O is in kmol O_2/m^3.

At the operating temperature of 160°C, the rate constant $k_1 = 1.05 \ s^{-1}$.

The reactor will operate at a pressure such that the sum of the partial pressures of the oxygen and nitrogen inside the vessel is 14 bar. Under steady-state conditions the reactor will contain 5 m^3 of liquid into which air will be dispersed at a flow rate of 450 m^3/h, measured at reactor conditions.

The quantities to be calculated are (1) the fraction of the oxygen supplied that reacts and (2) the corresponding rate of production of *o*-methylbenzoic acid in kilomole per hour. There is also the question of whether an agitated tank is the most suitable reactor for this process.

Further Data

　　Physical properties of system:

　　　　Henry's law coefficient for O_2 dissolved in *o*-xylene $H = 127 \ m^3$ bar/kmol

　　　　$(1.27 \times 10^7 \ Nm/kmol)$

Diffusivity of O_2 in liquid xylene $D_o = 1.4 \times 10^{-9}$ m^2/s
Equipment performance characteristics:
  Gas volume fraction in the dispersion $(1 - \varepsilon_g) = 0.34$
  Mean diameter of the bubbles present in the dispersion $= 1.0$ mm
  Liquid-phase mass transfer coefficient $k_L = 4.1 \times 10^{-4}$ m/s

Solution

Assumptions—Flow Patterns of Gas and Liquid

It will be assumed that, within the tank itself, both gas and liquid phases are well mixed. Fig. 4.5 shows a diagram of the reactor. The inflow of fresh *o*-xylene and the outflow of the products are shown as broken lines because they do not enter into the calculation. Also it is assumed that because the reaction is relatively fast and the solubility of oxygen is low, only a negligible amount of dissolved but unreacted oxygen leaves in the product stream.

General Assessment of Reaction Regime

To assess where the reactor fits into the general scheme of gas–liquid reactions (Fig. 4.3), the value of Hatta number β will first be calculated.

$$\beta = \sqrt{k_1 D_o}/k_L = \sqrt{(1.05 \times 1.4 \times 10^{-9})}/(4.1 \times 10^{-4}) = 0.0935$$

This value of β lies in region II, but fairly close to the boundary with region III. The value of a, the interfacial area per unit volume of dispersion, is 2040 m^2/m^3 (see below), which is quite large even for an agitated tank, so that the parameter $\varepsilon_L / \frac{aD_o}{k_L}$ on Fig. 4.3 is

$$(1 - 0.34)\left/\left(\frac{2040 \times 1.4 \times 10^{-9}}{4.1 \times 10^{-4}}\right)\right. = 95$$

Conditions in the tank will therefore be quite close to the curve drawn on Fig. 4.3 for $\varepsilon_L / \frac{aD_A}{k_L} = 100$. Rather than proceed by trying to read a reaction factor f_A from Fig. 4.3, it is better to set out the basic material balance for mass transfer and reaction as below.

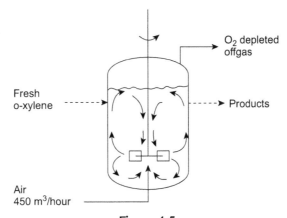

Figure 4.5
Oxidation of *o*-xylene. Diagram shows vaned-disc impeller to disperse gas, and circulation pattern of liquid and bubbles in tank.

Locating the position of β on Fig. 4.3 does, however, confirm that reaction will be occurring in the main bulk of the liquid and that an agitated tank is a suitable type of reactor.

Feed Rate of Oxygen

As the next step, the flow rate of the air (molar composition O_2 20.9%; N_2 79.1%) into the reactor will be calculated in kmol/s.

$$450 \text{ m}^3/\text{h is equivalent to } 450/3600 = 0.125 \text{ m}^3/\text{s of air.}$$

From the ideal gas law $PV = n\mathbf{R}T$, in 1 s the volume entering is equivalent to

$$n = \frac{14 \times 1.00 \times 10^5 \times 0.125}{8814 \times 433} = 0.0486 \text{ kmol air.}$$

This corresponds to an oxygen feed rate of $0.0486 \times 0.209 = 0.0102$ kmol/s.

Basis for Material Balance

Let f_u = Fraction of O_2 supplied that is unreacted.

1 kmol of air into reactor; i.e., 0.791 kmol N_2 and 0.209 kmol O_2

Therefore, kmol out of reactor: 0.791 kmol N2 and $0.209f_u$ kmol O_2

Therefore

$$\text{Mole fraction } O_2 = \frac{0.209f_u}{0.791 + 0.209f_u}$$

Therefore

$$\text{Partial pressure } O_2 = \frac{0.209f_u \times 14}{0.791 + 0.209f_u}$$

From Henry's law,

$$O_2 \text{ concentration at interface } C_{oi} = \frac{0.209f_u \times 14}{127(0.791 + 0.209f_u)} \text{ kmol/m}^3$$

Considering the rate r_{tank} for the whole tank and assuming a steady state,

Rate of mass transfer of O_2 from interface $\equiv$ Rate of reaction

$$-r_{tank} = k_L a V_d (C_{oi} - C_{OL}) = V_l k_i C_{OL}$$

where V_d is the volume of the gas–liquid dispersion in the tank and V_l is the volume of the liquid in the tank; note that $V_l = V_d \varepsilon_L$.

Therefore

$$C_{oi} - C_{OL} = \frac{V_l k_l}{k_L a V_d} C_{OL} = \frac{\varepsilon_L k_l}{k_L a} C_{OL}$$

Therefore

$$C_{OL} = \frac{C_{oi}}{(1 + \varepsilon_L k_1 / k_L a)}$$

Calculation of Interfacial Area a

To substitute numerical values, the interfacial area per unit volume of dispersion a needs to be calculated from the mean bubble diameter $d_b = 1.0$ mm and the gas volume fraction

$\varepsilon_G = (1 - \varepsilon_L) = 0.34$. Let there be n_b bubbles per unit volume of the dispersion, all of the same size.

Volume of the bubbles $\varepsilon_G = n_b \pi d_b^3 / 6$
Surface area of the bubbles $a = n_b \pi d_b^2$

Dividing and rearranging, $a = 6\varepsilon_G / d_b$

Substituting numerical values, $a = (6 \times 0.34)/0.001 = 2040 \ m^2/m^3$

Note that rearranging the above relationship to give $d_b = 6\varepsilon_G / a$ shows how a mean bubble size might be calculated from measurements of ε_G and a. A mean bubble diameter defined in this way (Volume 2, Chapter 1) is called a Sauter mean (i.e., a surface volume mean; see also Section 4.3.4).

Numerical Calculation of Reaction Rate
Substituting numerical values in the above equation for C_{oL},

$$C_{OL} = \frac{\dfrac{0.209 f_u \times 14}{127(0.791 + 0.209 f_u)}}{1 + \dfrac{1.05(1 - 0.34)}{4.1 \times 10^{-4} \times 2040}} = \frac{0.0603 \times 0.209 f_u}{0.791 + 0.209 f_u}$$

Rate at which O_2 reacts in whole vessel $-r_{tank} = V_l k_1 C_{oL}$

$$= 5 \times 1.05 \times \frac{0.0603 \times 0.209 f_u}{0.791 + 0.209 f_u}$$

This must be equal to the rate of O_2 removal from the air feed. That is,

$$0.0102(1 - f_u) = 5 \times 1.05 \times \frac{0.0603 \times 0.209 f_u}{0.791 + 0.209 f_u} = \frac{5 \times 1.05 \times 0.0603 f_u}{\left(\dfrac{0.791}{0.209} + f_u\right)}$$

That is,

$$31.04 f_u = (3.785 + f_u)(1 - f_u) = 3.785 - 2.785 f_u - f_u^2$$

Therefore

$$f_u^2 + 33.82 f_u - 3.785 = 0$$

$$f_u = \frac{-33.82 \pm \sqrt{33.82 + 4 \times 3.785}}{2} = \frac{-33.82 + 34.04}{2} = 0.112$$

the positive root being taken because f_u cannot be negative.
Thus, the fraction of the O_2 supplied that reacts is $(1 - 0.112) = \underline{0.888}$.

Therefore

$$\text{kmol of } O_2 \text{ reacting per second} = 0.0102 \times 0.888 = 0.0091$$

From the stoichiometry of the reaction, kilomoles per second of *o*-xylene reacting, and therefore kilomoles per second of the product *o*-methylbenzoic acid formed, will be $\frac{0.0091}{1.5} = 0.0060 \ kmol/s$. The hourly rate of production of the *o*-methylbenzoic acid will thus be $0.0060 \times 3600 = \underline{\underline{21.6 \ kmol/h}}$.

4.1.6.2.1 Further Comments

This reactor is typical of those used in a range of processes involving the direct oxidation of hydrocarbons with air or oxygen. Another example is the oxidation of cyclohexane to adipic acid, which is an intermediate in the manufacture of polyesters. Chemically, the reactions proceed by free radical chain mechanisms usually initiated by homogeneous catalysts, such as cobalt or manganese carboxylic acid salts or naphthenates.[16] From a chemical engineering point of view, some of these processes are extremely hazardous, involving quantities of volatile hydrocarbons maintained in the liquid phase at high temperatures and high pressures. Mechanical failure of the reactor could lead to the escape of a highly explosive vapor cloud with disastrous consequences. Understandably, measurements of agitation equipment performance characteristics, such as gas and liquid circulation patterns, bubble diameters, and interfacial areas, are much more difficult if not well-nigh impossible at reaction conditions, compared with experiments on air and water, which are very often used in laboratory measurements. Reactors of this type are not generally designed from first principles, but from extensive pilot plant development. Nevertheless a thorough understanding of fundamental principles is an invaluable guide to interpreting and extending the results of pilot plant studies.

4.1.6.3 Well-Mixed Bubble Column Reactors: Gas–Liquid Flow Patterns and Mass Transfer

In some cases, as in Example 4.4 that follows, a bubble column, which is relatively short in relation to its diameter, may be used (see Fig. 4.6). The bubbling gas will then generate sufficient circulation and turbulence in the liquid phase for the liquid to be assumed to be well mixed and uniform in composition[10] (except, in principle, in the thin films

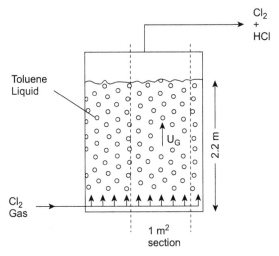

Figure 4.6
Batch chlorination of toluene in a well-mixed bubble column reactor (Example 4.4).

immediately surrounding the bubbles). The circulating liquid will also drag down smaller bubbles of gas, which can then mix with fresh gas. Under these circumstances, the gas phase can also be assumed to be well mixed, just as in the case of the agitated tank in Example 4.3. The question of taller bubble columns will be considered in the following section.

One of the purposes of giving Example 4.4 (on the chlorination of toluene) is to demonstrate the effect of different gas flow rates on the performance of a bubble column. The higher the gas flow rate, the larger the interfacial area α per unit volume of dispersion; gas–liquid mass transfer will take place more readily, and the concentration of the dissolved gas in the liquid will rise. Although the rate of reaction will increase, this is offset, as will be seen, by the disadvantage of a lower fractional conversion for the gas-phase reactant, i.e., in this example, the chlorine in the off-gas from the reactor. To show that this effect of gas flow rate on off-gas composition depends more on the height chosen for the column than on its diameter, the calculations will be carried out for a section of column of unit cross-sectional area (1 m^2) (see Fig. 4.6) and the gas flow rate will be expressed as a superficial gas velocity u_G (m/s).

Chlorination processes in bubble column reactors[10] are unusual in showing a significant gas-phase resistance to mass transfer. It will be seen from the low value of the Henry's law constant K in the list of data for the example below that the solubility of chlorine in toluene is much greater than the solubility of either the carbon dioxide or oxygen considered in the previous examples. This means that when the gas-phase mass transfer resistance is taken in combination with the liquid-phase resistance according to Eq. (4.27), which is derived in Volume 2, Chapter 12, then the gas-side contribution to the resistance is much greater if H is small.

$$\frac{1}{K_L a} = \frac{1}{k_L a} + \frac{1}{H k_G a} \tag{4.27}$$

where K_L is the overall mass transfer coefficient based on liquid-side concentrations.

Example 4.4
A Bubble Column Reactor—Batch Chlorination of Toluene[10]
The chlorination of toluene to give a mixture of *ortho-* and *para*-monochlorotoluenes is typical of many similar chlorinations carried out industrially.

As the reaction proceeds and the concentrations of the monochlorotoluenes increase, so further chlorination of the aromatic ring takes place to give dichlorotoluenes. The present example (see Fig. 4.6) is restricted to relatively low toluene conversions so that any subsequent reactions need not be considered. The process is carried out using a homogeneous (i.e., dissolved) catalyst such as ferric chloride $FeCl_3$; in this example, the catalyst used is stannic chloride $SnCl_4$, which, being more active than ferric chloride, is effective even at the relatively low temperature of 20°C. Note that the HCl formed as a by-product in the reaction is transferred back to the gas phase and leaves the reactor mixed with the unreacted chlorine. Thus, when the reaction is proceeding steadily, 1 mol of chlorine is replaced by 1 mol of HCl and the volume of gas passing through the reactor is unaffected by the reaction.

The example is concerned with a batch chlorination process. At the beginning, the fresh toluene charged to the reactor will contain no dissolved chlorine. After bubbling of the chlorine has commenced, a period of time will need to elapse before the concentration of the dissolved chlorine rises to a level that just matches the rate at which it is being removed from the solution by reaction. To avoid such a complication in this example, calculations are carried out for the stage when, after chlorine bubbling has continued at a steady rate, the fractional conversion of the toluene has reached a value of 0.10. It is then assumed that, at any instant in time, the rate of mass transfer of chlorine from the gas phase is just equal to the rate at which it reacts in the bulk of the liquid, i.e., the rate is given by Eq. (4.24).

The calculations required in this example may now be summarized as follows: Toluene is to be chlorinated in a batch-operated bubble column such that, for any particular gas rate chosen, the height of the dispersion in the column is 2.2 m. The reactor will operate at a temperature of 20°C and a pressure of 1 bar (any effect of hydrostatic pressure differences in the column may be neglected). The catalyst will be stannic chloride at a concentration of 5×10^{-4} kmol/m^3.

1. Determine whether a bubble column is a suitable reactor for this process.
2. Consider the stage when 10% of the toluene has been converted. For superficial gas velocities of 0.01, 0.02, 0.03, 0.04, 0.05, 0.06, 0.07, 0.08, 0.09, and 0.1 m/s, calculate:
   a. the mole fraction of chlorine in the outlet gas and the fractional conversion of the chlorine,
   b. the steady-state concentration of dissolved chlorine in the liquid phase, and
   c. the rate of the reaction in terms of kmol/s toluene reacting per cubic meter of dispersion.
3. For each of the superficial gas velocities in the range above, calculate the reaction time required for the fractional conversion of toluene to increase from 0.1 to 0.6.

Further Design Data Are as Follows
The reaction rate or the rate of consumption of chlorine at 20°C, catalyzed by stannic chloride, is

$$-r_{Cl} = k_3 C_{Sn} C_T C_{Cl}$$

where C_T, C_{Cl}, and C_{Sn} are the concentrations of toluene, chlorine, and stannic chloride, respectively, and $k_3 = 0.134$ (m^3/kmol)^{2}s^{-1} This equation may be written in the form of Eq. (4.1) as

$$-r_{Cl} = k_2 C_T C_{Cl}$$

where $k_2 = k_3 C_{Sn}$ incorporates the catalyst concentration C_{Sn} that remains unchanged during the course of the reaction (any small changes in the molar density of the liquid phase will be neglected).
Molar concentration of pure liquid toluene (calculated from the density of the liquid and the molecular weight of toluene): 9.36 kmol/m^3.

Henry's law constant H for dissolved chlorine at 20°C: 0.45 bar m^3/kmol.

Gas volume fraction in dispersion $\varepsilon_G = 4.0u_G$ where u_G is the superficial gas velocity (m/s).

Liquid-side volumetric mass transfer coefficient, $k_La = 0.078u_G$ (s^{-1}).

Gas-side volumetric mass transfer coefficient, $k_Ga = 0.104\ u_G$ (kmol/m^{3}s bar) ($=1.04 \times 10^{-6}$ kmol/N s m).

Diffusion coefficient of chlorine in toluene at 20°C, $D_{Cl} = 3.5 \times 10^{-9}$ m^2/s.

Liquid-side mass transfer coefficient, $k_L = 4.0 \times 10^{-6}$ m/s.

(Note: This experimental value appears to be remarkably low; the reason for the anomaly has yet to be fully resolved.)

Solution
1. Suitability of a bubble column reactor
   To assess the suitability of a bubble column using Fig. 4.3, the Hatta number $\beta = \sqrt{k_2 C_T D_{Cl}}/k_L$ will be evaluated. At the beginning of the reaction, the toluene concentration C_T is 9.36 kmol/m^3 (neglecting any dilution effect of the stannic chloride catalyst and the dissolved chlorine). The value of $k_2 = k_3 C_{Sn}$ is therefore $k_2 = 0.134 \times 5 \times 10^{-4} = 0.67 \times 10^{-4}$ m^3/kmol s.
   Thus,

   $$\beta = \sqrt{0.67 \times 10^{-4} \times 9.36 \times 3.5 \times 10^{-9}}\big/4.0 \times 10^{-6} = 0.37$$

   This value lies in the region of Fig. 4.3 corresponding to a moderately fast reaction occurring in the bulk of the liquid and not in the film. A bubble column and a stirred tank reactor, as sometimes used industrially, are therefore suitable types of reactor for the process.
2. Calculation of the chlorine conversion and the toluene reaction rate
   As in Example 4.3 with an agitated tank, let the fraction of chlorine passing through the reactor unreacted be f_u and, because Cl$_2$ is replaced by HCl, f_u is also the mole fraction of chlorine in the off-gas. Because the total pressure is 1 bar, the partial pressure of the chlorine will be f_u bar. Because the gas phase in the reactor is assumed to be well mixed, the equivalent interfacial chlorine concentration C_{Cli} is f_u/H, i.e., $f_u/0.45 = 2.22f_u$ kmol/m^3. Considering unit volume, i.e., 1 m^3 of dispersion, and following Eq. (4.24), the rate of mass transfer across the interface is now equated to the rate of the reaction in the bulk of the liquid where the concentration of the chlorine is C_{ClL}:

   $$J_{Cl} = K_La(C_{Cli} - C_{ClL}) = k_2 C_{ClL} C_T(1 - \varepsilon_G) \tag{A}$$

   where K_L is the overall mass transfer coefficient (there may be significant gas-side resistance) and $(1 - \varepsilon_G)$ has been substituted for ε_L. The value of C_T when a fraction α of the toluene has been converted, i.e., a fraction $(1 - \alpha)$ remains unconverted, will be $9.36(1 - \alpha)$ kmol/m^3. Using numerical values in Eq. (4.27),

   $$\frac{1}{K_La} = \frac{1}{0.078u_G} + \frac{1}{0.45 \times 0.104u_G} = \frac{1}{0.029u_G} \quad \text{or } K_La = 0.029u_G.$$

   Then,

   $$J_{Cl} = 0.029u_G(2.22f_u - C_{ClL}) = 0.67 \times 10^{-4}C_{ClL}9.36(1 - \alpha)(1 - 4.0u_G) \tag{B}$$

Solving this equation for C_{CIL} gives

$$C_{CIL} = \frac{0.0644 u_G f_u}{0.029 u_G + 0.627 \times 10^{-3}(1 - \alpha)(1 - 4.0 u_G)} \tag{C}$$

Now, using as a basis 1 m² cross-sectional area of the reactor and 1 s in time, the amount of chlorine *fed* to the reactor (from the ideal gas law $PV = nRT$) is $n = \frac{1.0 \times 10^5 u_G}{8314 \times 293} = 0.0411 u_G$ kmol; of this, a quantity $0.0411 u_G(1 - f_u)$ reacts. The rate at which chlorine is removed by reaction in a column 1 m² cross-sectional area and 2.2 m high, i.e., dispersion volume 2.2 m³, is $2.2 J_{Cl}$; thus in 1 s, using Eq. (B) for J_{Cl} and substituting for C_{CIL} from Eq. (C),

$$\text{Rate of removal of chlorine} = \frac{0.89 \times 10^{-4} u_G f_u(1 - \alpha)(1 - 4.0 u_G)}{0.029 u_G + 0.627 \times 10^{-3}(1 - \alpha)(1 - 4.0 u_G)}$$

$$= 0.041 u_G(1 - f_u).$$

Equating these two quantities and rearranging for f_u gives

$$f_u = \left(1 + \frac{0.00216(1 - \alpha)(1 - 4.0 u_G)}{0.029 u_G + 0.627 \times 10^{-3}(1 - \alpha)(1 - 4.0 u_G)}\right)^{-1} \tag{D}$$

Substituting this expression in Eq. (C) gives

$$C_{CIL} = \frac{0.0644 u_G}{0.029 u_G + 0.00279(1 - \alpha)(1 - 4.0 u_G)} \tag{E}$$

For a fractional conversion α of toluene of 0.1, Table 4.2 shows how f_u (which is the same as the mole fraction of chlorine in the outlet gas) and $(1 - f_u)$ (the fractional conversion of the chlorine) vary with the superficial gas velocity u_G over the range 0.01–0.05 m/s, as required in the problem. Table 4.2 also shows the concentration of the dissolved chlorine in the bulk liquid C_{CIL} calculated from Eq. (E). Shown also is the relative saturation of the liquid C_{C1L}/C^*_{C1L}, where C^*_{C1L} is the saturation chlorine concentration in the bulk liquid at the operating pressure of the reactor, i.e., 1.0 bar.

Thus, $C^*_{CIL} = P/H = 1.0/0.45 = 2.22$ kmol/m³.
The rate of the reaction in terms of kilomole of toluene reacting per unit volume of dispersion, i.e., J_T, which is equal to the rate at which chlorine reacts J_{Cl}, is given by Eq. (A). Substituting for C_T, this equation then becomes

$$J_T = J_{Cl} = k_2 C_{CIL} 9.36(1 - \alpha)(1 - \varepsilon_G) \tag{F}$$

Values of J_T over the required range of superficial gas flow rates with $\alpha = 0.1$, $\varepsilon_G = 4.0 u_G$, and C_{CIL} given by Eq. (E) are shown in Table 4.2.

3. Time required for conversion of the toluene from a mole fraction $\alpha = 0.1$ to $\alpha = 0.6$
   The reactor is operated batchwise with respect to the toluene. Considering now the change in the fractional conversion of the toluene with time, the rate of reaction J_T in terms of the rate of change in the number of moles of toluene per unit volume of dispersion is

$$-\frac{d[(1 - \varepsilon_G)9.36(1 - \alpha)]}{dt}.$$

Substituting this value of J_T into Eq. (F) and canceling the $9.36(1 - \varepsilon_G)$ term on each side,

$$\frac{d\alpha}{dt} = k_2 C_{CIL}(1 - \alpha) \tag{G}$$

Table 4.2: Results of calculations for batchwise chlorination of toluene.

At Toluene Fractional Conversion α = 0.1	Superficial Gas Velocity u_G (m/s)									
	0.01	0.02	0.03	0.04	0.05	0.06	0.07	0.08	0.09	0.1
Fraction chlorine unreacted f_u	0.308	0.381	0.444	0.500	0.550	0.595	0.635	0.672	0.705	0.735
Chlorine fractional conversion $(1 - f_u)$	0.692	0.619	0.556	0.500	0.450	0.405	0.365	0.328	0.295	0.265
Chlorine concentration in liquid (kmol/m³)	0.238	0.446	0.627	0.788	0.931	1.06	1.175	1.28	1.374	1.46
% chlorine saturation in liquid	10.7	20.1	28.3	35.5	41.9	47.7	52.9	57.6	61.9	65.8
Rate of reaction of toluene (kmol/m³ s × 10^4)	1.29	2.32	3.11	3.74	4.20	4.54	4.77	4.90	4.96	4.95
Time for reaction of toluene from α = 0.1 to α = 0.6 (h)	10.1	5.6	4.2	3.4	2.9	2.7	2.4	2.3	2.2	2.1

Substituting for C_{CIL} from Eq. (E),

$$\frac{d\alpha}{dt} = k_2 \frac{0.0644u_G(1-\alpha)}{0.029u_G + 0.00279(1-\alpha)(1-4.0u_G)}$$

Integrating,

$$k_2 t_r = \int_{0.1}^{0.6} \frac{0.029u_G + 0.00279(1-\alpha)(1-4.0u_G)}{0.0644u_G(1-\alpha)} d\alpha$$

where t_r is the time required for the reaction to proceed from a toluene conversion of 0.1 to a conversion of 0.6.

Simplifying,

$$k_2 t_r = \int_{0.1}^{0.6} \left[\frac{0.45}{1-\alpha} + 0.0433 \left(\frac{1}{u_G} - 4.0 \right) \right] d\alpha$$

That is,

$$k_2 t_r = \left[-0.45 \ln(1-\alpha) + 0.0433 \left(\frac{1}{u_G} - 4.0 \right) \alpha \right]_{0.1}^{0.6}$$

Substituting the value $k_2 = 0.67 \times 10^{-4}$ m^3/kmol-s and evaluating the integral, this equation for t_r becomes:

$$t_r = \frac{1}{0.67 \times 10^{-4}} \left[0.365 + 0.0217 \left(\frac{1}{u_G} - 4.0 \right) \right] \text{ s}$$

4.1.6.3.1 Further Comments

The results from Table 4.2 and Fig. 4.7 demonstrate how a bubble column reactor responds to increases in the gas feed rate. At low superficial gas velocities, the gas holdup and hence the gas–liquid interfacial area are small; the degree of saturation of the liquid is low and therefore the reaction rate is low. This is especially noticeable in the long period of time, 10.1 h, for the conversion of the toluene from 10% to 60%. On the other hand, the fractional conversion of the gaseous reactant, the chlorine, is greatest at the lowest superficial gas velocity. Importantly, the reaction rate and reaction time increases and decreases, respectively, with increasing superficial gas velocity, but tend to saturate after 0.8 m/s. Increasing the gas rate markedly increases the rate of conversion of the liquid reactant, toluene, but also increases the fraction of the gaseous reactant, chlorine, that passes through the column unreacted. As discussed in the following section, this has important consequences in the design of a bubble column reactor and in the design of the chemical process of which it forms a part.

The results given in Table 4.2 are based on the assumption that steady-state conditions have been reached and that, as indicated earlier, the gas flow rate through the reactor is

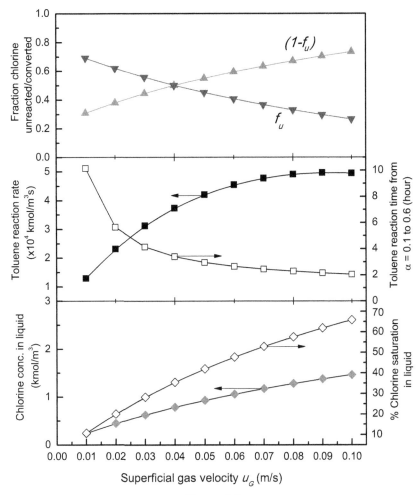

Figure 4.7

Results of calculations for batchwise chlorination of toluene. Variation of fraction of chlorine unreacted, fractional chlorine conversion, reaction rate and reaction time of toluene, chlorine concentration in liquid, and % chlorine saturation in liquid with superficial gas velocities from 0.01 to 0.1 m/s at a toluene fractional conversion of 0.1. The total time for reaction of toluene corresponds to conversion range from 0.1 to 0.6.

unaffected by the reaction. More detailed calculations show that, in the time taken for the toluene conversion to reach a value of 0.1, the concentration of chlorine reaches more than 99% of its steady-state value, thus justifying the assumption made in the solution.

4.1.7 High-Aspect Ratio Bubble Columns and Multiple-Impeller Agitated Tanks

The relatively low conversions of the gaseous reactant in many processes, such as the chlorination of toluene above, pose problems for the process design of such operations. One approach for the chlorine–toluene process would be simply to pass the Cl_2–HCl mixture through a downstream section of plant to separate the unreacted chlorine, which would then be recycled to the reactor. In an alternative design, chlorine might be passed from the first reactor into a second reactor in series, and then if necessary into a third reactor, and so on, as shown in Fig. 4.8. However, the problem arises because a bubble column with a low aspect ratio or a single-impeller agitated tank behaves essentially as a well-mixed reactor (aspect ratio is defined as L/d_c where L is the height of the gas–liquid dispersion in the column and d_c is the column diameter). A much more desirable situation would be to arrange for the gas to be more nearly in a state of plug flow. One of the reasons why the carbon dioxide concentration of the gas in the packed column of Example 4.2 could easily be reduced by a factor of 20 was that the gas could be assumed to be in plug flow.

Continuing this idea of trying to approximate to plug flow, the logical development is to use a bubble column that is tall in relation to its diameter, i.e., of high aspect ratio. As Fig. 4.9 shows, if instead of having a given volume of liquid in a shallow, large-diameter bubble column, the same volume of liquid is contained in a tall, smaller diameter column, there are two advantages. Firstly, there is the probability that the gas (and possibly the liquid) will be more nearly in a state of plug flow; and secondly, assuming that gas is supplied at the same volumetric flow rate, the superficial gas velocity through the column will be increased. Because the specific interfacial area a and the gas holdup ε_G increase with superficial gas velocity (as demonstrated by the equations used in Example 4.4), this will give an increase in the rate of reaction per unit volume of dispersion. (Note, however, that bubble columns generally operate at lower superficial gas velocities than packed columns; obviously, if the superficial gas velocity in the short column is already close to the limit above which the liquid would be entrained, no further increase is practicable.) One disadvantage of a tall column is the cost of compressing the gas to overcome the additional hydrostatic head.

For agitated tanks, a similar train of logic can be followed. Instead of the three separate stirred vessels shown in Fig. 4.8, all three impellers can be mounted on the same shaft[17] as shown in Fig. 4.10. In Chapter 2, Section 2.3, it was seen that as the number of stirred

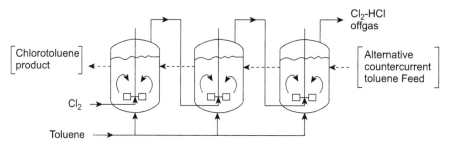

Figure 4.8

Three well-mixed stirred tank reactors (or low-aspect ratio bubble columns) for the chlorination of toluene showing the chlorine gas supplied in series to the reactors to improve the fractional conversion of chlorine.

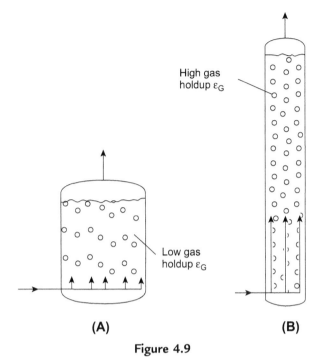

(A)   (B)

Figure 4.9
Alternative bubble columns each using the same volume of liquid and the same total gas flow rate. (A) Low-aspect ratio column: low superficial gas velocity u_G; small interfacial area a; low rate of reaction. (B) High-aspect ratio column: high superficial gas velocity u_G; large interfacial area a; higher rate of reaction. Note: For case (B) it must not be assumed that both gas and liquid phases are well mixed.

tanks in a series is increased, so does the overall flow behavior tend more nearly to plug flow. In biochemical reaction engineering (Chapter 6), aerobic stirred tank fermenters are often of the three impeller type of Fig. 4.10, to increase the fraction of oxygen usefully transferred from the compressed air supply to an economic level. Fig. 4.10 also illustrates that the similarity between a multiple-impeller stirred tank system and a bubble column extends further than might at first be imagined. The gas bubbles rising in a bubble column create circulation cells in which the mixing effect, at least for the liquid phase, is similar to that in the stirred compartments of the impeller-driven column. Both experimental data and theoretical considerations have indicated that the height of these mixing cells is approximately equal to the column diameter.

4.1.8 Axial Dispersion in Bubble Columns

4.1.8.1 Dispersion Coefficients for Gas and Liquid Phases

Although the mixing patterns in bubble columns do not obviously correspond to simple axial dispersion, the dispersed plug flow model has been found to hold reasonably well in

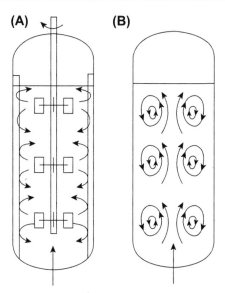

Figure 4.10
Multiple-impeller stirred vessel (A) and tall bubble column (B) showing similarity of liquid circulation patterns.

practice. For a two-phase gas–liquid system, the equation for gas-phase convection and dispersion (Chapter 2, Eq. 2.97) becomes

$$\frac{\partial C_G}{\partial t} = \varepsilon_G D_G \frac{\partial^2 C_G}{\partial z^2} - u_G \frac{\partial C_G}{\partial z} \tag{4.28}$$

where C_G is the concentration of dispersed species in the gas phase and D_G is the gas-phase dispersion coefficient. The gas holdup ε_G is introduced because it is equal to the fraction of the tube cross-sectional area occupied by the gas, i.e., the region in which gas dispersion occurs. The liquid-phase dispersion coefficient D_L is defined similarly.

Experimental measurements of dispersion coefficients[10] have shown that, unless the liquid velocity is unusually high, both gas and liquid phase dispersion coefficients depend primarily on the superficial gas velocity u_G and the column diameter d_c. An empirical equation for the gas-phase dispersion coefficient is

$$D_G = 50 \left(\frac{u_G}{\varepsilon_G} \right)^3 d_C^{1.5} \tag{4.29}$$

where the units are D_G(m²/s), u_G(m/s), and d_c(m). Note that 50 is a dimensional constant with unit s²/m²·⁵. u_G/ε_G in this expression is essentially the rising velocity of the bubbles relative to the liquid. This equation is recommended for use over a wide range of column diameters (from 0.20 to 3.2 m). An equation for the liquid-phase dispersion coefficient is

$$D_L = 0.35(u_G g)^{1/3} d_C^{4/3} \qquad (4.30)$$

where g is the gravitational acceleration. The constant 0.35 is here dimensionless. Baird and Rice[18] have shown that the form of this equation can be derived theoretically using dimensional analysis and the Kolmogoroff's theory of isotropic turbulence.

4.1.8.2 Modeling the Flow in Bubble Columns

Although the most realistic model for a bubble column reactor is that of dispersed plug flow in both phases, this is also the most complicated model; in view of the uncertainty of some of the quantities involved, such a degree of complication may not be warranted. Because the residence time of the liquid phase in the column is usually much longer than that of the gas, it is likely that the liquid phase will be well mixed even when the gas phase is not. Therefore, in modeling bubble column reactors, it is often acceptable to assume that the liquid phase is well mixed while the gas may be either well mixed, or in plug flow, or somewhere between these two extremes and described by the dispersed plug flow model.

One particular case of modeling a bubble column reactor will now be considered in more detail. This is the case where the reaction is very fast and occurs wholly within the liquid film surrounding the bubbles, i.e., in region I of Figs. 4.2 and 4.3. In this case, the concentration in the bulk of the liquid of the species A being transferred from the gas phase is zero, and the mixing pattern in the liquid has therefore no influence on the rate of transfer. The advantage of considering this case in the present context is that it will allow attention to be focused on the influence of gas-phase mixing on the performance of a bubble column.

To incorporate mixing by the dispersed plug flow mechanism into the model for the bubble column, we can make use of the equations developed in Chapter 2 for dispersed plug flow accompanied by a first-order chemical reaction. In the case of the very fast gas—liquid reaction, the reactant **A** is transferred and thus removed from the gas phase at a rate that is proportional to the concentration of **A** in the gas, i.e., as in a homogeneous first-order reaction. Applied to the two-phase bubble column for steady-state conditions, Eq. (2.121) becomes

$$\varepsilon_G D_G \frac{d^2 C_{Ag}}{dz^2} - u_G \frac{dC_{AG}}{dz} - k_{1G} C_{AG} = 0 \qquad (4.31)$$

where C_{AG} is the concentration of **A** in the gas phase, $\varepsilon_G D_G$ replaces D_L as in Eq. (4.28), and k_{1G} is now the first-order rate constant for the transfer of **A** from the gas phase. The solution to this equation gives the ratio of the concentrations of the reactant **A** at the exit (C_{Aout}) and inlet of the column (C_{Ain}): From Eq. (2.130),

$$\frac{C_{Aout}}{C_{Ain}} = \frac{4Ae^{\frac{u_G L}{2\varepsilon_G D_G}}}{(A+1)^2 e^{\frac{Au_G L}{2\varepsilon_G D_G}} - (A-1)^2 e^{\frac{Au_G L}{2\varepsilon_G D_G}}} \tag{4.32}$$

where

$$A = \left[1 + 4k_{1G}\varepsilon_G D_G / u_G^2\right]^{1/2}$$

The first-order constant k_{1G} will now be expressed in terms of the rate constant k_1 of the reaction in the liquid phase. From Eq. (4.17), the rate of transfer of **A** per unit area of gas−liquid interface is $\sqrt{(k_1 D_A)}C_{Ai}$; i.e., in terms of an enhanced mass transfer coefficient $k_L' = \sqrt{(k_1 D_A)}$, this rate of transfer is $k_L' C_{Ai}$. The rate of transfer per unit volume of dispersion J_A is thus

$$J_A = k_L' a C_{Ai} \tag{4.33}$$

where a is the interfacial area per unit volume of dispersion. Now C_{Ai} is the liquid-phase concentration of **A** at the interface. To convert this to a gas-phase concentration C_{AG}, first the Henry's law constant is introduced so that, assuming no gas-side mass transfer resistance, $C_{Ai} = P_A/H$ where P_A is the partial pressure of **A** in the gas. From the ideal gas law, $PV = nRT$; i.e., the gas-phase concentration $C_{AG} = n_A/V = P_A/RT$. Substituting in Eq. (4.33),

$$J_A = k_L' a \frac{RT}{H} C_{AG} = k_{1G} C_{AG}$$

That is,

$$k_{1G} = k_L' a \frac{RT}{H} = \sqrt{k_1 D_A} a \frac{RT}{H} \tag{4.34}$$

The influence of gas-phase mixing on the performance of a bubble column will now be demonstrated in Example 4.5.

Example 4.5
Effect of Gas-Phase Dispersion on the Gas Outlet Concentration From a Bubble Column: The Absorption of CO_2 From an Air Stream
Carbon dioxide at an inlet concentration of 0.1 mol% is to be removed from an air stream at a total pressure of 1 bar by bubbling through a 0.05 M (0.05 kmol/m^3) solution of NaOH at 20°C in a bubble column. The caustic soda solution passes through the column at such a rate that its composition is not significantly affected by the absorption. The superficial velocity of the gas will be 0.06 m/s. The ratio of outlet concentration of CO_2 in the air to inlet

concentration is to be calculated for the following cases; the height of the gas–liquid dispersion will be 1.5 m in each case:

1. The diameter of the column is 0.20 m and the gas is in plug flow.
2. The diameter of the column is 0.20 m and the dispersed plug flow model applies to the gas.
3. The diameter of the column is 1.5 m and the dispersed plug flow model applies to the gas.
4. The diameter of the column is 1.5 m and the gas phase is assumed to be well mixed.

Data: With respect to the chemical reaction, the conditions in this problem are very similar to those in Example 4.2 except that the concentration of the caustic soda solution here is 0.05 M (0.05 kmol/m^3); the same physicochemical data will therefore be assumed, i.e., a second-order rate constant for the reaction

$$CO_2 + OH^- = HCO_3^- \quad k = 9.5 \times 10^3 \ m^3/kmol \ s$$

Diffusivity of CO_2 in the liquid: $1.8 \times 10^{-9} \ m^2/s$.
Solubility of CO_2: Henry's law constant $H = 25$ bar m^3/kmol (2.5×10^6 Nm/kmol).Bubble *column mass transfer characteristics*: The following estimates for a superficial gas velocity of 0.06 m/s will be used for both columns:

Gas holdup	$\varepsilon_G = 0.23$
Gas–liquid interfacial area per unit volume of dispersion	$a = 195 \ m^2/m^3$
Liquid-side mass transfer coefficient	$k_L = 2.5 \times 10^{-4} \ m/s$

Note: Bubble column mass transfer parameters are difficult to estimate reliably.[10] The above figures are based on results from the sulfite oxidation method at a higher electrolyte concentration than the 0.05 M solution to be used in the present example, and therefore the values for ε_G and a may be overestimates.

Solution
First of all the value of Hatta number, $\beta = \sqrt{k^2 C_{OH} D_{CO_2}}/k_L$, will be calculated to check that the fast reaction regime will apply.

That is,

$$\beta = \sqrt{9.5 \times 10^3 \times 0.05 \times 1.8 \times 10^{-9}}/2.5 \times 10^{-4} = 3.7$$

From Fig. 4.3 it may be seen that this value, although smaller than that in Example 4.2, is still sufficiently large for all the reaction to occur in the film. In addition the concentration of dissolved CO_2 at the interface in contact with the incoming gas containing CO_2 at a partial pressure of 0.001 bar will be only $0.001/H$, i.e., $0.001/25 = 4 \times 10^{-5}$ kmol/m^3, which is much less than 0.05 kmol/m^3, the concentration of the OH^- ion, so that the reaction will still behave as pseudo-first order. The enhanced liquid-side mass transfer coefficient k_L' will thus be $k_L' = \sqrt{(9.5 \times 10^3 \times 0.05 \times 1.8 \times 10^{-9})} = 0.925 \times 10^{-3}$ m/s. The first-order rate constant k_{1G} is given by Eq. (4.34):

$$k_{1G} = 0.925 \times 10^{-3} \times 195 \times \frac{8314 \times 293}{25 \times 10^5} = 0.176 \ s^{-1}$$

(the factor 10^5 occurs to convert bar, which appears in the units of the Henry's law constant, to N/m^2).

1. Gas in plug flow in the small-diameter column
(Note that the actual diameter of the column does not appear in the calculation because the gas rate is expressed as a superficial velocity; however, plug flow for the large column would be extremely improbable.)
Very conveniently, Eq. (2.134) of Chapter 2 (which was derived as the limiting case of dispersed plug flow as the dispersion coefficient tends to zero) exactly describes this case if k_1 in that equation is replaced by k_{1G}.
That is,

$$\frac{C_{Aout}}{C_{Ain}} = \exp(-k_{1G}\tau)$$

where in the present case, τ $(=L/u_G)$ is the residence time of the gas in the column.
Thus

$$\frac{C_{CO_2out}}{C_{CO_2in}} = \exp\left(-0.176 \times \frac{1.5}{0.06}\right) = \underline{0.0123}.$$

This means that in a column with plug flow, the ratio of inlet to outlet concentration is $\frac{1}{0.0123} = \underline{81}$.

2. Gas in dispersed plug flow in the small-diameter column
For this case the gas-phase dispersion coefficient needs to be calculated from Eq. (4.29) and the result will now depend on d_c the diameter of the column. Inserting numerical values,

$$D_G = 50\left(\frac{0.06}{0.23}\right)^3 0.2^{1.5} = 0.0794 \text{ m}^2/s$$

Evaluating the parameters that appear in Eq. (4.32),

$$\frac{u_G L}{2\varepsilon_G D_G} = \frac{0.06 \times 1.5}{2 \times 0.23 \times 0.0794} = 2.464$$

$$A = \left[1 + \frac{4 \times 0.176 \times 0.23 \times 0.0794}{0.06^2}\right]^{1/2} = 2.138$$

Substituting in Eq. (4.32),

$$\frac{C_{CO_2out}}{C_{CO_2in}} = \frac{4 \times 2.138 \times e^{2.464}}{3.138^2 e^{2.138 \times 2.464} - 1.138^2 e^{-2.138 \times 2.464}} = \underline{0.0526}$$

This means that in the 0.20 m column with dispersed plug flow, the ratio of inlet to outlet concentration is $1/0.0526 = \underline{19}$.

3. Gas in dispersed plug flow in the large diameter column
Using again Eq. (4.29) but now with $d_c = 1.5$ m,

$$D_G = 50\left(\frac{0.06}{0.23}\right)^3 1.5^{1.5} = 1.63 \text{ m}^2/s$$

Again evaluating the parameters that appear in Eq. (4.32),

$$\frac{u_G L}{2\varepsilon_G D_G} = \frac{0.06 \times 1.5}{2 \times 0.23 \times 1.63} = 0.120$$

$$A = \left[1 + \frac{4 \times 0.176 \times 0.23 \times 1.63}{0.06^2}\right]^{1/2} = 8.63$$

Substituting in Eq. (4.32),

$$\frac{C_{CO_2 out}}{C_{CO_2 in}} = \frac{4 \times 8.62 \times e^{0.120}}{9.62^2 e^{8.62 \times 0.120} - 7.62^2 e^{-8.62 \times 0.120}} = \underline{0.162}$$

This means that in the 1.5-m column with dispersed plug flow, the ratio of inlet to outlet concentration is $1/0.162 = \underline{6.2}$.

4. Gas phase in the larger column is well mixed
   (In this case also, the actual diameter of the column does not appear in the calculation.)
   This is equivalent to a single stirred tank reactor with a first-order reaction and Eq. (1.124) of Chapter 1 can be used.
   That is,

$$\tau = \frac{(C_{in} - C_{out})}{k_1 C_{out}} = \frac{1}{k_1}\left(\frac{C_{in}}{C_{out}} - 1\right)$$

Therefore

$$\frac{C_{out}}{C_{in}} = \frac{1}{(1 + k_1 \tau)}$$

In the present context, k_1 is replaced by $k_{1G} = 0.176$, and

$$\tau = \frac{L}{u_G} = \frac{1.5}{0.06} = 25 \text{ s}$$

Thus

$$\frac{C_{CO_2 out}}{C_{CO_2 in}} = \frac{1}{(1 + 0.176 \times 2s)} = \underline{0.185}$$

This means that in the 1.5 m column with the gas phase well mixed, the ratio of inlet to outlet concentration is $1/0.185 = \underline{5.4}$.

4.1.8.2.1 Further Comments

The above example demonstrates that, due to increased gas-phase mixing, the performance of a bubble column can worsen dramatically as the column diameter is increased at the same superficial gas rate. This is an important point to remember when scaling up from a small-diameter pilot plant reactor to a full-scale production unit. One of the reasons why this effect is particularly marked in this example is that the plug flow reactor should, in theory, give a very low outlet concentration. Whenever the predicted ratio of inlet to outlet concentration for a reactor is very high (81 in this example), then the performance will be very susceptible to small departures from ideal plug flow. In cases where such a high ratio is definitely required, it is advisable to carry out the process in two separate stages.

4.1.9 Laboratory Reactors for Investigating the Kinetics of Gas–Liquid Reactions

To design a gas–liquid reactor on a fundamental basis, knowledge of the kinetics is essential. Some apparently straightforward, industrially important gas–liquid reactions, such as the absorption of nitrogen oxides NO_x in water to form nitric acid, are in fact extremely complex.[19] Although laboratory reactors can be very useful in investigating the mechanisms of complex reactions, discussion here will be limited to measuring the rate constants of simple first- or second-order reactions.

4.1.9.1 Types of Laboratory Gas–Liquid Reactors[3,10]

Choosing a laboratory reactor for the purpose of investigating a particular reaction is rather like choosing a reactor for an industrial-scale operation, in that the choice depends mainly on the intrinsic speed of the reaction—*fast*, *moderately fast*, or *slow*. As with large-scale reactors, the value of Hatta number, $\beta = \sqrt{k_2 C_{BL} D_A}/k_L$, is a useful guide, although β can only be calculated with hindsight *after* the rate constant has been measured. The aim for most types of laboratory reactor is to control the hydrodynamics so that each fresh element of liquid is exposed to the gas for a known contact time t_e (or a controlled spread of contact times). The different types of laboratory reactor can be broadly classified according to the range of contact times that each of them offers. The idea is to match approximately the contact time with the reaction time (which, for a first-order reaction, can be identified as the half-life of the reaction $(\ln 2)/k_1$ see Chapter 1). The contact times for the main types of laboratory reactor described below are shown in Table 4.3.[20]

In these types of laboratory reactors, the flow of the liquid is very carefully controlled so that, although the mass transfer step is coupled with the chemical reaction, the mass transfer characteristics can be decoupled from the reaction kinetics. For some reaction systems, absorption of the gas concerned may be studied as a purely physical mass transfer process in circumstances such that no reaction occurs. Thus, the rate of absorption

Table 4.3: Range of contact times for gas–liquid laboratory reactors.

Reactor Type	Contact Time (s)	Reaction Characteristic
Laminar jet	0.001–0.1	Very fast to fast
Laminar film	0.1–2	Fast to moderately fast
Stirred cell	1–10[a]	Moderately fast

[a]Average values calculated from the mass transfer coefficient k_L.

of CO_2 in water, or in nonreactive electrolyte solutions, can be measured in the same laboratory contactor as that used when the absorption is accompanied by the reaction between CO_2 and OH^- ions from NaOH solution. The experiments with purely physical absorption enable the diffusivity of the gas in the liquid phase D_L to be calculated from the average rate of absorption per unit area of gas–liquid interface $\overline{N}_A$ and the contact time t_e. As shown in Volume 1, Chapter 10, for the case where the incoming liquid contains none of the dissolved gas, the relationship is

$$\overline{N}_A = 2C_{Ai}\sqrt{\frac{D_A}{\pi t e}} \tag{4.35}$$

When the reaction is accompanied by a fast first-order or pseudo–first-order reaction (for example, CO_2 being absorbed into a solution of high OH^- ion concentration) with a rate constant k_1, then the average rate of absorption per unit area is given by[1]

$$\overline{N}_A = 2C_{Ai}\left(\frac{D_A}{k_1}\right)\left[\left(k_1 + \frac{1}{2t_e}\right)\operatorname{erf}\sqrt{k_1 t_e} + \sqrt{\frac{k_1}{\pi t_e}}\,e^{-k_1 t_e}\right] \tag{4.36}$$

from which k_1 may be calculated if D_A is known.

As an alternative to investigating the kinetics of a gas–liquid reaction on a laboratory scale, the mass transfer resistance may be minimized or eliminated so that the measured rate corresponds to the rate of the homogeneous liquid-phase reaction. This method of approach will be considered after first describing those reactors giving rise to controlled surface exposure times.

4.1.9.2 Laboratory Reactors With Controlled Gas–Liquid Mass Transfer Characteristics
4.1.9.2.1 Laminar Jet Contactor

The laminar jet apparatus[3,21] shown in Fig. 4.11 is simple in principle, but quite difficult to operate successfully in practice. The aim is to produce a vertical jet having a uniform velocity profile across any section. Various shapes of entry nozzles have been tried, but a simple orifice in a thin plate appears to provide the best way of avoiding a surface layer of more slowly moving liquid at the beginning of the jet. The jet tapers slightly toward the lower part as the liquid is accelerated under the influence of gravity. The receiving tube at the bottom of the jet must have a diameter that matches fairly closely the diameter of the jet; if the receiving tube is too small, liquid overflows, whereas if the receiver is too large, gas is entrained as bubbles and there is uncertainty about the amount of gas absorbed in the jet itself. A further complication is that ripples, which can be difficult to eliminate, may sometimes appear on the surface of the jet. If the jet is perfectly smooth and regular, then all elements of liquid moving down the jet have the same time of exposure t_e (i.e.,

---

[1] Derivation in Volume 1, Chapter 10 (fourth edition, 1994 revision).

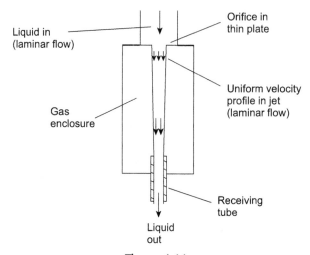

Figure 4.11
Laminar jet contactor.

the contact time), which can be calculated from the length and diameter of the jet, and the flow rate of the liquid. These same parameters can be varied to produce the range of contact times shown in Table 4.3. An advantage of the laminar jet is that the contact time is independent of the viscosity and density of the liquid. The rate of absorption of the gas is found by measuring the difference between the rates of inflow and outflow of the gas supplied to the enclosure.

4.1.9.2.2 Laminar Film Contactors

Two forms of laminar film contactor[3] are shown in Fig. 4.12. Both use a solid surface of regular geometry to support and guide the thin film of liquid, which is exposed to the gas. The wetted-tube contactor illustrated in Fig. 4.12A is very similar to the wetted-wall type of column described in Volume 2, Chapter 12, where the liquid flows down the inside wall of a tube. The advantage of using the outside wall of a tube, as shown in Fig. 4.12A, is that the distribution of liquid at the inlet can be more easily adjusted to give a film of uniform thickness around the whole circumference of the tube. A uniform thickness is important because the contact times between gas and liquid for different parts of the film will vary if the thickness varies. In addition a satisfactory liquid takeoff arrangement, such as the collar indicated in Fig. 4.12A, is more easily designed when the liquid flows on the outer wall of the tube. The hydrodynamic theory of laminar flow in a vertical film is treated in Volume 1, Chapter 3, where it is shown that the velocity at the free surface, and therefore the contact time t_e, in the present application, may be calculated from the flow rate of the liquid together with its viscosity and density. The thickness of the film can also be calculated from these quantities. When using a laminar film contactor, both extremes of

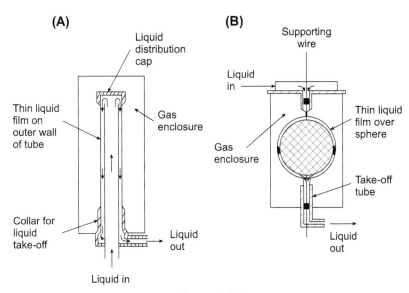

Figure 4.12
Laminar film contactors: (A) wetted wall and (B) wetted sphere.

unduly thick and unduly thin films are to be avoided. In thick films, any ripples that tend to develop on the surface of the film are less likely to be suppressed by the dampening effect of the solid wall. For thin films, there may be a tendency for imperfectly wetted areas to appear on the tube surface. Also, for a very thin film, the theory of gas absorption into a semiinfinite medium, which can be applied satisfactorily to thicker films, may need modification. For a film of medium thickness, which can be treated as semiinfinite in extent, Eqs. (4.35) and (4.36) above can be used to determine the diffusivity in the liquid D_A or the rate constant k_1, just as in the case of the laminar jet.

In the laminar film contactor shown in Fig. 4.12B, the supporting surface has the form of a sphere held in place by a wire.[3] This arrangement has the advantage that, in the regions where the liquid runs on to the sphere, and where it leaves the sphere, the surface areas exposed to the gas are relatively small so that, even if the hydrodynamics of the liquid flow in these regions is not ideal, the effect on the rate of absorption of the gas will be small. As in the case of the cylindrical tube, the contact time for each element of liquid as it flows around the sphere can be calculated from the liquid flow rate, although the mathematics of the analytical treatment is somewhat more complicated than in the case of the tube.

Instead of a single sphere, some authors[22] have used a vertical string of several adjoining spheres with the purpose of increasing the range of achievable contact times without the necessity of having to use a large-diameter single sphere; a disadvantage, however, is that it is necessary to make some assumptions about the mixing process that occurs in the liquid as it passes from one sphere to the next.

4.1.9.2.3 Stirred Cell Reactor

In this reactor the flow regime in the liquid phase is turbulent so that there is no unique exposure time that applies to all elements of liquid, but a spectrum of exposure times exists. Therefore, instead of attempting to specify contact times for the reactor, the mass transfer characteristics are described in terms of the liquid-phase mass transfer coefficient k_L.

The main features of the stirred cell reactor[3,23,24] shown in Fig. 4.13 are

1. the area of the surface between gas and liquid is clearly defined (the stirrer speed is so low that the surface remains flat), and
2. the stirrer in the liquid phase provides a means of maintaining steady and reproducible levels of turbulence and therefore consistent mass transfer characteristics.

The cell contains a relatively large volume of liquid so that, when operated as a stirred tank reactor with steady inflow and outflow of the liquid reagent, the composition of the liquid phase can be held constant. In fact, one of the original aims in developing this reactor was to simulate and maintain conditions steady at any particular point in a packed column.[3] The stirrer in the gas phase serves to keep the gas composition uniform and, in many cases, to eliminate any gas-phase mass transfer resistance.

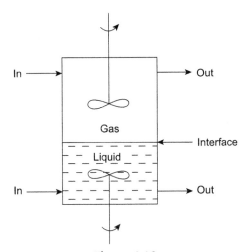

Figure 4.13

Stirred cell laboratory reactor. Note: In some designs a perforated plate is positioned in the interface to prevent excessive disturbance and to maintain a known area of contact between gas and liquid.

To investigate the kinetics of a reaction with the stirred cell, firstly experiments are carried out using a suitably related system in which gas absorption takes place by a purely physical mass transfer process (i.e., no reaction occurs). This establishes values of the physical mass transfer coefficient k_L for the range of stirrer speeds employed. Then the rate of gas absorption into the liquid with the reaction occurring is measured. Finding the rate constant for a fast first-order reaction, for example, is then a matter of working back through Eq. (4.16) to find the value of β and hence of k_1.

4.1.9.3 Reactors Eliminating Gas–Liquid Mass Transfer Resistance

For *slow* reactions, a rather different approach may be adopted whereby the mass transfer resistance is virtually eliminated altogether. This is achieved by dispersing the gas as small bubbles in a laboratory-scale stirred tank reactor as shown in Fig. 4.14, with the aim of saturating the liquid with the gas. If the bubbles are sufficiently small, i.e., if the interfacial area a is sufficiently large, then the concentration of the gas-phase reactant **A** in the bulk of the liquid C_{AL} becomes almost equal to the interface value, as shown in Fig. 1.2E. The rate of transformation is then the rate of the homogeneous reaction occurring in the liquid phase. As a test of whether the gas–liquid mass transfer resistance has been wholly eliminated, the agitator speed is increased, so increasing the gas–liquid surface area, until the overall rate of transformation levels off to an almost constant value. This type of reactor would be suitable for measuring the kinetics of the chlorine–toluene reaction, which is involved in Example 4.4.

For some *fast* reactions, there is another method of eliminating the gas–liquid mass transfer step and thereby of studying only the homogeneous liquid-phase reaction. A classic example is the measurement of the rate constant for the reaction between CO_2

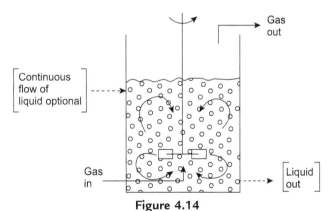

Figure 4.14
Laboratory stirred tank reactor for investigating the kinetics of slow reactions. Agitator disperses the gas with the aim of saturating the liquid with dissolved gas.

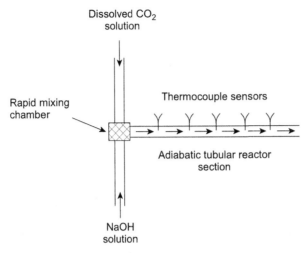

Figure 4.15

Rapid thermal flow reactor as used to determine the kinetics of the fast first-order reaction between CO_2 and OH^- ions: $CO_2 + OH^- = HCO_3^-$.

and OH^- ions using a fast flow reactor[25,26] (Fig. 4.15). The CO_2 is dissolved in water and a flowing stream of the solution is mixed very rapidly with a stream of the alkaline solution. Downstream from the mixing junction is situated a series of sensors that register the change in a property such as temperature, which is the property that has been used for monitoring the exothermic CO_2–OH^- reaction. Assuming that the reaction takes place adiabatically and under steady flow conditions, the temperature rise is a measure of the progress of the reaction as the mixed solutions proceed downstream. Other properties such as conductivity, pH, or light transmission can also be used to register the course of the reaction. A further development of the fast flow reactor is the "stopped-flow" technique[27] in which the flows are arrested soon after the start of the reaction. The progress of the reaction in the tube downstream of the mixing chamber is then followed as a function of time using spectroscopic techniques. In this method, much less of the liquid reagents are used than in the continuous steady flow method; it is therefore suitable for reactions involving enzymes or other biological fluids of limited availability.

4.2 Gas–Liquid–Solid Reactors

4.2.1 Gas–Liquid–Solid Reactions

In a surprisingly large number of industrially important processes, reactions are involved that require the simultaneous contacting of gas, liquid, and solid particles.[28] Very often the solid is a catalyst and it is on the surface of the solid that the chemical reaction occurs.

The need for three-phase contacting can be appreciated by considering, as an example, the hydrodesulfurization of a residual petroleum fraction, i.e., of the liquid taken from the base of a crude oil distillation column.

$$H_2 + R \cdot SH = R \cdot H + H_2S$$

where $R \cdot SH$ is representative of a wide range of sulfur compounds present in the oil. The catalyst is of necessity a solid, typically molybdenum disulphide (MoS_2) deposited on a highly porous alumina support.[29] The liquid, being a residue, cannot under any practical conditions be vaporized; and hydrogen, sometimes referred to as a "permanent" gas, cannot be liquefied except at extremely low temperatures. The reaction is normally carried out at high temperatures (300—350°C) and at high pressures (50—100 bar) to increase the solubility of the hydrogen in the oil.[30] Many industrial three-phase reactors involve hydrogenations or oxidations. In advanced oxidation processes such as photocatalytic degradation of toxic organic pollutants using visible radiation, oxygen/air is usually bubbled through an aqueous solution of waste water, whereas the TiO_2 catalyst is suspended.[31] Dissolved oxygen in aqueous solution and oxygen from the air bubbles are transferred to the catalyst surface, where it reacts with conduction band electrons generated via photoexcitation of the semiconductor photocatalyst to form superoxide radicals $\left(O_2 + e^- \rightarrow O_2^{\bullet-}\right)$. These then take part in a number of other reactions to form hydroxyl radicals in the aqueous phase, which acts as a precursor for degradation of organic compounds. Another reaction that has been extensively investigated, because of its importance if crude oil becomes scarce, is the conversion of carbon monoxide—hydrogen mixtures, derived from the gasification of coal, to hydrocarbon liquids in the Fischer—Tropsch process.[10] A finely divided solid catalyst based on iron is commonly used, the catalyst being suspended in a hydrocarbon liquid in a slurry-type reactor.

4.2.2 Mass Transfer and Reaction Steps

The individual mass transfer and reaction steps occurring in a gas—liquid—solid reactor may be distinguished as shown in Fig. 4.16. As in the case of gas—liquid reactors, the description will be based on the film theory of mass transfer. For simplicity, the gas phase will be considered to consist of just the pure reactant **A**, with a second reactant **B** present in the liquid phase only. The case of hydrodesulfurization by hydrogen (reactant **A**) reacting with an involatile sulfur compound (reactant **B**) can be taken as an illustration, applicable up to the stage where the product H_2S starts to build up in the gas phase. (If the gas phase were not pure reactant, an additional gas film resistance would need to be introduced, but, for most three-phase reactors, gas film resistance, if not negligible, is

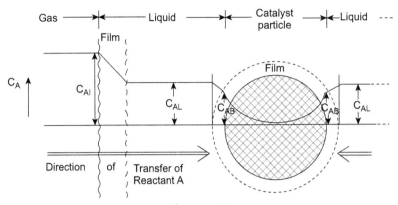

Figure 4.16

Mass transfer and reaction steps in gas–liquid–solids reactors (*solids* completely wetted by liquid), showing concentration gradients for reactant A being transferred from the gas phase.

likely to be small compared with the other resistances involved.) The reaction proceeds as follows:

1. Starting in the gas phase, the first step is that **A** (hydrogen) dissolves at the gas–liquid interface, where its concentration in the liquid will be C_{Ai} (equilibrium with the gas is assumed), and is transferred across the liquid film into the bulk of the liquid where its concentration is C_{AL}.

2. a. Reactant A is transferred, across the stagnant liquid film surrounding the particle, to the external surface of the solids where its concentration is C_{AS}.

   b. The second reactant **B** (mercaptan or similar sulfur compound) is also transferred from the bulk liquid to the surface of the solid.

3. Both reactants then diffuse through the porous structure of the catalyst and react at active sites on the surface of the solid. As in the case of gas–solid catalytic reactions, the driving force required for diffusion will reduce the concentrations of the reactants toward the interior of the particle and so lead to a reduction in the efficiency of the catalyst. To allow for this, an *effectiveness factor* η is introduced in just the same way as for gas–solid catalytic reactors (Chapter 3). An important difference, however, with three-phase reactors is that the pores of the catalyst will be filled with liquid, and diffusivities in liquids are much lower than in gases. This means that effectiveness factors are likely to be much below unity unless the high resistance to internal diffusion is offset by a very small particle size for the catalyst.

The final stage in the complete reaction process is the diffusion of the products of the reaction out of the particle and their eventual transfer back into the liquid or gas phase. However, only if the reaction is reversible will any buildup of products affect the rate of the reaction itself. For an irreversible reaction, generally the fate of the products is not so important and needs not be taken into account in determining the rate of the forward transfer and reaction steps.

4.2.3 Gas—Liquid—Solid Reactor Types: Choosing a Reactor

4.2.3.1 Significance of Particle Size

The design of a gas—liquid—solid reactor is very much dependent on the size of the solid particles chosen for the reaction, and the anticipated value of the effectiveness factor is one of the most important considerations. Generally, the smaller the particle size the closer the effectiveness factor will be to unity. Particles smaller than about 1 mm in diameter cannot, however, be used in the form of a fixed bed. There would be problems in supporting a bed of smaller particles; the pressure drop would be too great; and perhaps, above all, the possibility of the interstices between the particles becoming blocked too troublesome. There may, however, be other good reasons for choosing a fixed bed type of reactor.

When small particles are chosen, they have to be used in the form of a suspension in the liquid. The lower limit of particle size may be set by the nature of the catalyst but is most likely to depend on how easily the particles can be separated from the liquid products. To avoid filtration, the most convenient way of separating the particles is to allow them to settle under gravity, and the minimum particle size is set accordingly.

4.2.3.2 Types of Reactor

Three-phase reactors[28] can thus be divided into two main classes: (1) *suspended bed reactors* and (2) *fixed bed reactors*.

4.2.3.2.1 Suspended Bed Reactors

These may be further subdivided according to how the particles are maintained in suspension in the reactor.

1. Bubble columns: If the particles are very small, i.e., have relatively low sedimentation velocities, then they can quite easily be kept in suspension just by bubbling the gas through the liquid in a bubble column as shown in Fig. 4.1D. Special attention needs to be given to the design of the gas sparger to ensure that there are no stagnant regions where solids might be deposited.
2. Agitated tanks: A more positive way of suspending the particles is to use an agitated tank like that shown in Fig. 4.1F. For three-phase operation the impeller needs to be properly designed and positioned nearer to the bottom of the tank so that it will keep the solids in suspension as well as dispersing the incoming gas (see also Fig. 4.21). Information on the performance of such impellers can be found in the book *Mixing in the Process Industries*.[17] In some types of tall and narrow agitated tanks, two or three impellers are mounted on a single shaft (see Fig. 4.10A). Because of the higher local

velocities produced by the impeller in an agitated tank compared with those in a bubble column, an agitated tank can be used for larger sizes of particles.

Note that three-phase bubble columns and agitated tank reactors are sometimes referred to as *slurry reactors*.

3. Three-phase fluidized beds: A further method of suspending even larger particles is to fluidize the bed using an upward flow of liquid. If gas is then introduced at the bottom of the bed, three-phase contacting is achieved. As in bubble columns and agitated tank reactors, hydrodynamic interactions between the bubbles and the particles occur (see Fig. 4.17). Large particles can cause breakup of any large bubbles, increasing the interfacial area for gas—liquid transfer.

4.2.3.2.2 Fixed Bed Reactors

Apart from particle size, the main choice to be made with fixed bed reactors is the direction of flow, i.e., upward or downward for the gas and liquid phases. As shown in Fig. 4.18, there are three possibilities, any of which may be chosen for reasons particular to a given process.

1. Liquid and gas in cocurrent downflow—This configuration is sometimes called a *trickle bed* reactor[32] because, at low to moderate gas and liquid flows, the gas phase is continuous and the liquid flows as a thin film over the surface of the catalyst. In this regime, the hydrodynamic behavior of the liquid is influenced only to a small extent by changes in the gas flow rate. At higher gas flow rates there is more interaction between the liquid and gas flow regimes,[33] especially for liquids that have a tendency to foam.

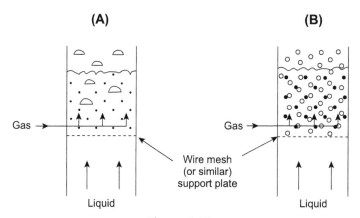

Figure 4.17

Effect of interaction between particles and bubbles in three-phase fluidized beds. (A) Fluidized bed of small particles giving large spherical cap bubbles. (B) Fluidized bed of large particles giving small, more nearly spherical bubbles. Note: Bubble sizes and breakup/coalescence behavior depend more on gas and liquid flow rates.

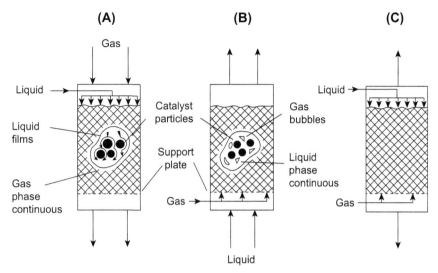

Figure 4.18

Flow regimes in three-phase fixed bed reactors. (A) Gas and liquid in cocurrent downward flow (trickle bed operation). (B) Gas and liquid in cocurrent upward flow (liquid floods bed). (C) Gas and liquid in countercurrent flow (not often used for catalytic reactors).

Fig. 4.19 is a flow map[34] showing the conditions under which trickling flow changes to a pulsing type of flow and, at very high gas rates, to spray flow.

Trickle bed reactors are widely used in the oil industry because of reliability of their operation and for the predictability of their large-scale performance from tests on a pilot plant scale. Further *advantages* of trickle bed reactors are as follows:

The flow pattern is close to plug flow, and relatively high reaction conversions may be achieved in a single reactor. If warranted, departures from ideal plug flow can be treated by a dispersed plug flow model with a dispersion coefficient for each of the liquid and gas phases.

Pressure drop through the reactor is smaller than that with upflow, and there is no problem with flooding, which might otherwise impose a limit on the gas and liquid flow rates used.

The particles of the bed are held firmly in place against the bottom support plate as a result of the combined effect of the forces attributable to gravity and fluid drag.

On the other hand, some *disadvantages* of trickle bed reactors are encountered:

Heat transfer in the radial direction is poor, so large reactors are best suited to adiabatic operation. If the temperature rise is significant, it may be controlled in some processes by recycling the liquid product or introducing a quench stream at an intermediate point.

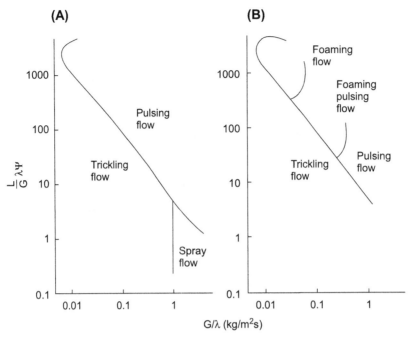

Figure 4.19

Flow regimes for cocurrent downward flow of gas and liquid through a fixed bed of particles. (A) Nonfoaming liquids. (B) Foaming liquids. L and G are liquid and gas flow rates per unit area $(\text{kg/m}^2 \text{ s})$; $\lambda = [(\rho_G/\rho_{air})(\rho_L/\rho_{water})]^{1/2}$; $\psi = [\sigma_{water}/\sigma_L]\left[(\mu_L/\mu_{water})\big/(\rho_{water}/\rho_L)^2\right]^{1/3}$ where ρ_L, ρ_G, ρ_{air}, and ρ_{water} are densities of liquid, gas, air, and water; μ_L and μ_{water} are viscosities of liquid and water; σ_L and σ_{water} are surface tensions of liquid and water.

At low liquid flow rates, incomplete wetting of the catalyst by the liquid may occur, as illustrated in Fig. 4.20, leading to channeling and a deterioration in reactor performance.[35]

2. Liquid and gas in cocurrent upflow: At low flow rates the packed bed reactor (see Fig. 4.18B) behaves rather like the packed bubble column shown in Fig. 4.1E with the liquid phase continuous and the gas dispersed as bubbles. The higher liquid holdup compared with a trickle bed reactor is likely to be the main reason for choosing the upflow mode of operation; it has the added advantage that higher gas—liquid mass transfer coefficients are also obtained. However, at high flow rates, there is the possibility that the catalyst particles will be lifted and fluidized unless constrained from above by a perforated plate pressing down on the bed. In any case the pressure drop will be higher than for a downflow reactor.

3. Liquid in downflow—gas in upflow: This situation is similar to that in the packed column shown in Fig. 4.1A with countercurrent flow of gas and liquid as in a

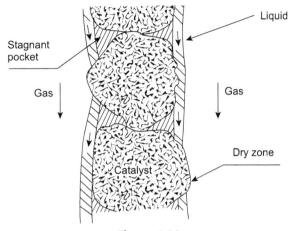

Figure 4.20
Trickle flow of liquid over a bed of particles, showing possible complications of stagnant pockets of liquid and dry zones due to poor wetting of the solid surface.

conventional gas absorption process. In a three-phase reaction, consideration of the concentration driving forces for mass transfer shows that countercurrent operation is usually of little advantage unless the reaction is reversible. Because, with small catalyst particles, countercurrent operation can easily lead to overloading and flooding of the column, it is rarely used in reactors.

4.2.4 Combination of Mass Transfer and Reaction Steps

4.2.4.1 Suspended Bed Reactor

The individual mass transfer and reaction steps outlined in Fig. 4.16 will now be described quantitatively. The aim will be firstly to obtain an expression for the overall rate of transformation of the reactant and then to examine each term in this expression to see whether any one step contributes a disproportionate resistance to the overall rate. For simplicity we shall consider the gas to consist of just a pure reactant **A**, typically hydrogen, and assume the reaction that takes place on the interior surface of the catalyst particles to be first order with respect to this reactant only, i.e., the reaction is pseudo-first order with rate constant k_1. In an agitated tank suspended bed reactor, as shown in Fig. 4.21, the gas is dispersed as bubbles, and it will be assumed that the liquid phase is "well mixed," i.e., the concentration C_{AL} of dissolved **A** is uniform throughout, except in the liquid films immediately surrounding the bubbles and the particles. (It will be assumed also that the particles are not so extremely small that some are present just beneath the surface of the liquid within the diffusion film and are thus able to catalyze the reaction before **A** reaches the bulk of the liquid.)

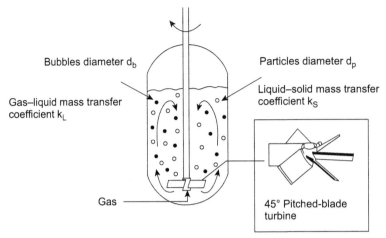

Figure 4.21
Suspended bed agitated tank reactor: Combination of mass transfer and reaction steps. Impeller used would typically be a pitched blade turbine, pumping downward as shown, serving both to suspend particles and to disperse gas.

Choosing a clearly defined basis is an all-important first stage in the treatment of three-phase reactors. In this case the overall reaction rate r_t, will be based on unit volume of the whole dispersion, i.e., gas + liquid + solid. On this basis let:

a = gas–liquid interfacial area per unit volume of *dispersion*,
ε_G = volume fraction of gas, i.e., bubbles,
ε_L = volume fraction of liquid, and
ε_P = volume fraction of particles.

$$\varepsilon_G + \varepsilon_L + \varepsilon_P = 1 \tag{4.37}$$

For the particles, the external surface area is most conveniently defined as

a_p = external surface area per unit volume of *particles*.

For both bubbles and particles, there will be a distribution of sizes in the dispersion. The above quantities can be related to the volume/surface or Sauter mean bubble and particle diameters d_b and d_p (see Volume 2, Chapter 1). Thus, for bubbles (as in Example 4.3),

$$d_b = \frac{6\varepsilon_G}{a} \tag{4.38}$$

and for the particles,

$$d_p = \frac{6}{a_P} \tag{4.39}$$

Note that, per unit volume of dispersion, the surface area of the particles is $a_P \varepsilon_P$.

As Fig. 4.16 demonstrates, the two mass transfer steps, gas to liquid and liquid to solid, and then the chemical reaction, take place in series. This means that, in the steady state, each must proceed at the same rate as the overall process. Continuing then on the basis of unit volume of dispersion, and using the reactant concentrations shown in Fig. 4.16, with the gas–liquid and liquid–solid film mass transfer coefficients k_L and k_s shown in Fig. 4.21, the overall rate r_t may be written as

$$r_t = k_L a (C_{Ai} - C_{AL}) \text{ corresponding to gas–liquid mass transfer} \tag{4.40}$$

or

$$r_t = k_s a_P \varepsilon_P (C_{AL} - C_{As}) \text{ corresponding to liquid–solid mass transfer} \tag{4.41}$$

or

$$r_t = k_1 C_{As} \eta \varepsilon_P \text{ from the rate of reaction within the particle} \tag{4.42}$$

The reaction term follows from the assumption of a first-order reaction with a rate constant k_1 defined on the basis of unit volume of *particle* (see Chapter 3); ε_P allows for the change in basis to unit volume of dispersion. The effectiveness factor η is also included to take into account any diffusional resistance within the pores of the particle.

Eqs. (4.40)–(4.42) are then rearranged to give

$$\frac{r_t}{k_L a} = C_{Ai} - C_{AL}$$

$$\frac{r_t}{k_s a_P \varepsilon_P} = C_{AL} - C_{AS}$$

$$\frac{r_t}{k_1 \eta \varepsilon_P} = C_{AS}$$

On adding these equations, C_{AL} and C_{AS}, which are unknown, cancel out so that

$$r_t = \left(\frac{1}{k_L a} + \frac{1}{k_s a_P \varepsilon_P} + \frac{1}{k_1 \eta \varepsilon_P} \right)^{-1} C_{Ai} \tag{4.43}$$

Eq. (4.43) expresses the overall rate r_t in terms of the overall driving force C_{Ai}, the concentration existing at the gas–liquid interface, which is known from the solubility of the gas in the liquid. Thus, if the pressure of the reactant **A** in the gas phase is P_A, then C_{Ai} is given by

$$C_{Ai} = P_A / K \tag{4.44}$$

where K is the Henry's law constant. The three terms in parentheses on the right-hand side of Eq. (4.43) can be regarded as the effective resistances associated with each of the three individual steps, $(k_L a)^{-1}$ for the gas–liquid mass transfer step, $(k_S a_P \varepsilon_P)^{-1}$ for the liquid–solid mass transfer step, and $(k_1 \eta \varepsilon_P)^{-1}$ for the pore diffusion with reaction step. Eq. (4.43) thus has the form, familiar in other cases of mass transfer and heat transfer where resistances are in series, and the overall resistance is obtained by adding terms involving the reciprocals of transfer coefficients. In the numerical example that follows, we will evaluate each of these resistances and discuss how the process might be changed to minimize the contribution of the step offering the largest proportion of the resistance.

Example 4.6

The Hydrogenation of α-Methyl Styrene in an Agitated Tank Slurry Reactor

Note: The hydrogenation of α-methyl styrene, although not itself a commercially important reaction, has been studied as a model for other more complex catalytic hydrogenations carried out on an industrial scale. The kinetics of the hydrogenation reaction catalyzed by palladium deposited within porous alumina particles were investigated by Satterfield et al.[36] The particles were crushed to a very small size to eliminate diffusional resistance within the pores; i.e., the effectiveness factor in the experiments was effectively unity. In addition, very intense agitation was used to disperse the hydrogen and so eliminate interphase mass transfer resistances. The result of these experiments was that the reaction appeared to be pseudo-first order with respect to hydrogen with a rate constant k_1 equal to 16.8 s^{-1} at 50°C. Also from measurements of the porosity of the larger sized catalyst particles, the effective diffusivity D_e of hydrogen in the particles, with α-methyl styrene filling the pores, was estimated to be $0.11 \times 10^{-8} \text{ m}^2/\text{s}$.

We will now consider the design of an agitated tank slurry reactor, which might be used industrially for this reaction. We will choose a particle size of 100 μm rather than the very small size particles used in the laboratory experiments, the reason being that industrially we should want to be able to easily separate the catalyst particles from the liquid products of the reaction. If we use spherical particles (radius r_0, diameter d_p), the Thiele modulus ϕ (see Chapter 3) is given by

$$\phi = \lambda \frac{r_0}{3} = \lambda \frac{d_p}{6} \quad \text{(from Eq. 3.19)}$$

where $\lambda = \sqrt{\frac{k_1}{D_e}}$

and the effectiveness factor η by

$$\eta = \frac{3}{\lambda r_0} \left\{ \coth(\lambda r_0) - \frac{1}{\lambda r_0} \right\} \quad \text{(Eq. 3.18)}$$

Substituting for λr_0,

$$\eta = \frac{1}{\phi} \left(\coth 3\phi - \frac{1}{3\phi} \right)$$

Using the above values for d_p, k_1, and D_e,

$$\phi = 2.06 \quad \text{and} \quad \eta = 0.407$$

The fact that η, which can be regarded as a catalyst efficiency, has a value of 0.407, rather than unity, is the price we pay for choosing catalyst particles of such a size that they can be separated easily.

The size of the bubbles produced in the reactor and the gas volume fraction will depend on the agitation conditions, and the rate at which fresh hydrogen is fed to the impeller, as shown in Fig. 4.21. (Some hydrogenation reactors use gas-inducing impellers to recirculate gas from the headspace above the liquid, whereas others use an external compressor.) For the purposes of the present example, the typical values, $d_b = 0.8$ mm and $\varepsilon_G = 0.20$, will be taken. The values of the transfer coefficients k_L and k_s are also dependent to some extent on the agitation conditions, although these will depend mainly on the physical properties of the system such as the viscosity of the liquid and the diffusivity of the dissolved gas. The values taken here will be $k_L = 1.23 \times 10^{-3}$ m/s and $k_s = 0.54 \times 10^{-3}$ m/s. (These have been estimated for typical reactor conditions from the correlations described in textbooks such as those by Tatterson[9] and Harnby et al.[17])

The last important design parameter to be specified is the catalyst loading to be used in the process, i.e., the ratio of catalyst to liquid charged to the reactor. There is little point in using more catalyst than is really necessary so, as a first trial, a solid/liquid ratio of 0.10 by volume will be used. The pressure of the hydrogen gas in the reactor also needs to be decided. Hydrogen is one of the least soluble gases so that industrial hydrogenations are usually carried out at high pressures to increase the solubility under reactor conditions, i.e., to increase C_{Ai} in Eq. (4.44). In this example, a pressure of 30 bar will be used compared with which the vapor pressure of the α-methyl styrene is negligible. At $50°C$ the value of the Henry's law constant for hydrogen dissolved in α-methyl styrene is 285 m^3 bar/kmol.

Using the above values, from Eqs. (4.38) and (4.39), the gas–liquid interfacial area a is 1500 m^2/m^3 and the particle surface area is 6×10^4 m^2/m^3. In the dispersion, where the gas volume fraction ε_G is 0.20, the solid-to-liquid ratio of 0.10 for the feed charged to the reactor corresponds to the ratio $\varepsilon_P/\varepsilon_L$ so that, from Eq. (4.37), ε_P is 0.073 and ε_L is 0.727. The resistance terms in Eq. (4.43) can now be examined in turn:

$$\frac{1}{k_L a} = \frac{1}{1.23 \times 10^{-3} \times 1500} = \underline{0.54\ s}$$

$$\frac{1}{k_s a_p \varepsilon_P} = \frac{1}{0.54 \times 10^{-3} \times 6 \times 10^4 \times 0.073} = \underline{0.42\ s}$$

$$\frac{1}{k_1 \eta \varepsilon_P} = \frac{1}{16.8 \times 0.407 \times 10^4 \times 0.073} = \underline{2.00\ s}$$

It is clear from these calculations that the term $(k_1 \eta \varepsilon p)^{-1}$ arising from pore diffusion with reaction is considerably larger than either of the others. Unless a more active catalyst can be found, or possibly the operating temperature can be varied, little can be done to increase k_1. The effectiveness factor η is considerably less than unity and could be increased by choosing a smaller particle size, with the disadvantage of a more difficult separation. However, a distinct improvement can be made by using a higher catalyst loading, i.e., a higher ratio of solid to liquid in the charge to the reactor. There will be a limit to the extent that the amount of catalyst can be increased because, as the solids content rises, the slurry will become non-Newtonian, dispersion of the gas will become more difficult, and the bubble size d_b will increase to the detriment of the gas—liquid interfacial area a in the $(k_L a)^{-1}$ resistance term. This example demonstrates the importance of achieving an optimized balance in the choice of conditions for a gas—liquid—solid reactor.

Finally a value for the overall rate r_t of the hydrogenation process can be calculated from Eq. (4.43). For a pressure of 30 bar, Eq. (4.44) gives a value for C_{Ai} of 0.105 kmol/m³. Using the above values for the individual resistances, the overall rate becomes

$$r_t = (0.54 + 0.42 + 2.00)^{-1} \times 0.105 = 0.036 \; \text{kmol}/\text{m}^3\text{s}.$$

4.2.4.2 Three-Phase Fluidized Suspended Bed Reactor—Combination of Mass Transfer and Reaction Steps

At the present time, three-phase fluidized beds are not often chosen for gas—liquid—solid reactions despite their advantages of good heat and mass transfer and, in principle, freedom from the blockages that can occur with fixed beds.[30] The reason may be that, because of the pronounced hydrodynamic interactions between the phases as indicated in Fig. 4.17, development of a three-phase fluidized bed reactor for a new process would necessarily require extensive experimental work, which would be both time-consuming and costly. Also in the present state of knowledge, scale-up from pilot plant operation to a full-scale design would still be uncertain. By contrast, scale-up of trickle bed reactors is relatively well understood so that a trickle bed reactor might be chosen for a process despite any of the advantages that, in theory, a three-phase fluidized bed reactor might have to offer.

The three-phase fluidized bed reactors so far developed commercially for processes such as the hydrogenation of petroleum residues use beds of various particle sizes depending on the type of feedstock and the particular experience of the company that developed the process. Reactors with small particles will tend to behave in the way shown in Fig. 4.17A, producing comparatively large bubbles. Apart from having an upward flow of liquid, these are somewhat similar to the bubble column type of slurry reactor. A three-phase fluidized bed reactor using relatively large particles, as shown in Fig. 4.17B, requires higher liquid flow rates to maintain the particles in the fluidized state, but has the advantage of producing smaller bubbles and therefore a larger gas—liquid surface area for mass transfer. To compare this type of three-phase fluidized bed reactor with the agitated slurry type of reactor using small particles, as previously considered in Example 4.6, the hydrogenation of α-methyl styrene will again be taken as a model reaction in the example that follows.

One of the areas of uncertainty in three-phase fluidized bed design is the mixing that occurs within each of the liquid and gas phases. As the particle size increases, so the flow patterns approximate more nearly to plug flow. For the hydrogenation of α-methyl styrene using pure hydrogen, gas phase mixing will not affect the process and, because the reaction rate is independent of the α-methyl styrene concentration, mixing in the liquid phase will have an effect only through the dissolved hydrogen concentration. In the example, it will be assumed that a steady state has been established between the rates of mass transfer and the rate of the reaction (which is the case for the agitated tank suspended bed reactor in Example 4.6). Unless the liquid is already saturated with dissolved hydrogen before entering, this steady state may not apply near the liquid inlet. (If the liquid is in plug flow, the variation of dissolved hydrogen concentration near the liquid inlet will be similar to that shown in Fig. 4.23 where, in Example 4.8, the effect is considered in more detail for a trickle bed reactor.)

Although bubble sizes for large particles with the α-methyl styrene system have not been measured directly and any prediction must be regarded as highly speculative, Fig. 4.22 shows an estimate of the variation of mean bubble diameter with size of catalyst particles.[37] This estimate is based on measurements with glass beads in water, with subsequent adjustments to allow for the different densities of the particles and differences in the viscosity and surface tension of the liquids.

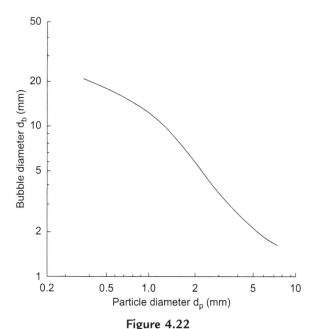

Figure 4.22

Estimated bubble sizes in α-methyl styrene with catalyst particles in a three-phase fluidized bed; particle volume fraction $\varepsilon_P = 0.5$, gas volume fraction $\varepsilon_G = 0.2$.

Example 4.7

The Hydrogenation of α-Methyl Styrene in a Three-Phase Fluidized Bed Reactor
The performance of a three-phase fluidized bed is to be assessed in relation to a model reaction, the hydrogenation of α-methyl styrene at 50°C. The same catalyst will be used as in Example 4.6, excepting that the diameter of the spherical particles will be 2.0 mm. The liquid and gas flow rates will be such that the volume fraction of particles in the bed will be 0.50 and the volume fraction of gas 0.20; i.e., the relationship shown in Fig. 4.22 will apply, from which the estimated bubble size is 5.0 mm.

Solution

Because the conditions for the reaction are the same as in Example 4.6, the pseudo—first-order rate constant k will again be 16.8 s^{-1}, and the effective diffusivity of hydrogen in the liquid filled pores of the catalyst D_e will be 0.11×10^{-8} m^2/s. Also because the transfer coefficients k_L and k_s depend mainly on the physical properties of the system, the same values, namely, $k_L = 1.23 \times 10^{-3}$ m/s and $k_s = 0.54 \times 10^{-3}$ m/s, will be used even though the hydrodynamics of the three-phase fluidized bed will be different and the particle size is larger.

As in Example 4.6 the aim is again to evaluate the individual resistance terms in Eq. (4.43) and to examine them in turn. In addition the overall rate of hydrogenation at the same hydrogen pressure of 30 bar will be calculated.

1. Gas—liquid mass transfer resistance $1/k_L a$:
   Because $\varepsilon_G = 0.20$ and $d_b = 0.005$ m, from Eq. (4.38) the gas—liquid interfacial area $a = 6\varepsilon_G/d_b = 6 \times 0.20/0.005 = 240$ m^2/m^3.

   $$\frac{1}{k_L a} = \frac{1}{1.23 \times 10^{-3} \times 240} = \underline{3.39\ s}$$

2. Liquid—solid mass transfer resistance $1/k_s a_p \varepsilon_p$:
   From Eq. (4.39), $a_P = 6/d_P = 6/0.002 = 3000$ m^2/m^3:

   $$\frac{1}{k_s a_p \varepsilon_p} = \frac{1}{0.54 \times 10^{-3} \times 3000 \times 0.50} = \underline{1.23\ s}$$

3. Reaction within the particle $(k_1 \eta \varepsilon p)^{-1}$:
   As in Example 4.6, the effectiveness factor η for spherical particles is given by

   $$\eta = \frac{1}{\phi}\left(\coth 3\ \phi - \frac{1}{3\phi}\right)$$

   where

   $$\phi = \frac{d_P}{6}\sqrt{\frac{k_l}{D_e}}$$

   Thus, in the present case, $\phi = 41.2$ and $\eta = 0.0241$.

   $$\frac{1}{k_1 \eta \varepsilon p} = \frac{1}{16.8 \times 0.0241 \times 0.5} = \underline{4.94\ s}$$

   The overall rate of reaction r_t is therefore obtained from Eq. (4.43), using the value for C_{Ai} of 0.105 kmol/m^3, from Example 4.6

   $$r_t = (3.39 + 1.23 + 4.94)^{-1} \times 0.105 = \underline{0.0110\ kmol/m^3 s}.$$

If even larger particles are considered, repeating the above calculation shows that, although the gas—liquid resistance (1) decreases, both the liquid—solid resistance (2) and the reaction term (3) increase. As a result of these opposing tendencies, there must exist an optimum particle size for which the overall resistance is a minimum and therefore the overall reaction rate r_t is a maximum. More detailed calculations[37] show that for the value of the rate constant $k_1 = 16.8 \text{ s}^{-1}$ used above, the optimum particle size is indeed approximately 2 mm, which is the size taken for the example. However, for lower values of the rate constant, the reason for the optimum shifts toward smaller particle sizes is that the gas—liquid interfacial area for mass transfer becomes less important.

The overall rate of reaction calculated for the three-phase fluidized bed reactor above is approximately one-third of the rate calculated for the agitated tank slurry reactor in Example 4.6. The main reasons are the very poor effectiveness factor and the relatively smaller external surface area for mass transfer caused by using the larger particles. Even the gas—liquid transfer resistance is greater for the three-phase fluidized bed, in spite of the larger particles being able to produce relatively small bubbles; these bubbles are not, however, as small as can be produced by the mechanical agitator. Although the three-phase fluidized bed does not appear to perform very favorably compared with the agitated tank slurry reactor, it should be remembered that the fluidized bed does not have the sealing and other mechanical complications associated with an agitator. In addition the gas and liquid will be more nearly in plug flow in the fluidized bed, although this advantage may to some extent be lost if the liquid has to be recycled to provide a high liquid velocity for fluidization.

4.2.4.3 Trickle Bed Reactor—Combination of Mass Transfer and Reaction Steps

One of the advantages of the cocurrent downward flow, trickle bed type of three-phase reactor is that both gas and liquid move nearly in plug flow. For the gas, any departure from ideal plug flow is usually unimportant, but for the liquid phase, the departure from plug flow may be sufficient in some cases to warrant using the dispersed plug flow model. In the example that follows, it will be assumed for simplicity that the reactor operates under isothermal conditions, that the distribution of both gas and liquid is uniform, and that there are no radial concentration gradients. The reaction involved is the hydrogenation of thiophene to remove the sulfur as hydrogen sulfide. This reaction has been studied as a model for industrial hydrodesulfurization processes involving feedstocks consisting of complex mixtures of sulfur-containing hydrocarbons. The catalyst for the reaction is sulfided cobalt—molybdenum oxide on an alumina support.

In general, for a trickle bed reactor, a material balance is required for each of the components present, taken over each of the gas and liquid phases. In the example, however, pure hydrogen will be used and the volatility of the other components will be assumed to be sufficiently low that they do not enter the vapor phase. This means that the material balance on the gas phase can be omitted. It also means that there will be no gas film resistance to gas—liquid mass transfer. In the liquid phase, material balances are required for (1) the hydrogen (reactant **A**) and (2) the thiophene (reactant **B**). The amounts of each of the reactants consumed will be linked by the stoichiometric equation:

$$\text{thiophene} + 4\,H_2 = C_4H_{10} + H_2S$$

In deriving the material balance equations, the dispersed plug flow model will first be used to obtain the general form but, in the numerical calculations, the dispersion term will be omitted for simplicity. As used previously throughout, the basis for the material balances will be unit volume of the whole reactor space, i.e., gas plus liquid plus solids. Thus in the equations below, for the transfer of reactant **A**, $k_L a$ is the volumetric mass transfer coefficient for gas—liquid transfer and $k_s a_s$ is the volumetric mass transfer coefficient for liquid—solid transfer.

Consider an element height δz of a reactor of unit cross-sectional area, as shown in Fig. 4.23A. Let the gas phase consist of a pure reactant **A** (hydrogen in the example) at a

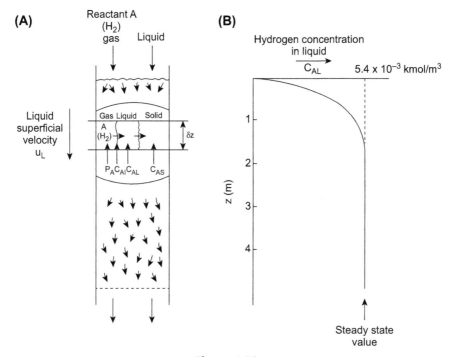

Figure 4.23
Trickle bed reactor. (A) Schematic diagram showing individual mass transfer steps for a hydrogenation process; (B) profile of dissolved hydrogen in the liquid flowing down the column, calculated in Example 4.8.

pressure P_A. The concentration of **A** dissolved in the liquid at the gas–liquid interface will therefore be $C_{Ai} = P_A/K$, where K is the Henry's law constant. We will now consider each of the terms in the mass balance equation for **A**, in the liquid phase, over the element δz:

Rate of transfer of **A** into the element from the gas phase: $k_L a(C_{Ai} - C_{AL})\delta z$.

Rate of transfer of **A** out of the element to the solid: $k_s a_s(C_{AL} - C_{AS})\delta z$.

Rate of convection of **A** into the element with the liquid flow: $u_L C_{AL}$.

Rate of convection of **A** out of the element by liquid flow: $u_L\left(C_{AL} + \frac{dC_{AL}}{dz}\delta z\right)$.

Rate of transport of **A** into element by axial dispersion: $-D_L\frac{dC_{AL}}{dz}$.

Rate of transport of **A** out of element by axial dispersion: $D_L\left(\frac{dC_{AL}}{dz} + \frac{d^2C_{AL}}{dz^2}\delta z\right)$.

Because conditions in the reactor do not change with time, taken over 1 s, the net sum of all the transfer terms above, on canceling out δz, gives the following equation:

$$D_L\frac{d^2C_{AL}}{dz^2} - u_L\frac{dC_{AL}}{dz^2} + k_La(C_{Ai} - C_{AL}) - k_Sa_S(C_{AL} - C_{AS}) = 0 \qquad (4.45)$$

$$\text{Dispersion} \qquad \text{Convection} \qquad \text{Gas to liquid} \qquad \text{Liquid to solid}$$

The concentration C_{AS} is determined by the rate of reaction within the solid, which must be equal to the rate of transfer from the liquid:

$$k_Sa_S(C_{AL} - C_{AS}) = k_1 C_{AS}\eta\epsilon_P \qquad (4.46)$$

In this equation the reaction is taken to be pseudo-first order with respect to **A** (as is the case for hydrogen in the reaction with thiophene) with an effectiveness factor η. The factor ϵ_P appearing in this equation converts the basis of the reaction rate constant from unit volume of particles to unit volume of bed to be consistent with the basis used for the volumetric mass transfer coefficients.

From Eq. (4.46), solving for C_{AS},

$$C_{AS} = \frac{k_Sa_S}{(k_sa_S + k_1\eta\epsilon_P)}C_{AL} \qquad (4.47)$$

In addition, because C_{Ai} is given by $C_{Ai} = P_A/H$, C_{Ai} and C_{AS} from Eq. (4.47) may be substituted into Eq. (4.45) to give a differential equation in C_{AL}:

$$D_L\frac{d^2C_{AL}}{dz^2} - u_L\frac{dC_{AL}}{dz} + k_La\left(\frac{P_A}{K} - C_{AL}\right) - \left(\frac{1}{k_sa_s} + \frac{1}{k_1\eta\epsilon_P}\right)^{-1}C_{AL} = 0 \qquad (4.48)$$

The equation corresponding to Eq. (4.45) for the material balance on the reactant **B** (thiophene in the example), present in the liquid phase only, can be derived in the same way giving:

$$D_L \frac{d^2 C_{BL}}{dz^2} - u_L \frac{dC_{BL}}{dz} - k_{sB} a_S (C_{BL} - C_{BS}) = 0 \qquad (4.49)$$

where k_{sB} is the liquid–solid mass transfer coefficient for reactant **B**, which will differ from k_s owing to the different physical properties of **B**. The concentration C_{BL} is determined similarly to C_{AL} by the rate of reaction within the solid, but because the reaction is pseudo-first order and the reactant **B** (thiophene) is the one in considerable excess, the rate of reaction term involves just the concentration of the reactant **A**, C_{As}.

That is, the equation for transfer and reaction of reactant **B** is

$$k_{sB} a_S (C_{BL} - C_{BS}) = \frac{1}{4} k_1 C_{As} \eta \varepsilon_P \qquad (4.50)$$

so that Eq. (4.49) becomes

$$D_L \frac{d^2 C_{BL}}{dz^2} - u_L \frac{dC_{BL}}{dz} - \frac{1}{4} k_1 C_{As} \eta \varepsilon_P = 0$$

where the factor of $\frac{1}{4}$ arises because 4 mol of hydrogen is needed to react with 1 mol of thiophene. Substituting for C_{As} from Eq. (4.47), the concentration C_{BL} is given by the solution of the differential equation:

$$D_L \frac{d^2 C_{BL}}{dz^2} - u_L \frac{dC_{BL}}{dz} = \frac{1}{4} \left(\frac{1}{k_{sas}} + \frac{1}{k_1 \eta \varepsilon_P} \right)^{-1} C_{AL} = 0 \qquad (4.51)$$

which requires Eq. (4.48) to be solved first so that C_{AL} can be expressed as a function of z.

When dispersion is neglected, i.e., ideal plug flow is assumed, the two equations, 4.48 for C_{AL}, and 4.51 for C_{BL}, are simplified and become

$$-u_L \frac{dC_{AL}}{dz} + k_L a \left(\frac{P_A}{H} - C_{AL} \right) - k^0 C_{AL} = 0$$

or

$$-u_L \frac{dC_{AL}}{dz} - (k_L a + k^0) C_{AL} + \frac{P_A}{H} k_L a = 0 \qquad (4.52)$$

and

$$-u_L \frac{dC_{BL}}{dz} - \frac{1}{4} k^0 C_{AL} = 0 \qquad (4.53)$$

where k^0 is given by

$$\frac{1}{k^0} = \frac{1}{k_s a_s} + \frac{1}{k_1 \eta \varepsilon_P} \tag{4.54}$$

4.2.4.4 Trickle Bed Reactor—Simplified Steady-State Treatment

A further simplification of the case of the hydrogenation reaction above can be obtained if it is assumed that, across any section through the reactor, the rate of hydrogen transfer from gas to liquid is equal to the rate of transfer from liquid to solid, which in turn is just equal to the rate of reaction of hydrogen within the solid catalyst. This assumption implies that as soon as the liquid enters the reactor, the steady-state concentration of hydrogen in the liquid C_{AL} is instantly established. Because the rate of the reaction is independent of the thiophene concentration and pure hydrogen is used, this concentration of hydrogen in the liquid C_{AL} is the same throughout the reactor. In this case the overall rate of the reaction r_t can be expressed as a combination of the individual mass transfer and reaction steps in exactly the same way as for the suspended bed reactors previously considered. Thus, in terms of $k_s a_s$ the volumetric coefficient for mass transfer to the catalyst particles (a_s is equivalent to $\alpha_P \varepsilon_P$), Eq. (4.43) becomes

$$r_t = \left(\frac{1}{k_L a} + \frac{1}{k_s a_s} + \frac{1}{k_1 \eta \varepsilon_P} \right)^{-1} C_{Ai} \tag{4.55}$$

This is the rate at which the hydrogen reacts. From the stoichiometry of the reaction, the rate at which the thiophene reacts is $\frac{1}{4} r_t$. Thus this term replaces the $\frac{1}{4} k^0 C_{AL}$ term in Eq. (4.53) for the thiophene material balance, which then becomes

$$-u_L \frac{dC_{BL}}{dz} - \frac{1}{4} r_t = 0 \tag{4.56}$$

Although the steady-state concentration of hydrogen in the liquid phase does not appear in Eq. (4.55) for the overall rate, if the value of this concentration denoted $C_{\overline{AL}}$ is required, it is most easily obtained from Eq. (4.52) by setting dC_{AL}/dz equal to zero because there will then be no variation of C_{AL} with z.

Thus,

$$C_{\overline{AL}} = \frac{P_A}{H} \frac{k_L a}{(k_L a + k^0)} \tag{4.57}$$

Alternatively, this equation can be obtained by eliminating r_t and C_{As} from Eqs. (4.40)–(4.42), which are the basic equations for the trickle bed reactor with the steady-state assumption.

The numerical results from these two treatments will now be compared in the following example.[38]

Example 4.8

The Hydrodesulfurisation of Thiophene in a Trickle Bed Reactor

A trickle bed reactor is to be designed to remove 75% of the thiophene present in a hydrocarbon feed, which contains 0.012 kmol/m^3 thiophene. The bed of catalyst particles will operate at a temperature of 200°C and a pressure of 40 bar. Under these conditions the rate of the reaction may be assumed to be first order with respect to dissolved hydrogen and independent of the thiophene concentration. The superficial velocity of the liquid in the reactor will be 0.05 m/s.

The *objective* of the calculation is to determine the depth of bed required to give a fractional conversion of thiophene at the outlet of 0.75. As part of the calculation, a graph of the concentration of hydrogen in the liquid phase C_{AL} versus the distance z from the top of the catalyst bed is to be drawn. Further data are as follows:

First-order rate constant with respect to hydrogen (here expressed as k_1^m, the rate constant per unit mass of catalyst, which is often the practice for heterogeneous reactions catalyzed by solids, as in Chapter 3):

$$k_1^m = 0.11 \times 10^{-3} \text{ m}^3/(\text{kg catalyst}) \text{ s.}$$

Bulk density of catalyst: $\rho_b = 960 \text{ kg/m}^3$.
Effectiveness factor: $\eta = 1$.

Volumetric mass transfer coefficients for hydrogen:

Gas to liquid: $k_L a = 0.030 \text{ s}^{-1}$. Liquid to solid: $k_s a_s = 0.50 \text{ s}^{-1}$.
Henry's law constant for dissolved hydrogen under the above conditions:
$H = 1940 \text{ bar m}^3/\text{kmol}$.

Solution

The rate constant given, k_1^m has, as its basis, unit mass of catalyst, whereas the term $k_1 \varepsilon_P$ in Eq. (4.54) is the rate constant k_1 per unit volume of particles converted to a basis of *unit volume of bed*. To convert k_1^m similarly it has to be multiplied by ρ_b the bulk density of the bed. That is,

$$k_1 \varepsilon_P = \rho_b k_1^m = 960 \times 0.11 \times 10^{-3} = 0.106 \text{ s}^{-1}$$

Then from Eq. (4.54) with $\eta = 1$,

$$\frac{1}{k^0} = \frac{1}{0.05} + \frac{1}{0.106 \times 1} \qquad \therefore \quad k^0 = 0.085 \text{ s}^{-1}$$

Proceeding now to the differential equation whose solution will give C_{AL}, the concentration of hydrogen in the liquid phase as a function of z, i.e., Eq. (4.52), this becomes

$$-0.05 \frac{dC_{AT}}{dz} - (0.030 + 0.085)C_{AL} + \frac{40}{1940} 0.030 = 0$$

That is,

$$\frac{dC_{AL}}{dz} = 0.0124 - 2.30C_{AL}$$

At the inlet to the bed, when $z = 0$, $C_{AL} = 0$, integrating with this condition,

$$\int_0^{C_{AL}} \frac{1}{(0.0124 - 2.30C_{AL})} dC_{AL} = z$$

That is,

$$-\frac{1}{2.30} \ln(0.0124 - 2.30C_{AL}) + \frac{1}{2.30} \ln 0.0124 = z$$

or

$$\frac{0.0124 - 2.30C_{AL}}{0.0124} = e^{-2.30z}$$

or

$$(1 - 185C_{AL}) = e^{-2.30z}$$

Therefore

$$C_{AL} = 5.4 \times 10^{-3} (1 - e^{-2.30z}) \tag{A}$$

Eq. (A) gives the required dependence of the hydrogen concentration in the liquid phase upon the distance from the inlet to the reactor and is plotted in Fig. 4.23B.

The corresponding differential equation representing the material balance on thiophene is Eq. (4.53). Substituting numerical values and C_{AL} from Eq. (A), this becomes

$$-0.05 \frac{dC_{LB}}{dz} - \frac{0.085}{4} 5.4 \times 10^{-3} (1 - e^{-2.30z})$$

That is,

$$-\frac{dC_{LB}}{dz} = 0.0023 (1 - e^{-2.30z})$$

Integrating,

$$(C_{LB})_{in} - (C_{LB})_{out} = 0.0023 \left(z + \frac{1}{2.30} e^{-2.30z} - \frac{1}{2.30} \right)$$

Thus the fractional conversion of the thiophene α_T is

$$\alpha_T \frac{(C_{LB})_{in} - (C_{LB})_{out}}{(C_{LB})_{out}} = \frac{0.0023}{0.012} \left(z + \frac{1}{2.30} e^{-2.30z} - 0.435 \right) \tag{B}$$

It can be seen from this expression that if $z > 1$, $\frac{1}{2.30} e^{-2.30z} < 0.044$ and if $z > 4$, $\frac{1}{2.30} e^{-2.30z} < 0.00004$ so that neglecting this term in Eq. (B) corresponds to an error of less than 0.0012% when $z > 4$. When this term is neglected, Eq. (B) becomes explicit for z, so that with $\alpha_t = 0.75$, the desired fractional conversion for the thiophene, Eq. (B) becomes

$$0.75 = \frac{0.0023}{0.012} (z - 0.435)$$

Thus, $z = \underline{4.35 \text{ m,}}$ which is the answer required.

Calculation Using Simplified Steady-State Treatment

In this case the rate of reaction with respect to hydrogen is given by Eq. (4.55) with C_{Ai} equal to P_A/H.

That is,

$$r_t = \left(\frac{1}{0.030} + \frac{1}{0.50} + \frac{1}{0.106}\right)^{-1} \frac{40}{1940} = 0.00046 \text{ kmol/m}^3 \text{ s}$$

Eq. (4.56) for the thiophene material balance therefore becomes

$$-0.05\frac{dC_{LB}}{dz} - \frac{0.00046}{4} = 0$$

Integrating,

$$(C_{LB})_{in} - (C_{LB})_{out} = \frac{0.00046}{0.05 \times 4}z = 0.0023z$$

and

$$\alpha_T = \frac{(C_{LB})_{in} - (C_{LB})_{out}}{(C_{LB})_{out}} = \frac{0.00023}{0.012}z = 0.192z$$

Thus, when $\alpha_t = 0.75$, $z\frac{0.75}{0.192} = \underline{0.39 \text{ m}}$, which is the answer required. The steady-state concentration of hydrogen $C_{\overline{AZ}}$ is given by Eq. (4.57), so that

$$C_{\overline{AL}} = \frac{40}{1940}\frac{0.030}{(0.030 + 0.085)} = 5.4 \times 10^{-3} \text{ kmol/m}^3.$$

This value, which is independent of z, is plotted in Fig. 4.23B for comparison with the more exact variation with z obtained in the first calculation.

Nomenclature

		Units in SI System	Dimensions M, N, L, T, θ
A	$[1 + 4k_{1G}\varepsilon_G D_G/u_G^2]^{\frac{1}{2}}$	—	—
a	Interfacial area per unit volume of dispersion	m^{-1}	L^{-1}
a_P	Surface area per unit volume of solid particle	m^{-1}	L^{-1}
a_S	Surface area per unit volume of fixed bed	m^{-1}	L^{-1}
C_A, C_B	Molar concentrations of **A**, **B**	$kmol/m^3$	NL^{-3}
C_{AG}	Value of C_A in bulk gas	$kmol/m^3$	NL^{-3}
C_{AL}, C_{BL}	Value of C_A, C_B in bulk liquid	$kmol/m^3$	NL^{-3}
$C_{\overline{AL}}$	Concentration of **A** in liquid phase with steady-state assumption	$kmol/m^3$	NL^{-3}
C_G	Molar concentration of dispersed species in gas phase	$kmol/m^3$	NL^{-3}
C_L^*	Saturation concentration in bulk liquid	$kmol/m^3$	NL^{-3}

		Units in SI System	Dimensions M, N, L, T, θ
D_A, D_B	Diffusivity of **A**, **B** in mixture	m²/s	L^2T^{-1}
D_{AB}	Diffusivity of **A** in **B**	m²/s	L^2T^{-1}
D_e	Effective diffusivity	m²/s	L^2T^{-1}
D_L	Liquid-phase diffusivity	m²/s	L^2T^{-1}
D_G	Gas-phase dispersion coefficient	m²/s	L^2T^{-1}
D_L	Liquid-phase dispersion coefficient	m²/s	L^2T^{-1}
d_b	Bubble diameter	m	L
d_c	Column diameter	m	L
d_p	Particle diameter	m	L
f_A	Reaction factor for **A** (Eq. 4.10)	—	—
f_u	Fraction of gas unreacted	—	—
H	Henry's law constant	Nm/kmol	$MN^{-1}L^2T^{-2}$
G	Molar gas flow rate per unit area	kmol/m² s	$NL^{-2}T^{-1}$
h	Height of packing	m	L
J_A	Transformation rate of **A** per unit volume of reactor	kmol/m³ s	$NL^{-3}T^{-1}$
K_G	Overall mass transfer coefficient based on gas phase (cf k_G)	kmol/N s	$NM^{-1}L^{-1}T$
K_L	Overall mass transfer coefficient based on liquid phase (cf k_L)	m/s	LT^{-1}
k_G	Gas film mass transfer coefficient, driving force as partial pressure difference	kmol/N s	$NM^{-1}L^{-1}T$
k_L	Liquid film mass transfer coefficient, driving force as molar concentration difference	m/s	LT^{-1}
k'_L	Reaction-enhanced liquid film mass transfer coefficient	m/s	LT^{-1}
k_S	Mass transfer coefficient from liquid to solid	m/s	LT^{-1}
k_1	Reaction constant for first-order reaction	s⁻¹	T^{-1}
k_2	Reaction constant for second-order reaction	m³/kmol s	$N^{-1}L^3T^{-1}$
k_3	Reaction constant for third-order reaction	(m³/kmol)²/s	$N^{-2}L^6T^{-1}$
k_{1G}	First-order rate constant for transfer from gas phase	s⁻¹	T^{-1}
k^0	Combination of liquid–solid transfer coefficient and rate constant, defined by Eq. (4.46)	s⁻¹	T^{-1}
k_1^m	First-order rate constant defined on basis of unit mass of catalyst	m³/kg s	$M^{-1}L^3T^{-1}$

Continued

		Units in SI System	Dimensions M, N, L, T, θ
L	Height of column	m	L
N_A	Molar rate of transfer of **A** per unit area	kmol/m^2 s	$NL^{-2}T^{-1}$
$\overline{N}_A$	Mean value of N_A over exposure time	kmol/m^{2}s	$NL^{-2}T^{-1}$
n	Number of moles	kmol	N
n_b	Number of bubbles per unit volume dispersion	m^{-3}	L^{-3}
P	Total pressure	N/m^2	$ML^{-1}T^{-2}$
P_A	Partial pressure of **A**	N/m^2	$ML^{-1}T^{-2}$
P_{Ae}	Partial pressure of **A** in equilibrium with liquid concentration C_A	N/m^2	$ML^{-1}T^{-2}$
R	Universal gas constant	J/kmol K	$MN^{-1}L^2T^{-2}\theta^{-1}$
r_A, r_B	Reaction rate of **A**, **B** per unit volume	kmol/m^{3}s	$NL^{-3}T^{-1}$
r_t	Overall reaction rate per unit volume	kmol/m^{3}s	$NL^{-3}T^{-1}$
r_{tank}	Reaction rate in whole volume of tank	kmol/s	NT^{-1}
r_0	Radius of sphere	m	L
T	Absolute temperature	**K**	θ
T	Time	s	T
t_e	Exposure time	s	T
t_r	Reaction time	s	T
u_G	Superficial gas velocity	m/s	LT^{-1}
V	Volume	m^3	L^3
V_d	Volume of dispersion	m^3	L^3
V_L	Volume of liquid	m^3	L^3
x	Distance in direction of transfer	m	L
y	Mole fraction	—	—
α	Fractional conversion	—	—
α_T	Fractional conversion of thiophene	—	—
β	Hatta number $\sqrt{k_1 D_A}/k_L$ or $\sqrt{(k_2 C_{BL} D_A)}/k_L$ (Eq. 4.11)	—	—
δ	Film thickness	m	L
ε_G	Volume fraction of gas	—	—
ε_L	Volume fraction of liquid	—	—
ε_P	Volume fraction of particles	—	—
ν_A, ν_B	Stoichiometric coefficients for **A, B**	—	—
η	Effectiveness factor	—	—
ρ	Density	kg/m^3	ML^{-3}
σ	Surface tension	N/m	MT^{-2}
γ	Residence time	s	T
μ	Viscosity	Ns/m^2	$ML^{-1}T^{-1}$
ϕ	Thiele modulus	—	—

	Units in SI System	Dimensions M, N, L, T, θ
	Suffixes	
A	Component **A**	
B	Component **B**	
e	Equilibrium value	
i	Value at interface	
L	Liquid	
s	Solid	
out	Value at outlet	
in	Value at inlet	
air	Value for air	
water	Value for water	

References

1. Doraiswamy LK, Sharma MM. Heterogeneous reactions. In: *Fluid—fluid-solid reactions*, vol. 2. Wiley; 1984.
2. van Swaaij WPM, Versteeg GF. Mass transfer accompanied with complex reversible chemical reactions in gas—liquid systems: an overview. *Chem Eng Sci* 1992;**47**:3181.
3. Danckwerts PV. *Gas-liquid reactions*. McGraw-Hill; 1970.
4. Resnick W, Gal-or B. Gas—liquid dispersions. In: Drew TB, Cokelet GR, Hoopes JW, Vermeulen T, editors. *Advances in chemical engineering*, vol. 7. Academic Press; 1968. p. 295.
5. Froment GF, Bischoff KB, De Wilde J. *Chemical reactor analysis and design*. 3rd ed. John Wiley and Sons, Inc.; 2011.
6. Lightfoot EN. Steady state absorption of a sparingly soluble gas in an agitated tank with simultaneous first order reaction. *AIChE J* 1958;**4**:499.
7. Van Krevelen DW, Hoftijzer PJ. Kinetics of gas—liquid reactions. Part 1. General theory. *Rec Trav Chim Pays-Bas* 1948;**67**:563.
8. Reid RC, Prausnitz JM, Poling BE. *The properties of gases and liquids*. 4th ed. McGraw-Hill; 1987.
9. Tatterson GB. *Fluid mixing and gas dispersion in agitated tanks*. McGraw-Hill; 1991.
10. Deckwer W-D. *Bubble column reactors*. Wiley; 1992.
11. Kohl AL, Riesenfeld FC. *Gas purification*. Gulf Publishing; 1979.
12. Astarita G, Savage DW, Bisio A. *Gas treating with chemical solvents*. Wiley-Interscience; 1983.
13. Danckwerts PV, Sharma MM. The absorption of carbon dioxide into solutions of alkalis and amines (with some notes on hydrogen sulphide and carbonyl sulphide). *Chem Eng* 1966;(202). CE 244.
14. Bruijn W, van't Riet K, Smith JM. Power consumption with aerated Rushton turbines. *Trans Inst Chem Eng* 1974;**52**:88.
15. Sittig M. *Organic chemical process encyclopedia*. Noyes Development Corporation; 1969.
16. Gates BC. *Catalytic chemistry*. Wiley; 1992.
17. Harnby N, Edwards MF, Nienow AW. *Mixing in the process industries*. 2nd ed. Butterworth-Heinemann; 1992.
18. Baird MHI, Rice RG. Axial dispersion in large unbaffled columns. *Chem Eng J* 1975;**9**:171.
19. Sherwood TK, Pigford RL, Wilke CR. *Mass transfer*. McGraw-Hill; 1975.
20. Westerterp KR, van Swaaij WPM, Beenackers AACM. *Chemical reactor design and operation*. Wiley; 1984.
21. Duda JL, Ventras JS. Laminar liquid jet diffusion studies. *AIChE J* 1968;**14**:286.

22. Davidson JF, Cullen EJ, Hanson D, Roberts D. The hold-up and liquid film coefficient of packed towers. Part 1: behaviour of a string of spheres. *Trans Inst Chem Eng* 1959;**37**:122.

23. Levenspiel O, Godfrey JH. A gradientless contactor for experimental study of interphase mass transfer with/without reaction. *Chem Eng Sci* 1974;**29**:1723.

24. Sharma MM. Some novel aspects of multiphase reactions and reactors. *Trans Inst Chem Eng A* 1993;**71**:595.

25. Pearson L, Pinsent BRW, Roughton FJW. Study of fast reactions. *Faraday Soc Discuss* 1954;**17**:141. The measurement of the rate of rapid reactions by a thermal method.

26. Pinsent BRW, Pearson L, Roughton FJW. The kinetics of combination of carbon dioxide with hydroxide ions. *Trans Faraday Soc* 1956;**52**:1512.

27. Hague DN. *Fast reactions*. Wiley-Interscience; 1971.

28. Shah YT. *Gas−liquid−solid reactor design*. McGraw-Hill; 1979.

29. Topsøe H, Clausen BS, Topsøe N-Y, Pedersen E. Recent basic research in hydro-desulfurization catalysts. *Ind Eng Chem Fundam* 1986;**25**:25.

30. Fan L-S. *Gas−liquid−solid fluidization engineering*. Butterworth; 1989.

31. Vinu R, Madras G. Energy efficiency and renewable energy through nanotechnology. In: Zang L, editor. *Green energy and technology*. London: Springer-Verlag; 2011.

32. Herskowitz M, Smith JM. Trickle-bed reactors: a review. *AIChE J* 1983;**29**(1).

33. Ng KM. A model for flow regime transitions in cocurrent down-flow trickle-bed reactors. *AIChE J* 1986;**32**:115.

34. Charpentier JC, Favier M. Some liquid hold-up experimental data in trickle-bed reactors for foaming and non-foaming hydrocarbons. *AIChE J* 1975;**21**:1213.

35. Gianetto A, Specchia V. Trickle-bed reactors: state of art and perspectives. *Chem Eng Sci* 1992;**47**:3197.

36. Satterfield CN, Pelossof AA, Sherwood TK. Mass transfer limitations in a trickle-bed reactor. *AIChE J* 1969;**15**:226.

37. Lee JC, Sherrard AJ, Buckley PS. Optimum particle size in three phase fluidized bed reactors. In: Angelino H, Couderc JP, Gibert H, Laguerie C, editors. *Fluidization and its applications*. Toulouse: Société de Chimie Industrielle; 1974. p. 407.

38. Smith JM. *Chemical engineering kinetics*. 3rd ed. McGraw-Hill; 1981.

Further Reading

Astarita G. *Mass transfer with chemical reaction*. Elsevier; 1967.

Danckwerts PV. *Gas−liquid reactions*. McGraw-Hill; 1970.

Kastanek F, Zahradnik J, Kratochvil J, Cermak J. *Chemical reactors for gas−liquid systems*. Ellis Horwood; 1991.

Ramachandran PA, Chaudhari RV. *Three-phase catalytic reactors*. Gordon and Breach; 1983.

CHAPTER 5

Fundamentals of Biological Sciences

Sathyanarayana N. Gummadi

Indian Institute of Technology Madras, Chennai, India

Learning Outcomes

- The fundamentals of biological sciences including properties of biomolecules such as proteins, nucleic acids, carbohydrates, and lipids.

- Familiarize the basic concepts of microbiology necessary to understand the importance of working with microorganisms for commercial production.

- Understand the mechanism of various cellular metabolic pathways, which is vital in designing bioprocesses for various industrially important metabolites.

- Theory behind employing various strategies to enhance the microbial production of commercially important metabolites.

- An introduction to biological reactions and enzyme kinetics with emphasis on control and regulation of metabolic pathways.

5.1 Introduction

Previous chapters in this volume have been concerned with chemical reaction engineering and refer to reactions typical of those common in the chemical process industries. There is another class of reactions, often not thought of as being widely employed in industrial processes, but which are finding increasing application, particularly in the production of *fine chemicals*. These are biochemical reactions, which are characterized by their use of enzymes or whole cells (mainly microorganisms) to carry out specific conversions. The exploitation of such reactions by man is by no means a recent development—both the fermentation of fruit juices to make alcohol and its subsequent oxidation to vinegar are examples of biochemical reactions, which have been used since antiquity.

Living organisms are chemically complex and are capable of carrying out a wide range of transformations, which can often be manipulated by controlling their environmental conditions or by changing their genetic constitution. While the primary interest of organisms is to replicate themselves and the formation of other products is incidental to

Coulson and Richardson's Chemical Engineering. http://dx.doi.org/10.1016/B978-0-08-101096-9.00005-4

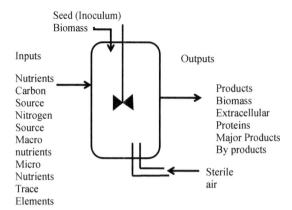

Seed (Inoculum)
Biomass

Inputs

Outputs

Nutrients
Carbon
Source
Nitrogen
Source
Macro
nutrients
Micro
Nutrients
Trace
Elements

Products
Biomass
Extracellular
Proteins
Major Products
By products

Sterile
air

Figure 5.1

Schematic diagram showing the method for microbial growth and product formation. Microbial growth is considered as an autocatalytic reaction where initial amount of biomass is added to the fermenter as seed culture. These microbes utilize the inputs (nutrients and air) and increase its population thereby producing higher concentration of biomass (productive biomass). In addition to its growth, microorganisms produce and secrete various metabolites and products.

this, the biochemical engineer can often modify the environment to harness the biological transformations to suit his own ends (Fig. 5.1). It may be that it is in the engineer's interests to minimize the proportion of the nutrient or reactant (henceforth called as "substrate") supply, which goes toward replication so that a greater yield of his desired product is obtained. On the other hand, the *biomass* itself might be the desired product so that the engineer would be concerned with enhancing the natural inclination of the organism to replicate itself.

The active biochemical constituents of cells are a particular group of proteins that have catalytic properties. These catalytic proteins, or *enzymes,* are in some ways similar to inorganic catalysts but are distinctive in other, quite important respects. Enzymes are very powerful catalysts, capable of enhancing the overall rates of reactions much more markedly; they are much more specific than the average inorganic catalyst (Fig. 5.2).

Furthermore, the conditions under which the reactions proceed are typically mild, temperatures, for example, being generally under 100°C and frequently below 50°C. Enzymes are water soluble but are frequently bound to membranes within the cells or retained in the microbe by the cell walls. The structure of those cell walls is such that they permit the uptake of substrates and secretes out the by-products of the cell's growth. Enzymes usually retain their catalytic activity when isolated from the cell and are often used as such, thus removing the need for providing for the maintenance of the cell itself while retaining the desirable specificity and activity of the protein. However, because the enzymes are soluble in water, it is often necessary to immobilize them, perhaps by attaching them to a suitable support, so that they are not carried away with the product.

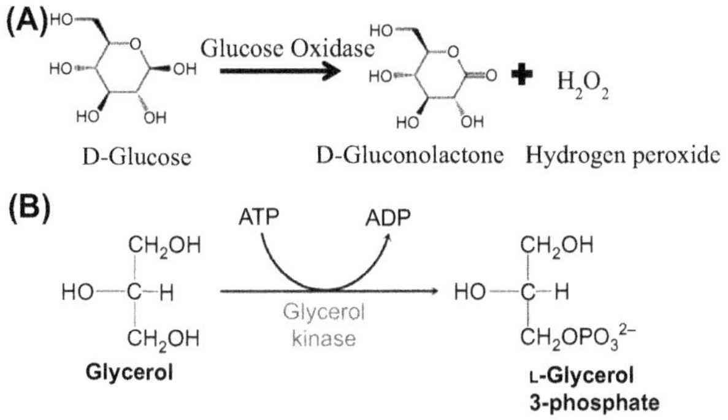

Figure 5.2

Specificity of enzyme. (A) Glucose oxidase enzyme reacts only with ᴅ-glucose and converts to gluconolactone and do not react with ʟ-glucose, (B) Enzymatic reactions are highly specific. For example, glycerol kinase phosphorylates only at $3'$ position forming ʟ-glycerol 3-phosphate. *ADP, adenosine diphosphate; ATP, adenosine triphosphate.*

There is an increasing number of areas where bioreactors are serious alternatives to conventional chemical reactors, particularly when their mild conditions and high selectivity can be exploited. In the pharmaceutical industry, microorganisms and enzymes can be used to produce specific stereoisomers selectively, a very desirable ability because it can be that only the one isomer (possibly an optical isomer) may possess the required properties (Fig. 5.3). In such applications, the limitations of bioreactors are clearly outweighed by the advantages in their use. The fact that the products are formed in rather dilute aqueous solution and at relatively low rates may be of secondary importance, and it may then be economically feasible to employ multiple separation stages in their purification.

Many of the problems encountered in the processing of biological materials are similar to those found in other areas of chemical engineering, and the separation processes used are frequently developments from counterparts in the chemical industry. However, biological

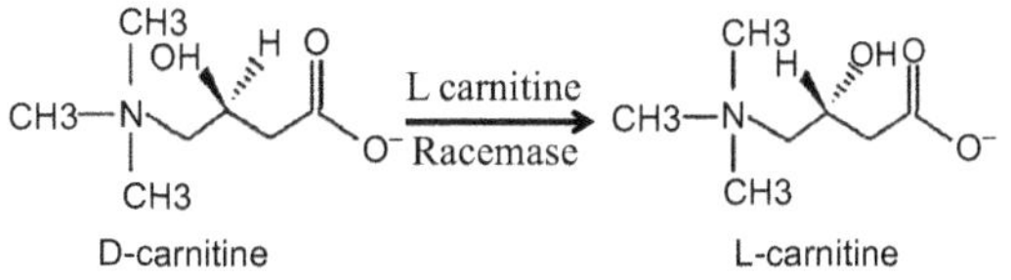

Figure 5.3

Stereospecificity of enzyme. Carnitine racemase is an enzyme which acts on ᴅ-carnitine and converts it to ʟ-carnitine. This enzyme acts specifically on ᴅ-carnitine in a racemic mixture and converts it in to ʟ-carnitine. Degradation of ᴅ-carnitine involves its initial racemization to ʟ-carnitine. It is essential for the transport of long-chain fatty acids through the inner mitochondrial membrane.

Table 5.1: Process engineering and its relations to biological sciences and biochemical engineering.

Biological Sciences	Process	Chemical Engineering Science	
Microbiology	Culture choice		
Genetics	Mass culture	Reactor design	
Biochemistry	Cell responses	Control	Biochemical
Physiology	Process operation	Unit operations	engineering
Chemistry	Product recovery	Energy & material utilization	

materials frequently have rheological properties, which make then difficult to handle, and the fact that their density differs little from that of water and the interfacial tensions are low can give rise to difficulties in *physical* separation of product.

Table 5.1 shows the major areas of the biological sciences that are of significance to the process engineer. Systematics, genetics, biochemistry, and physiology are all important when considering the applications of biological processes. These areas all impinge on the choice of microbe, the type of culture/reactor used, how the culture will react to such an environment, and how the process is operated. The utility of these systems is legion, and the ingenuity of man is the limiting factor in the exploitation of this resource. It is the challenge to the biochemical engineer to exploit these materials in a variety of processes, central to which are biological reactions. However, it should be noted that the underlying principles and disciplines associated with the industrialization of biological systems are those of chemical engineering.

Fig. 5.4[1] shows an outline of a biological process and at its center is the reaction. Clearly, there are three important aspects to reactors; the nature and processing of the raw materials, the choice and manipulation of the catalysts, and the control of the reaction process from which the products must be recovered. Consideration has to be given to the behavior and properties of the biological materials employed if the process is to be successfully designed.

This chapter covers general background of the nature of living systems for the benefit of those without a background of biological science, and next chapter will discuss the characteristics of biochemical reactions and the design features of reactors.

5.1.1 Cells as Reactors

To survive, an organism must grow and reproduce itself using resources from the surrounding environment. Living systems capture and utilize energy from their environment to produce highly ordered structures so as to give rise to autocatalytic processes. Another notable, if not more important property is the fundamental variation

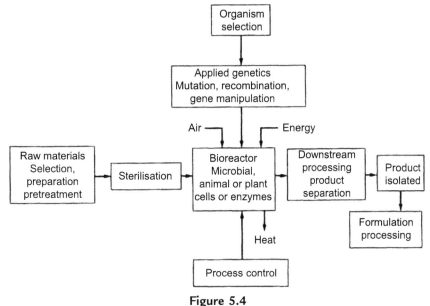

Figure 5.4
Schematic overview of a biotechnological processes.[1]

within living systems on which the principles of natural selection (selection of the fittest) may act. This selection allows the natural optimization of living processes for the evolution of new biological information structures and even new life forms.

The basis of life lies in chemistry, albeit of a rather specialized form, which creates a variety of novel solutions to a set of problems associated with life. These include the following:

1. The acquisition of substrate (nutrients), which will provide the energy and other nutrients that make up the structure of the cell,
2. The conversion of the substrates into the structure of the living system,
3. The retention of information such that the chemical structures may be reproduced when required, and
4. The introduction of variation in the biological information so as to encourage the change or adaptation of the organisms to the environment.

The complexity of the bacterial cell (for example) is thus a match for any chemical plant in the power and sophistication of its chemistry. Inside every cell there is a controlled environment in which several thousand catalyzed reactions occur and where their products and reactants may be found. Cells also contain the control systems to match and a store of information to coordinate and reproduce the whole system. All of these activities take place at relatively mild temperatures and pressures and, overall, these reaction processes make up the *metabolism* of the cells.

5.1.2 The Biological World and Ecology

Living systems, as far as is known, are limited to this planet. The power of living systems is immense if their activities are considered on the global scale. The biosphere, or the environment in which life exists, is a rather limited part of the planet, the most important of which is that near or on surface, or in the oceans.

Evidence for activity of living processes appears to expand steadily. There is good evidence that life can successfully exist in pressurized water at $120°C$, in ice at $-10°C$, and at extremes of pH $(1-11)$ in saturated salt solutions, and even when subjected to high levels of radiation and high levels of toxic compounds.

Living systems are important in the control of rates of flow of many elemental materials in the environment. Indeed, much of environmental chemistry is dependent upon the activities of living systems. The nutrient cycles of carbon, oxygen, nitrogen, sulfur, and phosphorus (the major constituents of living cells) are good examples of processes mediated by living systems. In all of these cycles, cells obtain materials from the environment and transform them by oxidation and reduction processes. There are two basic types of nutrition which are associated with the carbon cycle. The first involves the fixation and reduction of carbon dioxide to carbohydrate and the oxidation of water to oxygen in the presence of an external inorganic energy source; when it involves light, it is called *photosynthesis*. This form of nutrition, called *autotrophy,* is the ability to use materials derived from inorganic sources. Plants, for example, are *autotrophs*. The second basic mode of nutrition is the reverse process and the most important reaction involved is the oxidation of organic materials to carbon dioxide and water in the presence of oxygen (*respiration*). This form of nutrition is called *heterotrophy,* being able to live on other organic materials in the environment. Animals, for example, are *heterotrophs*.

The carbon cycle is illustrated in Fig. 5.5. All living systems require an external source of energy, either in the form of chemical bond energy, as chemical (redox) potential, or as some form of electromagnetic radiation usually in or near the visible light region.

Energy is initially captured from the sun's radiation and is then used to fix carbon dioxide and convert it into the organic molecules that constitute living tissue. The energy is stored in organic molecules, which may be oxidized to release part of that energy to drive other processes. It is this flow of materials and energy that drives all living systems. By interacting with one another, various cell types can become associated and heavily interdependent. The net result is that the energy absorbed from the sun is rapidly dissipated, whereas the materials associated with the biosphere are rapidly circulated.[2]

The biosphere has a complex chemistry, and many aspects of the processes and their consequences are not fully understood. However, pollution arises from the cyclic processes

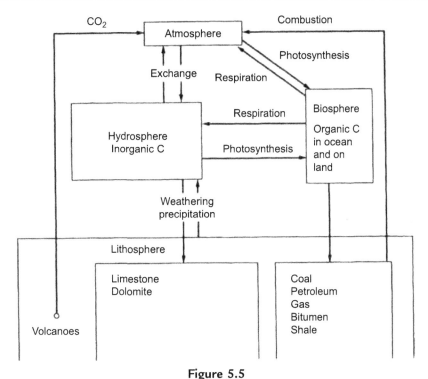

Figure 5.5
The circulation of carbon in nature.

being out of balance, or from the accumulation of other undesirable materials. Indeed, a major perception of pollution is the disruption of these cycles. The activities of man are now on such a scale that they significantly affect the environment and its chemistry.

5.1.3 Biological Products and Production Systems

Table 5.2 [3] shows the markets that are associated with the products of biotechnology. Not shown here is the impact of biological systems on waste treatment and environmental systems. The products illustrated in the table are a mixture of traditional products such as alcoholic beverages and food materials, commodity chemicals, which are produced by large-scale fermentation (e.g., citrates), in addition to specialized compounds (e.g., enzymes and vaccines). More recently the products of so-called genetic engineering, such as insulin and other therapeutic proteins, have begun to make an impact. The advent of genetic engineering (manipulation of cells' genetic material) is now providing a powerful tool, not only for the fundamental understanding of biological systems but also in the manipulation and exploitation of biological materials and processes in medicine and industry. It should also be noted that the production systems are primarily based on microorganisms. This is mainly due to their rapid growth rates and the ease with which they can be manipulated in the laboratory and on the plant.

Table 5.2: World sales of major products of biotechnology.[3]

Product	Quantity (1000 tonne)	Price ($ per tonne)	Value ($ millions)
Fuel/Industrial Ethanol			
USA	1000	576	576
Brazil	4500	576	2600
India	400	576	230
Others	400	576	230
Total			3636
High Fructose Syrups			
USA	3150	400	1260
Japan	600	400	240
Europe and others	200	400	80
Total			1580
Antibiotics			
Penicillin and synthetic penicillins			3000
Cephalosporins			2000
Tetracyclines			1500
Others			1500–2000
Total			8000–8500
Other Products			
Citric acid	300	1600	480
Monosodium glutamate	220	2500	550
Yeast biomass	450	1000	450
Enzymes			400
Lysine	40	4000	160
Total			2040
Total of all products listed			15,256–15,756

5.1.4 Scales of Operation

The scales of bioreactions are listed in Table 5.3. In today's economic environment there are three distinct scales of operation. On the large scale the emphasis of research is generally on fermentation technology and process engineering, while this emphasis changes as the scale of operation is reduced and ultimately the research is dominated by work in genetic manipulation and catalyst development.

5.1.4.1 Large-Scale Processes

Typically, large-scale processes are dominated by the cost of raw materials and the downstream processes are usually well understood. However, the treatment of effluent is a significant problem and can make an appreciable contribution to the cost of production. Surplus biomass, spent slurry, and either utilization or disposal of feed residues represent problems that must be solved economically. This is the major problem associated with the production of biofuels.

Table 5.3: Classification of industrial microbial processes by scale of operation.[2]

Process	Phase of Product in Slurry	Typical Broth Volume	Typical Scale or Form	Typical Volume or Weight Reduction Ratio From Broth
Large scale				
Ethanol and soluble products (conventional process)	Liquid	Widely varying scale	5%–12% in slurry	5–10
Single-cell protein from methanol	Solid	>4000 m^3/d	170 t/d	
Single-cell protein + lipids from methanol	Solid	Unknown	3%–5% in slurry	5–7
Single-cell protein from gas oil	Solid	>2000 m^3/d		
Medium scale				
Polysaccharides	Liquid + solid	40–200 m^3 per batch	3% in slurry	Unknown[a]
Stable organic acids (e.g., citric acid)	Liquid	50–250 m^3 per batch	10% in slurry	4
Small scale				
Antibiotics Cephalosporin	Liquid	50–200 m^3 per batch	3% in slurry	5–10
Penicillin G (K salt)	Liquid	40–200 m^3 per batch	3% in slurry	7–10
Streptomycin	Liquid	40–200 m^3 per batch	1.5% in slurry	5–10
Vitamins				
Cyanocobalamin (Vitamin B$_{12}$)	Solid	40–100 m^3 per batch	3% in slurry	7
Enzymes				
Extracellular	Liquid	10–200 m^3 per batch	<10.5% pure enzyme	
Intracellular Conventional route product	Solid	Not widely established		8
Glucose isomerase[b]	Solid	40–200 m^3 per batch	3% in slurry	5–10
Two-phase aqueous process	Solid	Not widely established		
Vaccines and Antibodies				
Poliomyelitis vaccine	Liquid + solid	>1 m^3 per batch	250,000 doses in 1 m^3	5–10

[a]Viscosity effects limit the concentration that can be achieved.
[b]Product is dried whole cells that exhibit enzymatic activity.

5.1.4.2 Medium-Scale Processes

Medium-scale processes are usually associated with commodity organic chemicals and with the substitution of natural products in place of materials from other sources. These processes generally produce stable products and utilize conventional process operations for the recovery of the product. The economic viability of such processes is sensitive to the concentration of the product in the fermenter broth.

5.1.4.3 Small-Scale Processes

Small-scale processes are concerned with the production of fine chemicals, pharmaceuticals, and enzymes. Generally, products from these fermentations are not available from other sources. A significant cost of production is the development of the process and catalyst, together with the safety testing of the products. The materials produced are chemically and physically diverse in form, and consequently, a wide variety of separation processes is employed. In many cases the capacity of the production equipment may be less than 1 m^3, yet the potency of the product may be such that 300,000 doses can be produced, and only a few batches a week will satisfy the world market for such a product. As a consequence, various recovery operations operate, which are economically acceptable on a small scale.

5.1.4.4 The Role of Biochemical Engineering

While it may be said that biochemical engineering involves the application of chemical engineering principles to biological systems and the manufacture of biologically derived products in general, there is, however, a considerable emphasis on processes involving the growth of microorganisms because either the organisms themselves represent the product or the formation of product is in some way related to the growth process. The typical rates of growth of microbes are such that they double their total mass in a few hours and, in some cases, in a matter of minutes. When compared with the growth rates of more complex life forms with doubling times of weeks or months, this makes them attractive systems on which to base commercial process, even though their care and manipulation present their own problems and difficulties.

Batch reactions, or semicontinuous operations, have in the past tended to dominate processes involving microbial growth. This has largely been due to problems in maintaining sterility (maintenance of a pure culture) in the case of microbial reactors. However, sterility is not crucial in case of enzyme reactors. Apart from those cases where the desired product is formed in a manner not simply related to the growth of the organism, there can be a clear economic attraction in making a process continuous rather than batch. A large-scale process that does not require pure culture conditions is wastewater treatment, where the objective is to consume feed substrates as rapidly as possible. It was the first to be developed as a widespread, successful continuous operation.

Within the constraints imposed, particularly by the requirements of sterility, for example, when monocultures are used, a major role of biochemical engineering is to provide appropriate mathematical descriptions of biochemical reaction processes and from these to devise suitable design criteria for economically viable processes.

5.2 Biomolecules

All living organisms are made of four major biomolecules: proteins, nucleic acids, carbohydrates, and lipids. Amino acids, sugars, and nucleotides are building blocks of proteins, carbohydrates, and nucleic acids. This section will cover the properties and basic functions of these four biomolecules in a living cell.

5.2.1 Amino Acids and Their Properties

The monomers from which proteins are synthesized are the α L-amino acids. The amino acids are differentiated by the various radical groups attached to the α carbon (or C2). Since there are generally four different functional groups attached to the C2 atom, it is asymmetrical and thus will have two possible optical isomers. However, with few exceptions, only the L-amino acids are found in proteins.

Amino acids have at least two groups, which can become ionized in aqueous solution.

$$
\begin{array}{ccccc}
\underset{\substack{| \\ NH_2}}{R-CH-COO^-} & \underset{\substack{-H^+ \\ pK_a\ 10}}{\overset{+H^+}{\rightleftharpoons}} & \underset{\substack{| \\ NH_3^+}}{R-CH-COO^-} & \underset{\substack{-H^+ \\ pK_a\ 3}}{\overset{+H^+}{\rightleftharpoons}} & \underset{\substack{| \\ NH_3^+}}{R-CH-COOH} \\
\text{Anion} & & \text{Zwitterion} & & \text{Cation}
\end{array}
$$

Net charge   − 1   0   + 1

An amino acid therefore can have a number of charged forms, it is an anion at high pH and a cation at low pH; at neutral pH, it normally has no net charge, but exists as a dipolar ion or *zwitterion* with both a positive and negative charge.

The 20 common naturally occurring amino acids are shown in Fig. 5.6. They are subdivided into three groups depending on the nature of the side chains, i.e., nonpolar with alkyl and hydroxy side chains, polar with basic groups, and polar with acid groups.

Non-polar side chains		Polar side chains	
–R	Amino-acid	–R	Amino-acid

Non-polar side chains –R	Amino-acid	Polar side chains –R	Amino-acid
— CH_3	Alanine (Ala)	**Negative charge at pH 7**	
— $CH.CH_3$ with CH_3	Valine (Val)	— CH_2C $\nwarrow O$ $\diagdown O^-$	Aspartic Acid (Asp) or Asparate
— $CH_2CH.CH_3$ with CH_3	Leucine (Leu)	— CH_2CH_2C $\nwarrow O$ $\diagdown O^-$	Glutamic Acid (Glu) or Glutamate
— $CH.CH_2CH_3$ with CH_3	Isoleucine (Ile)	**Positive charge at pH 7**	
		— $(CH_2)_4\overset{+}{N}H_3$	Lysine (Lys)
		— $(CH_2)_3NHC.NH_2$ with $\|$ $+NH_2$	Arginine (Arg)
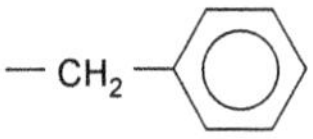 — CH_2—⬡	Phenylalanine (Phe)	**Uncharged at pH 7**	
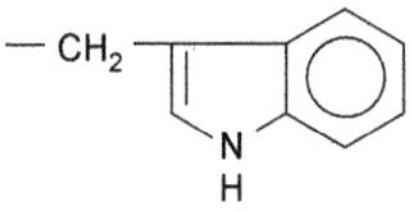 — CH_2—(indole) N H	Tryptophan (Trp)	—H	Glycine (Gly)
		— CH_2OH	Serine (Ser)
— CH_2CH_2—S—CH_3	Methionine (Met)	— $CH.CH_3$ with OH	Threonine (Thr)
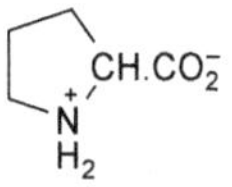 (pyrrolidine ring) $CH.CO_2^-$ $\overset{+}{N}$ H_2 (complete structure)	Proline (Pro)	— CH_2SH	Cysteine (Cys)
		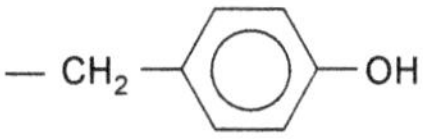 — CH_2—⬡—OH	Tyrosine (Tyr)
		— CH_2C $\nwarrow O$ $\diagdown NH_2$	Asparagine (Aspn)
		— CH_2CH_2C $\nwarrow O$ $\diagdown NH_2$	Glutamine (Glun)
		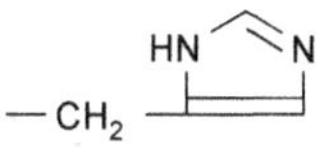 HN⌢N — CH_2—(imidazole)	Histidine (His)

Figure 5.6
The side chains of the 20 amino acids commonly found in proteins.

5.2.2 Proteins and Their Structure

The simple condensation between two amino acids to form one molecule yields a dipeptide. Polypeptides are thus large molecules of amino acids linked together via peptide bonds.

$$\underset{\text{NH}_2\text{CHCOOH}}{\overset{\text{R}_1}{|}} + \underset{\text{H}_2\text{N CH COOH}}{\overset{\text{R}_2}{|}} \longrightarrow \underset{\text{H}_2\text{N CH CONH CHCOOH}}{\overset{\text{R}_1 \qquad \text{R}_2}{|\qquad\quad|}} + \text{H}_2\text{O}$$

Simple proteins are composed only of amino acids linked together via a *peptide bond* as illustrated above. Note that the molecule as such now has an *N-terminal end* (free amino group) and a *C-terminal end* (free carboxylic acid group). Peptides and proteins are differentiated by size and interaction of the molecule—proteins have a peptide backbone, which is sufficient for long-distance intermolecular interactions to take place.[4] This phenomenon occurs at about 2000–3000 molecular weight and above.

There are a vast number of proteins in living systems. It has been estimated that at least 1000 different proteins are present at any one time in the bacterium *Escherichia coli*. The molecular composition of a typical bacterium is shown in Appendix A.1. They form a vital part of living systems and have a wide variety of uses. However, for the function of individual proteins to be clearly understood, they usually have to be purified (Appendix A.1).

The structure of proteins, as with the structure of carbohydrates, has various levels—primary, secondary, tertiary, and quaternary. The tertiary and quaternary structures and their subtleties are most important in the biological function of the molecule. Consider an enzyme (a protein-based catalyst)—its structure allows the binding of specific molecules, which then react catalytically to give products. Conversely, enzymes are very susceptible to environmental conditions, which alter their tertiary structure.

Protein structure may be further complicated by the inclusion of materials such as metal ions and porphyrin rings and by the addition of carbohydrates, lipids, and nucleic acids. Such compounds are called *conjugated proteins*.

5.2.2.1 Primary Structure

A major problem, until recently, was the determination of the protein primary structure, but with the advent of modern analysis of DNA this has become comparatively easy. One of the first structures to be described was that of insulin, which contains 60 amino acids and has a molecular weight of 12,000. Once the primary structure is known, it is possible to predict the secondary and tertiary structures using additional information obtained through X-ray crystallography of the crystallized protein.

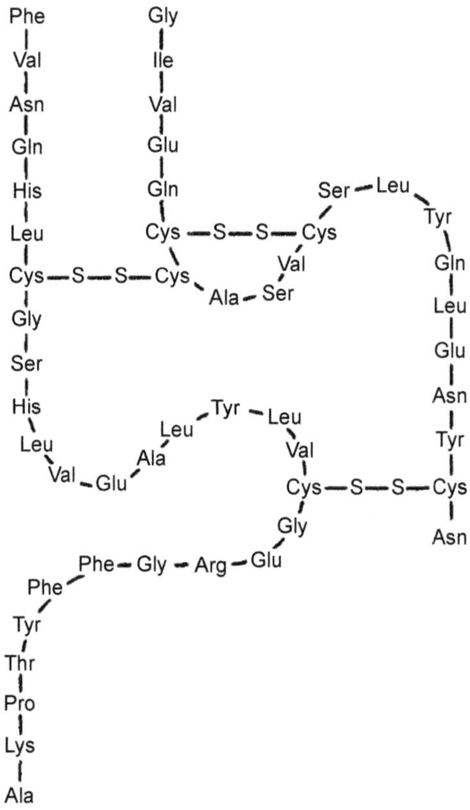

Phe
|
Val
|
Asn
|
Gln
|
His
|
Leu
|
Cys—S—S—Cys
|
Gly
|
Ser
|
His
|
Leu
|
Val—Glu

Gly
|
Ile
|
Val
|
Glu
|
Gln
|
Cys—S—S—Cys
|
Ala—Ser

Ser—Leu
|
Tyr
|
Gln
|
Leu
|
Glu
|
Asn
|
Tyr
|
Cys—S—S—Cys
|
Asn

Val

Tyr—Leu
|
Leu   Val
|
Ala
|
Cys—S—S—Cys
|
Gly
|
Phe—Gly—Arg—Glu
Phe
|
Tyr
|
Thr
|
Pro
|
Lys
|
Ala

Figure 5.7
Protein primary structure. The amino acid sequence of ox insulin.

Unlike polysaccharides, proteins do not have branched chains, but several chains may be linked together via disulfide bridges rather than peptide bonds. The primary structure of ox insulin is shown in Fig. 5.7. The protein consists of two peptide chains, which are linked via the formation of the disulfide bridges. Disulfide bridges are formed by the condensation of the thiol groups of two cysteine residues.

5.2.2.2 Secondary Structure

Secondary structure arises from the way in which the primary structure is folded, maximizing the number of hydrogen bonds, and in effect lowering the free energy of the molecule and its interaction with water. Generally, these interactions occur over relatively short range between different parts of the molecule.

A major subdivision due to different secondary structures is shown by the differences between those of globular and fibrous proteins. Fibrous proteins are generally structural proteins that are insoluble in water—for example, keratin and collagen, which are major components of skin and connective tissues. Globular proteins are usually soluble in water and have functional roles, enzymes being typical examples. Secondary structure also generates a number of structural interactions and conformations. These structures are largely determined by the side groups of the individual amino acids which will hinder or aid certain conformational structures. There are two forms: the α *helix* and the β *pleated sheet.*

The α helix structure is a right-handed helix, meaning that the chain spirals clockwise as it is viewed down the peptide chain. Because of the limited rotation around either side of the peptide bond, only a few angles are energetically stable. In the case of the helix, a full turn is achieved by 3.6 amino acid units. The structure is stabilized by the weak hydrogen bonding as shown in Fig. 5.8. The β pleated sheet structure occurs commonly in insoluble structural proteins and only to a limited extent in soluble proteins. It is characterized by hydrogen bonding between polypeptide chains lying side by side, as illustrated in Fig. 5.9. Fig. 5.10 shows the structure of hemoglobin, a conjugate protein where there are extensive regions of α helix around the prosthetic group, which is an iron porphyrin.

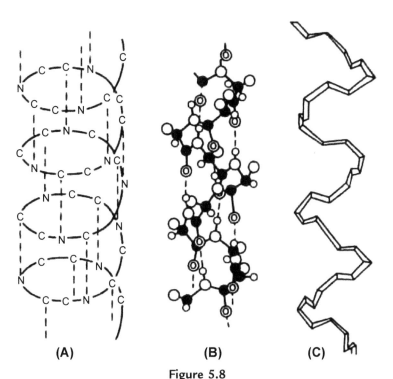

(A)　　　　**(B)**　　　　**(C)**

Figure 5.8

Protein secondary structure: α-helix and the interactions that stabilize its structure. Structure is shown in three ways as (A) hydrogen bonding, (B) ball and stick, and (C) ribbon structure.

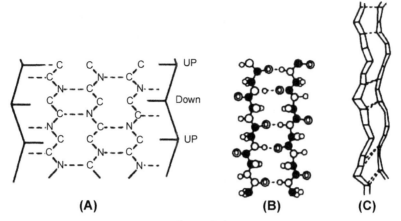

Figure 5.9
Protein secondary structure: β-pleated sheet and the interactions that stabilize its structure. Structure is shown in three ways as (A) hydrogen bonding, (B) ball and stick, and (C) ribbon structure.

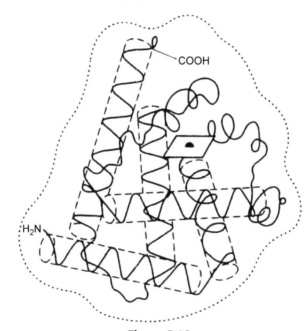

Figure 5.10
The structure of hemoglobin and its iron porphyrin prosthetic group. This diagram is highly simplified and the dotted line indicates the overall shape of the folded molecule, and the zones of the α helical structure within the molecule are also shown.

5.2.2.3 Tertiary Structure of Proteins and the Forces That Maintain It

Although steric effects and hydrogen bonding maintain the secondary structure of proteins, these same factors also contribute to the stability of the tertiary structure. These involve long-range interactions between the amino acids of the polymer. Tertiary structure plays a key role in the function of proteins, including enzymes in biological systems. The forces responsible for maintaining the tertiary structure of proteins are the so-called weak forces, consisting of van der Waals forces, hydrogen bonding, and weaker electrostatic interactions (Fig. 5.11).

The behavior of the side chains in an aqueous environment is important in determining the tertiary folding (structure) of the protein. Nonpolar groups are *hydrophobic* (showing aversion to water) while polar groups are *hydrophilic* (showing affinity for water). Alteration of these interactions can therefore markedly affect the activity of enzymes. A change in the pH, for example, will affect the charge on the protein molecule, thence its structure and function. Table 5.4 compares the bond energies of various types of interaction in macromolecules, and it should be noted that these are 5%+10% of that of a covalent bond.

Changing the environmental conditions can easily provide sufficient energy to alter the tertiary structure of proteins significantly.

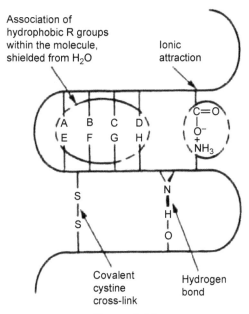

Figure 5.11

Forces maintaining the tertiary structure of globular proteins.

Table 5.4: Bond energies of types of bond occurring in macromolecules.

Type of Bond	Energy (MJ/kmol)
Covalent	200—400
Hydrogen	Up to 20
Van der Waals	Up to about 4
Hydrophobic	4—8
Ionic	Up to 4

5.2.3 Physical and Chemical Properties of Proteins

The chemical and physical behavior of proteins is very complex and the prediction of these characteristics has met with limited success as studies on these materials are based upon empirical measurements. The difficulties lie in the lack of understanding of protein structures; only a few three-dimensional structures are known in detail, and the systems in which they have been studied are very simple. Studies of complex systems of mixtures or intermolecular interactions are limited and the appropriate techniques have yet to be developed. However, with advances in the molecular modeling of proteins, it is becoming possible to predict more of their physical and chemical behavior, and modeling is now a major area of research activity.[4]

5.2.3.1 Chemical Properties of Proteins

The chemical properties of a protein are dictated by the chemical properties of the amino acids of which it is composed. There are many functional groups, which may participate in reaction processes; these include alkyl, hydroxyl, carboxyl, amino, and thiol groups. Some of these groups can be used in protein immobilization, for example, where the protein is covalently fixed to other materials using such functional groups.

Some important reactions are used to estimate the concentration of proteins. These assays all assume that the reacting groups of a protein have an average distribution through the protein molecule. In the *Biuret reaction*, a purple color develops when the protein is treated with alkaline copper sulfate. This reaction is dependent on peptide bonds and not on the side chains of individual amino acids present. In the *Folin—Ciocalteu reaction*, the protein is treated with tungstate and molybdate under alkaline conditions and the formation of a complex such as phenylalanine and tyrosine gives rise to a blue color. Lowry developed one of the most widely used protein assays in which a combination of the above reactions is involved.[4,5]

5.2.3.2 Physical Properties of Proteins

A rational understanding of the physical properties of proteins can be based on a sound knowledge of protein structure.[4] The more important properties are outlined below.

5.2.3.2.1 Acid—Base Properties

By definition, a base is a proton acceptor while an acid is a proton donor, and the general equation for any acid—base reaction can be written as

$$HA \rightleftharpoons H^+ + A^-$$

The strength of an acid is determined by its ability to give up protons, whereas the strength of a base is determined by its ability to take up protons. This strength is indicated by the dissociation or equilibrium constant, (pK_a), for the acid or base; strong acids have a low affinity for protons, while weak acids have a higher affinity and only partially dissociate (e.g., HCl (strong) and acetic acid (weak)).

Thus, for weakly acidic groups, as found in amino acids:

$$R-COOH \rightleftharpoons H^+ + RCOO^-$$

and for weakly basic groups, as found in amino acids:

$$R-NH_3 \rightleftharpoons H^+ + R-NH_2$$

The side groups of amino acids consist of carboxylic acid, amino groups, thiols, alcohols, and alkyl groups. Proteins will thus have a charge, which will also be dependent on pH (Table 5.5). For example, at higher pH values a protein will be negatively charged and will move toward the anode when exposed to an electric field. When the charges on a molecule are balanced, then there is no movement in an electric field. The pH at which this occurs is known as the *isoelectric point* of the protein. This is an important physical property, which is used to characterize a protein, and it depends on the amino acid content and the

Table 5.5: The ionizable groups which contribute to the acid—base properties of proteins, shown with their approximate pK$_a$ values.

Ionizable Group	Dissociation Reaction	Approximate pK$_a$
α-Carboxyl	$-COOH \rightleftharpoons H^+ + -COO^-$	3.0
Aspartyl carboxyl	$-CH_2COOH \rightleftharpoons H^+ + -CH_2COO^-$	3.9
Glutamyl carboxyl	$-CH_2CH_2COOH \rightleftharpoons H^+ + -CH_2CH_2COO^-$	4.1
Histidine imidazole	$-CH_2 \rightleftharpoons H^+ + -CH_2$	6.0
α-Amino	$-NH_3^+ \rightleftharpoons H^+ + -NH_2$	8.0
Cysteine sulfydryl	$-CH_2SH \rightleftharpoons H^+ + -CH_2S^-$	8.4
Tyrosyl hydroxyl	$-OH \rightleftharpoons H^+ + -O^-$	10.1
Lysyl amino	$-(CH_2)_4NH_3^+ \rightleftharpoons H^+ + -(CH_2)_4NH_2$	10.8
Arginine guanidine	$-(CH_2)_3NH.C.NH_2 \rightleftharpoons H^+ + -(CH_2)_3NH.C.NH_2$ $\quad\quad\quad \| \quad\quad\quad\quad\quad\quad\quad\quad \|$ $\quad\quad +NH_2 \quad\quad\quad\quad\quad\quad\quad +NH_2$	12.5

pK_a is equilibrium constant.

three-dimensional molecular structure of protein. Generally, proteins with a high isoelectric point have a high concentration of basic amino acids, while those with a low isoelectric point will be high in acidic amino acids.

The conformation of a protein is dependent on weak intramolecular interactions and is also likely to change with varying pH. Therefore its function and its catalytic activity will be highly dependent on pH.[4]

5.2.3.2.2 Solubility of Globular Proteins

Another property of proteins, which is important in the understanding of the limits of their catalytic activity, as well as being useful in their recovery, is solubility. The solubility of globular proteins in aqueous solution is enhanced by weak ionic interactions, including hydrogen bonding between solute molecules and water. Therefore, any factor, which interferes with this process, must influence solubility. Electrostatic interactions between protein molecules will also affect solubility, since repulsive forces will hinder the formation of insoluble aggregates.[4]

5.2.3.2.2.1 Salt Concentration The addition of a small amount of neutral salt usually increases the solubility of a protein, and changes the interaction between the molecules as well as changing some amino acid charges. The overall effect is to increase the solubility. This phenomenon is known as *salting in.* However, at high concentrations of salts the solvating interactions between protein and water are reduced, and the protein may be precipitated from solution—a process termed *salting out.* Heavy metals (such as Hg and Zn) combine irreversibly with protein to form salts. Similarly, acid-insoluble salts can also be formed; this effect is often used to deproteinise solutions.

5.2.3.2.2.2 pH At extremes of pH, the charge on the side chains of amino acids will be different from that at the physiological pH (pH 5 to 8). The change in charge on the side chains will cause disruption of the tertiary structure, usually irreversibly. The solubility of the proteins also changes considerably around the isoelectric point; this is due to the fact that protein molecules will interact causing aggregation, in contrast to the repulsion, which occurs when molecules have an overall charge.

5.2.3.2.2.3 Organic Molecules in the Solvent The introduction of a water-miscible organic component lowers the dielectric constant of water. This has the net effect of increasing the attractive forces between groups of opposite charge and therefore of reducing the interactions between the water and the protein molecules; consequently, the solubility of the protein will decrease.

5.2.3.2.2.4 Temperature The range of temperature over which proteins are soluble depends on their source. Typically, the solubility of globular proteins increases with temperature up to a maximum. Above this critical temperature, thermal agitation within the molecule disrupts the tertiary structure and gives rise to denaturation. In enzymes, this behavior is paralleled to some extent by changes in activity.

5.2.3.3 Stability of Proteins

One of the major problems that a biochemical engineer will encounter is that of the stability of protein materials. The biological function of the molecule is determined by its secondary and tertiary structures, and if these are upset irreversibly then the protein becomes denatured. Denaturation can occur under relatively mild conditions, as proteins are usually stable only at very narrow ranges of pH (e.g., 5–8) and temperature (e.g., 10–40°C). The boiling of an egg illustrates this point. In its uncooked form the white is a slimy clear protein solution, but under acidic conditions, or when put in boiling water, the solution gels to white solid. The stability of tertiary structure can be appreciated by the relative strengths of these associations as compared with that of a covalent bond.

5.2.4 Nucleic Acids

The basic monomers of nucleic acids are nucleotides, which are made up of heterocyclic nitrogen-containing compounds, purines, and pyrimidines, linked to pentose sugars. There are two types of nucleic acids, and these can be distinguished on the basis of the sugar moiety of the molecule. Ribonucleic acids (RNA) contain ribose, whereas deoxyribonucleic acid (DNA) contains deoxyribose. The bases cytosine (C), adenine (A), and guanine (G) are common in both RNA and DNA. However, RNA molecules contain a unique base, uracil (U), whereas the unique DNA base is thymidine (T).

5.2.4.1 Deoxyribonucleic Acid Structure

DNA is a linear polymer of nucleotides linked via phosphodiester bonds, which are formed between the 3' and 5' hydroxy groups of the ribose moiety of the nucleotide. Fig. 5.12 shows the covalent backbone of DNA. These polymers are normally found as an aggregated pair of strands held together by hydrogen bonding (Fig. 5.13). DNA is composed of four nucleotides: adenine (A), thymine (T), guanine (G), and cytosine (C). The primary structure of DNA is the linear sequence of nucleotide residues comprising the polydeoxyribonucleotide chain. DNA varies in length and in nucleotide composition, the shortest molecules being in the order of 10^3 bases in viruses, 10^6 in bacteria, and 10^8 in humans.

The secondary structure of DNA was deduced by Watson and Crick in 1953.[5] The key piece of data came from the results of their interpretation of some X-ray diffraction patterns of DNA crystals. This, together with Chargaff's data[5] showing that base ratios A:T and G:C are constant, some model building, and a large amount of intuition and inspiration, enabled Watson and Crick to deduce that native DNA was a double-helix

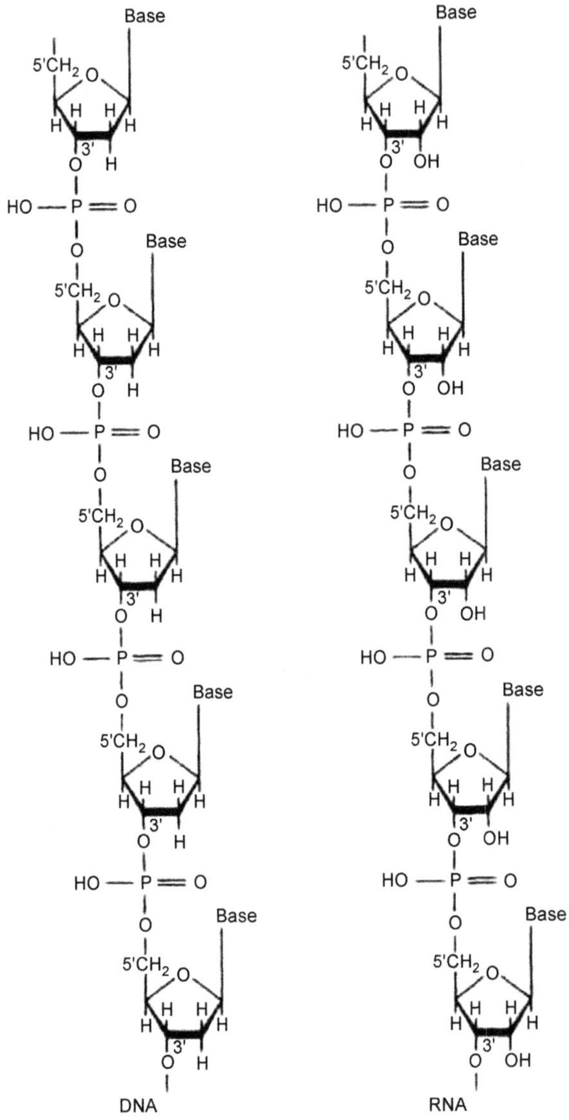

Figure 5.12
The covalent backbone structure of nucleic acids.

structure. The two strands of DNA are held together by a number of weak forces, the most important being hydrogen bonding. Fig. 5.14 shows the basic double-helix structure and the nature of the hydrogen bonding between A and T and G and C base pairs.

The tertiary structure of DNA is complex. DNA does not normally exist as a straight linear polymer, but as a supercoiled structure. Supercoiling is associated with special proteins in eukaryotic organisms. Prokaryotic organisms have one continuous molecule,

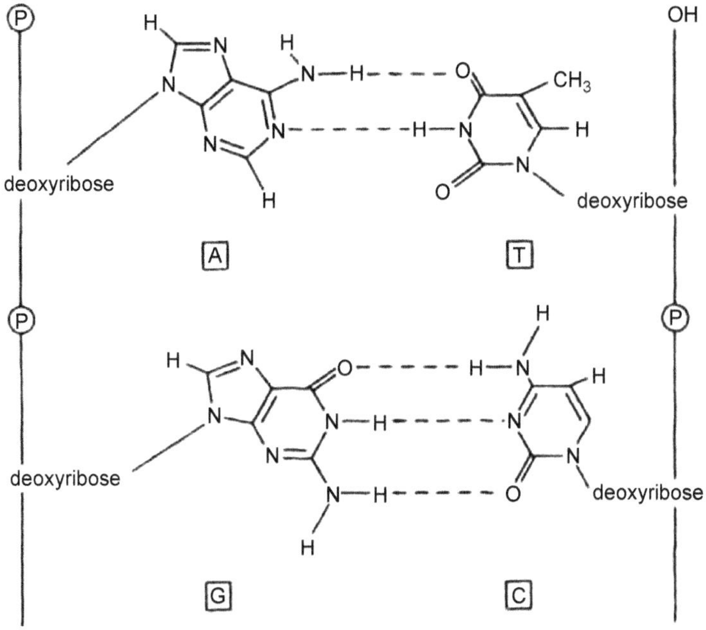

Figure 5.13
AT and GC base pairs held together by hydrogen bonding.

whereas eukaryotes have many (e.g., humans have 46). Viruses also contain nucleic acids and their genetic material can be either DNA or RNA. DNA stores the complete genetic information required to specify the structure of all proteins and RNAs of each organism; program in space and time the orderly biosynthesis of cells or tissues; determine the activities of the organism throughout its life cycle; and determine the individuality of the organism.

5.2.4.2 Ribonucleic Acid

RNA consists of long strings of ribonucleotides, polymerized in a similar way to DNA, but the chains are considerably shorter than those of DNA. RNA contains ribose rather than deoxyribose and also contains uracil instead of thymidine. This has important connotations in the secondary structure of RNA, which does not form the long helices found in DNA. RNA is usually much more abundant than DNA in the cell and its concentration varies according to cell activity and growth. This is because RNA has several roles in protein synthesis. There are three major classes: messenger RNA (mRNA), ribosomal RNA (rRNA), and transfer RNA (tRNA).

mRNA is the template used by ribosomes for the translation of genetic material into an amino acid sequence, and it is derived from a specific DNA sequence. The genetic code is made up of trinucleotide sequences, or *codons*, on mRNA. Each mRNA has a unique

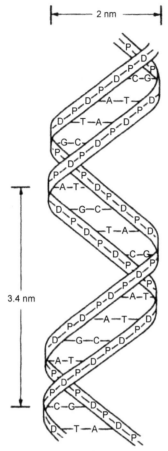

Figure 5.14
The double-helix structure of DNA. *A*, Adenine; *C*, cytosine; *D*, deoxyribose; *G*, guanine; *P*, phosphate ester bridge; and *T*, thymine.

sequence coding for each protein (polypeptide chain). tRNA is a single strand of RNA but it is in a highly folded conformation. The molecules are usually 70–95 ribonucleotides long (equivalent 23,000 to 30,000 MW). Each of the 23 amino acids has one or more tRNAs to which it is able to bind. In this bound form the amino acids are transported into the ribosome. tRNA molecules therefore serve as adapters for translating the genetic code or codons of the mRNA into the sequence of amino acids or proteins. Each tRNA contains a trinucleotide sequence called an *anticodon sequence,* which is complementary to a codon, the trinucleotide sequence of mRNA that codes for one amino acid.[5] rRNAs are the major components of ribosomes and make up 65% by weight of the structure. The role of ribosomal RNA is very complex, but it is essential for protein synthesis that occurs within the ribosome organelle.

5.2.5 Carbohydrates

Carbohydrates form the major structural components of the cell walls. The elemental analysis of carbohydrates showed that it contains C, H, and O with empirical formula of $(CH_2O)_n$. Carbohydrates are nothing but polyhydroxy aldehydes (aldoses) or ketones (ketoses). The simplest aldose is glyceraldehyde and other examples are D-erythrose, D-threose, and D-galactose (Fig. 5.15). The simplest ketose is dihydroxyacetone, and other examples are xylulose and fructose (Fig 5.16). Aldoses are called reducing sugars because they have reducing property. Both aldoses and ketoses have chiral carbon atom and hence they exist as isomers. For example, glucose can exist in two possible isomers (enantiomers) as D-glucose and L-glucose when it exists in linear structure (Fig. 5.17). The structures shown in Fig. 5.16 and 5.17 are called linear structure or open-chain structure and they exist in solution. However, sugars exist in ring structures. For example, glucose exists as α- and β-D-glucose in crystalline form (Fig. 5.18).

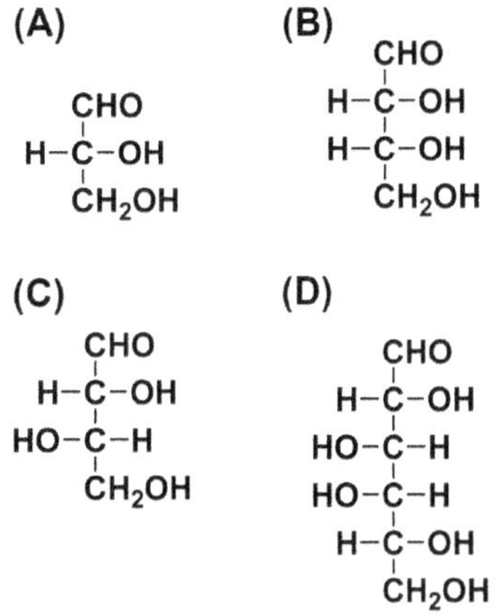

Figure 5.15
Chemical structure of aldoses. (A) Glyceraldehyde; (B) D-Erythrose; (C) D-Threose; (D) D-Galactose.

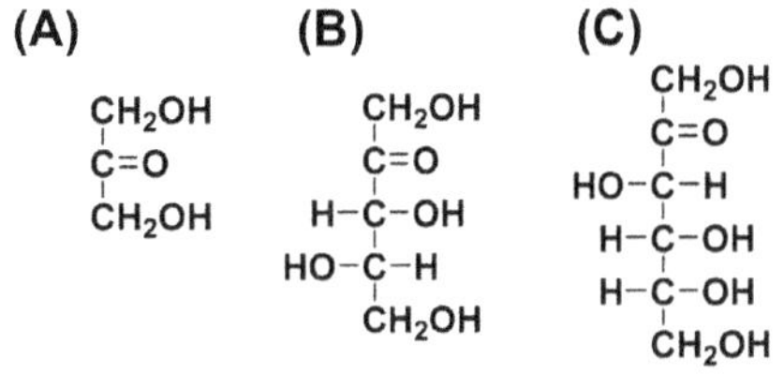

Figure 5.16
Chemical structure of ketoses. (A) Dihydroxyacetone; (B) Xylulose; (C) Fructose.

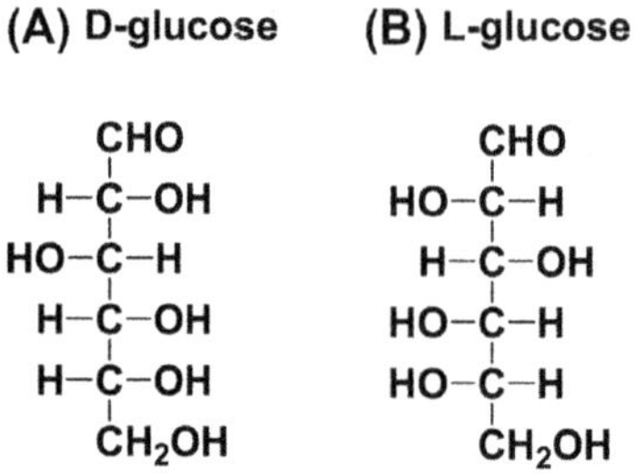

(A) D-glucose
(B) L-glucose
CHO
H-C-OH
HO-C-H
H-C-OH
H-C-OH
CH2OH
CHO
HO-C-H
H-C-OH
HO-C-H
HO-C-H
CH2OH

Figure 5.17: Enantiomers of Glucose.

The fifth carbon is the chiral carbon. When the carbon molecule is bonded to four different groups, it is termed as chiral carbon. If the hydroxyl group on the fifth carbon is to the right of molecule, then it is called D-glucose (A) and when the hydroxyl group of the fifth carbon is to the left of the molecule, it is called L-glucose (B). More naturally occurring sugars are D isomers.

(A)
(B)
(i) Boat form
CH2OH
O
OH
OH
OH
OH
CH2OH
O
OH
OH
OH
OH
(ii) Chair form
OH
O
HO
HO
OH
OH
OH
O
HO
HO
OH
OH

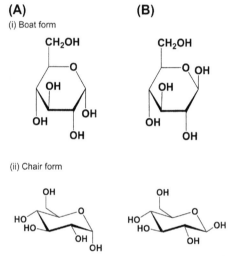

Figure 5.18: α and β structures of glucose in (i) boat configuration and in (ii) chair configuration.

The D-glucose can exist in two forms: (A) α-D-glucose and (B) β-D-glucose. They differ only in the direction that −H and −OH groups point on carbon 1. When alpha-glucose molecules are joined chemically to form a polymer starch is formed. When beta-glucose molecules are joined to form a polymer, cellulose is formed.

The molecular weight of carbohydrates ranges between 10^2 and 10^6 Da. Smaller molecules, which contain 3 to 9 carbon atoms, are referred to monosaccharides. Disaccharides are formed by condensation reaction between two monosaccharides, and two monomers are linked by glycosidic bond (Fig. 5.19). Similarly, polysaccharides are formed by condensation reaction between several monosaccharides. Polysaccharides are often called as oligosaccharides if the number of monomer units is less than 10.

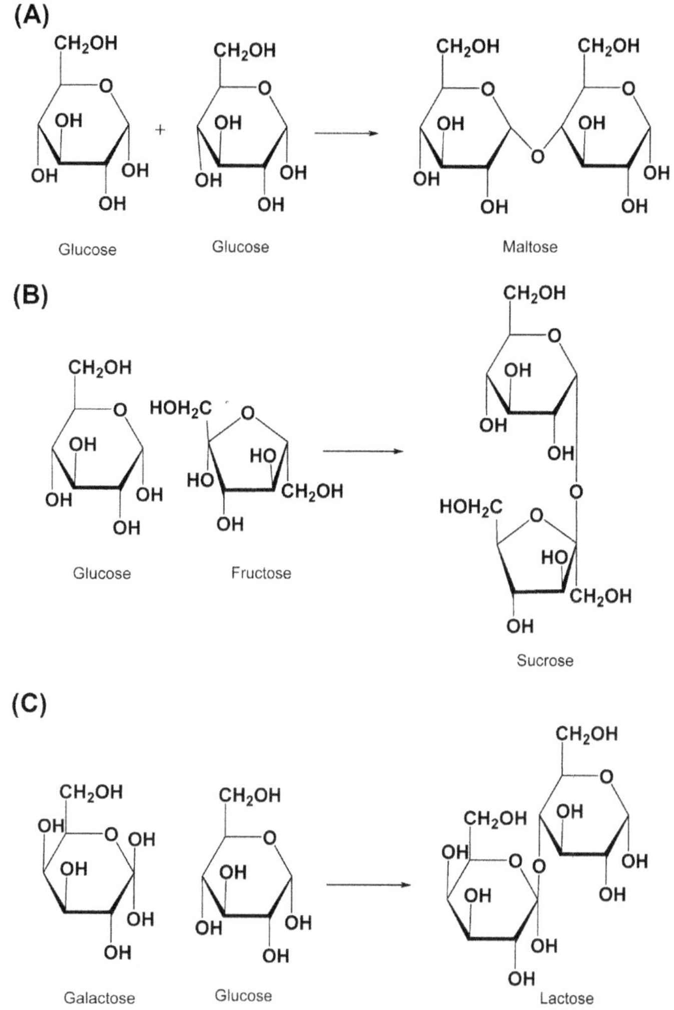

Figure 5.19

Formation of disaccharides from monosaccharides. (A) Maltose; (B) Sucrose; (C) Lactose.

The most commonly found polysaccharide is cellulose, which makes up over 30% of the dry weight of wood. Other structural forms are hemicellulose (a mixed polymer of hexose and pentose sugars), pectins, and chitin (Fig. 5.20). Apart from contributing to the structure, some polymers also act as energy storage materials in living systems. Glycogen and starch form the major carbohydrate stores of animals and plants, respectively. Carbohydrate structure, like that of nucleic acids and proteins, is complex, and various levels of structure can be identified.

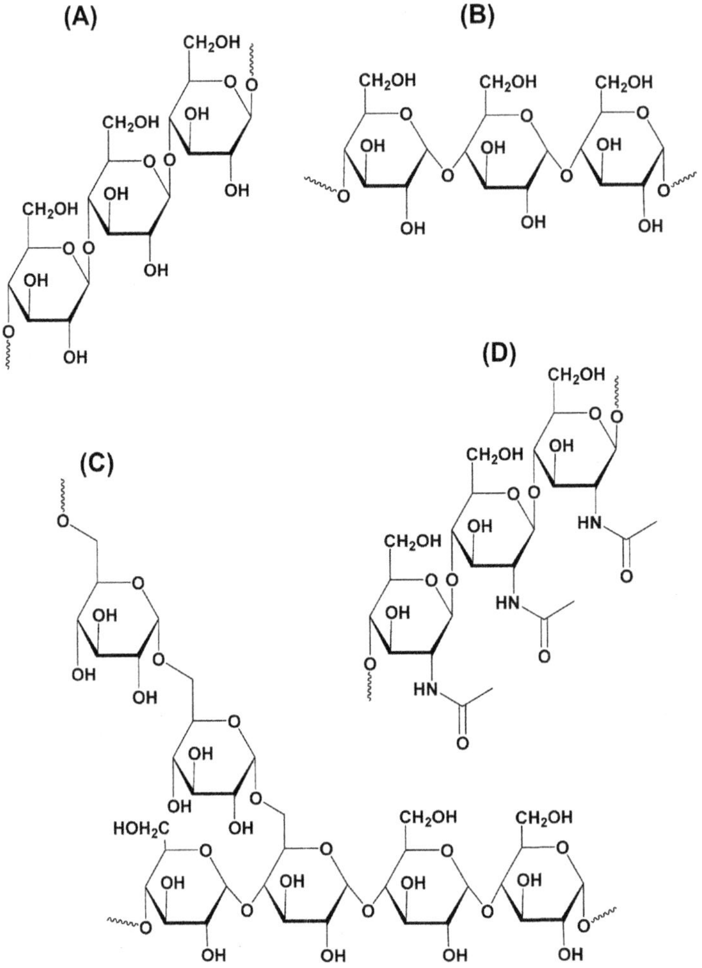

Figure 5.20
Chemical structure of polysaccharides. (A) Cellulose—β(1→4) linked D-glucose units;
(B) Starch—α(1→4) linked D-glucose units; (C) Glycogen—α(1→4) linked D-glucose units and
branched to new chain by α(1→6) glycosidic bonds; (D) β(1→4) linked D-glucose units with
one hydroxyl group on each monomer replaced with an acetyl amine group.

5.2.5.1 Cell Walls

Cell walls are derived from carbohydrates and additionally contain a number of polymeric
materials, ranging from peptidoglycan and other complex polymers in bacteria to glycans/
mannans in yeast and to chitin and cellulose in fungi. The most important function of
these walls is to provide the cell with mechanical strength, in the same way that a tire
reinforces the inner tube and the leather of the rugby ball supports the bladder. It therefore
provides a rigid structure against which the membrane sits and, in consequence, the cells
are sufficiently robust to cope with osmotic stress and other environmental shocks.[6,7]

5.2.6 Lipids

Lipids are the major components of membranes; they have complex structures comprising fatty acids esterified with alcohols to form glycerides, and other lipids based on esters of phosphatidylethanolamine. Other important lipid components are based on sterols. Within this hydrophobic structure, proteins provide ports of entry and exit from the interior of the cell and distinguish the inside from the outside of the cell.

Lipids are defined as water-insoluble compounds extracted from living organisms by weakly polar or nonpolar solvents. The selective solubility of lipids in nonpolar solvents is due to the high proportion of carbon and hydrogen, and thus they are insoluble in water. The term lipid covers a structurally diverse group of compounds, and there is no universally accepted scheme for classifying lipids. A few examples of lipids are given in Fig. 5.21.

In biological systems, the hydrocarbon content of the lipids is derived from polymerization of acetate followed by reduction of the chain so formed. Polymerization of acetate can give rise to (1) fatty acids—long linear hydrocarbon chains, (2) terpenes—branched chain hydrocarbons via a five-carbon intermediate, isopentene, (3) acetogenins—linear or cyclic structures that are partially reduced (Fig. 5.22).

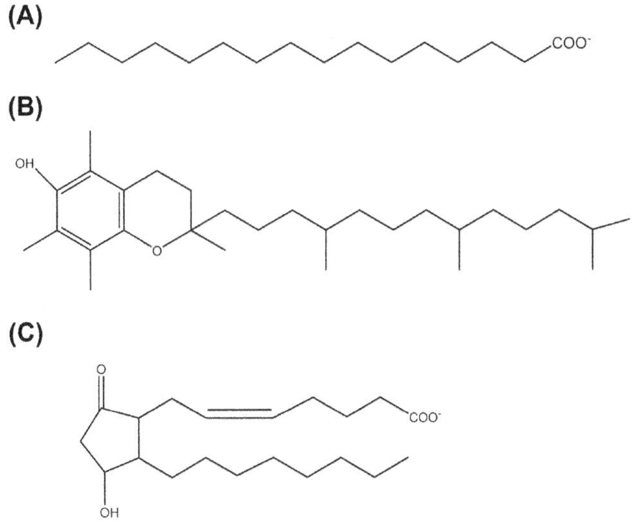

Figure 5.21
Examples of lipids. (A) Palmitate—the most common saturated fatty acid; (B) α-Tocopherol form of vitamin E; (C) Prostaglandin E_2.

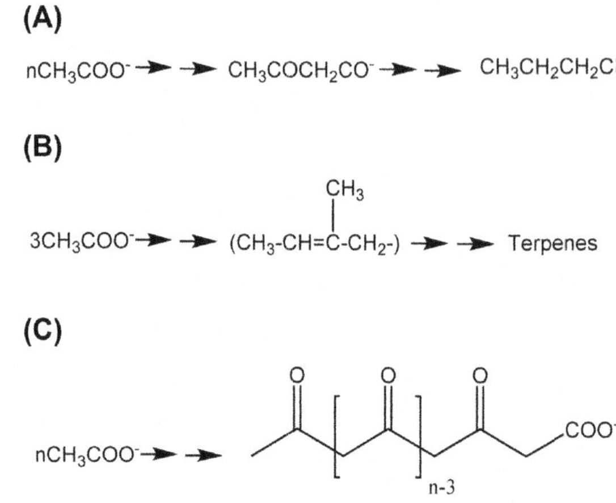

(A)

nCH_3COO^- ⇢ → $CH_3COCH_2CO^-$ ⇢ → $CH_3CH_2CH_2CH_2$---

(B)

$$3CH_3COO^- ⇢ → (CH_3-CH=\overset{\overset{\textstyle CH_3}{|}}{C}-CH_2-) → → \text{Terpenes}$$

(C)

nCH_3COO^- ⇢ →

$$\text{...}COO^-$$
$n-3$

Figure 5.22
Classes of lipids. (A) Fatty acids; (B) Terpenes; (C) Acetogenins.

Fatty acids have the general molecular formula of $CH_3(CH_2)_nCOOH$ and vary in chain length and degree of unsaturation. Most consist of linear carbon chains but a few have branched chains. Fatty acids occur in very low quantities in the free state and are found mostly in an esterified state as components of other lipids. Lipid-containing fatty acids include glycerolipids, sphingolipids, and waxes.

Glycerolipids are lipids containing glycerol in which the hydroxyl groups are substituted. Glycerolipids are the most abundant lipids in animals and can be classified as mono-, di-, triacylglycerols; phosphoglycerides; and glycoglycerolipids. Triacylglycerols are neutral glycerolipids and are mainly used by the cell as a food store. TAGs are the most abundant lipids among the glycerolipids. Although found in most cells, TAGs are particularly present in adipose tissue as fat depot. Diacyl- and monoacylglycerols are found in very small amounts in cell and are usually the metabolites of TAGs and phosphoglycerides. Phosphoglycerides or phospholipids are polar glycerolipids derived from sn-glycerol-3-phosphoric acid in which the phosphoric acid moiety is esterified with certain alcohols and the hydroxyl groups on C-1 and C-2 are esterified with fatty acids. A few examples of phospholipids include phosphatidylcholine, phosphatidylethanolamine, phosphatidylserine, and phosphatidylglycerol. In some instances, C-1 is etherified with a long-chain fatty alcohol leading to the formation of plasmalogen. About 1% of the total phosphoglycerides are in the form of lysophosphoglycerides, in which one of the acyl substituents is not present. Glycoglycerolipids have a hydrophobic and a hydrophilic part similar to phosphoglycerides but the polar part is provided by a carbohydrate moiety rather than by an esterified phosphate. A few examples for glycerolipids are given in Fig. 5.23.

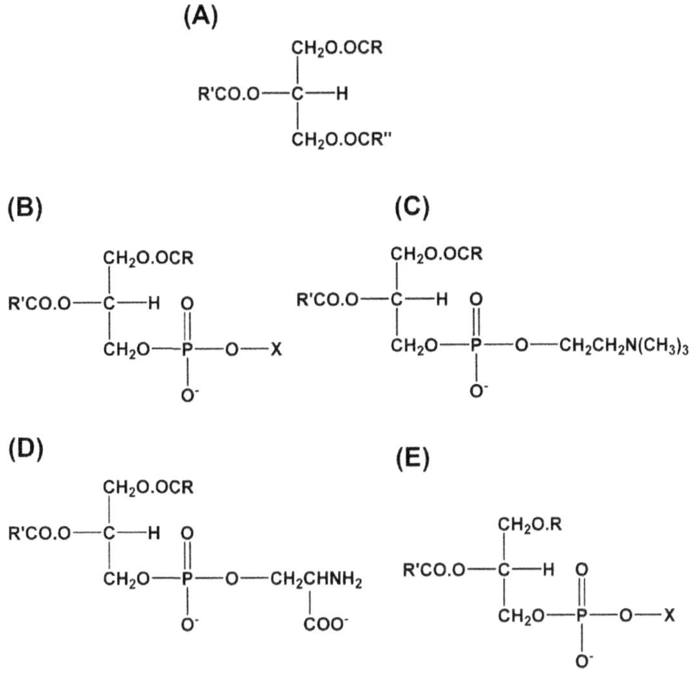

Figure 5.23

Examples of glycerolipids. (A) 1,2,3-Triacyl-sn-glycerol; (B) Phosphoglyceride structure; (C) Phosphatidylcholine; (D) Phosphatidylserine; (E) Plasmalogen.

Sphingolipids are built from long-chain hydroxylated bases rather than glycerol. Sphingosine and sphinganine are two such bases present in animals (Fig. 5.24A and B). Sphingolipids can exist as phosphosphingolipids and glycosphingolipids based on the substitution at the primary hydroxyl position. In phosphosphingolipids, the primary hydroxyl group is esterified with choline phosphate to form sphingomyelin (Fig. 5.24C). In glycosphingolipids, the primary hydroxyl group is glycosylated. More than 50 types of glycosphingolipids are known based on the differences in the carbohydrate portion of the molecule (Fig. 5.24D).

Terpenes are lipids derived from isoprene, are a broad group of unsaturated hydrocarbons found in essential oils of plants, and are based on a cyclic molecule having the formula $C_{10}H_{16}$. Terpenes are classified as monoterpenes (10 carbon atoms), sesquiterpenes (15 carbon atoms), diterpenes (20 carbon atoms), triterpenes (30 carbon atoms), and tetraterpenes (40 carbon atoms). Triterpenes and tetraterpenes give rise to carotenoids and steroids. Carotenoids are hydroxylated derivatives of 40-carbon hydrocarbons called carotenes. Carotenoids absorb visible light due to the high level of conjugation. Most of

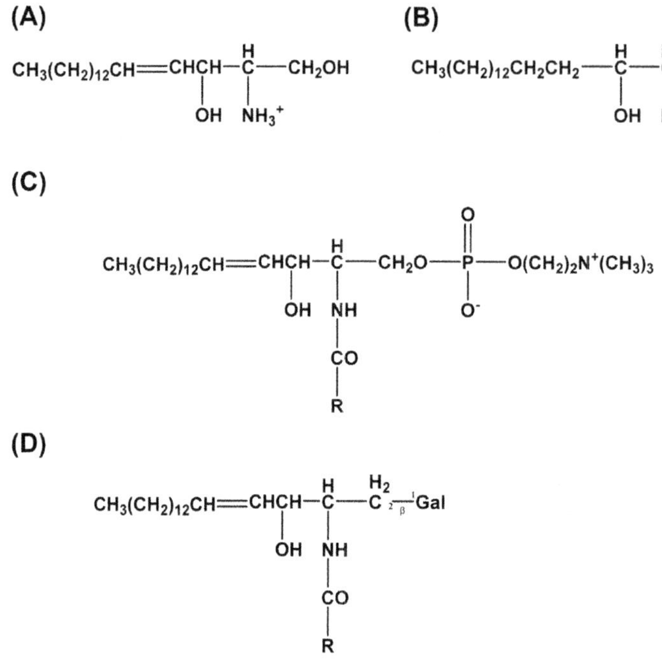

(A)

$CH_3(CH_2)_{12}CH=CHCH-\overset{H}{\underset{OH}{C}}-CH_2OH$
$\quad\quad\quad\quad\quad\quad\quad\quad\quad OH\quad NH_3^+$

(B)

$CH_3(CH_2)_{12}CH_2CH_2-\overset{H}{\underset{OH}{C}}-\overset{H}{\underset{NH_3^+}{C}}-CH_2OH$

(C)

$CH_3(CH_2)_{12}CH=CHCH-\overset{H}{\underset{OH}{C}}-CH_2O-\overset{O}{\overset{\|}{P}}-O(CH_2)_2N^+(CH_3)_3$
$\quad\quad\quad\quad\quad\quad\quad\quad OH\quad NH\quad\quad\quad O^-$
$\quad\quad\quad\quad\quad\quad\quad\quad\quad\quad\quad CO$
$\quad\quad\quad\quad\quad\quad\quad\quad\quad\quad\quad R$

(D)

$CH_3(CH_2)_{12}CH=CHCH-\overset{H}{\underset{OH}{C}}-\overset{H_2}{\underset{NH}{C}}\overset{|}{\underset{\beta}{}}Gal$
$\quad\quad\quad\quad\quad\quad\quad\quad OH\quad NH$
$\quad\quad\quad\quad\quad\quad\quad\quad\quad\quad CO$
$\quad\quad\quad\quad\quad\quad\quad\quad\quad\quad R$

Figure 5.24
Examples of sphingolipids. (A) Sphingosine; (B) Sphinganine; (C) Sphingomyelin; (D) Galactoceramide.

the yellow and red pigments in nature are due to carotenoids. Examples include β-carotene, zeaxanthin, lutein, etc. (Fig. 5.25A). Steroids are derivatives of perhydrocyclopentanophenanthrene. Sterols are steroids containing one or more hydroxyl groups. Some examples of sterols include cholesterol, testosterone, and cholic acid (Fig 5.25B and C).

5.3 Essential Microbiology for Engineers

The concept of the cell, or cell theory, was developed by the middle of the 19th century when living structures had been observed with the microscopes, which were then available. The theory stated that all living systems are composed of cells, which with their products, form the basic building blocks of living systems. It was later found that the cells were made up of various chemicals shared by all cells. This was a very important observation, as the study of simple single-cell systems such as bacteria has led into the understanding of more complex systems. J. Monod once said "What is true for *Escherichia coli* is true for elephants, only more so." Organisms can thus consist of either

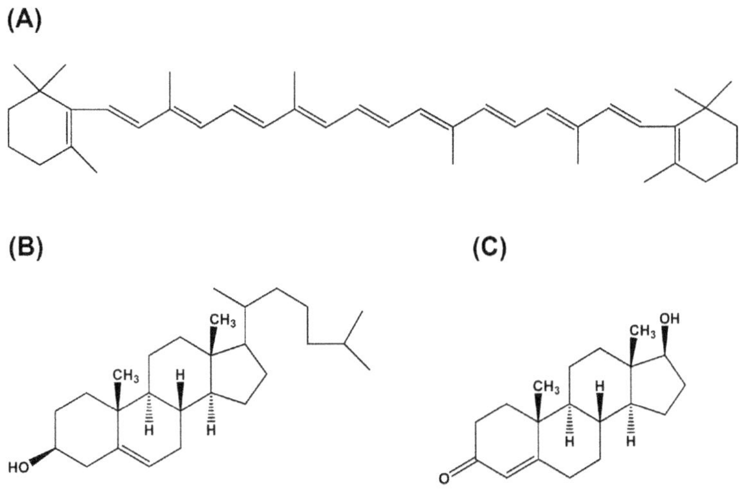

(A)

(B)   (C)

Figure 5.25

Examples of carotenoids. (A) β-Carotene and sterols; (B) Cholesterol; (C) Testosterone.

of a single cell or of a series of specialized cell types, which make up multicellular organisms. Cells are between 1 and 50 μm in diameter and cannot normally be seen with the naked eye; free-living forms are called microorganisms.

The cells of most multicellular organisms are differentiated both in structure and function. This differentiation is most highly developed in plants and animals, where there is profusion of cell types, each performing a different function within the collective whole of the organism. Generally, once a cell has become differentiated to perform a specific activity it is rarely able to change its function; it is committed until it dies.

Cells, or systems of cells, may be derived from microorganisms (protists), animals, and plants; the differences between these cell types are principally ones of organization. For growth and survival, the cell must contain the basic metabolic systems and an information store for the coordination of its synthesis. Various types of cells are used in biological processes and these can be derived from almost any living material. Fortunately, there is a common biochemical system found in all living organisms. All that is required is the knowledge of their chemical and physical requirements for growth, from which the growth and reaction conditions may be formulated.

5.3.1 Classification

Organisms are classified according to structure and function. The biological world is divided into three kingdoms: plants, animals, and protists. Animals and plants are generally classified according to their visible structure. Protists do not have visible structural differences, and they are therefore classified according to chemical differences and biochemical properties. Organisms are named in Latin, or in Latinized terms, using a binary nomenclature where the first name represents the group or *genus,* whereas the second represents the *species.* Usually, a species may be subdivided into strains, which nowadays correspond to strain numbers associated with international culture collections, such as the National Collection of Industrial and Marine Bacteria (NCIMB) or the American Type Culture Collection (ATCC). These collections hold many thousands of strains for research and industrial applications; many hold the same strains, e.g., *E. coli* NCIMB 9481, which is equivalent to ATCC 12435 and is a mutant strain derived from K12 (NCIMB 10214, ATCC 23716). Similarly, yeasts and fungi and other microorganisms are obtainable from culture collections.[8]

Microorganisms, the major form of organisms for biotechnology, can be split into several groups based upon biochemical activity and structure. The basic classification of the protist is shown in Fig. 5.26. There are two major types, the *Prokaryotes* and the *Eukaryotes,* and Table 5.6 illustrates the main distinguishing features between the two cell types. The major differences are reviewed in the following sections.

5.3.2 Prokaryotic Organisms

Microbes are relatively simple in structure in that, when viewed under the light and electron microscope, few complex structures are observed. The general structure of prokaryotic cells is shown in Fig. 5.27.[9]

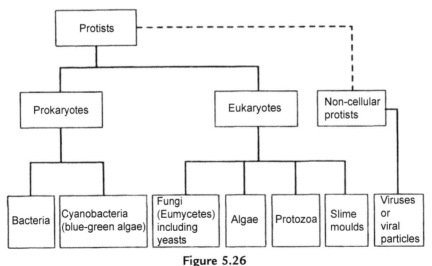

Figure 5.26
Classification within the kingdom of Protists.[2]

Table 5.6: Features distinguishing prokaryotic and eukaryotic cells.

Feature	Prokaryotic Cells	Eukaryotic Cells
Size of organism	<1–2 × 1–4 μm	>5 μm in width or diameter
Genetic system location	Nucleoid, chromatin body, or nuclear material	Nucleus, mitochondria, chloroplasts
Structure of nucleus	Not bounded by nuclear membrane	Bounded by nuclear membrane
	One circular chromosome	One or more linear chromosomes
Sexuality	Chromosome dues not contain histones	Chromosomes have histones
	No mitotic division	Mitotic nuclear division
	Nucleolus absent	Nucleolus present
	Functionally related genes may be clustered	Functionally related genes not clustered
	Zygote partially diploid (merozygotic)	Zygote diploid
Cytoplasmic nature and structures		
Cytoplasmic streaming	Absent	Present
Pinocytosis	Absent	Present
Gas vacuoles	Can be present	Absent
Mesosome	Present	Absent
Ribosomes	70S distributed in the cytoplasm[a]	80S[a] arrayed on membranes (e.g., endoplasmic reticulum), 70S in mitochondria and chloroplasts
Mitochondria	Absent	Present
Chloroplasts	Absent	May be present
Golgi structures	Absent	Present
Endoplasmic reticulum	Absent	Present
Membrane-bound (true) vacuoles	Absent	Present
Outer cell structures		
Cytoplasmic membranes	Generally do not contain sterols	Sterols present
	Contain part of respiratory and, in some, photosynthetic machinery	Do not carry out respiration and photosynthesis
Cell wall	Peptidoglycan (murein or mucopeptide) as component	Absence of peptidoglycan
Locomotor organelles	Simple fibril	Multifibrilled with microtubules
Pseudopodia	Absent	May be present
Metabolic mechanisms	Varied, particularly that of anaerobic energy-yielding reactions, some fix atmospheric N_2, some accumulate poly-β-hydroxybutyrate as reserve material	Glycolysis is pathway for anaerobic energy-yielding mechanism
DNA base ratios $(C + G)^b$	28%–73%	About 40%

[a]S refers to the Svedberg unit, the sedimentation coefficient of a particle in the ultracentrifuge.
[b]C refers to cytosine and G refers to guanine.

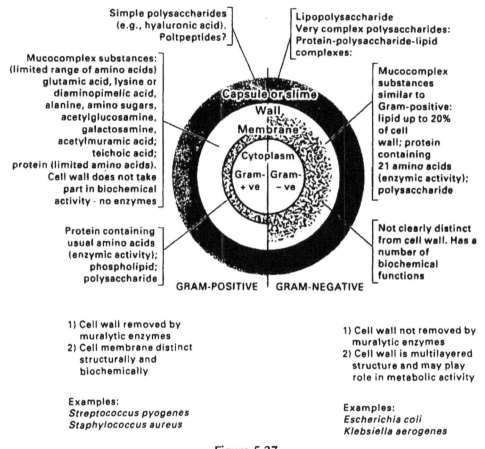

Figure 5.27
Summary of the morphology of Gram-positive and Gram-negative bacteria.[9]

Prokaryotic organisms do not include nuclear material bounded by a membrane and therefore have no true nucleus. The membrane is the most important structural component; Fig. 5.28[10] is a diagrammatic representation of a bacterial membrane. This not only contains lipid which partitions the interior of the cell from the environment, but also is packed with proteins and carrier molecules. It is these molecules that make it an energized selective barrier for importing and exporting materials. The membrane also provides an essential component in the mechanism of energy generation, by electron transport processes (Zone A, Fig. 5.28) in which differences of chemical potential (usually in the form of pH gradient) over the membrane are exploited to generate ATP (Zone D, Fig. 5.28), to cause the movement of *flagella* (Zone B, Fig. 5.28) and to drive *active transport* processes (Zone C, Fig. 5.28).

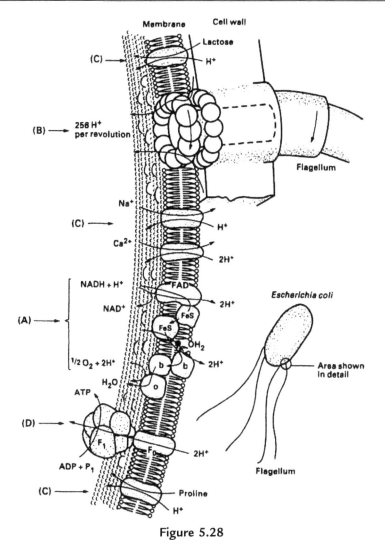

Figure 5.28
The structure and activities of the cytoplasmic membrane involving proton transfer.[10]

The most important cell types of the prokaryotes are bacteria. Bacterial morphology is
rather limited and, only under the light microscope, cocci (spherical forms), bacilli (rod or
cylindrical forms), spirilla (spiral forms), and aggregations thereof, may be observed. The
structures of these systems can be further subdivided on the basis of capsules, flagella, and
other subcellular materials, which may be present. Table 5.7 summarizes the structures
found in various bacteria. An important division between prokaryotic types is made on the
basis of cell wall structure. The *Gram stain* (named after its inventor) is capable of
showing this distinction and the bacteria are divided into two main groups on the basis of
this stain, i.e., Gram-positive (no lipopolysaccharide or LPS layer) and Gram-negative
(extensive LPS layer) (Fig 5.27).[6,7]

Table 5.7: Size and composition of various parts of bacteria.

Part	Size	Comments
Slime Layer Microcapsule Capsule Slime	5–500 nm	Complex materials that vary in composition mainly polysaccharide but may contain significant proteins. Responsible for antigenic properties of cells.
Cell Wall	10–20 nm	20% of cell dry wt.
Gram-positive organisms		Mainly a mixed polymer of muramic acid and peptide teichoic acids and polysaccharide.
Gram-negative organisms		Have outer semipermeable structure of lipopolysaccharide with inner structure of muramic polymer, between these structures there is a space containing protein.
Cell Membrane	10–20 nm	Doubled-layered membrane; main semipermeable barrier of cell; 5%–10% of cell dry wt: 50% protein, 30% lipid and 20% carbohydrate.
Flagellam	0.1 × 12,000 nm	Protein structure arises from membrane. Responsible for motility.
Inclusions Spores	1–2 μm dia	Specialized resistant intracellular structures.
Storage granules	0.05–2 μm	Consist of polysaccharide, lipid, polyhydroxybutyrate and sulfur.
Chromatophores	50–100 nm	Specialized structures containing photosynthetic apparatus.
Ribosomes	10–30 nm	Organelles for protein synthesis; consist of RNA and protein and make up to 20% dry wt of the cell and is a function of growth rate.
Nuclear material		Poorly aggregated materials but can occupy up to 50% cell volume. Consist of DNA usually as a single molecule and makes up to 3% cell dry wt.
Cytoplasm		Free proteins (enzymes); about 50% of dry wt.

Adapted from Atkinson B, Mavituna F. Biochemical engineering and biotechnology handbook, *2nd ed. Basingstoke: Macmillan Publishers; 1991.*

Prokaryotic organisms are metabolically the most diverse of all living systems and are responsible for most degradative processes in the biosphere. They can be grown in the presence or absence of oxygen and form a wide range of organic products. This property has both positive and negative impacts on society. On the positive side, they represent a massive resource of biocatalysis for the biotransformation of organic materials and the degradation of herbicides, insecticides, and other artificial chemicals. On the negative side, they represent the principal agents causing the deterioration of biomaterials, e.g., food and wood, and are major hazards to public health (food poisoning and other diseases).

The ability of bacteria to grow rapidly in a wide range of environments is a very important property, which is exploited in many processes both in mono and mixed cultures.

5.3.2.1 Life Cycles of Prokaryotes

Most bacteria divide by *binary fission*: the cell cycle involves a mature cell in which the nuclear material of the cell first separates, and then an ordered and usually symmetric division of the remaining material occurs to produce two daughter cells.

One form of asymmetric division caused by unfavorable conditions is sporulation, which is a highly coordinated process ultimately producing spores that are resistant to adverse environmental conditions. When conditions are favorable, outgrowth to vegetative cells then occurs. Streptomycetes and the blue-green bacteria are exceptions to this relatively simple cellular morphology and each of these types of bacteria shows various forms of cellular differentiation, both in terms of biochemical and morphological features.[6,7]

5.3.3 Eukaryotic Organisms

There also exist structurally more complex organisms in the microbial world; these, exemplified by the yeast and the fungi, are called eukaryotes. Animals and plants are even more complex in that the cells are organized in complex interacting multicellular structures, e.g., leaves, stems, and roots in plants, or muscles, nerves, organs, and the cardiovascular system in animals. It is possible to grow individual cellular components of multicellular organisms or organs by the use of specialized media and culture conditions in reactors. The most important eukaryotic cells in biotechnology are the yeasts and fungi, although animal cells are becoming increasingly used. Most eukaryotic cells undergo a form of asymmetric division whereby the parent and daughter cells are distinguishable and the cultures have population profiles. With the exception of immortalized cancer cells, and cell fusions thereof, it is thought that most cells have a finite life.

The cell cycles of eukaryotic organisms are complex and not only involve changes in morphology but also variations in the genetic complement of the cell. Typically, the simplest organisms contain a single set of chromosomes for most of their life cycle, and in this condition, cells are termed *haploid*. Cell division in this state is termed *mitosis* and involves only replication of DNA. However, in some situations cells can fuse to contain two or more sets of chromosomes. When two sets of chromosomes are present, the cells are termed *diploid*. In the diploid state, a specialized form of cell division called *meiosis* can take place, where the chromosomes are segregated between two cells to return the cell to the haploid state. Generally, this is associated with a sexual process where specialized cell types fuse to form the diploid state before returning to the more usual haploid state on meiotic cell division. The time spent in one or other of these two states varies considerably from organism to organism; however, in most organisms of industrial importance, cells multiply by mitotic processes.

5.3.3.1 Yeast

Yeasts reproduce by budding, fission, and sporulation, the most common process being budding. In budding, the nucleus, which is present in the parent cell, enlarges and then extends into a bud forming a dumbbell-shaped structure that later divides giving rise to a nucleus in both the parent and daughter cell. On division, an inactive surface scar remains on the mother cell surface, and ultimately after several more divisions the surface becomes covered with the scar material; it is thought that this eventually brings about the death of the cell. Sporulation can occur in yeast and usually involves the formation of spores via either asexual or sexual processes; the spores are called conidia and ascospores, respectively.

The overall life cycle of a particular yeast, *Saccharomyces cerevisiae,* is summarized in Fig. 5.29, which shows how it is possible for the cells to fuse to form various cell and spore types. The figure shows the possible types of reproduction in yeast. Generally, industrial strains of *S. cerevisiae*, brewers' yeast, reproduce by budding/fission processes and only sporulate under specialized conditions. However, many strains of yeast are capable of cell fusion to form spores or cells with increased genetic complements. Such strains have many sets of chromosomes and are termed *polyploid.* Active fermentation of industrial strains involves growth by mitotic division and nutrient depletion, which results in stationary cells with little or no spore formation.

5.3.3.2 Fungi

The name, fungi, is derived from their most obvious representatives, the mushrooms (Greek mykes and Latin fungus). They share many properties with plants, possessing a cell wall and liquid-filled intracellular vacuoles; microscopically they exhibit visible streaming of the cytoplasm and they lack motility. However, they do not contain photosynthetic pigments and are heterotrophic. Nearly all grow aerobically and obtain

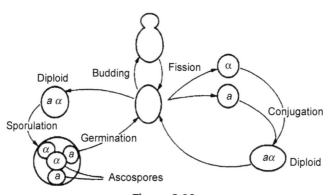

Figure 5.29
The yeast cell cycle.[2]

Table 5.8: Size and properties of parts of fungi.[2]

Part	Size (μm)	Comments
Outer fibrous layer	0.1–0.5	Very electron-dense material
Cell wall	0.1–0.25	Zygomycetes, ascomycetes, and basidiomycetes contain chitin (2%–26% dry wt); oomycetes contain cellulose, not chitin; yeast cells contain glucan (29%), mannan (31%), protein (13%), and lipid (8.5%)
Cell membrane	0.007–0.01	Much-folded, double-layered membrane; semipermeable to nutrients
Endoplasmic reticulum	0.007–0.01	Highly invaginated membrane or set of tubules, probably connected with both the cell membrane and the nuclear membrane and concerned in protein synthesis and probably other metabolic functions
Nucleus	0.7–3	Surrounded by a double membrane (10 nm), containing pores (40–70 nm wide); flexible and contains cytologically distinguishable chromosomes Nucleolus about 3 nm
Mitochondria	0.5–1.2 × 0.7–2	Analogous to those in animal and plant cells containing electron transport enzymes and bounded by an outer membrane and an inner membrane forming cristae; probably develop by division of existing mitochondria
Inclusions		Lipid and glycogen-like granules are found in some fungi Ribosomes in all fungi

their energy by oxidation of organic substances. Table 5.8 shows the properties and parts of fungi which are typical of eukaryotic cells, the most interesting component is the cell wall with its great diversity of structure and composition. Fungi do show a limited morphological differentiation, which is associated with spore formation or fruiting bodies. Some species have forms of sexual reproduction.

The vegetative body is a *thallus*. It consists of filaments about 5 μm in diameter, which are multibranched or spread over or into the nutrient medium. The filaments or *hyphae*, can be present without cross walls as in lower fungi or divided into cells by *septa* in higher fungi. The total hyphal mass of the fungal thallus is called the *mycelium*. In certain situations, during transition between asexual and sexual reproduction, various other tissue structures are formed, e.g., *Plectrenchyma* (mushroom flesh).

Growth usually involves elongation of hyphae at their tips or apices. In most fungi, every part of the mycelium is capable of inoculation and a small piece of mycelium is sufficient to produce a new thallus (Fig. 5.30).

There are many forms of structure and mechanism involved in the reproductive processes of fungi. Most fungi are capable of reproduction in two ways: asexual and sexual. Asexual

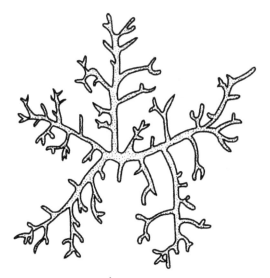

Figure 5.30
The mycelium structure of molds.

methods are by budding, fragmentation, and formation of spores, while the sexual process involves cell fusion. Spore formation is the most widely distributed and highly differentiated method. Further details of the biology and industrial uses of fungi are given by Berry[11] and Onions et al.[12] The structural diversity of sporulating bodies is used as a basis for classification. Fig. 5.31 shows the structural differences in fruiting bodies between Penicillium and Aspergillus.

5.3.3.3 Algae and Protozoa

Algae and protozoa are now being used in industrial fermentations; generally, they are used for the production of glycerol, pigments, and their derivatives. Algae are autotrophic and generally require light for growth. Some of the most useful forms are, however, heterotrophic, which means that in the presence of an organic carbon source they will grow in traditional fermenters in the absence of light. Usually, algal fermentations are not enclosed and take place in ponds having a significant impact on waste treatment processes.[13]

Protozoa, again, are found in waste treatment processes where these generally motile forms graze on bacterial populations. The most important and interesting application concerns the use of these organisms as biological control agents.

5.3.3.4 Animal and Plant Cells

Animal and plant cells are also used in fermentations but rarely in fully differentiated forms. Most differentiated cells have a finite life and so, to grow cells successfully from

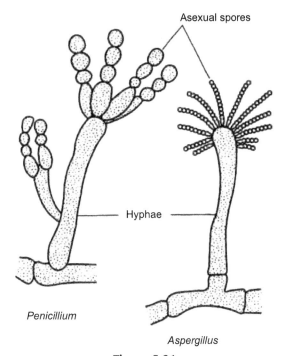

Figure 5.31
The hyphae of *Aspergillus* and *Penicillium*, two industrially important molds.

plants, undifferentiated cells or callus are used. Animal cells for fermentations are derived from the fusion products where the cell is immortalized by fusing the differentiated cell with a cancer cell. The most interesting types of cells are those which produce antibodies; these are highly specific reagents and of high value. The most important considerations with these specialized fermentations are the expense of the medium, the relatively slow growth rates, the low productivity, and shear sensitivity. These problems are being overcome: shear sensitivity can be obviated by the use of correct cell selection methods, and media are being reformulated to reduce the content of expensive components. Specialized reactor systems, such as airlift reactors and immobilized cell reactors, have now been developed[2,14] for this purpose.

5.3.4 General Physical Properties of Cells

There are many important physical properties of cells of significance in biological processing, but there is a lack of detailed information concerning the physical properties of cells and their components. For example, the colloidal and surface properties of cells, cell walls, and proteins, which are of considerable importance in separation and purification processes, are not fully understood.

The size and shape of cells, as shown in Table 5.9, can be used as the basis of methods to separate cells from a liquid medium. Cells are only slightly denser than water and separation from liquids by centrifugation requires the use of intense fields. The shape of cells is important because of its effect on the rheology of the fermentation broth. Generally, bacteria and yeast present few problems until very high concentrations are encountered, e.g., 50 g dry wt/1. However, filamentous organisms at relatively low concentrations can cause significant changes in theological properties and many broths quickly become non-Newtonian in character.[2]

Table 5.9: Shape and size of biological particles.[2]

Microorganisms	Shape	Size
Bacteria		
Bacillus megaterium	Rod shaped	2.8–1.2–1.5 µm
Serratia marcescens	Rod shaped	0.7–1.0 × 0.7 µm
Staphylococcus albus	Spherical	1.0 µm
Streptomyces scabies	Filamentous	0.5–1.2 µm dia
Fungi		
Botrytis cinerea conidiophores	Filamentous	11–23 µm dia
B. cinerea, conidia	Ellipsoidal	9–15 × 5–10 µm
Rhizopus nigricans sporangiopores	Filamentous	24–42 µm dia
R. nigricans sporangiospores	Almost spherical	11–14 µm dia
R. nigricans zygospores	Spherical	150–200 µm dia
Saccharomyces uvarum (*Saccharomyces carlsbergensis*) vegetative cells	Ellipsoidal	5–10.5 × 4–8 µm
Algae		
Chlamydomonas kleinii	Ovoid	28–32 × 8–12 µm
Ulothrix subtilis	Filamentous	4–8 µm dia
Protozoa		
Amoeba proteus	Amorphous	<600 µm dia
Euglena viridis	Spindle shaped	40–65 × 14–20 µm
Paramecium caudatum	Slipper shaped	180–300 µm length
Slime molds		
Badhamia utricularis Plasmodia	Amorphous	Indefinite, extensive
B. utricularis spores	Spherical	9–12 µm dia
Protein molecules		
Egg albumin		4.0 nm
Serum albumin		5.6 nm
Serum globulin		6.3 nm
Hemocyanin		22.0 nm

Another important parameter is the surface charge associated with microbial particles. This reflects the composition of the cell wall surrounding the cell membranes; it is complex and dependent on environmental conditions and on the stages of the life cycle.[6]

5.3.5 Tolerance to Environmental Conditions

Although individual microbial cell systems usually grow under only a narrow range of physical conditions, there are many cell types, which are adapted to extreme but specific environmental conditions. Cells have a series of cardinal conditions, i.e., a maximum, a minimum, and an optimum with respect to physical conditions. Environmental conditions such as temperature, pressure, pH, and ion concentration can all affect the growth of cells. For the more detailed explanation of the physical conditions of growth the reader is referred to specific texts on the subject.[15,16]

5.3.6 Chemical Composition of Cells

5.3.6.1 Elemental Composition

The elemental composition of various microorganisms is shown in Table 5.10. Six major elements—carbon, hydrogen, nitrogen, oxygen, phosphorus, and sulfur—are associated with cells and make up over 90% of their weight. The remaining elemental material consists mainly of the alkali metals and transition metals. While the compositions shown in Table 5.10 are typical of living cells, the composition of the cell can vary considerably during the growth cycle and with the type of cell. Materials associated with energy storage functions, such as carbohydrates (starch) and fats (polyhydroxybutyrate), are noted in this respect and are dependent on environmental conditions. Similarly, materials related to growth rate, such as RNA, can undergo significant changes in their concentration during the growth cycle.

There is a molecular hierarchy within cells. It is possible for some organisms, *autotrophs*, to synthesize all their structure from simple chemicals, such as carbon dioxide, water, and ammonia, using light energy. On the other hand, *heterotrophs* use organic materials to form the building blocks of cells, carbohydrates, nucleotides, amino acids, and fatty acids. These basic building blocks are then polymerized into polysaccharides, nucleic acids, proteins, and fats. These in turn form macromolecular structures, *organelles* (small membrane-bound structures), and ultimately cells and multicellular structures, as illustrated in Fig. 5.32.

A typical molecular analysis of various microorganisms is shown in Table 5.11.[18] Most of the elemental composition of cells is found in three basic types of materials—proteins, nucleic acids, and lipids. In Table 5.A1, the molecular composition of a bacterium is shown in more detail. Water is the major component of the cell and accounts for

Table 5.10: Elemental composition of microorganisms.[2]

Microorganism	Limiting Nutrient	Dilution rate[a] (h^{-1})	Composition (% by wt)							Empirical Chemical Formula	Formula "Molecular" Weight
			C	H	N	O	P	S	Ash		
Bacteria			53.0	7.3	12.0	19.0			8	$CH_{1.666}N_{0.20}O_{0.27}$	20.7
			47.0	4.9	13.7	31.3				$CH_2N_{0.25}O_5$	25.5
Enterobacter (Aerobacter) aerogenes			48.7	7.3	13.9	21.1			8.9	$CH_{1.78}N_{0.24}O_{0.33}$	22.5
Klebsiella aerogenes	Glycerol	0.1	50.6	7.3	13.0	29.0				$CH_{1.74}N_{0.22}O_{0.43}$	23.7
Yeast	Glycerol	0.85	50.1	7.3	14.0	28.7				$CH_{1.73}N_{0.24}O_{0.43}$	24.0
			47.0	6.5	7.5	31.0			8	$CH_{1.66}N_{0.13}O_{0.49}$	23.5
			50.3	7.4	8.8	33.5				$CH_{1.75}N_{0.15}O_5$	23.9
			44.7	6.2	8.5	31.2	1.08	0.6		$CH_{1.64}N_{0.16}O_{0.52}P_{0.01}S_{0.005}$	26.9
Candida utilis	Glucose	0.08	50.0	7,6	11.1	31.3				$CH_{1.82}N_{0.19}O_{0.47}$	24.0
	Glucose	0.45	46.9	7.2	10.9	35.0				$CH_{1.84}N_{0.2}O_{0.56}$	25.6
	Ethanol	0.06	50.3	7,7	11.0	30.8				$CH_{1.82}N_{0.19}O_{0.46}$	23.9
	Ethanol	0.43	47.2	7.3	11.0	34.6				$CH_{1.84}N_{0.2}O_{0.55}$	25.5

[a]Dilution rate in continuous fermentation, generally equal to the specific growth rate μ and is given by the ratio of volumetric throughput to liquid volume in the fermenter (See Section 5.11.3).

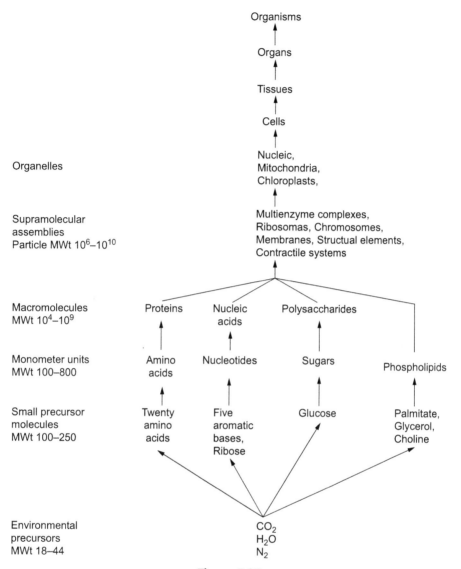

Figure 5.32

The hierarchy of biological structure. The approximate relative sizes as molecular weights (MWt) or equivalent.[17]

80%—90% of the total weight, while proteins form the next most abundant group of materials, and these have both structural and functional properties. Most of the proteins present will be in the form of enzymes. Nucleic acids are found in various forms—RNA and DNA. Their primary function is the storage, transmission, and utilization of genetic information. Polysaccharides and lipids are associated with wall and membrane structures and they can also act as energy storage materials within the cell. Small molecules,

Table 5.11: Chemical analyses, dry weights, and the populations of different microorganisms obtained in culture.[18]

Microorganism	Composition (% Dry wt)			Population in Culture (Numbers mL^{-1})	Dry Weight in Culture (g (100 mL)$^{-1}$)	Comments
	Protein	Nucleic Acid	Lipid			
Algae (small unicellular)	10–60 (50)	1–5 (3)	4–80 (10)	$4 \times 10^7 - 8 \times 10^7$	0.4–0.9	Figure in parentheses is a commonly found value; composition varies with the growth conditions
Bacteria	40–50	13–25	10–15	$2 \times 10^8 - 2 \times 10^{11}$	0.02–2.9	Mycobacterium may contain 30% lipid
Fungi (filamentous)	10–25	1–3	2–7		3–5	Some Aspergillus and Penicillium species contain 50% lipid
Viruses	50–90	5–50	<1	$10^8 - 10^9$	0.0005^a	Viruses with a lipoprotein sheath may contain 25% lipid
Yeasts	40–50	4–10	1–6	$10^8 - 10^9$	1–5	Some Rhodotorula and Candida species contain 50% lipid

^{a}For a virus 200 nm in diameter.

monomers of the polymers described above together with metal salts make up 0.5% weight of the cell; however, within the cell, this pool of small molecules is constantly and rapidly turning over.

Much of the chemistry of the cell is common to all living systems and is directed towards ensuring growth and cell multiplication, or at least the survival of the cell. Organisms also share various structural characteristics. They all contain genetic material (DNA), membranes (the boundary material between the cell and the environment), cytoplasm (small particulate materials, ribosomes, and enzyme complexes), and cell walls or surfaces (complex structures external to the membrane). In addition, there are various distinct membrane-bound organelles in eukaryotic organisms, which have specialized functions within the cell (Tables 5.6–5.8).[5,7]

5.4 Cellular Metabolism

5.4.1 The Roles of Metabolism

In Section 5.1 it is stated that the main purpose of a living system is to ensure its survival and reproduction. This is achieved through the chemistry of the cell and there are two primary activities.

5.4.1.1 The Synthesis of Materials for Cell Structure

Synthesis within the cell is a complex process. Fig. 5.33 illustrates how the general metabolic system of the cell is used to acquire and create the basic building blocks of the cell. Once assembled, the building blocks are then used to produce polymeric materials, most importantly proteins and nucleic acids (see also Fig. 5.32).

5.4.1.2 The Generation of Energy for Growth, and for Chemical and Mechanical Work

To maintain a highly ordered structure in relation to the surrounding environment, the cell must consume energy. The major source of energy for living systems is ultimately traceable to light (electromagnetic radiation), which is converted into various forms of chemical energy. This captured energy can then drive synthetic processes, such as the reduction of carbon dioxide to glucose (i.e., photosynthesis). Glucose and other organic compounds can also be considered as sources of energy as they can be oxidized to carbon dioxide and water with the release of energy.

Enzymes are organized into metabolic pathways, which collectively constitute metabolism. Two types of metabolism are found in cells: *catabolism* (breakdown pathways) and *anabolism* (synthetic pathways). Linking these two types of metabolic reactions are the intermediary reactions of central metabolism. Cells, which contain many complex polymers, thus have the means to generate and convert monomeric materials into the

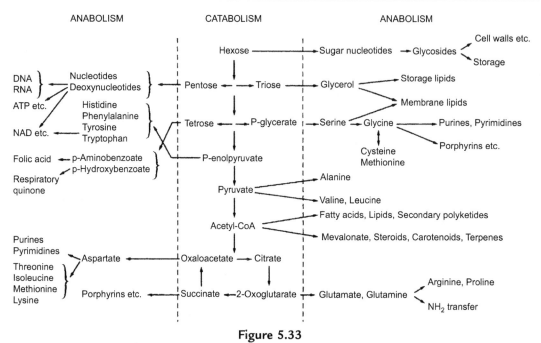

Figure 5.33

Anabolic pathways (synthetic) and central catabolic pathways. Only the main biosynthetic routes and their principal connections to catabolic routes, are shown, in simplified version. The relationships between energy (ATP) and redox (NAD$^+$, NADP$^+$) metabolism, nitrogen metabolism, etc. are omitted.[19]

complex biological structure. The sources of these materials are the simpler components from the cell's environment, such as inorganic salts and glucose (Fig. 5.32).

5.4.2 Types of Reactions in Metabolism

5.4.2.1 Catabolic Metabolism

The breakdown of materials from the environment and their conversion into usable forms take place through a series of catabolic pathways and the process is one type of catabolism. The general pattern of metabolism is one of converging pathways in that many materials are broken down to the common materials of central metabolism. A good analogy is that of the flow of materials and resources to a central point for conversion. For example, a steel work receives coal and iron ore which is smelted to form molten iron, converted to steel, and then formed into rod, sheet, and bar. The metabolic breakdown processes of herbicides, insecticides, and natural polymeric materials follow an analogous pattern. The breakdown of glucose along the glycolytic pathway is another example (Section 5.4.6). These pathways provide useful routes for the isolation of new enzymes for biotransformation processes.[20]

5.4.2.2 Anabolic Metabolism

Anabolism concerns the synthetic pathways of metabolism, which are involved in the generation of the building blocks and the subsequent synthesis of macromolecules. The pattern of these pathways is generally divergent. They start from the components of central metabolism and then diverge into the many pathways associated with synthesis, the polymer building blocks, and then the polymers themselves. An analogue of this would be the distribution of materials, e.g., steel to a component manufacturer, nuts and bolts to an automotive manufacturing plant, and the final distribution of the car to the end user. Fig. 5.33 shows the synthesis of amino acids, sugars, and nucleotides to polymeric materials such as proteins, carbohydrates, and nucleic acids (see also Fig. 5.32).

5.4.2.3 Intermediary Metabolism

Anabolism and catabolism are complementary to one another and may be compared with the biologically mediated cycling of carbon (and other materials) in the environment (Fig. 5.5). However, there is an area of metabolism where the two types of metabolism coincide and become indistinguishable. These pathways are known as intermediary metabolism and are the central part of cellular metabolism where the basic requirements for growth are met. At this metabolic cross-roads, materials are either rearranged into synthetic precursors or are oxidized to generate energy.

5.4.2.4 Primary and Secondary Metabolism

The metabolism of an organism is in most cases related to growth and reproduction and it is then termed *primary* metabolism. In many cases, however, metabolism can be uncoupled from growth by limiting some nutrient or by environmental control. This nongrowth-associated metabolism of an organism is called *secondary metabolism.*

The most obvious product of primary metabolism is the biomass itself or the fermentation products of anaerobic organisms. Table 5.12 shows primary *metabolites* (biochemical intermediates and products) of industrial interest. The yields from wild-type strains are not commercially viable, and this can be improved by various strain improvement techniques (Appendix A.2).

Secondary metabolism is commonly achieved by uncoupling the anabolic from the growth pathways. The subsequent overflow metabolites are then channeled toward *secondary products,* which may include antibiotics such as penicillin, tetracycline, and streptomycin. Fig. 5.34 shows the relationship between secondary products and central anabolic pathways.

Table 5.12: Some primary metabolites of industrial interest and some overproducing species.[21]

Metabolite	Organism	Uses (Present and Future)
Glutamate	*Corynebacterium glutamicum, Brevibacterium* spp.	Food industry as flavor-accentuating agent
Lysine, threonine	*Corynebacterium glutamicum, Brevibacterium flavum, Escherichia coli*	Essential amino acids, added to supplement low-grade protein
Guanylic, inosinic, and xanthylic acids	*Bacillus subtilis, Corynebacterium glutamicum*	Food industry as flavor-accentuating agents
Citric acid	*Aspergillus niger, Candida* spp.	Acidulant in food industry; pharmaceuticals (effervescent powders); esters used as plasticizers
Itaconic acid	*Aspergillus terreus, Aspergillus itaconicus*	Synthetic fiber and resin manufacture; copolymers (e.g., styrene-butadiene)
Fumaric acid	*Rhizopus arrhizus, Rhizopus nigricans*	Plastics and food industries
Cyanocobalamin (vitamin B_{12})	*Propionibacterium shermanii, Pseudomonas dentrificans*	Food and animal feed supplement
Riboflavin (vitamin B_2)	*Eremothecium ashbyii, Ashbya gossypii*	Food and animal feed supplement
β-Carotene	*Blakeslea* spp., *Choanephora* spp.	Precursor of vitamin A: coloring agent in food industry; food supplement
Xanthan gum, dextran	*Xanthomonas campestris, Leuconostoc mesenteroides*	Food, pharmaceutical, textile industries. Useful for thickening, stiffening, and setting properties
Acetic acid	*Saccharomyces ellipsoideus* or *Saccharomyces cerevisiae*, plus *Acetobacter* spp.; *Clostridium* spp.	Vinegar, preservative in food industry; chemical feedstock; polymer industry
Acetone, butanol	*Clostridium acetobutylicum*	Solvents in chemical industry; thinners; synthetic polymers
Ethanol	*Saccharomyces cerevisiae, Zymomonas mobilis, Clostridium thermocellum*, and other *Clostridium* spp.	Alcoholic beverages; solvent in chemical industry; fuel extender

5.4.3 Energetic Aspects of Biological Processes

In most engineering processes, thermal energy balances form one of the most important criteria in design, and the heat balance frequently approximates to the energy balance of the system. However, in biological systems the total energy balance must be considered and heat transfer within living cells is relatively unimportant compared with the transport, storage, and utilization of chemical energy. This is because enzymes operate over a narrow range of temperature, typically 15−40°C.

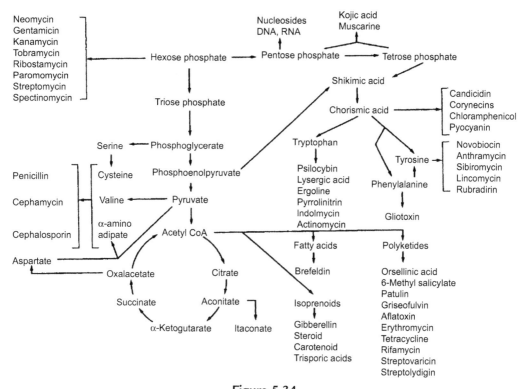

Figure 5.34
Summary of different metabolic routes leading to secondary metabolites.[22]

Utilizable forms of free energy in biological systems occur at several levels—as stored energy, e.g., polysaccharides; as energy from the environment, e.g., food; as energy at the intermediate level, e.g., glucose in the cell; as utilizable chemical energy, e.g., adenosine triphosphate.

In all organisms, the free energy released in redox reactions is conserved in the energy-carrier molecule adenosine triphosphate (ATP), which is the universal carrier of free energy in biological systems. ATP functions as an energy carrier by virtue of its two "high-energy" phosphate anhydride bonds. When these bonds are hydrolyzed:

$$ATP + H_2O \rightleftharpoons ADP + P_i + H^+ \quad \Delta G^{0'} = -30.55 \text{ MJ/kmol}$$
$$ADP + H_2O \rightleftharpoons AMP + P_i + H^+ \quad \Delta G^{0'} = -30.55 \text{ MJ/kmol}$$

The high, negative $\Delta G^{0'}$ values indicate that hydrolysis of the phosphate bond is strongly favored thermodynamically.

ATP formation from adenosine diphosphate (ADP) and inorganic phosphate (P_i) is driven by the energy generated when the fuel molecules are oxidized in heterotrophs, or when

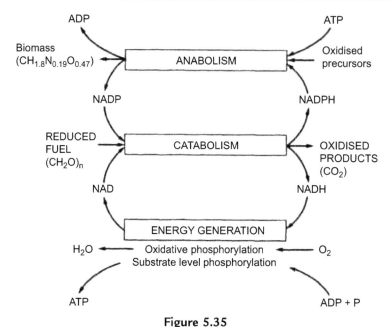

Figure 5.35
The interrelationships among metabolism, energy, and redox processes.

light energy is trapped in *phototrophs*. The energy in ATP is utilized, via breakdown to ADP + P$_i$, to drive thermodynamically unfavorable processes such as biosynthesis, active transport, and motility. ATP is not used in energy storage, but rather as a carrier between storage compounds such as fats and carbohydrate and the energy-consuming processes including biosynthesis and transport. Thus, ATP and ADP cycle between energy-requiring and energy-generating reactions as shown in Fig. 5.35.

Energy transfer via ATP is accomplished by the transfer of phosphate groups. In the breakdown of fuel molecules (catabolism), energetically favorable reactions produce the high-energy phosphorylated compounds for which the $\Delta G^{0'}$ value for hydrolysis is greater than that of ATP. Such compounds can spontaneously transfer phosphate groups to ADP, producing ATP, with a net negative $\Delta G^{0'}$. In biosynthesis, ATP is used to phosphorylate intermediates; thereby producing activated derivatives capable of further reaction. The detailed mechanisms by which ATP is generated are discussed in Section 5.4.6.

Another important set of reactions in living systems comprises those that involve redox processes. Reduced pyridine nucleotides, nicotinamide adenine dinucleotide (NAD), and nicotinamide adenine dinucleotide phosphate (NADP) are the universal carriers of hydrogen and electrons, or metabolic reducing power. Many redox processes may be linked together in electron transport chains.

All organisms use the same pair of pyridine nucleotides as carrier molecules for hydrogen and electrons. Both of these molecules accept hydrogen and electrons in the redox reactions of catabolism and become reduced. The oxidative half-reactions of catabolism generally produce two H^+ and two electrons. The nicotinamide ring can accept two electrons and one H^+ and, since the second H^+ is released into the solution, most redox reactions in biological systems take the form:

$$AH_2 + NAD(P) \rightleftharpoons A + NAD(P)H + H^+$$

NADH and NADPH have different metabolic functions. NADPH generally transfers H^+ and $2e^-$ to oxidized precursors in the reduction reactions of biosynthesis. Therefore, NADPH cycles between catabolic and biosynthetic reactions and serves as a carrier in the same way that ATP serves as the energy carrier. NADH is used almost exclusively in redox processes associated with energy metabolism.

The roles of NADH and NADPH in the overall strategy of metabolism are shown in Fig. 5.35. Fuel molecules, such as glucose, are oxidized in catabolism; they lose electrons, and these reducing equivalents are transferred to an environmental acceptor such as oxygen, with concomitant ATP production (see oxidative phosphorylation, Section 5.4.6). However, some reducing equivalents are conserved and reutilized in the synthesis of cellular components, with the consumption of ATP, as oxidized intermediates are reduced to synthetic precursors with subsequent polymerization. The pyridine nucleotides thus have roles in both synthetic and energy-generation process.

5.4.4 Energy Generation

Energy-evolving reactions (exogenic) are coupled as described above to a reaction requiring energy (endogenic). It was noted that energy in the cell may be stored in various forms (storage chemicals, e.g., fats and carbohydrates; and glucose) and converted rapidly to the universal energy chemical ATP, which is then employed to do osmotic, chemical, and mechanical work. There are two basic systems by which the synthesis of ATP can be achieved—*substrate-level phosphorylation* and *electron transport—linked phosphorylation* of which *oxidative phosphorylation* is the more important process.

5.4.5 Substrate-Level Phosphorylation

Substrate-level phosphorylation refers to those reactions associated with the generation of energy by the transfer of phosphate groups in metabolism and is exemplified by fermentative metabolism where it is the sole source of energy (e.g., yeast and bacteria growing in anaerobic conditions).

5.4.5.1 Glycolysis

The sequence of biochemical steps by which glucose is degraded to pyruvic acid is called glycolysis or the Embden—Meyerhof—Parnas (EMP) pathway and is shown in Fig. 5.36. This is the best-known metabolic pathway and is central to the metabolism of carbohydrates, having roles in energy generation and in providing the carbon skeletons used in biosynthetic reactions. This pathway occurs in both anaerobic and aerobic systems, but environmental conditions and the type of organism determine the fate of pyruvate produced by this pathway. For example, pyruvate is completely oxidized to carbon dioxide and water in most aerobic systems, but under anaerobic conditions a variety of organic acids and solvents can be produced.

The net products of the oxidation of glucose, two molecules each of NADH, ATP, and pyruvate, are formed by a series of simple reactions, either ring splitting or the transfer of a small group such as phosphate or hydrogen. Two molecules of ATP are hydrolyzed to ADP when used to activate glucose to fructose 1, 6 di-phosphate via hexokinase, and phosphofructokinase enzymes (area 1 of Fig. 5.36). Once activated, the six-carbon-atom molecule is split into two three-carbon compounds, following which two reactions are effective in transferring phosphate to ADP to form ATP, for each three-carbon moiety catalyzed by 3-phosphoglycerate kinase and pyruvate kinase (area 2 of Fig. 5.36). Overall, the pathway results in the formation of a net two molecules of ATP, as two molecules of ATP are consumed in the activation stage. The other remaining feature of the pathway is the net production of reducing equivalents in the form of NADH from NAD.

The overall stoichiometry of the EMP pathway is thus:

$$C_6H_{12}O_6 + 2\ P_i + 2\ ADP + 2\ NAD^+ \rightarrow 2\ C_3H_4O_3 + 2\ ATP + 2\ NADH + H^+$$

Apart from the energy and reducing power derived from these reactions, the pathway also provides carbon skeletons for the synthesis of cellular structures. The energy provided by this pathway is used to drive the synthetic processes of the cell, and the EMP pathway is the major route for glucose metabolism. However, there are specialized alternative routes by which glucose may be metabolized and the reader is referred to specialized biochemical texts for an account of these.[5,23]

5.4.5.2 The Metabolic Fates of Pyruvate in Anaerobic Conditions
5.4.5.2.1 Lactic Acid Fermentation

Another feature of the EMP pathway is that there is only a small pool of NAD/NADH within the cell and NADH must be oxidized back to NAD before it can participate in another set of glycolytic reactions. There is a whole series of routes by which this can be achieved. The simplest way is by the reduction of pyruvate to lactate (Fig. 5.37). This type of metabolism is widespread and is encountered in both lactic acid bacteria and human

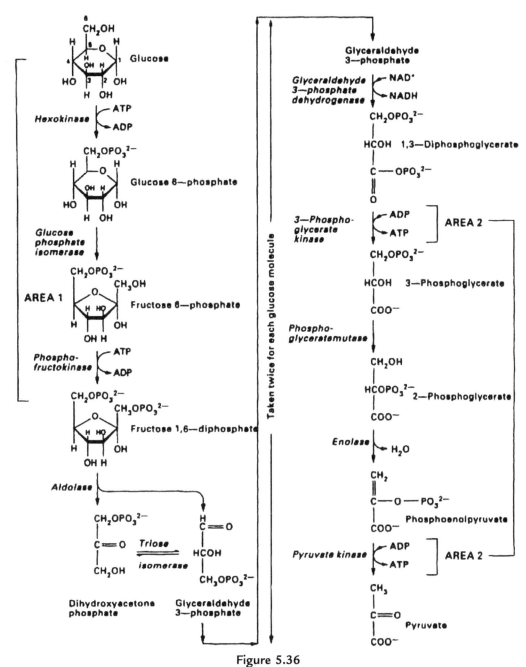

Figure 5.36
The Embden—Meyerhof—Parnas pathway. The six-carbon substrate, (glucose) yields two three-carbon intermediates to produce two moles of pyruvate.

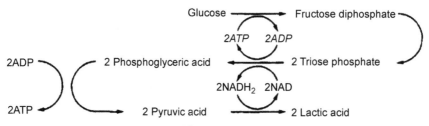

Figure 5.37

The lactic acid fermentation. The reaction sequence results in two molecules of lactic acid, two molecules of ATP, and a balanced redox process.

muscle. Thus, the pyruvate and the reducing equivalents in $NADH + H^+$ react to produce lactic acid:

$$C_3H_4O_3 + NADH + H^+ \rightarrow C_3H_6O_3 + NAD^+$$

The overall glycolytic sequence from glucose to lactic acid can be written as follows:

$$C_6H_{12}O_6 + 2\ Pi + 2\ ADP \rightarrow 2\ C_3H_6O_3 + 2\ ATP + 2\ H_2O\ (\Delta G^{0'} = -135.7\ \text{MJ/kmol})$$

The free energy change for glucose breakdown alone is

$$\text{Glucose} \rightarrow 2\ C_3H_6O_3\ \Delta G^{0'} = -196.8\ \text{MJ/kmol}$$

The difference between the free energy changes of the two reactions shows how much of the energy is retained by the molecules in the pathway (i.e., $-196.8 + 135.7 = -61.1$ MJ/kmol) as 2 mol of ATP ($30.55 \times 2 = 61.1$ kJ/mol). The efficiency of free energy transfer from the glucose molecule to the phosphate bonds of ATP is thus $(61.1/196.8) \times 100 = 31\%$. Correction of the standard free energy data for the concentrations and the pH found in living cells suggests that this estimate is low and should more realistically be above 50%.[5]

5.4.5.2.2 Other Fermentations

The lactic acid fermentation represents the simplest fermentative metabolism but, economically, the most important anaerobic fermentation is the alcohol fermentation of yeast. The major difference between alcoholic and the homolactic fermentation is that pyruvate is decarboxylated to acetaldehyde in the alcohol fermentation and this is then reduced to ethanol, as illustrated in Fig. 5.38.

Other products can be produced in fermentative bacteria but the central feature of all these pathways is the strict maintenance of the oxidation–reduction balance within the fermentation system. This gives rise to another important tool in assessing fermentation pathways—a mass balance of the substrate and products. The amount of carbon, hydrogen,

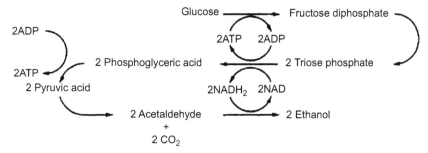

Figure 5.38
The alcoholic fermentation sequence results in two molecules of ethanol from glucose, two molecules of ATP, and a balanced redox process.

and oxygen in the fermentation products (including cells) must correspond to the quantities in the substrate utilized.

e.g.,

$$\text{Homolactic fermentation: } C_6H_{12}O_6 \rightarrow 1.9\ C_3H_6O_3 + \text{cells}$$
$$\text{Alcohol fermentation: } C_6H_{12}O_6 \rightarrow 1.8\ C_2H_5OH + 1.8\ CO_2 + \text{cells}$$

The diversity of fermentative metabolism, summarized in Fig. 5.39, results in the production of a variety of products such as neutral compounds (ethanol, isopropanol acetone, butanediol, and butanol), organic acids (formic, acetic, propionic, butyric, succinic), and gases (hydrogen and carbon dioxide).[23] There are various physiological advantages in producing this diverse collection of products but the most important is the generation of additional energy via the acid–phosphate intermediates as shown:

$$\text{Acetyl CoA} + P_i \rightarrow \text{Acetyl} - P + \text{CoA}$$
$$\text{Acetyl} - P\ \text{ADP} \rightarrow \text{Acetate} + \text{ATP}$$

Although acids are generally more highly oxidized than the original substrates, if the fermentation can at the same time reduce another compound or release its reducing equivalents in the form of hydrogen, then energy-generating acids can be produced. For example, a mixed acid fermentation of the enteric bacteria produces lactate, acetate, ethanol, formate, hydrogen, and carbon dioxide.

5.4.5.2.3 Fermentation of Other Materials

Many other carbohydrates can be fermented, including most sugars (hexoses and pentoses), disaccharides (maltose), and polysaccharides (starch and cellulose). Extracellular enzymes are used to hydrolyze the polymeric materials down to sugar monomers. Once in monomeric forms, hexose and pentose sugars can be taken up and metabolized in the cell where they eventually feed into the glycolytic pathway. Some of

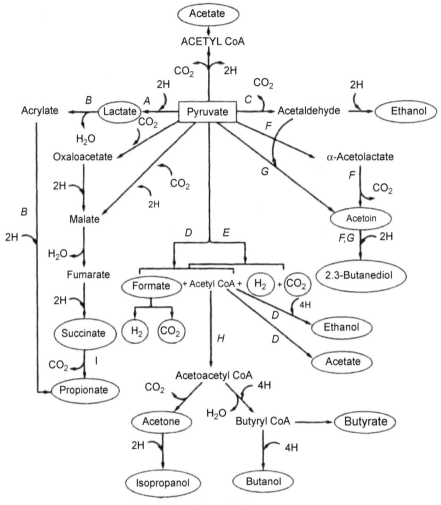

Figure 5.39

The end products (circled) of microbial fermentations of pyruvate. Letters indicate the organisms able to perform these reactions. (A) Lactic acid bacteria (*Streptococcus, Lactobacillus*); (B) *Clostridium propionicum*; (C) Yeast, *Zymomonas mobilis, Sarcina ventriculi*; (D) *Enterobacteriaceae* (Coli-aerogenes); (E) Clostridia; (F) Aerobacter, Bacillus polymyxa; (G) Yeast; (H) Clostridia; (I) Propionic acid bacteria.[24]

these sugars may be more highly reduced than glucose and accordingly the product spectrum reflects this and the redox balance of the system is maintained.

Anaerobic fermentations are thus capable of producing a wide variety of products from carbohydrates. Energy generation and chemicals production by fermentation are becoming attractive possibilities. At present these fermentations are being investigated for the possible conversion of wastes into useful chemicals, particularly in the western world. In developing countries, especially in the tropics and where there are no fossil fuel reserves, this type of process is being used for the production of fuels and chemicals, e.g., the "gasohol" program of Brazil where ethanol is produced from sugar cane. Ethanol can either be used as a motor fuel or it can be dehydrated to ethylene by chemical means for the production of other bulk chemicals. In Europe and America, the production of bulk chemicals from biomass is more expensive than by petrochemical routes, but the situation will change as oil becomes more difficult to find and recover. By utilizing the anaerobic fermentation pathways, it is possible to produce many of the basic precursor molecules of the modern petrochemical industry and thereby to maintain a sustainable technology for chemicals production.

5.4.6 Aerobic Respiration and Oxidative Phosphorylation

Phosphorylation linked to electron transport is the other major mechanism of generation of ATP in cells. Carbon substrates are completely oxidized in the presence of oxygen to carbon dioxide and reducing equivalents, the latter usually in the form of NADH. NADH is a source of electrons, which are then used to generate via electron transport processes before they ultimately reduce oxygen to form water. This process is commonly called *aerobic respiration.*

Respiration is a broad term for the oxidation of organic materials to carbon dioxide. When the electron acceptor is not oxygen they are called *anaerobic respirations.* Table 5.13 illustrates possible types of respiration. Nitrate, sulfate, and carbon dioxide are significant electron acceptors in anaerobic respiration and many are of environmental significance.

Aerobic respiration can be subdivided into a number of distinct but coupled processes, such as the carbon flow pathways resulting in the production of carbon dioxide and the oxidation of $NADH + H^+$ and $FADH_2$ (flavin adenine dinucleotide) to water via the electron transport systems or the respiratory chain.

Table 5.13: Reductants and oxidants in bacterial respirations.

Reductant	Oxidant	Products	Organism
H_2	O_2	H_2O	Hydrogen bacteria
H_2	SO_4^2	$H_2O + S^{2-}$	*Desulfovibrio*
Organic compounds	O_2	$CO_2 + H_2O$	Many bacteria, all plants and animals
NH_3	O_2	$NO_2^- + H_2O$	Nitrifying bacteria
NO_2^-	O_2	$NO_3^- + H_2O$	Nitrifying bacteria
Organic compounds	NO_3^-	$N_2 + CO_2$	Denitrifying bacteria
Fe^{2+}	O_2	Fe^{3+}	*Ferrobacillus* (iron bacteria)
S^{2-}	O_2	$SO_4^{2-} + H_2O$	*Thiobactilus* (sulfur bacteria)

5.4.6.1 Carbon Flow and the Generation of Reducing Power in Oxidative Phosphorylation

Glucose is converted to pyruvate by the glycolytic pathway (Fig. 5.38). Pyruvate is then oxidized to acetyl—CoA and CO_2 by the pyruvate decarboxylase enzyme:

$$CH_3COCOOH + NAD^+ + CoA-SH \rightarrow CH_3CO-S-CoA + CO_2 + NAD + H^+$$

The transformation of pyruvate to carbon dioxide is achieved by the several steps in a cyclical series of reactions known as the tricarboxylic acid (TCA) cycle. The name of the cycle comes from the first step where acetyl—CoA is condensed with oxaloacetic acid to form citric acid, a TCA. Once citrate is formed the material is converted back to oxaloacetate through a series of 10 reactions, as illustrated in Fig. 5.40, with the net production of two molecules of carbon dioxide and reducing equivalents in the form of four molecules of $NADH + H^+$ and one molecule of $FADH_2$, together with 1 mol of ATP. The overall stoichiometry of the TCA cycle from pyruvate is

$$CH_3COCOOH + ADP + P_i + 2H_2O + FAD + 4NAD \rightarrow$$
$$3CO_2 + ATP + FADH_2 + 4(NADH + H^+)$$

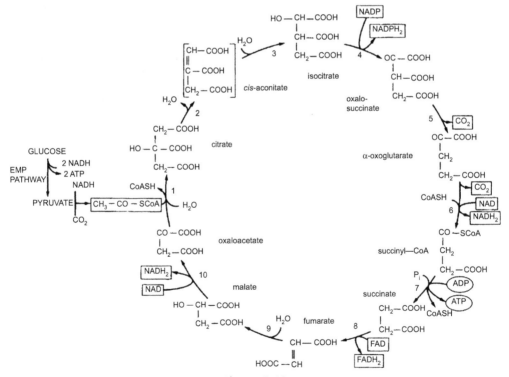

Figure 5.40

Oxidation of acetyl-CoA via the tricarboxylic acid cycle. Individual enzymes of the pathway are marked. 1, Citrate synthase; 2 and 3, cis-aconitate hydratase; 4 and 5, isocitrate dehydrogenase; 6, α-oxo glutarate dehydrogenase; 7, succinate thiokinase; 8, succinate dehydrogenase; 9, fumarase; 10, malate dehydrogenase.[23]

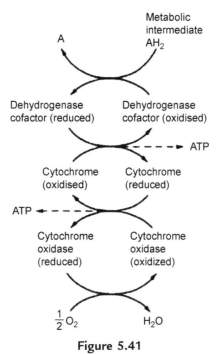

Figure 5.41

Simplified scheme showing route taken by electrons. At each step, the reduced form of one carrier reduces (or passes electrons to) the oxidized form of the next.

5.4.6.2 The Respiratory Chain

Carbon from many sources can be successfully oxidized to carbon dioxide, and the reduced electron carriers produced are ultimately oxidized by oxygen to form water. It is possible to pass electrons through a series of proteins, which participate in a series of interlinked redox reactions, known collectively as the *electron transport chain* or *system*. Thus, if redox proteins are ordered correctly, electrons will shuttle down the series, as illustrated in Fig. 5.41, giving rise to chain-like transport behavior. Fig. 5.41 shows quite clearly how the term electron transport chain originated. As an electron is passed down through the redox carriers, at first it is reduced by electrons and then reoxidized as the electron is passed from it to the next carrier. Several carriers may be linked together, each one more highly oxidized than the previous one, so that the electron is passed along a gradient of redox potential to a terminal electron acceptor, which, in the case of aerobic respiration, is oxygen (Fig. 5.42).

5.4.6.3 Electron Transport Processes Linked to Phosphorylation

Energy can be generated by electron transport processes, the operation of which can be understood by looking at the orientation and the location of the electron transport systems.

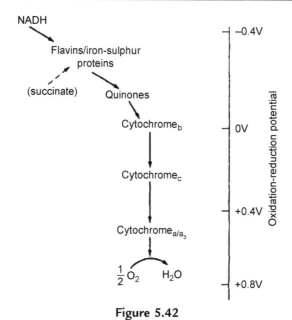

Figure 5.42

The passage of electrons through an electron transport chain involving the passing of electrons down a potential gradient.

The electron transport systems are localized in the cell membranes of bacteria whereas in yeast and animal cells (eukaryotes), it is localized in the membranes of *mitochondria* and other specialized organelles. The redox carriers of the electron transport chain straddle the membrane and, as electrons are passed down the chain, they also transport protons across the membrane, as shown in Fig. 5.43. The result is the passage of protons from the inside to the outside of the membrane and the creation of a proton or pH gradient. The pH gradient represents a source of potential energy (a concentration gradient and a gradient of charge), which can now be utilized to generate ATP. This generation of ATP is achieved by channeling the transport of the protons across the membrane from the outside back into the inside of the cell through a specific protein pore called ATPase. The energy released by this process is coupled to phosphorylation of ADP + P_i to ATP. The high-energy electron released from oxidation of carbohydrate can thus drive ADP phosphorylation, as illustrated in Fig. 5.43, but the precise biochemical coupling of proton transport and ATP generation is yet to be elucidated.

The nature of this process means that the stoichiometry of these reactions is variable, although it is thought that the process will produce a maximum of 3 mol of ATP from the oxidation of 1 mol NADH + H^+ and 2 mol ATP from 1 mol $FADH_2$.

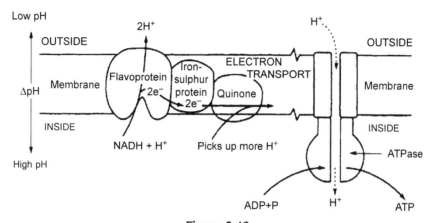

Figure 5.43
Energy generation linked to electron transport process.

i.e.,

$$NADH + H^+ + 1/2\ O_2\ 3\ ADP + 3\ P_i \rightarrow NAD + 4\ H_2O + 3\ ATP$$

$$FADH_2 + 1/2\ O_2\ 2\ ADP + 2\ P_i \rightarrow FAD + 4\ H_2O + 2\ ATP$$

Thus, from 1 mol of glucose the following numbers of moles of product are generated:

Process	Product		
	NADH + H$^+$	FADH$_2$	ATP
Glycolysis	2		2
Oxidation of two pyruvates	8	2	2
Oxidation of NADH + H$^+$	−10	−2	34
	0	0	38

That is, overall 38 mol of ATP may be generated from the metabolism of 1 mol of glucose. The overall process is summarized in Fig. 5.44.

5.4.6.4 Energy Efficiency of Aerobic Respiration (Oxidative Phosphorylation)

Aerobic respiration potentially makes available much more energy for use by the cell than glycolysis (19 times). Now, for the oxidation of glucose:

$$C_6H_{12}O_6 + 6\ O_2 \rightarrow 6\ CO_2 + 6\ H_2O\ \Delta G^{0'} = -2871\ \text{MJ/kmol}$$

whereas in glycolysis and respiration:

$$C_6H_{12}O_6 + 38\ ADP + 38\ P_i \rightarrow 6\ CO_2 + 38\ ATP + 44\ H_2O$$

Because ATP hydrolysis has a standard free energy change of −30.55 MJ/kmol, the free energy change of the reaction is approximately:

$$\Delta G^{0'} = 38 \times 30.55 = -1161\ \text{MJ/kmol}$$

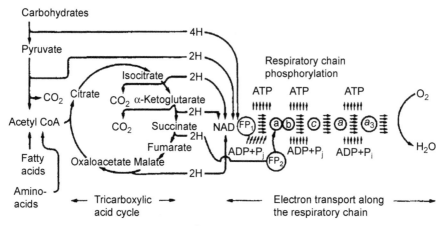

Figure 5.44
Summary of the potential energy released from oxidation of carbon materials to carbon dioxide and water via the tricarboxylic acid cycle and electron transport in the respiratory chain.[5]

$$\text{Therefore, the energy capture efficiency} = \frac{1161}{2871} \times 100 = 40\%$$

Again, corrections, which take pH and the nonstandard concentration of reactants into account, suggest that the energy capture is in fact greater than 70%, with most of the remaining energy being dissipated as heat.

5.4.6.5 Energy Capture and Electron Transport Processes

One of the most significant problems associated with electron transport—mediated processes is that they are not directly linked to the energy generation process. The amount of energy generated by these processes is variable, primarily due to leakage of the protons against a pH gradient over the membrane. Alternatively, some electron transport processes may avoid or nullify the transport of protons across the membrane and thus reduce the number of protons transported and reduce the membrane potential. Energized membranes also do work connected with the biosynthetic process by maintaining *homeostasis* (meaning self-maintenance of a constant environment) within cells. As a consequence, the yield of cell material from a substrate is variable and depends on a number of factors, including growth rate, environmental conditions, and the organism employed.

If energy is generated solely by substrate-level phosphorylation, as with anaerobic fermentative metabolism of bacteria and yeast, then the yield is more tightly linked to the amount of energy generated. Generally, 10–12 kg of cell dry matter can be synthesized per kilomole of ATP generated in metabolism.

5.4.7 Photosynthesis

Glucose is a major source of energy for heterotrophs; however, the main source of energy on the planet is sunlight. This ultimate source of energy can be captured by organisms (e.g., plants, algae, and bacteria) in the process called photosynthesis. Electromagnetic radiation in the wavelength range 300–800 nm (visible light) drives the fixation of carbon dioxide from the atmosphere into glucose. As is common with synthetic processes, the fixation requires reducing power, NADPH (CO_2 is very highly oxidized compared with glucose $C_6H_{12}O_6$), and chemical energy (ATP). It can therefore be considered as an energy capture processes and is, in effect, the reverse of respiration. It also completes the cycling of carbon material in the environment; carbon dioxide, water, and light are utilized to reduce organic materials and oxygen, while glucose and oxygen are then oxidized back to carbon dioxide and water.

5.4.7.1 An Overview of the Photosynthetic Process

The radiant energy of the sun is captured and converted to chemical energy by photosynthetic organisms by the overall stoichiometry:

$$6CO_2 + 6H_2O \xrightarrow{\text{light}} C_6H_{12}O_6 + O_2$$

The photosynthetic reactions can be split into two major groups, the light and the dark reactions. The *light* reactions are those involved with light energy captured in a variety of pigments including chlorophyll. This light energy is used to split water and to create reducing power in the form of NADPH and energy (ATP), with the simultaneous formation of molecular oxygen.

$$H_2O + NADP + HPO_4^- + ADP \rightarrow O_2 + ATP + NADPH + H^+$$

The resultant NADPH and ATP provide the reducing power and free energy to drive the reduction of carbon dioxide (*the dark reactions*) via the pentose phosphate pathway (or Calvin cycle[5]) and lead ultimately to the synthesis of glucose according to the overall stoichiometry given below.

$$6CO_2 + 12(NADPH + H^+) + 18\,ATP \rightarrow C_6H_{12}O_6 + 12\,NADP + 18\,ADP + P_i + 6\,H_2O$$

Both the "light" and "dark" reactions occur in the *chloroplast*, a specialized membrane-bound organelle. For further details of this process the reader is referred to a general biochemistry textbook such as that by Lehninger.[5]

5.5 Cellular Control Mechanisms and Their Manipulation

In previous sections, the basic aspects of cell structure, cell biochemistry, and important biological pathways have been discussed. These biological reactions in the cell are mediated by highly specific biocatalysts called enzymes. Enzymes act similar to catalysts that act by accelerating chemical reactions but at milder conditions.

5.5.1 Biological Versus Chemical Reaction Processes

Biological reactions are like any other chemical processes, except that they are catalyzed by a variety of sophisticated catalysts whose structure is based upon proteins and ancillary associated materials, such as metal ions, carbohydrates, and nucleotides. Biological catalysts, or enzymes, have a very precise architecture, and this is reflected in their reactivity with both their products and substrates (e.g., reactants, food materials for growth); usually these are stereospecific. Enzymes provide the means to generate and reproduce the complex structure of living systems. Whole cells may be used to provide catalytic materials, or if the cells are ruptured, the catalyst may be extracted and purified. Table 5.14 compares biological and chemical catalysts and it may be seen that most enzymes compare well with their chemical counterparts in terms of *turnover number* (see Section 5.5.2) and operational temperature range. With all proteins, there is, however, a major problem in relation to their stability, although strategies for obtaining more stable

Table 5.14: Some turnover numbers for enzyme and inorganic catalyzed reactions.[19]

Catalyst	EC Number	Reaction	Turnover Number (s^{-1})	Temperature $(°C)$
Enzymes				
Bromelain	3.4.22.4	Hydrolysis of peptides	4×10^{-3} -5×10^{-1}	0–37
Carbonic anhydrase	4.2.1.1	Hydration of carbonyl compounds	8×10^{-1} -6×10^{5}	0–37
Fumarase	4.2.1.2	L − Malate $\rightleftharpoons$ fumarate + H_2O	10^{3} (forward) 3×10^{3} (backward)	0–37
Papain	3.4.22.2	Hydrolysis of peptides	8×10^{-2}–10	0–37
Ribonuclease	3.1.4.22	Transfer phosphate of polynucleotide	2–2×10^{-3} $\times 10^{2}$	0–37
Trypsin	3.4.21.4	Hydrolysis of peptides		0–37
Inorganic catalysts				
Aluminum trichloride/ alumina		n-Hexane isomerization	10^{-2} 1.5×10^{-2}	25 60
Copper/silver		Formic acid dehydrogenation	2×10^{-7}	25
Cu_3Au			3×10^{10}	327
Silica–alumina		Cumene cracking	3×10^{-8} 2×10^{4}	25 420
Vanadium trioxide		Cyclohexene dehydrogenation	7×10^{-11} 10^{2}	25 350

catalysts are now being developed through the study of protein structure and its manipulation (now called *protein engineering*) and the isolation of organisms from extreme environments.

5.5.2 Properties of Enzymes

As with normal chemical reactions, there are many reasons for studying the kinetics of enzyme-catalyzed reactions: to be able to predict how the rate of reaction will be affected by changes in reaction conditions; to aid in the determination of the mechanism of the reaction—that is to identify the sequence of reaction events that intervene between reactants and terminal products; to identify and characterize the enzyme molecule in functional terms, as protein molecules are structurally too complex for the application of conventional structural analysis; to explain aspects of metabolic regulation and cell differentiation in terms of mechanisms that control either the variety or the quantity of enzymes in the cell; to understand the kinetic aspects of reactions, which are critical in the design and operation of bioreactors.

An understanding of the influence of environmental conditions on the kinetics of enzyme reactions is essential for the design of processes based on the use of these materials as catalysts. Their growth kinetics is also governed by similar kinetic equations (cf. Monod growth equations, Section 6.4.3).

In five respects, the catalytic activity of enzymes differs from that of other catalysts.

5.5.2.1 Efficiency

Turnover number (molecules reacted per catalytic site per unit time) at room temperature is usually much higher than that for industrial chemical catalysts (see Table 5.14).

5.5.2.2 Specificity

A characteristic feature of enzymes is that they are specific in action, some showing complete (or absolute) specificity (or discrimination) for only one type of molecule. If a substance exists in two stereochemical forms, L and D isomers, enzymes may recognize only one of the two forms. For example, *glucose oxidase* will oxidize D(+) glucose only and no other hexose isomer. Other enzymes have group specificity, showing activity towards a series of closely related compounds (e.g., branched-chained amino acids, keto acids, alcohols, etc.).

The stereochemical specificity of enzymes depends on the existence of at least three different points of interaction, each of which must have a binding or catalytic function. A catalytic site on the molecule is known as an *active site* or *active center* of the enzyme. Such sites constitute only a small proportion of the total volume of the enzyme and are located on or near the surface. The active site is usually a very complex physicochemical space, creating microenvironments in which the binding and catalytic areas can be found. The forces operating at the active site can involve charge, hydrophobicity, hydrogen bonding, and redox processes. The determinants of specificity are thus very complex but are founded on the primary, secondary, and tertiary structures of proteins.

Several models have been put forward to explain specificity. The Fischer "lock and key" hypothesis was the earliest hypothesis proposed to explain the interaction between the substrate and the active site, which is considered rigid throughout the binding process. Here the active site of the protein is matched and fitted by a specific substrate and the hypothesis explains in simple terms the matching of the two complex architectures. Later, as protein structure became better understood, the Koshland "induced fit" hypothesis was put forward to explain the possible conformational change associated with the formation of the enzyme—substrate complex in structural enzymes. Here, both the enzyme and the substrate change their structure slightly to accommodate each other. Finally, the Haldane and Pauling "transition state" hypothesis represents yet a further development. It takes into account the way the active site might actually catalyze the reaction by forcing the formation and stabilization of a transitional state between substrates and enzymes. This assists the reaction by which the desired product is formed.[5]

5.5.2.3 Versatility

The versatility of enzyme catalysis is shown by the types of reactions that can be catalyzed. There are six major groups of enzymes arranged according to their reactivity:

1. *Oxidoreductase*—oxidation—reduction reactions,
2. *Transferases*—transference of an atom or group between two molecules,
3. *Hydrolases*—hydrolysis reactions,
4. *Lyases*—removal of a group from a substrate (not by hydrolysis),
5. *Isomerases*—isomerization reactions,
6. *Ligases*—the synthetic joining of two molecules, coupled with the breakdown of a pyrophosphate bond in a nucleoside triphosphate.

The International Union of Biochemistry has drawn up a numeric system of classification based on the type of reaction catalyzed. Enzyme numbers can be generated using the first digit to represent the enzyme group indicated above, then three other numbers added to identify the enzyme uniquely. The additional numbers are based on properties such as reaction mechanism, substrate, and *cofactor* type (i.e., secondary substrate, suppling reducing power, energy, or catalytic intermediate). The number conventions are unique within each major reaction group.

For example,

> Lactate dehydrogenase, L-lactate: NAD^+ oxidoreductase (E.C. 1.1.1.27),
> Lactate dehydrogenase, D-lactate: NAD^+ oxidoreductase (E.C. 1.1.1.28),
> Alcohol dehydrogenase, ethanol: NAD^+ oxidoreductase (E.C. 1.1.1.1.).

5.5.2.4 Controlled Expression

Enzyme expression and activity can be controlled in a number of ways at the metabolic level by specialized enzymes (e.g., allosteric enzymes), whose activity can be modulated by environmental conditions, and at the genetic level by the environmentally controlled synthesis of the enzymes. These processes are discussed further in Section 5.5 and Appendix A.2.

5.5.2.5 Stability

Enzymes are relatively unstable as only weak molecular forces hold the functional tertiary structure together. The kinetic aspects of stability are discussed below, and the structure of protein is discussed in more detail in Section 5.2.2.

5.5.3 Enzyme Kinetics

5.5.3.1 Factors Responsible for Enzyme Catalysis

Enzymes, like all catalysts, enhance the rate of reaction but do not alter the thermodynamic equilibrium. The reactant molecules (substrates) must collide for a reaction to take place and the collision must both be in the correct orientation and have sufficient energy (the activation energy) for the reaction to take place.

The Arrhenius equation expresses the relationship between the rate constant k and the activation energy $\mathbf{E}_a$:

$$k = Ae^{-E_a/RT} \tag{5.1}$$

where $\mathbf{R}$ is the universal gas constant, T is the absolute temperature, and A is the Arrhenius constant. In enzyme-catalyzed reactions, the rate constant k may be many orders of magnitude larger than that of the uncatalyzed reaction. To accommodate the activity of an enzyme into the model, the concept of the formation of enzyme—substrate and enzyme—product complexes is introduced. Thus, the first step of an enzyme-catalyzed reaction is the formation of an enzyme—substrate (ES) complex, which is generally a more stable entity than the enzyme alone. The stability is attributed to the hydrogen bonding, van der Waals forces, and ionic interactions between the enzyme and the substrate. The enhancement of the rate at which the reaction takes place depends on the lowering of the activation energy occurring as a result of the intra- and intermolecular interactions between the enzyme and the substrate. The products, once formed, are released when the enzyme and product dissociate. A more detailed and extensive explanation of enzyme catalysts is given by Fersht.[26]

For most enzymes, the rate of reaction can be described by the Michaelis—Menten equation, which was originally derived in 1913 by Michaelis and Menten.[27] The model is governed by the equation describing the rate of enzymatic reactions by relating the reaction rate to the concentration of substrate and the formula is given as

$$R_0 = \frac{V_{max}S}{S + K_m}$$

where, R_0 is the initial rate of the reaction, V_{max} represents the maximum reaction rate at saturating substrate concentration, S is the substrate concentration, and K_m is the Michaelis—Menten constant (Fig 5.45). The detailed derivation and discussion of the Michaelis—Menten kinetics will be discussed in Chapter 6.

5.5.4 The Control of Enzyme Activity

The control and integration of cellular processes to produce appropriate structures and functions is vital for the efficient reproduction and use of resources. The activity of the cell may be controlled in a number of ways. Fig. 5.46 illustrates the interactions that have been shown to take place within cells. Two major systems of control exist.

The first involves various feedback and feed-forward control mechanisms associated with metabolic pathways. Here a chemical present in the cell, usually an end product of a metabolic sequence, will influence the activity of an enzyme at the beginning of the

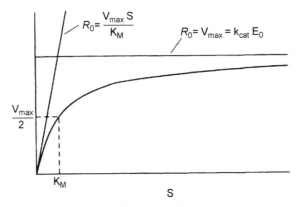

Figure 5.45
Initial reaction rate R_0 plotted against substrate concentration S for a reaction obeying Michaelis—Menten kinetics.

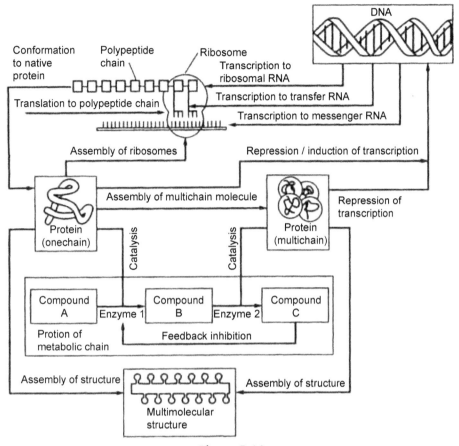

Figure 5.46

A hierarchy of control and information transmission in the cell.[25]

pathway. This is usually achieved by the presence of an *allosteric enzyme* whose properties are significantly changed by effector molecules (Fig. 5.46).

A second major set of controls is associated with the expression or synthesis of the enzyme itself, so that cells adapt to environmental conditions with an optimized spectrum of enzymes. In bacteria, this control is usually achieved by changing the rate of transcription of DNA to mRNA. Studies of such systems have shown that the genome (or complete genetic material) is partitioned into sections relating to groups of enzymes (*structural genes*) and their control systems (*regulatory sequences*), which are collectively known as *operons*. Knowledge of these control systems may be exploited to produce systems capable of enhanced production of metabolic products and enzymes.

5.5.5 The Control of Metabolic Pathways

5.5.5.1 Regulatory Enzymes

All enzymes exhibit various features that could conceivably be elements in the regulation of their activity in cells. All have a characteristic pH optimum, which makes it possible for their catalytic rates to be altered by changes in intracellular pH (e.g., in ribulose diphosphate carboxylase[5]). The activities also depend on the concentration of substrates, which may vary according to intracellular conditions. Moreover, many require metal ions or vitamins, and the activity of enzymes may be a function of the concentrations in which such materials are present (e.g., the effect of iron limitation on the citric acid fermentation of *Aspergillus niger*[2]). However, over and above these factors, some enzymes have other properties that can endow them with specific regulatory roles in metabolism. Such specialized forms are called *regulatory* enzymes, of which there are three major types.

In *allosteric enzymes*, the activity of the enzyme is modulated by a noncovalently bound metabolite at a site on a protein other than the catalytic site. Normally, this results in a conformational change, which makes the catalytic site inactive or less active. *Covalently modulated enzymes* are interconverted between active and inactive forms by the action of other enzymes, some of which are modulated by allosteric-type control. Both of these control mechanisms are responsive to changes in cell conditions and typically the response time in allosteric control is a matter of seconds as compared with minutes in covalent modulation. A third type of control, the control of enzyme synthesis at the transcription stage of protein synthesis (see Appendix A.3), can take several hours to take effect.

5.5.5.2 Allosteric Enzymes and the Regulation of Biosynthetic Pathways

The concept of control of metabolic activity by allosteric enzymes or the control of enzyme activity by ligand-induced conformational changes arose from the study of metabolic pathways and their regulatory enzymes. A good example, the multienzymatic sequence catalyzing the conversion of L-threonine to L-isoleucine, is shown in Fig. 5.47.

The first enzyme in the sequence, L-threonine dehydratase, is strongly inhibited by L-isoleucine, the end product, but not by any other intermediates in the sequence. The kinetic inhibition is not competitive with the substrate, nor is it noncompetitive or uncompetitive. Isoleucine is quite specific in this characteristic, and other amino acids or other analogues do not inhibit. This type of inhibition is called end-product inhibition or feedback inhibition.

This first enzyme, whose activity is modulated by an end product, is an allosteric enzyme where, in addition to the active site, it has another space specific for binding the ligand, which modulates the active site. Some negative modulators inhibit, as shown above with isoleucine on threonine hydratase, whereas others may stimulate or positively modulate the

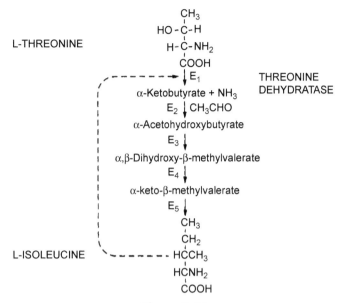

Figure 5.47

Feedback inhibition of the formation of isoleucine from threonine. This enzymatic pathway (E_1 to E_2) is inhibited by its product isoleucine at the first step E_1 (threonine dehydratase).

enzyme. Some enzymes have only one modulator and are called *monovalent*, whereas others have several and are called *polyvalent* modulators. Moreover, some allosteric enzymes have both negative and positive modulators. Fig. 5.48 illustrates some patterns of allosteric modulation. The advantage of these control systems is that cellular materials are economically used.

Allosteric enzymes have unique structural characteristics: they are much larger in molecular weight than average enzymes and are structurally more complex. This leads to a difficulty in purification because they are oligomeric (have more than one peptide chain) and are sensitive to low as well as high temperatures. They also show atypical kinetic characteristics, which fail to conform to the classic Michaelis–Menten relationship.

There are two types of control shown by allosteric enzymes. They can be modulated by a molecule other than a substrate of the enzyme (termed *heterotrophic enzymes,* e.g., threonine dehydratase) or by the substrate itself (termed *homotropic enzymes,* e.g., oxygen binding to hemoglobin). They contain two or more binding sites for the substrate, and activity is modulated by the number of binding sites, which are filled.

5.5.5.3 Kinetics of Allosteric Enzymes

Allosteric enzymes do not follow the Michaelis–Menten kinetic relationships between substrate concentration V_{max} and K_m because their kinetic behavior is greatly altered by

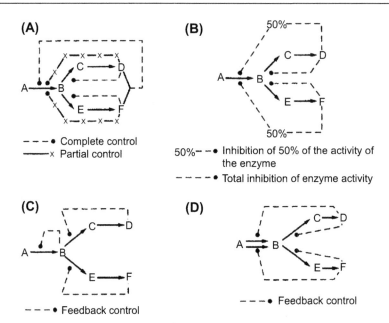

(A)

----• Complete control
——x Partial control

(B)

50%----• Inhibition of 50% of the activity of
          the enzyme
- - - -• Total inhibition of enzyme activity

(C)

----• Feedback control

(D)

- - -• Feedback control

Figure 5.48
Patterns of allosteric modulation in anabolic pathways. (A) The control of a biosynthetic pathway by the cooperative control by end products D and F; (B) The control of a biosynthetic pathway by sequential feedback control; (C) The control of a biosynthetic pathway by the cumulative control of products D and F; (D) The control of two isoenzymes (catalyzing the conversion of A to B) by end products D and F.

variations in the concentration of the allosteric modulator. Generally, homotropic enzymes show sigmoidal behavior with reference to the substrate concentration, rather than the rectangular hyperbolae shown in classical Michaelis–Menten kinetics. Thus, to increase the rate of reaction from 10% to 90% of maximum requires an 81-fold increase in substrate concentration, as shown in Fig. 5.49A. *Positive cooperativity* is the term used to describe the substrate concentration–activity curve, which is sigmoidal; an increase in the rate from 10% to 90% requires only a ninefold increase in substrate concentration (Fig. 5.49B). *Negative cooperativity* is used to describe the flattening of the plot (Fig. 5.49C) and requires over 6000-fold increase to increase the rate from 10% to 90% of maximum rate.

Some allosteric enzymes are also classified by the way in which they are affected by the binding of a modulator; some affect the value of K_m without affecting that of V_{max}. They are classed as K-series enzymes while others, which affect V_{max} without affecting K_m, are called M-series enzymes. Fig. 5.50 shows the characteristic kinetic patterns observed for

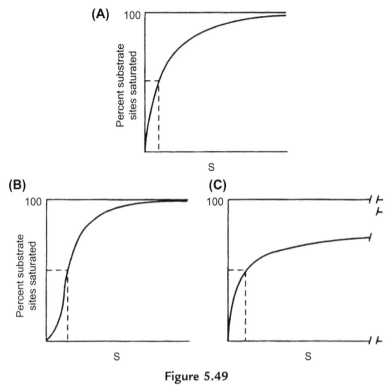

Figure 5.49
Comparison of idealized plots of percent maximum reaction rate as a function of substrate concentration for (A) A nonregulatory enzyme obeying Michaelis—Menten kinetics; (B)Regulatory enzyme following positive cooperativity; (C) Regulatory enzyme showing negative cooperativity

K-series and M-series enzymes. There are, of course, exceptions to these two extremes of kinetic behavior.

Altering the control of metabolic pathways can also be achieved by genetic manipulation. As proteins are generated using the template stored as a fragment of DNA, the structure of the allosteric enzyme may be altered so that there is little or no regulatory control. It is therefore possible to generate mutants that overproduce metabolites, and techniques based on this principle have been most widely exploited in amino acid and nucleotide production.[2]

Amino acid fermentations are usually performed with mutants that remove the control by feedback inhibition. This is achieved by chemical mutation and screening for the appropriately modified organisms. The result of such a mutation is the channeling of the feedstock carbon to the desired product. Consider the system illustrated in Fig. 5.51. The production of lysine, threonine, and methionine is controlled by a number of mechanisms,

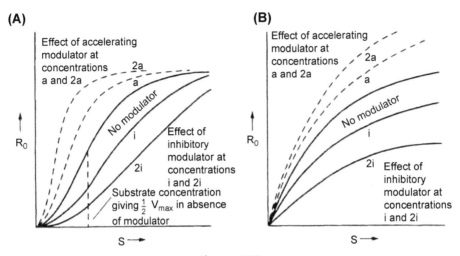

Figure 5.50

Effect of positive and negative modulators of reaction rate. Substrate concentration–rate curves for (A) K enzymes that modulate activity by altering K_m; (B) M enzymes that modulate activity by altering the maximum rate.

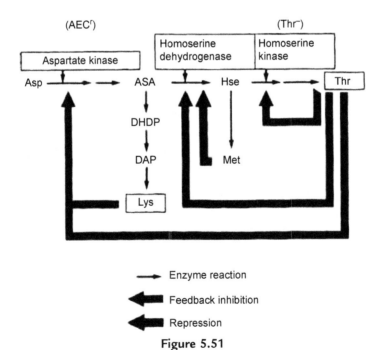

Figure 5.51

Regulation of lysine (Lys), threonine (Thr), and methionine (Met) in *Brevibacterium flavum*.[28]

the most notable of which is the feedback control of aspartate kinase by threonine (Thr) and lysine (Lys). Two basic strategies are employed. The first uses an *auxotrophic mutant* that requires threonine (designated Thr$^-$), for growth. In this situation, only limiting amounts of threonine are fed to the organism and this results in very low levels of threonine in the cell, and thus a reduction of feedback inhibition by threonine. The organism will now produce high levels of lysine. The second strategy uses regulatory mutants of aspartate kinase. If the structure of aspartate kinase is altered so that it becomes insensitive to the molecules that cause feedback inhibition, then the organism will overproduce lysine. This is achieved by feeding the organism with an analogue of lysine, S-2-aminoethyl-L-cysteine (AEC), a powerful feedback inhibitor of the aspartate kinase. In the presence of AEC, complete feedback inhibition occurs and no growth is possible, as neither lysine nor threonine will be synthesized. If the enzyme structure is altered so that they become resistant to AEC or AEC^r, the AEC no longer binds and then the synthesis of threonine and lysine will take place and growth will occur. As a consequence of the enzyme alteration, there is a good chance that lysine will also no longer act as a feedback inhibitor and so an enhanced production of lysine is achieved in a fermentation involving such a mutant.[28]

5.5.5.4 Covalently Modulated Enzymes

In a second class of regulatory enzymes the active and inactive forms are interconverted by covalent modifications of their structures by enzymes. The classic example of this type of control is the use of glycogen phosphorylase from animal tissues to catalyze the breakdown of the polysaccharide glycogen yielding glucose-1-phosphate, as illustrated in Fig. 5.52.

There are two forms of the enzyme: phosphorylase *a* and the less active form phosphorylase *b*. Phosphorylase *a* is an oligomeric protein with four major subunits. Phosphorylation of a serine hydroxy group produces an active form of the enzyme. Removal of the phosphate causes a breakdown of the tetramer to a dimeric form which is the less active phosphorylase *b*. Reactivation is achieved by the enzyme phosphorylase kinase which catalyzes the phosphorylation at the expense of ATP.

Another important characteristic of these enzymes is the amplification cascade involving phosphorylase kinase and phosphorylase. The chemical signal, here phosphorylase kinase, can react with many molecules of phosphorylase *b*, which in turn produces phosphorylase *a*, which converts glycogen into glucose 1-phosphate.

A different type of covalent regulation of enzyme activity is the enzyme-catalyzed activation of inactive precursors of enzymes (zymogens) to give catalytically active forms. The best examples are the digestive enzymes, e.g., trypsin. Proteolytic enzymes would digest the inside of the cells that produce the enzyme, so they are produced in an inactive

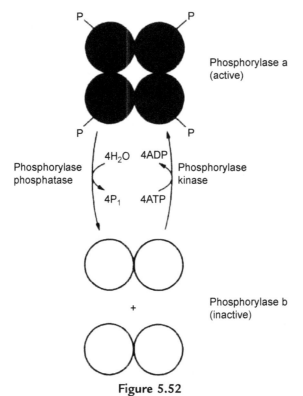

Figure 5.52

Modulation of glycogen phosphorylase activity by covalent modulation.

form which is activated to the true enzyme once they have entered the digestive system of the animal.

5.5.5.5 Isoenzymes

The final form of metabolic regulation is effected by the use of isoenzymes, which are multiple forms of an enzyme. For example, lactate dehydrogenase exists in five forms in a rat. They differ in primary structure and have different isoelectric points, but they all catalyze the reversible reduction of pyruvate to lactate.

$$CH_3COCOO^- + NADH + H^+ \rightleftharpoons CH_3CHOHCOO^- + NAD$$

A careful kinetic study has shown that, although all catalyze the same reaction, both the kinetic constants K_m and V_{max} differ. The kinetic characteristics match the requirements of the tissues, e.g., V_{max} is high in skeletal muscle but low in heart muscle.

Nomenclature

		Units in SI System	Dimensions in M, N, L, T, θ
A	Arrhenius constant	s^{-1}	T^{-1}
E_a	Activation energy	$J/kmol$	$MN^{-1}L^2T^{-2}$
$\Delta G^{0\prime}$	Gibbs free energy at pH 7	$J/kmol$	$MN^{-1}L^2T^{-2}$
K_m	Michaelis constant	kg/m^3	ML^{-3}
k	Rate constant	s^{-1}	T^{-1}
R_0	Initial rate of reaction	kg/m^3s	$ML^{-3}T^{-1}$
R	Universal gas constant	$8314\,J/kmol/K$	$MN^{-1}L^2T^{-2}\theta^{-1}$
S	Substrate concentration	kg/m^3	ML^{-3}
T	Absolute temperature	K	θ
V_{max}	Enzyme velocity constant	kg/m^3s	$ML^{-3}T^{-1}$

Appendices

A.1 Proteins

Of all the macromolecules present in living systems, proteins and nucleic acids are the most significant. Nucleic acid in the form of DNA forms the stored structural and regulatory material for the organism, whereas proteins and their expression represent the functional form of this information. A typical composition of a bacterial cell is given in Table 5.A1.

Protein is formed mainly of polymerized amino acids. The primary structure, unlike that of synthetic polymers, is nonrepetitive and, for its production, requires a chemical template stored in the structure of the DNA molecule. The sizes of proteins vary considerably (in a range of molecular weights from 6000 to 1,000,000). Proteins fulfill many roles within the cell, the most important of which is that of catalysis. Proteins that have catalytic activity are called enzymes, while other proteins have important roles in storage, transport, protection (antibodies), as chemical messengers (hormones), and in structure.[4,5]

Protein Purification and Separation

Protein separation processes generally utilize differences in one or more of the physical properties of the materials, which can be purified by the use of three or four basic operations, including the following:

Gel filtration (size),
Ultrafiltration (size),
Ultracentrifugation (density and size),
Electrophoresis (charge and size),
Ion exchange (charge),
Precipitation by salts or solvent (hydrophobicity and charge).

Table 5.A1: Molecular composition of a typical bacterium.

Component	Mass ($\times 10^{13}$ g)	Percentage of Total Mass	Molecular Weight, M_r	Molecules Per Cell[a]
Entire cell	15	100		
Water	12	80	18	4×10^{10}
Dry weight	3	20		
Protein				
Ribosomal	0.22	1.5	4×10^4	3.3×10^5
Nonribosomal	1.5	10	5×10^4	1.8×10^6
RNA				
Ribosomal, 16S	0.15	1	6×10^5	1.5×10^4
Ribosomal, 23S	0.30	2	1.2×10^6	1.5×10^4
tRNA	0.15	1	2.5×10^4	3.5×10^5
mRNA	0.15	1	10^6	9×10^5
DNA	0.15	1	4.5×10^9	2
Polysaccharides	0.15	1	1.8×10^2	5×10^7
Lipids	0.15	1	10^3	9×10^6
Small molecules	0.08	0.5	4×10^2	1.2×10^7

[a]Calculated by dividing the weight of a component in grams by its molecular weight to give the number of gram molecular weights (moles) of the component. Since there are approximately 6×10^{23} molecules in one mole of any compound, the number of moles multiplied by 6×10^{23} gives the number of molecules. For example, the number of moles of water in 12×10^{-13} g of water is $12 \times 10^{-13}/18$ or about 7×10^{-14}. The number of molecules of water per cell is then $7 \times 10^{-4} \times 6 \times 10^{23}$ or about 4×10.

Generally speaking, these methods result in very mild treatments, which will maintain the biological function of proteins that are usually associated with their tertiary structure. If the configuration of the protein is altered, the process is generally reversible. A discussion of many of these unit operations can be found in Volume 2, Chapters 17–20.

A.2 Strain Improvement Methods

The improvement of the biocatalyst, in the form of either whole cell or isolated enzyme, can be achieved by a change in the structure of the DNA of the cell. This can be effected either by random selection of mutants or by a concerted development of a strain by careful genetic manipulation. There is a good analogy between modern information technology using computers and data storage and the storage of biological information, its utilization and manipulation. Indeed some of the terms used in computing are derived from biological processes, for example, computer viruses and their ability to infect machines and programs.

The manipulation of the stored information in the DNA can result in the creation of new strains of organisms with new or enhanced capabilities, and this may be achieved by a variety of methods including mutation and in vivo and in vitro recombination (recombinant DNA technology). "Genetic engineering" covers the manipulation of DNA in the laboratory and can be utilized for the improvement of traditional processes and provide the enabling technology for the generation of new ones; for example, the production of human insulin by bacteria.

The structure of nucleic acids is discussed in Section 5.2.4, whereas Appendix A.3 gives details of how information can be stored in the cell in the form of DNA, genes, plasmids, and the mechanisms of protein synthesis.

Mutation and Mutagenesis

Chromosomes are by no means inert, stable structures, holding dead information in storage. They are constantly undergoing changes of various kinds, some of which are involved in genetic recombination. However, other changes are accidental and random in nature and, if not repaired, result in a permanent genetic mutation. In Section 5.2.4 and Appendix A.3, some of these repair mechanisms are discussed (e.g., DNA repair), but the results of these phenomena will now be discussed further.

DNA is subject to damage by electromagnetic radiation or reactive chemicals, many of which have been introduced into the environment as a by-product of industrial activity. Some chemicals may not be dangerous per se, but may be metabolized to products, which are dangerous. There are three major classes of such chemicals.[5]

1. Deaminating agents, such as nitrous acid or compounds that can be metabolized to nitrous acid or nitrites. These reagents are capable of removing amino groups from cytosine, adenine, and guanine.
2. Alkylating agents that can cause methylation and ethylation of bases. Guanine, for example, reacts with dimethyl sulfate to form o-methylguanine.
3. Analogues of bases, which mimic the normal bases present in DNA. Thus, many aromatic and polyaromatic compounds produced by industry are potential mutagens.

Although DNA proofreading and repair mechanisms within cells are very effective, inevitably some errors in replication remain uncorrected, and as a result become perpetuated in the DNA of the organism. Such permanent changes are called *mutations*.

Mutations caused by the replacement of a single base with an incorrect one are called *substitution mutations*. Such mutations will have the effect of changing one codon-triplet and, depending on the code, this may alter an amino acid in the polypeptide chain. Examples of hypothetical mutations are shown in Table 5.A2. *Silent mutations* are changes in the base sequence, which cause no change to the functionality product, e.g., an enzyme activity. Other mutations are so catastrophic that the organisms can no longer function. If, for example, the enzyme is on a central pathway, and becomes critically mutated, this

Table 5.A2: Effect of some hypothetical single-base mutations on the biological activity of the resulting protein products.

Mutation		Wild type (unmutated DNA triplet)	Mutated triplet
A single-base substitution causing no change in the amino-acid sequence; a silent mutation	DNA template RNA codon Amino acid	(3′)–GGT–(5′) (5′)–CCA–(3′) ┤ Pro ├	–GGA– –CCU– ┤ Pro ├
A single-base mutation resulting in an amino-acid change that may not alter the biological activity of the protein because the amino-acid replacement is in a noncritical position and also resembles the normal amino-acid; also a silent mutation		(3′)–TAA–(5′) (5′)–AUU–(3′) ┤ Ile ├	–GGA– –CUU– ┤ Leu ├
A lethal single-base mutation in which a serine residue essential for enzyme activity is replaced by phenylalanine to give an enzymatically inactive product		(3′)–AGA–(5′) (5′)–UCU–(3′) ┤ Ser ├	–AAA– –UUU– ┤ Phe ├
A leaky mutation in which the amino-acid change results in a protein that retains at least some of its normal activity		(3′)–CGT–(5′) (5′)–GCA–(3′) ┤ Ala ├	–CCT– –GGA– ┤ Gly ├
A hypothetical beneficial mutation, in which the amino-acid replacement yields a protein with improved biological activity, giving the mutated organism an advantage: it is not possible to predict advantageous amino-acid replacements		(3′)–TTC–(5′) (5′)–AAG–(3′) ┤ Lys ├	–TCC– –AGG– ┤ Arg ├

process is termed a *lethal mutation. Leaky mutants* are those in which the alteration in amino acids changes the kinetic characteristics of the enzyme so that, for example, it does not function very well or, conversely (but rarely), it may be an improvement on the original enzyme.

Substitution mutation is only one type; *insertion* and *deletion mutations* are much more numerous and more lethal. Insertion or deletion of nucleotides causes *frameshift* mutations, in which a base pair is inserted in or deleted from a gene, giving rise to a more extensive type of mutation damage. The consequence of such a mutation is the disruption of the linear order of the codons in the DNA and thence of the amino acid sequence for which it codes. The disruption begins at the site where the base has been gained or lost, and the result of such a mutation is a garbled protein sequence. For example, consider the analogy of words in a sentence. Consider the following sentence of three-letter words:

THE CAT WAS RED AND BAD

On deletion of W, it would read:

THE CAT ASR EDA NDB ADX

and the meaning is lost. Sometimes, one frameshift mutation can be canceled out by a second one; e.g., when the first adds a base while the second takes another base out. Such a mutation is called a *suppressor mutation* as it suppresses the first mutation. Frameshift mutations are caused by molecules that intercalate between two adjacent base pairs and, in effect, add an additional base to the sequence. Thus, on replication, these molecules cause an additional base to be added to the sequence; for example, acridine is a mutagen that intercalates with DNA and causes such a phenomenon.

Mutation is a rare event as far as an individual human being is concerned; the probability that a mutation will occur in any one protein is about 1 in 10^9. For a human cell it is probably 1 in 10^5 (calculated from the incidence of hemophilia), but mutagenic agents in effect increase the mutation frequency. Furthermore, a specific mutation in the enzymes involved in DNA repair will also increase the level of mutation.

Statistical evidence strongly suggests that continued exposure of human beings to certain chemical agents, especially in the work place, results in an increased incidence of cancer. For example, workers involved in the production of naphylamines have a much higher incidence of bladder cancer than the general population and, interestingly, many cancers are the result of loss of the growth control mechanisms of the cell. Dyes and many other aromatic compounds, exhaust gases, medicines, and cosmetics all contain chemicals that are possibly carcinogenic. To protect humans from exposure to such chemicals, extensive animal tests have had to be developed. This is now a significant part of the cost (£100,000 to millions) of developing a drug and takes a considerable time (2–3 years).

On the positive side, the use of mutagens to increase the degree of variation in a culture, followed by selecting and screening individuals with enhanced capabilities, has been one of the most successful methods in improving microbial strains. A classic case is the development of penicillin G fermentation where the productivity of culture strains was improved by a factor of more than a 1000 over a period of about 20 years, using mutant screening program.

Genetic Recombination in Bacteria

There are several mechanisms by which genetic recombination occurs in nature. Genes or sets of genes can also be recombined in the test tube to produce new combinations that do not normally occur in nature. It is possible to isolate the genes necessary for the formation of a specific protein and to splice them with other forms of DNA to yield these new combinations. Such artificial recombinant DNAs are extremely useful as tools in genetic research and in biotechnology. The development of methods for isolating and splicing genes into recombination has been a major scientific and technological advance.

Biological exchange of genes to form a modified chromosome is called *genetic recombination* and can occur between DNA molecules that are similar. Their similarity allows the molecules to associate closely with one another and to exchange genetic material. Recombination occurs in many situations and has been studied in great depth for bacteria. A summary of the basic types of recombination in bacteria is shown in Fig. 5.A1. All require the new DNA to enter the recipient cell before the recombination event can take place. The simplest is called *transformation* and involves the uptake by the cell of a small piece of DNA from the environment. Once absorbed it may then undergo recombination with the chromosome. *Transduction* involves the virus-mediated

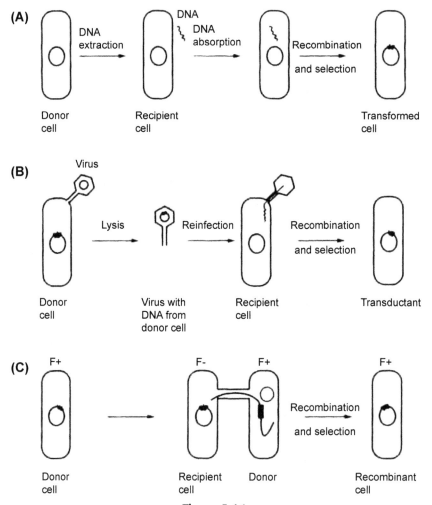

Figure 5.A1
Summary of the principal mechanism of genetic recombination. (A) Transformation (the absorption of DNA from the environment); (B) Transduction (the viral-mediated transfer of DNA from one cell to another); (C) Conjugation (the transfer of DNA between touching cells and involving specialized plasmid).

transfer of DNA from one cell to the other. The bacterial virus is able to take up small fragments of donor cell DNA during its reproductive cycle. If the modified virus is then introduced to another culture, then DNA may be transferred into the recipient culture. The final form of bacterial recombination is the concerted passage of material between touching cells. DNA from one organism (designated F^+) containing a sex plasmid is able to pass the replicated chromosome from one organism to another, which does not contain the plasmid (F^- strain). This process occurs in a linear fashion and, if it is allowed to proceed for long enough, a complete transfer of material will take place. This process, called *conjugation*, has been a very useful tool for the mapping of genes on the chromosome.

Genetic Engineering

The increase in knowledge about the structure and properties of DNA, the discovery of the genetic systems of bacteria, and the development of practical methods in the manipulation of DNA have led to the development of gene cloning (clone is derived from the Greek word *klon = cutting*, as in plant propagation). This concept has been extended to the cellular and molecular levels so that material is either derived from a single cell or from a piece of DNA. Another term used to describe these methods is recombinant DNA (rDNA) technology.

In pure research, gene cloning has very important applications in the understanding of gene function (e.g., control of gene expression or protein synthesis); control of DNA replication and repair; obtaining DNA sequence information; determining structure—function relationships in proteins (protein engineering).

In applied research and industry, gene cloning offers a mechanism for producing many valuable products from animals, employing some of the more appropriate techniques used for the production of bacteria. Typical of the premolecular biology products are antibiotics, organic acids, amino acids, and solvents. The introduction of the techniques of molecular biology to industrial biology has led to the development of processes for animal and plant proteins using bacteria and yeast, which are more easily used on a large scale than the original animal and plant cells. Examples of such products are hormones and specialized proteins such as vaccines and antibodies.

Physical and Biochemical Techniques Used in the Manipulation of DNA

Following the discovery of DNA structure and the genetic code, various associated techniques involving the manipulation and separation of DNA fragments have been developed. Several of the physical methods are useful in nucleic acid biochemistry; these include centrifugation, electrophoresis, and chromatographic separation of DNA fragments.

The production of DNA fragments can be achieved by specific enzymatic hydrolysis. Enzymes that hydrolyze DNA, usually at the phosphodiester bonds, are collectively known as nucleases. Those specific to RNA are ribonucleases (RNases), and those specific to DNA are deoxyribonucleases (DNases). Nucleases are also classified according to how they attack the polymer chain. Exonucleases attack the polymer chains from the end; some are specific to the ends of the polymer (see Section 5.2.4), while others, endonucleases, attack the polymer from within the chain. There is a very important subgroup of the endonucleases, the *restriction endonucleases*, which are very useful reagents in the study of DNA. Although their normal physiological functions involve DNA replication, DNA repair, and DNA recombination, they can be used in the laboratory to cut DNA selectively, yielding several products. A single cut in a double strand called a *NICK*; cutting several bases from a strand is called a *GAP*; if both strands are cut, then the result is called a *BREAK*.

Recombinant DNA Technology

The development of the means to isolate and splice genes into new combinations was a major scientific and technological advance.[29,30] Fig. 5.A2 outlines how an animal gene may be transferred to a bacterium. Recombinant DNA (rDNA) technology follows a basic outline (Fig. 5.A3). A fragment of DNA containing the gene to be cloned is isolated, and

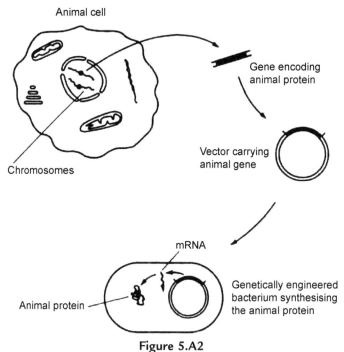

Figure 5.A2
A possible scheme for the production of an animal protein by a bacterium.

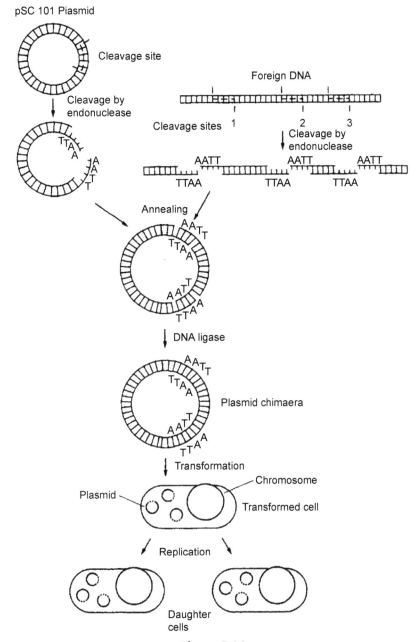

Figure 5.A3

The basic steps of recombinant DNA technology. Using this sequence of steps, a fragment of foreign DNA can be annealed and sealed into a plasmid which in turn can be introduced into a living cell where the foreign DNA is replicated and expressed.

then inserted into a circular DNA molecule called a *vector* to form a recombinant DNA molecule or a *chimaera* (an organism made up of two distinct pieces of genetic material; in Greek mythology a chimaera was a fire-spouting monster with a lion's head, a goat's body, and a serpent's tail). The vector acts as a vehicle that transports genes into the host cell, which is usually a plasmid or a virus. Within the cell the vector multiplies, producing numerous identical copies of itself including the cloned gene. As the organism multiplies, further vector replication takes place. After a large number of cell divisions, a colony or clone of identical host cells, each containing one or more copies of the cloned DNA, is produced.

There are several specific techniques for the following:

(a) *Cutting and splicing DNA*. The discovery of restriction endonucleases gave the first clue as to how genes can be recombined in the laboratory. The analogy of film or tape editing can be used to illustrate the principle of these techniques in which a story can be retold in a number different ways. A piece of DNA is cut at a specific site with a restriction endonuclease to leave a staggered two-strand break to give "sticky" or cohesive ends, which enables DNA from two sources to be mixed and joined together. Other enzymes may be used to produce a sticky end from a blunt end, as shown for example in Fig. 5.A3.

(b) *Introduction of DNA into living cells*. The next step in developing recombinant DNA technology was finding a means of introducing foreign DNA into the host cell. This is now done by using either plasmids (via transformation) or viruses (via transduction). Plasmids (see Appendix A.3) are small circular, autonomous DNA units, which can contain 2000—1,000,000 bases; the smaller plasmids may have 20 or so copies per cell. In each plasmid, several genes are replicated, transcribed, and translated independently, but simultaneously, with chromosomal genes. They are readily isolated from chromosomal DNA as they are smaller and less dense. They have two important properties:

(i)   They are able to pass from one cell to another, indeed from one species to another. For example, *Salmonella typhimurium* can acquire permanent resistance to certain antibiotics, such as penicillin, when it is mixed with a strain of *E. coli* cells resistant to penicillin.

(ii)   Foreign genes can be spliced into the plasmids which can then be carried as passengers into *E. coli* cells and become part of the host cell *genome* (meaning genetic complement).

A good cloning vehicle must generally have a number of features:

(i)   It must be able to replicate in the foreign organism;

(ii)   It must be small, so as to ease purification;

(iii)   It must occur at high copy numbers (e.g., 10—50 copies of the plasmid per cell).

Considerable effort has been spent on developing the vectors (for example, pBR322) so that they incorporate desirable features. These include:

(i)   Ease of purification;

(ii) Enzyme manipulation (i.e., specific manipulation of DNA by restriction endonucleases);

(iii) Antibiotic resistances for ease of selection (i.e., if pieces of DNA are added to the sites within a resistance gene, they become sensitive to the antibiotic—this is called *insertional inactivation*);

(iv) High copy number of the plasmid is possible and can reach 1000 copies under the right conditions.

The plasmid pBR322 has such desirable features (Fig. 5.A4) and is a good example of how such plasmids are developed. It has 4363 base pairs and is thus small and, when

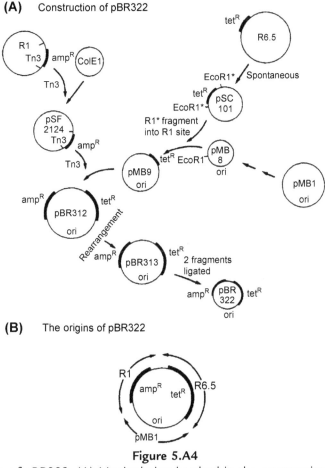

Figure 5.A4

The pedigree of pBR322. (A) Manipulation involved in the construction of pBR322; (B) Summary of the origins of pBR322[29].

pieces of cloned DNA have been added, the molecule is still small enough to handle. It has antibiotic resistances to two materials, ampicillin (amp^r) and tetracyclin (tet^r), and has many sites where restriction enzymes can cut the molecule. Furthermore, it is produced in high copy numbers. Such plasmid cloning vectors do not arise by chance but are themselves products of gene manipulation. The construction of pBR322 is shown in Fig. 5.A4. It is constructed from three natural plasmids isolated from drug-resistant organisms and then manipulated to the appropriate size and configuration.

An Example of Cloning Using a Plasmid

The plasmid can be used, not only to transfer and express foreign genes in *E. coli* as with pBR322, but also to study the control of gene expression, etc. Figs. 5.A2 and 5.A3 show some of the procedures, each of which involves a number of steps:

(1) *Isolation of genes and the preparation of cloned DNA.* There are two general methods for obtaining cloned DNA:
   (i) The "shotgun" approach where the complete cell DNA is treated with endonucleases, which generate staggered ends. These fragments are then combined with plasmids, which have been "opened" by the same exonuclease. The product is an exceedingly complex mixture of recombinant plasmid-containing organisms, which are then screened for an organism with the correct plasmid composition.
   (ii) Specific methods used to select the required pieces of DNA by isolating its complementary mRNA. In rapidly growing cells, mRNA can make up a large proportion of cell mass, and it is possible to isolate the mRNA and use it as a template for selection of DNA fragments. Several other methods are used, but for the sake of simplicity these will not be considered here.
(2) *Construction of the gene-bearing vector.* Once the clone of DNA has been isolated, this is put into a vector (in this case a plasmid) which will transfer it from the host to the recipient organism. Again, by the use of restriction enzymes to cut the plasmid vector and the cloned DNA, the cloned DNA can now be inserted into the vector.
(3) *Insertion of the loaded plasmid.* The plasmid is then introduced into the host in one of the number of ways, either in conjugation, or more commonly by transformation using the purified plasmid and the desired host.
(4) *Expression of cloned genes.* Once inside the microbe, the genes are then expressed by the normal protein synthesis equipment of the cell. The expression or production of the cloned material can be a significant problem as the promoters and operators may not be compatible with the control systems of these cells. Therefore, careful manipulation of these control sequences can enhance the production of the protein required. Similarly, the subsequent folding of the protein and/or excretion of the protein may well not be accomplished, and this is a symptom of a basic incompatibility between the host cell and the production of the foreign protein.

Enhanced gene expression can be achieved by the addition of a promoter to the cloned gene. For example, the *lac promoter* may be added upstream of the protein DNA sequence, giving rise to the expression of the protein in the presence of lactose and in the absence of glucose (cf. operon hypothesis Appendix A.3).

Genetically Engineered Products

An example of the use of genetic engineering is the production of somatostatin. This is a growth hormone, which regulates the secretion of insulin from the pancreas and is a polypeptide with 14 amino acid residues. As the gene is so small, it can be synthesized chemically and then joined to the end of the β-galactosidase gene. Thus, when the β-galactosidase gene is expressed, the protein chimaera (or hybrid) is produced and excreted into the medium and, after selective enzymatic hydrolysis, it yields a biologically active somatostatin molecule.

The engineering of eukaryotic genes in eukaryotic organisms (yeast) is still in its infancy, and its application is not as well developed as that of the bacterial systems. This is due to the increased complexity encountered in the structure and function of eukaryotic chromosomes. However, many advances have been made in the development of this system, particularly for the production of materials that require posttranslational modification of the protein, and where other additional materials must be added before a fully functional molecule is produced (e.g., glycoproteins).

Many chemicals important to genetic research have been produced at high levels, e.g., DNA ligase production in *E. coli* can be enhanced several hundred fold. Other enzymes can also be produced for industrial use and enhanced degradative ability has been generated; for instance, enhanced petroleum degraders can be added to oil spills to accelerate clean-up operations. However, it is in the production of medically related compounds that this technology has been most successfully applied as, for example, in the production of insulin.

Historically, insulin has been produced by extracting it from the pancreas of pigs and oxen, but the protein is not precisely the same as that found in humans and, although it functions in the same manner, its use can produce unwelcome side effects. Insulin produced from cloned DNA is identical to human insulin and is consequently considered safer. Another major advantage is that the number of pigs slaughtered does not then limit the production of insulin. Using *E. coli* or yeasts, the process can be far more easily controlled and matched to demand so that this, and many other hormones, is now produced by this method.

Other proteins, which are being made using genetically engineered organisms, include vaccines for animals and humans. Table 5.A3 shows some examples of products that have used rDNA technology.

Table 5.A3: Products of recombinant DNA technology.

Substance	Application
Blood coagulation factors VIII IX	Hemophilia
Human gonadotropin	Sterility
Human insulin	Diabetes
Human serum albumin	Blood substitute
Interferon alpha 2	Antiviral/Antitumor
Interferon alpha	Antiviral/Antitumor
Interferon beta	Antiviral/Antitumor
Hepatitis B vaccine	Vaccine
Human growth hormone	Dwarfism
Lymphokinines (interleukin 2)	Stimulation of immune system
Tissue plasminogen activator	Myocardial infarction

A.3 Information Storage and Retrieval in the Cell

Introduction

Biological information is stored in the nucleotide base structure of DNA. By manipulating the structure of DNA it is possible to improve and understand the structure and functional relationships of catalysts (termed *protein engineering*). Considerable improvements have been made to products derived from enzyme systems, both in the quality and number of metabolites, enzymes and proteins produced by organisms (often termed *genetic engineering*).

Today, knowledge of the molecular aspects of genetics has arisen from the convergence of three disciplines: (1) genetics; (2) biochemistry; and (3) molecular physics. This is epitomized by the discovery of the structure of DNA as a double helix by Watson and Crick. This structure was verified using data obtained from these three fields: (1) genetic coding in the form of genes; (2) X-ray analysis of DNA crystal structure; and (3) the chemical composition of DNA. The Watson and Crick model not only accounted for the structure but also showed how replication could be performed with precision.

The central dogma of molecular genetics: The "central dogma" was based upon the findings of Watson and Crick and states that the flow of information is essentially in one direction from DNA to protein. Three major steps can be defined in the process: replication, transcription, and translation of genetic material, as shown in Fig. 5.A5.

The control of processes involving biological systems should start from a sound knowledge of their biological properties and characteristics. It has been shown that metabolism consists of many reactions, which supply both the energy and the chemical materials for the synthesis and reproduction of the cells. Central to the control of these systems is the utilization and retrieval of information stored within the cell's DNA.

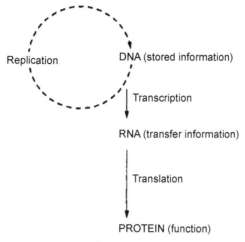

Figure 5.A5

Information storage and flow in cell systems, showing the major processes involved.

Two major areas are covered: (1) the transmission and translation of genetic material to produce protein and (2) mutation, genetic recombination, and manipulation. The reader is referred to Section 5.2.4 for the nature, dimension, and conformation of genetic material.

DNA and Gene Structure

Two molecules of DNA make up the single chromosome of bacteria. In eukaryotes, chromosome structure is very complex and there can be a large number of chromosome within the cell, e.g., some yeasts have 7 while humans have 23 pairs.

The first evidence that DNA is the bearer of genetic information came in 1943, through the work of Avery McCloud and McCarty. They found that DNA extracted from a virulent (disease causing) strain of the bacterium *Streptococcus pneumoniae* permanently transformed a nonvirulent strain of this organism into a virulent form.[5]

Genes: Genes are segments of DNA and are defined in the classical biological sense as a portion of a chromosome, which determines or specifies a single character, for example, eye color. This was later redefined to mean that one gene is equivalent to one protein, i.e., a segment of DNA equals a protein. Today the definition has been further refined, to take account of the two function types of DNA sequences; first there are *structural genes*, and second these are *regulatory genes* or sequences. The regulatory sequences consist of several components, denoting the stop and start of the structural genes and others that control the rate of transcription of structural genes. There are many genes on a single chromosome; in bacteria such as *E. coli* there are over 1100, which can be identified on the only chromosome. However, taking into account the regulatory sequences, 3000–5000 genes are thought to be a better estimate. Various viral and bacterial chromosomes have been mapped and the task of mapping the human chromosomes systematically (the Human Genome Project) started seriously in 1992.

Because in *E. coli* there are at least 3000 genes and the chromosome has 4×10^6 nucleotide pairs the average size of a gene is 4,000,000/300 = 1300, or less, nucleotide pairs. The true value is more likely to be 900–1000 pairs per gene.

In prokaryotes, there is only one copy of DNA per cell, and in nearly all cases there is only a single copy of a gene. Apart from regulatory DNA and signaling sequences, there are very few *silent*, or nontranslated DNA sequences, in prokaryotes. Moreover, each gene is typically collinear with the amino acid sequence for which it codes. This is in contrast to the organization of eukaryotic DNA, which is structurally, and functionally far more complex; silent DNA splits the structural parts of genes and there are multiple copies of the chromosomes.

Plasmids: Apart from chromosomes, prokaryotes and eukaryotes also contain small pieces of autonomous DNA. Autonomous extrachromosomal fragments of DNA in bacteria are called plasmids. Most plasmids are very small and contain only a few genes as compared with the chromosome. Plasmids carry genetic information and, like chromosomes, are passed from each cell into daughter cells at division, or they can be taken up from the environment. They can apparently lead separate lives from the chromosomal DNA for many generations but some plasmids can, however, integrate with the chromosome DNA and leave again in a coordinated manner. The best examples of plasmids are those which carry resistance to various antibiotics, such as tetracycline and streptomycin. Bacterial cells that contain such plasmids become resistant to antibiotics, and they can become a significant problem in bacterial infections, especially in hospitals where the use of antibiotics is prevalent. Further, such behavior is significant as it is possible for plasmids to be transferred to antibiotic sensitive strains. From the biotechnological point of view, plasmid DNA is easily isolated from bacterial cells and is small enough to be easily manipulated. A major strategy in the transfer of foreign DNA into bacteria is putting the foreign DNA into the plasmid and then transferring the plasmid into the host cell. There the plasmid DNA is transcribed together with the foreign DNA and synthesis takes place of the protein coded by the foreign DNA. Eukaryotes also have extrachromosomal DNA, which can be found in the cytoplasm, mitochondria, and chloroplasts.

Processes Involving the Nucleic Acid Synthesis

Replication and Transcription: DNA is replicated to yield daughter molecules. The enzymes and other proteins participating in replication and transcription of DNA are among the most remarkable biological catalysts known. Not only do they form enormously long macromolecules from the mononucleotide precursors using guanosine triphosphate (GTP and the energy equivalent ATP) but they also synthesize genetic information from the template strand to an extraordinary fidelity. In addition, these enzymes must solve complex mechanical problems, since the parental duplex DNA must be unwound in advance of the replication enzymes, so that they can gain access to the information stored in the base sequences of the molecule. The transcription enzymes also have extraordinary properties. Not only can they make a large assortment of different RNAs, but they also start and stop the translation at specific parts of the DNA molecule. They also respond to various regulatory signals, so that different proteins are expressed at various specific times in the life cycle. DNA and RNA polymerases and other proteins that help carry out replication and transcription of DNA are thus vitally important materials in the perpetuation of genetic information.

The Watson and Crick hypothesis for DNA replication proposed that each strand of DNA is used as a template for the production of one of the daughter DNA molecules. Thus the result of replication would be that one strand of DNA is present in each daughter molecule of DNA. This is a semiconservative mechanism of replication. A simplified diagram of replication is shown in Fig. 5.A6; however, the replication patterns are different in bacteria and in eukaryotes.

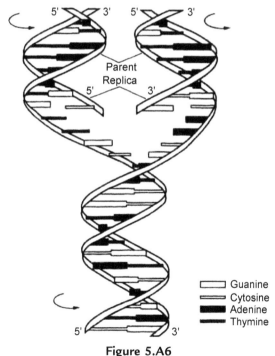

Figure 5.A6
A simplified diagram of DNA replication.

DNA Transcription

The first step in the utilization of information held within the DNA structure is the transcription of DNA into RNA. The product of transcription is a complementary strand of RNA, produced with high accuracy. There are three types of RNA products from the transcription of DNA.

Transfer (tRNA) and ribosomal RNA (rRNA) are transcribed and used in protein synthesizing processes. Messenger RNA (mRNA) codes the amino acid sequence of proteins, and 95% of the total DNA transcribed is used for this purpose. In prokaryotes, a single mRNA molecule may code for a single polypeptide or for two or more polypeptide chains. There is a triplet code for each amino acid; 300 ribonucleotides code for a 100-amino acid sequence. Fig. 5.A7 shows the relationship between the nucleotide sequence on DNA and RNA and the amino acid sequence of protein.

The most important difference between replication and transcription is that not all the DNA is used in the transcription. Usually, only small groups of genes are transcribed at any one time. Thus, the transcription of DNA is selective, turned on by specific regulatory

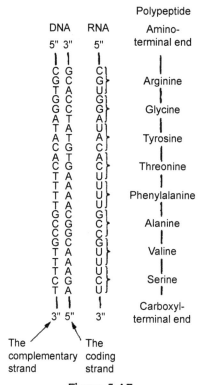

Figure 5.A7
The collinearity of the nucleotide sequences of DNA, mRNA, and amino acid sequence.

sequences, which indicate the beginning and ending of the region to be transcribed. The control of protein synthesis is discussed further.

The discovery of DNA polymerase and its dependence on a DNA template led to the search for enzymes, which could make an RNA molecule complementary to the DNA. RNA synthesis does not require a primer strand; it does, however, require a specific initiation signal on the DNA template strand to allow binding and initiation. As the RNA strand is synthesized it forms a temporary helix with the template DNA, but when complete the mRNA breaks off at the stop site on DNA. Once released from DNA, some of the RNA is processed further, for the specific structures of rRNA and tRNA.

Protein Synthesis (Translation)

An overview of protein synthesis is shown in Fig. 5.A8. The linear sequence in mRNA that is translated to protein contains four bases: adenine, uracil, guanine, and cytosine. The four "letters" A, U, G, and C constitute the mRNA "alphabet." This basic alphabet is used in triplets of bases called codons. The codons on mRNA pair up with anticodon or complementary triplets on the tRNA, thus matching the mRNA code to an amino acid sequence.

Protein synthesis occurs in five major stages:

1. *Activation of the amino acids.* This stage takes place in the cytoplasm. Each of the 20 amino acids is covalently attached to a specific tRNA at the expense of ATP hydrolysis (i.e., it is an energy-driven process). Each amino acid has a specific enzyme for this reaction to ensure that the correct amino acid is linked to the tRNA molecule.
2. *Initiation of the polypeptide chain.* mRNA bearing the code for the polypeptide is bound to the small subunit of RNA, followed by the initiating amino acid, and is attached to its tRNA to form an initiation complex. The tRNA of the initiating amino acid base pairs with a specific nucleotide triplet or codon on the mRNA that signals the beginning of the polypeptide chain. This process requires GTP (ATP equivalent) plus three proteins called initiation factors.
3. *Elongation.* The polypeptide chain is now lengthened by covalent attachment of successive amino acid units, each of which has been carried to the ribosome by a tRNA, which is base paired to the corresponding codon. Two molecules of GTP are required for each amino acid residue added to the peptide.
4. *Termination and release.* The completion of the polypeptide is signaled by a termination codon in the mRNA and is followed by the release of the polypeptide from the ribosome.
5. *Folding and processing.* To achieve its native, biologically active form, a polypeptide must undergo folding into its proper three-dimensional conformation. Before or after folding, the new polypeptide may undergo processing by enzymes so that prosthetic groups, carbohydrates, or nucleotides, may be added to the protein.

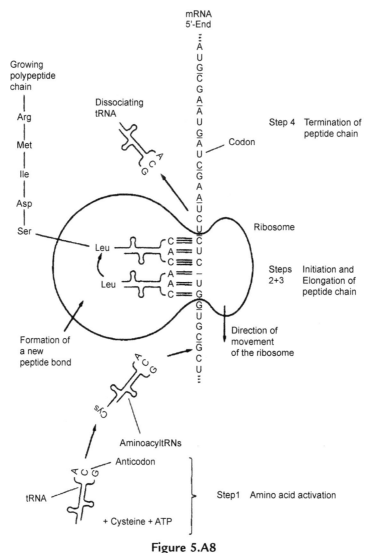

Figure 5.A8
The major steps involved in protein synthesis.

Energy consumption in protein synthesis. The energy consumed by the process is the guarantee of fidelity of protein synthesis. Four molecules (or more) of GTP (ATP equivalent) are required for the synthesis of each peptide bond.

Two high-energy bonds in the synthesis of aminoacyl tRNA:
        One is consumed in the elongation step (GTP),
        One is consumed in the translocation step,
        plus additions for corrections.

Thus $(4 \times 30.55 =)$ 122.2 MJ/kmol of phosphate group energy is required to generate a peptide bond, a bond energy of -20.6 MJ/kmol. This represents an enormous thermodynamic push as net energy of 101.6 MJ/kmol remains from synthesis. This process may appear wasteful but it is the price that is paid for near perfect fidelity. In bacteria, there is a very tight coupling between transcription and translation, as shown in Fig. 5.A8. Another feature is the fact that mRNA is very short lived and may only last a few minutes, before being degraded by the cell nucleases. The short life of the mRNA makes for a rapid control of synthesis of protein when it is no longer needed.

The Genetic Code

In Section 5.2.4, the structure of DNA was discussed and the basis for storage of information is shown to be in the base sequence of the DNA molecule. There are four types of bases in nucleic acid structure and these represent a four-letter alphabet. Combinations of two bases give only 4^2 or 16 unique combinations; while a three-base code gives 4^3 or 64 combinations and a four-base code will give 4^4 (256) combinations. The code must contain at least three letters to give enough bases to code for all the amino acid and stop sequences. A problem in a three-letter code, however, is that there are too many codes for all the possible amino acids and stop/start sequences. In fact, different codes have been found to have the same meaning, i.e., some amino acids have more than one codon and the code is thus said to be degenerated.

The genetic code is shown in Table 5.A4. All the possible 64 combinations are shown and there are codes for either start and stop signals or for amino acids. There are up to six

Table 5.A4: The genetic code.

First Position	Second Position				Third Position
	U	**C**	**A**	**G**	
U	Phe	Ser	Tyr	Cys	U
	Phe	Ser	Tyr	Cys	C
	Leu	Ser	Stop	Trp	A
	Leu	Ser	Stop	Trp	G
C	Leu	Pro	His	Arg	U
	Leu	Pro	His	Arg	C
	Leu	Pro	Gin	ArS	A
	Leu	Pro	Gin	Arg	G
A	Ile	Thr	Asn	Ser	U
	Ile	Thr	Asn	Ser	C
	Ile	Thr	Lys	Arg	A
	Met	Thr	Lys	Arg	G
G	Val	Ala	Asp	Gly	U
	Val	Ala	Asp	Gly	C
	Val	Ala	Glu	Gly	A
	Val	Ala	Glu	Gly	G

codes for some amino acids, e.g., arginine, leucine, and serine, while tryptophan and methionine have unique codes. The major features of the genetic code include the following:

1. Methionine and tryptophan have only one codon. All the others have at least two, and serine and leucine have six. Many have four codons;
2. Those amino acids which have four codons are coded by the first two bases of the triplet while the third can be variable;
3. Similarly, coded acids have similar chemistry, e.g., serine and threonine. U (uracil) in the second position will inevitably code for hydrophobic amino acids; a mutation here will result in the replacement of one hydrophobic acid by another; and
4. The code is the same for all organisms.

The Control of Protein Synthesis

Living cells have many control mechanisms for regulating the synthesis of proteins so that each cell has the correct amount of each protein to carry out its metabolic activities smoothly and efficiently. For example, although *E. coli* contains 3000 proteins, it can have different amounts of each protein under different conditions. For example, when growing on glucose, *E. coli* has very high copy numbers of the glycolytic enzymes while it will only have five or so copies of β-galactosidase. However, when grown on lactose, the copy number of this enzyme increases rapidly. Regulation of enzyme synthesis provides each cell with the proper ensemble of enzymes for balanced growth. It also allows the organism to economize on the highly energy-expensive process of protein synthesis.

The control and expression of protein in eukaryotic cells is more complex than in bacteria. Research into this field is ongoing and, as such, is beyond the scope of this text. Many systems of control are known in bacteria and can be used to illustrate the type of control mechanisms and the importance of the environmental control of protein synthesis. The first and best documented example is that of the lactose (lac) operon is *E. coli.*

It has been established that some enzymes are produced only in the presence of their substrates and they are then said to be *inducible.* The quantity of such enzymes will vary considerably with changing environmental conditions. Other enzymes that are always detectable at constant levels, irrespective of nutrition and environmental conditions, are called *constitutive* enzymes.

In the induction of enzymes of galactose metabolism in *E. coli*, three enzymes are involved: β-galactosidase (which catalyzes the hydrolysis of the β-glycosidic bonds of lactose); galactose permease (which is responsible for transport of lactose across the cell membrane); and a third enzyme, A-protein, apparently not directly involved in galactose metabolism. The system has an environmental *inducer,* galactose, and in its presence the number of β-galactosidase molecules rises from $5-10$ to 10,000 within the cell. The addition of the inducer can increase the protein production in less than 5 min after its addition. Protein synthesis of these enzymes stops almost immediately in the absence of lactose.

Another important change in the concentration of an enzyme in bacteria, seemingly the opposite to that observed in enzyme induction, is *enzyme repression.* An example of enzyme repression is demonstrated when *E. coli* is grown in the presence of ammonium salts which act as the sole source of nitrogen for growth. Such systems require all the biosynthetic apparatus to be present to make all the amino acids required by the cell. If, for example, histidine is added to the medium, a whole set of enzymes is no longer synthesized. Repression, like induction, is a reflection of the principle that the cell uses energy as economically as possible. In general, repression operates in anabolic pathways (biosynthetic), while in catabolic pathways (degradation for energy) induction usually operates.

The Operon Hypothesis

The molecular and genetic relationship between enzyme induction and repression was clarified by the genetic research of Jacob and Monod at the Pasteur Institute, Paris (see Ref. 9). Their classic work led them to develop the *operon hypothesis* for the control of protein synthesis in prokaryotes, which has since been verified by direct biochemical experiments.

Jacob and Monod proposed that the operon consists of a sequence of genes, which are partitioned into *structural genes* and *regulatory genes.* In the case of the lactose, or *lac,* operon there are *z, y,* and α coding for the three enzymes galactosidase, galactoside permease, and A-protein. It was proposed that an area referred to as the i-region was responsible for the *regulatory gene,* coding the amino acid sequence for a regulatory protein called the *repressor* protein. The repressor protein in the absence of the inducer will bind to another regulatory site called the *operator* or the *o*-site. If the repressor protein is bound to the *o*-site, then the RNA polymerase, which is bound to the promoter site (*p*), is unable to transcribe the structural genes. However, in the presence of the inducer the repressor protein is inactivated and no longer binds to the operator. The RNA polymerase can now transcribe the structural genes. Fig. 5.A9 illustrates the control mechanisms associated with an operon; the mechanisms of both induction and repression are illustrated.

An operon thus consists of a series of functionally related structural genes, which are turned on and off together, plus their regulatory gene, the operator. This overall process is called *transcriptional control* since control is primarily directed at the rate of transcription of genes, then at their corresponding mRNAs and ultimately at the amount of protein synthesized.

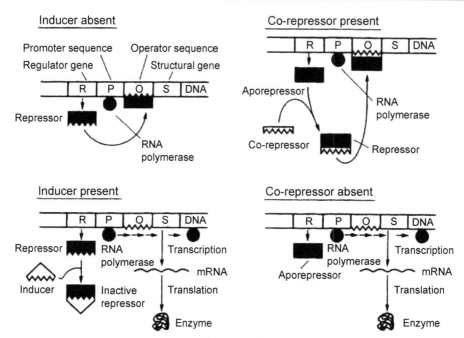

Figure 5.A9

The control of protein (enzyme) synthesis at the level of DNA transcription. Two types of control are illustrated. On the left, the inducer acts by inactivating a repressor protein allowing transcription of the structural genes. On the right, a repressor combines with an aporepressor protein to form an active repressor complex which blocks transcription of the structural genes.

Another way in which gene expression is regulated is by translational control, where the rate of protein synthesis is controlled at the point of transcription of mRNA into polypeptides. Generally, the majority of the control mechanisms in bacteria is at the transcriptional level. Translational control is less well understood and appears to be a secondary mechanism in bacteria, but it is thought to be very important in eukaryotic organisms.

Catabolic Repression

The operon theory has to be further elaborated when considering another common phenomenon, catabolite repression. If glucose and lactose are present in the growth medium, then *E. coli* will utilize only the glucose and ignore the lactose until the glucose is exhausted. In the presence of glucose the cells no longer make the proteins required for lactose metabolism. This repression of the lactose, or lac, proteins is called *catabolite repression.*

The organism senses whether glucose is available by another regulatory mechanism, which cooperates with the lac repressor and the lac operator. The promoter is therefore

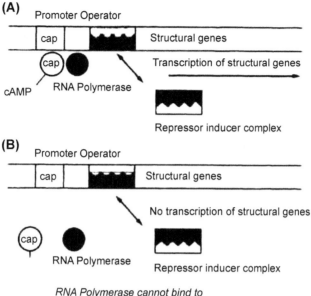

Figure 5.A10

The control of enzyme synthesis by catabolite repression. A control region of the lac operon contains the catabolite activator protein (CAP) binding site within the promoter region. (A) At low energy levels, the CAP protein binds enhancing protein synthesis; (B) When glucose is present and energy levels are high the CAP protein does not bind, so reducing the rate of protein synthesis.

subdivided into two specific regions, each of distinctive function. One is the RNA polymerase entry site, where RNA polymerase first becomes bound to DNA and the other is the protein-binding site for the catabolite activator protein (CAP) (Fig. 5.A10). The CAP protein-binding site controls the polymerase site which, when bound to the DNA, allows successful transcription provided that the repressor is not bound. When the CAP protein is not bound, then RNA polymerase cannot bind and transcription cannot take place.

The binding of the CAP protein to the promoter site on DNA can be modulated by the chemical, cyclic AMP (cAMP). When this is bound to the CAP protein, the CAP protein will bind to the promoter site on the DNA, which allows the RNA polymerase to bind to DNA at the promoter site. Now, if glucose is absent but lactose is present, then the operator site will be open and, where these two conditions are met, the RNA polymerase will function to produce the lac proteins required. On the other hand, when ample glucose is available, then the concentration of cyclic AMP is low and the CAP–cAMP complex cannot form. The net result of this is that the RNA polymerase cannot enter the promoter site for the lac genes; thus the lac genes are only synthesized when glucose is unavailable. The lac operon thus has both positive and negative control characteristics. The CAP "senses" whether glucose is available by having two binding sites, one for the binding of DNA at the promoter site, and the other for cAMP. In *E. coli,* cAMP is a chemical messenger (or hunger signal) which signals the presence or absence of glucose for use as cell fuel. Fig. 5.A10 illustrates this type of control mechanism.

As research has progressed in this area, many systems have been shown to follow a similar behavior. Such phenomena are important in the production of enzymes and metabolites. In industrial applications, these very important control mechanisms usually have to be bypassed for the reliable generation of high concentrations of enzymes. Good examples may be seen in carbohydrase production systems, such as those for amylases and cellulases.[2] Similarly, inducible enzymes may be made constitutive by altering the structure of the control site of the operon. Alternatively, the structure of the repressor/inducer proteins may be altered so as to change the binding characteristics of sites which can then no longer bind to DNA, a similar principle to that outlined when using control mutants of allosteric enzymes (Section 5.5.5).

References

1. Smith JE. *Biotechnology principles.* Wokingham: Van Nostrand Reinhold (UK) Co Ltd; 1985.
2. Atkinson B, Mavituna F. *Biochemical engineering and biotechnology Handbook.* 2nd ed. Basingstoke: Macmillan Publishers; 1991.
3. Hacking AJ. *Economic aspects of biotechnology.* Cambridge: Cambridge University Press; 1986.
4. Creighton TE. *Proteins: structures and molecular properties.* 2nd ed. New York: W. H. Freeman and Company; 1993.
5. Nelson NL, Cox MM. *Lehninger principles of biochemistry.* 6th ed. Basingstoke: Macmillan Publishers; 2012.
6. Schelgel HG. *General microbiology.* 7th ed. Cambridge: Cambridge University Press; 1993.
7. Stanier R. *General microbiology.* 5th ed. London: Macmillan Press; 1987.
8. Kirsop BE, Doyle A. *Maintenance of microorganisms—a manual of laboratory methods.* 2nd ed. London: Academic Press, Inc.; 1991.
9. Hibbert DB, Jones AM. *Macmillan dictionary of chemistry.* London: Macmillan Press; 1987.
10. Hinckle PC, McCarthy R. How cells make ATP. *Sci Am* 1978;**238**:104.
11. Berry R, editor. *Physiology of industrial fungi.* Oxford: Blackwell Scientific Publications; 1988.
12. Onions AHS, Allsopp D, Eggins HOW. *Smith's introduction to industrial mycology.* 7th ed. London: Edward Arnold; 1981.
13. Horan NJ. *Biological waste water treatment systems.* Chichester: John Wiley and Sons; 1989.
14. Stanbury PF, Hall S, Whitaker A. *Principles of fermentation technology.* 3rd ed. Oxford: Pergamon Press; 2016.
15. Ingraham JL, Maaloe O, Neidhardt FC. *Growth of the bacterial cell.* Sunderland Massachusetts: Sinauer Associates, Inc.; 1983.
16. Baldtz RH, Demain AL, Soloman NA, editors. *Manual of industrial microbiology and biotechnology.* 3rd ed. Washington: American Society for Microbiology; 2010.
17. Wood WB, Wilson JH, Benbow RM, Hood LF. *Biochemistry—a problems approach.* 2nd ed. Menlo Park, California: W. A. Benjamin Inc.; 1981.

18. Aiba S, Humphrey AE, Millis NF. *Biochemical engineering*. 2nd ed. New York: Academic Press; 1973.
19. Ratledge C. Biochemistry of growth and metabolism. In: Bu'lock JD, Kristiansen B, editors. *Basic biotechnology*. London: Academic Press; 1987. p. 39.
20. Kieslech K. Biotechnology. In: Rehm HJ, Reed G, editors. *Biotransformations*vol. 6A. Weinheim, Germany: Verlag Chemie; 1984.
21. Britz JE, Demain AL. Regulation of metabolite synthesis. In: Moo-Young M, editor. *Comprehensive biotechnology*vol. 1. Oxford: Pergamon; 1985. p. 617.
22. Malik VS. Microbial secondary metabolism. *Trends Biochem Sci* 1980;**5**:68.
23. Gottschalk G. *Bacterial metabolism*. New York: Springer Verlag; 2012.
24. Mandlestam J, McQuillan K, Dawes I. *Biochemistry of bacterial growth*. 3rd ed. Oxford: Blackwell Scientific Publications; 1985.
25. Bailey JE, Ollis DF. *Biochemical engineering fundamentals*. 2nd ed. New York: McGraw-Hill; 1986.
26. Fersht A. *Structure and mechanism in protein science: a guide to enzyme catalysis and protein folding*. New York: W. H. Freeman and Co.; 1999.
27. Michaelis L, Menten ML. Die Kinetik der Invertinwirkung [The kinetics of invertin action]. *Biochem Z* 1913;**49**:333.
28. Hirose Y, Sano K, Shibia H. Amino acids. In: Perlman D, editor. *Ann. Reports on fermentation process*vol. 2. New York: Academic Press; 1978. p. 155.
29. Brown TA. *Gene cloning and DNA analysis—an introduction*. Hoboken: John Wiley and Sons; 2016.
30. Primrose SB, Twyman R, Old RW. *Principles of gene manipulation—an introduction to genetic engineering*. 6th ed. Hoboken: Wiley; 2002.

Further Reading

Cano RJ, Colome JS. *Microbiology*. St. Paul, Minnesota: West Publishing Co.; 1986.
Palmer T. *Understanding enzymes*. 4th ed. Chichester: Ellis Horwood Limited; 1995.
Segal IH. *Biochemical calculations*. 2nd ed. New York: John Wiley; 1976.
Berg JM, Tymoczko JL, Stryer L. *Biochemistry*. 5th ed. New York: Freeman; 2002.

CHAPTER 6

Biochemical Reaction Engineering

Sathyanarayana N. Gummadi
Indian Institute of Technology Madras, Chennai, India

Learning Outcomes

1. Understand the underlying mechanisms of enzyme action, enzyme kinetics including inhibition kinetics, immobilization, and protein denaturation kinetics.

2. Familiarize the basic concepts of microbial stoichiometry and their importance in bioprocess engineering.

3. Overview of microbial growth kinetics and the factors affecting microbial growth including temperature, pH, and product formation.

4. Explaining the basic concepts of bioreactor configurations and the theory essential for design of bioreactors.

5. Learning the significance of important kinetic parameters to be considered while designing a process for industrial production.

6.1 Enzymes: Biocatalysts

Enzymes are usually proteins, which act as catalysts and increase the rate of chemical reactions in biological systems without themselves undergoing any change. Most of the enzymes require a nonprotein component called as cofactor (or coenzyme) for efficient catalytic activity. The protein component without any cofactor is termed as apoenzyme, and the protein—cofactor complex is termed as holoenzyme. A classic example of an enzyme-catalyzed reaction is given below.

$$\text{Urea} \rightarrow \text{NH}_3 + \text{CO}_2$$

The conversion of urea to ammonia and carbon dioxide by chemical catalysis in the presence of palladium(II) complex (rate constant for formation of CO_2—1.7×10^{-3} min^{-1}) is 10^5 times faster than the uncatalyzed decomposition of urea. Alternatively, when the enzyme urease is used for decomposition of urea, the ratio of the catalyzed rate to the uncatalyzed rate is 10^{14}. The high rate of enzymatic reactions at milder operating conditions such as temperature and pH can easily trump the chemically catalyzed reactions

Coulson and Richardson's Chemical Engineering. http://dx.doi.org/10.1016/B978-0-08-101096-9.00006-6

that require raucous temperature and pressure. The high catalytic rate of enzymes is attributed to its extreme specificity toward the catalyzed reaction.

6.1.1 Specificity of Enzymes

One of the significant features of enzyme biocatalysts is that they are highly specific in catalytic action. Specificity of the enzymes can be of different types including group specificity, absolute specificity, and stereochemical specificity. In group specificity, the enzyme acts on a particular chemical group of different closely related substrates (e.g., hexokinases catalyzes the phosphorylation of different hexose sugars). Enzyme exhibiting absolute specificity acts only on a particular substrate (e.g., glucokinase catalyzes the phosphorylation of glucose). Enzymes can also show stereochemical specificity where they can predominantly catalyze either D-isomer or L-isomer of a substrate (e.g., oxidation of L-amino acids to oxo acids is catalyzed by L-amino acid oxidase, whereas the oxidation of D-amino acids is catalyzed by D-amino acid oxidase).

6.1.1.1 Active Site

The high specificity of the enzymes can be attributed to the structural and functional properties of the enzyme–substrate interactions. Ogston claimed that there must be at least three different points of interactions in the enzyme for exhibiting specificity. The interactions occur at a specific site in the enzyme termed as active center or active site, which might consist of binding and catalytic sites. Binding sites interact with specific groups in the substrate, thereby aligning the substrate and enzyme at a specific orientation in the proximity of the catalytic sites. For example, the enzyme glycerol 3-phosphate dehydrogenase specifically acts on glycerol to form glycerol 3-phosphate. In this reaction, the glycerol should bind to the enzyme in a specific alignment so that the third carbon of glycerol is phosphorylated. If the orientation of the binding is changed, the binding of the substrate to the enzyme is not strong enough for the reaction to proceed (Fig. 6.1).

6.1.1.2 Fischer Lock and Key Hypothesis

The specificity of the enzyme with the substrate can be explained using the lock and key analogy, hypothesized by Emil Fischer in the year 1894. According to the lock and key hypothesis, the substrate fits into the active site of the enzyme as a key fits into the lock (Fig. 6.2). The structure of the enzyme and the substrate remains unchanged throughout the reaction process.

6.1.1.3 Koshland's Induced-Fit Hypothesis

The change in three-dimensional structure of enzyme on binding to the substrate was not considered in the lock and key hypothesis. In 1958, Koshland hypothesized that the structure of the substrate may be complementary to the enzyme–substrate complex but not

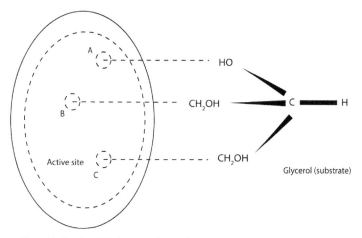

Figure 6.1
Schematic of three-point interaction between enzyme and substrate. A, B, C are the sites on the enzyme, and glycerol was taken as the substrate. Each point of interaction might be involved in binding or catalysis.

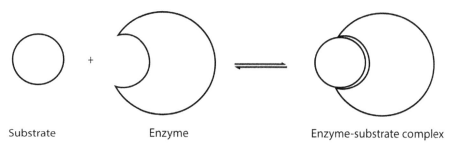

Figure 6.2
Schematic of the interaction between enzyme and substrate according to the Fisher lock and key model.

the free enzyme. The enzyme undergoes an induced conformational change during binding to the substrate. In induced-fit hypothesis, the substrate is rigid but the active site of the enzyme is flaccid. The floppy nature of the active site ensures the proper fitting of catalytic sites of the enzyme and reacting groups in the substrate (Fig. 6.3).

6.1.1.4 Transition State Stabilization

Though specificity of the enzyme was described by lock and key and induced-fit hypothesis, the mechanism of enzyme catalysis was not considered. In transition state stabilization hypothesis, it was assumed that the substrate is bound in an undistorted form, but various unfavorable interactions exist in the enzyme—substrate complex. As the

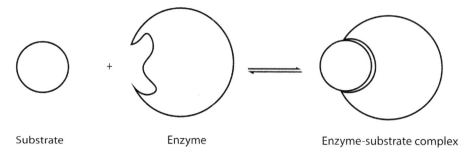

Substrate Enzyme Enzyme-substrate complex

Figure 6.3
Schematic of the interaction between enzyme and substrate according to the Koshland induced fit model.

reaction progresses, the substrate distorts and favors the following reaction scheme: enzyme—substrate complex → transition state → products, by diminishing unfavorable interaction (Fig 6.4).

6.1.2 Enzyme Kinetics: Michaelis—Menten Equation

For most enzymes, the rate of reaction can be described by the Michaelis—Menten equation that was originally derived in 1913 by Michaelis and Menten.[1] The Michaelis—Menten assumption was that an equilibrium between enzyme, substrate, and enzyme—substrate complex was set up instantly and the formation of products from the enzyme substrate complex is too slow to disturb the equilibrium.

The simplest general equation for a single-substrate enzyme— catalyzed reaction would be

$$E + S \underset{k_{-1}}{\overset{k_1}{\rightleftharpoons}} ES \underset{k_{-2}}{\overset{k_2}{\rightleftharpoons}} E + P$$

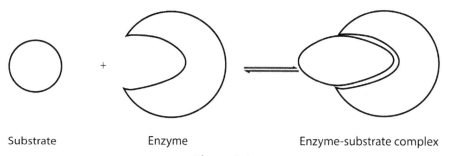

Substrate Enzyme Enzyme-substrate complex

Figure 6.4
Schematic of the interaction between enzyme and substrate according to the transition state stabilization model.

By considering the initial period of the reaction, the concentration of product is negligible and the reverse reaction of enzyme—substrate complex formation can be ignored. The reaction is simplified to

$$E + S \underset{k_{-1}}{\overset{k_1}{\rightleftharpoons}} ES \overset{k_2}{\rightarrow} E + P$$

Using the equilibrium assumption,

$$k_1[E][S] = k_{-1}[ES] \tag{6.1}$$

The dissociation constant of $[ES]$ is given by K_s,

$$\frac{[E][S]}{[ES]} = \frac{k_{-1}}{k_1} = K_s \tag{6.2}$$

The total enzyme concentration $[E_0]$ will be equal to the sum of free enzyme concentration $[E]$ and the bound enzyme concentration $[ES]$.

$$[E] = [E_0] - [ES] \tag{6.3}$$

$$\frac{([E_0] - [ES])[S]}{[ES]} = K_s \tag{6.4}$$

$$K_s[ES] = [E_0][S] - [ES][S] \tag{6.5}$$

$$[ES]([S] + K_s) = [E_0][S] \tag{6.6}$$

$$[ES] = \frac{[E_0][S]}{[S] + K_s} \tag{6.7}$$

$[ES]$ governs the rate of product formation: $R_0 = k_2[ES]$.

Substituting the expression of $[ES]$ in the above equation,

$$R_0 = \frac{k_2[E_0][S]}{[S] + K_s} \tag{6.8}$$

The limiting initial velocity V_{max} is reached when the substrate concentration is very high and all the enzymes are present as enzyme—substrate complex, $V_{max} = k_2[E_0]$.

$$R_0 = \frac{V_{max}[S]}{[S] + K_s} \tag{6.9}$$

If we assume that the initial substrate concentration $[S_0]$ is much greater than the initial enzyme concentration $[E_0]$, the formation of enzyme substrate concentration will result in insignificant change in free substrate concentration. Thus, R_0 can be written as

$$R_0 = \frac{V_{max}[S_0]}{[S_0] + K_s} \tag{6.10}$$

The Michaelis–Menten equilibrium assumption cannot be applicable for all enzyme-catalyzed reactions because the reaction proceeds at a faster rate and disturbs the equilibrium.

6.1.3 The Steady-State Approximation

Consider the simplest reaction scheme between an enzyme and a substrate, Scheme 6.1. It is assumed that the decomposition of the enzyme–substrate complex to yield product is not reversible, whereas its decomposition to substrate **S** and free enzyme **E** is reversible. The decomposition of enzyme–substrate complex (**ES**) is assumed to be rate limiting. The rate constants for the formation and breakdown of the enzyme–substrate (**ES**) complex are given by

$$E + S \underset{k_{-1}}{\overset{k_1}{\rightleftharpoons}} ES \overset{k_2}{\longrightarrow} E + P$$

Scheme 6.1

The steady-state assumption is valid when the concentration ES of the enzyme–substrate complex **ES** is constant and when the total enzyme concentration, E_0, is small relative to that of the substrate, i.e., $E_0 \ll S$. In most cases this assumption holds over a long period of the reaction, as illustrated in Fig. 6.5, which shows the significance of this assumption during reaction processes.

Because the concentration of the enzyme–substrate complex is constant, its rate of change is zero. That is

$$\frac{d[ES]}{dt} = 0 \tag{6.11}$$

Rate of formation of $[ES]$ at any time $t = k_1[E][S]$
Rate of breakdown of $[ES]$ at the time $t = k_{-1}[ES] + k_2[ES]$
Using the steady state assumption, $k_1[E][S] = k_{-1}[ES] + k_2[ES] = [ES](k_{-1} + k_2)$

Separating constants and variables:

$$\frac{[E][S]}{[ES]} = \frac{k_{-1} + k_2}{k_1} = K_m \tag{6.12}$$

The composite term $((k_{-1} + k_2)/k_1)$ in the above equation is the Michaelis constant K_m.

Because

$$[ES] = [E_0] - [E] \tag{6.13}$$

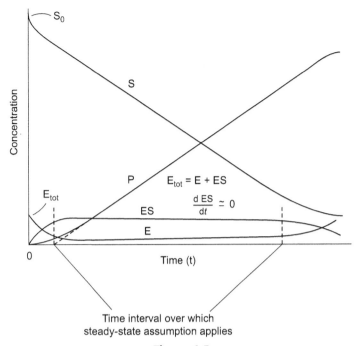

Figure 6.5
The steady-state assumption. Concentration of substrate, **S**; free enzyme, **E**; product **P**;
and enzyme-substrate complex, **ES** over time *t*. After an initial burst, the concentration of
enzyme–substrate complex is almost constant and the steady-state assumption is applicable.

substituting Eq. (6.12) into Eq. (6.13) then gives

$$ES = \frac{[E_0][S]}{K_m + [S]} \tag{6.14}$$

and

$$R_0 = k_2[ES] \tag{6.15}$$

where R_0 is the initial rate of reaction. Substituting Eq. (6.14) into Eq. (6.15) yields the
rate law for the reaction:

$$R_0 = \frac{k_2[E_0][S]}{K_m + [S]} \tag{6.16}$$

where $k_2[E_0] = V_{max}$, the maximum rate of reaction, and therefore

$$R_0 = \frac{V_{max}[S]}{[S] + K_m} \tag{6.17}$$

Usually, substrate concentration $[S_0]$ will be much greater than the enzyme concentration $[E_0]$.

$$R_0 = \frac{V_{max}[S_0]}{[S_0] + K_m} \tag{6.18}$$

From the derivation shown here, it should be noted that nothing is implied concerning the mechanism of action of the enzyme, and different enzyme reaction mechanisms cannot be distinguished kinetically. Fig. 6.6 shows the relation between rate of the reaction and substrate concentration. At low substrate concentrations, it is first-order with respect to the substrate, while at high (saturating) concentration the reaction is zero-order and is limited by the enzyme concentration and the turnover number k_{cat}. (In the simple situation above, $k_2 = k_{cat}$, but in more complex situations this may not be so and k_{cat} will be a sum of decomposition terms.[2]) The K_m value for the reaction is the concentration of substrate required to give half the maximum rate. Typically values for K_m are in the range of $0.01-20$ mM ($10^{-5} - 0.02$ kmol/m^3).

6.1.4 The Significance of Kinetic Constants

6.1.4.1 K_m and K'_s

The derivation using the steady-state assumptions shows that

$$K_m = \frac{k_{-1} + k_2}{k_1} \quad \text{(Eq. 6.12)}$$

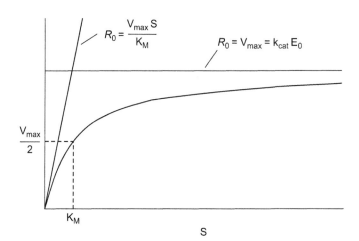

Figure 6.6
Initial reaction rate R_0 plotted against substrate concentration S for a reaction obeying Michaelis−Menten kinetics.

The rapid equilibrium hypothesis of Michaelis—Menten shows that the relationship between K_m and K_s' is

$$K_m = \frac{k_2}{k_1} + K_s' \tag{6.19}$$

(as the dissociation of **ES**, $K_s' = k_{-1}/k_1$).

It would appear, therefore, that the view of the Michaelis—Menten situation envisaged by Briggs and Haldane is a special case in which k_2 is so much smaller than k_{-1} that the value of the ratio is negligible compared with K_s', and $K_m = K_s'$ when $k_2 \gg k_1$.

The apparent dissociation constant may be treated as an overall dissociation constant for much more complex situations involving several linked rate constants.[1]

6.1.4.2 The Maximum Rate V_{max}

According to the interpretation of the two-step mechanism offered above, the definition of V_{max} reflects the fact that the decomposition of the **ES** complex is rate limiting so that when all the enzymes are in the form of **ES** a maximum rate will be observed, i.e., $[ES] = E_0$. Thus, $V_{max} = k_2[E_0]$.

It should also be noted that k_2 is equivalent to k_{cat} in the Michaelis—Menten rapid equilibrium hypothesis when the decomposition rate of the enzyme—substrate complex is fast, as described above in Scheme 6.1. However, the value of k_{cat} may be attributable to more complex situations involving several decomposition terms.

It is also feasible that by following changes in the value of V_{max} under different reaction conditions, it might be possible to obtain information concerning the kinetics of the rate-limiting step in the decomposition of **ES**. The catalytic constant or turnover number (k_{cat}) is a first-order rate constant that refers to the properties and reactions of the enzyme—substrate, enzyme—intermediate, and enzyme—product complexes. The unit of k_{cat} is $time^{-1}$, and $1/k_{cat}$ is the time required to turn over a molecule of substrate on an active site.

6.1.4.3 The Ratio k_{cat}/K_m

This ratio is of fundamental importance in the relationship between enzyme kinetics and catalysis. The constant K_{cat} is called the turnover number, which is obtained from the expression,

$$V_{max} = k_{cat}[E_0] \tag{6.20}$$

Turnover number denotes the maximum number of substrate molecules that can be converted to products per molecule of enzyme per unit time. For reactions discussed above, $k_{cat} = k_2$; for complex reactions, k_{cat} will involve several individual rate constants.

For a simple reaction,

$$R_0 = \frac{k_2[E_0][S]}{K_m + [S]} \tag{6.21}$$

We know that $[E_0] = [E] + [ES]$ and $[E][S]/[ES] = K_m$.

Substituting in equation,

$$R_0 = \frac{k_2}{K_m}[E][S] = \frac{k_{cat}}{K_m}[E][S] \tag{6.22}$$

The term k_{cat}/K_m is called the catalytic efficiency of the enzyme. A high value indicated that the limiting factor for the overall reaction is the frequency of enzyme and substrate collisions. A low value could be compared to the equilibrium assumption. A comparison of catalytic efficiency with different substrates could be used to understand the specificity of the enzyme. K_m is independent of enzyme concentration and is the characteristic of the system being investigated. Because the catalytic step is the rate-limiting step, k_1 and k_{-1} are much greater than k_2; in such case, the equilibrium assumption is valid, $K_m \approx K_s$ which gives the indication of the affinity of an enzyme toward the substrate. A low K_s value indicates high affinity of the enzyme toward the substrate, and a high K_s value indicates low affinity of the enzyme toward the substrate.

6.1.5 Estimation of Kinetic Constants

6.1.5.1 The Lineweaver–Burk Plot (Double Reciprocal Plot)

Graphical transformation of the representation of enzyme kinetics is useful as the value of V_{max} is impossible to obtain directly from practical measurements. A series of graphical transformations/linearizations may be used to overcome this problem. Lineweaver and Burk (see Ref. 3) simply inverted the Michaelis–Menten equation (Eq. 6.18). Thus:

$$R_0 = \frac{V_{max}[S_0]}{[S_0] + K_m} \tag{6.23}$$

$$\frac{1}{R_0} = \frac{[S_0]}{V_{max}[S_0]} + \frac{K_m}{V_{max}[S_0]} \tag{6.24}$$

Therefore:

$$\frac{1}{R_0} = \frac{K_m}{V_{max}}\frac{1}{[S_0]} + \frac{1}{V_{max}} \tag{6.25}$$

A plot $\frac{1}{R_0}$ against $\frac{1}{[S_0]}$ gives a straight-line graph with an intercept of $\frac{1}{V_{max}}$ and a slope of $\frac{K_m}{V_{max}}$ as shown in Fig. 6.7.

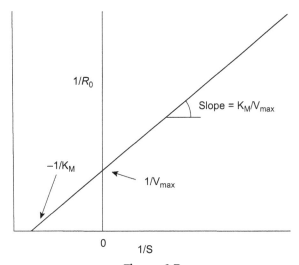

Figure 6.7
The Lineweaver–Burk plot.

This transformation suffers from a number of disadvantages. The data are reciprocals of measurements, and small experimental errors can lead to large errors in the graphically determined values of K_m, especially at low substrate concentrations. Departures from linearity are also less obvious than on other kinetic plots such as the Eadie–Hofstee and Hanes plots (see Ref. 3).

6.1.5.2 Eadie–Hofstee Plot and Hanes Plot

If the Lineweaver–Burk (double reciprocal) equation (Eq. 6.25) is multiplied by V_{max} throughout, and after simplification it yields

$$R_0 = -K_m \frac{R_0}{[S_0]} + V_{max} \tag{6.26}$$

The Eadie–Hofstee plot of R_0 against R_0/S has an intercept of V_{max} on the R_0 axis, while on the other axis the intercept is V_{max}/K_m (Fig. 6.8).

The Hanes plot also starts with the Lineweaver–Burk transformation (Eq. 6.25) of the Michaelis–Menten equation which in this instance is multiplied by S throughout; on simplification, this yields

$$\frac{[S_0]}{R_0} = \frac{1}{V_{max}}[S_0] + \frac{K_m}{V_{max}} \tag{6.27}$$

A plot of S/R_0 versus S is linear with a slope of $1/V_{max}$. The intercept on the S/R_0 axis gives K_m.

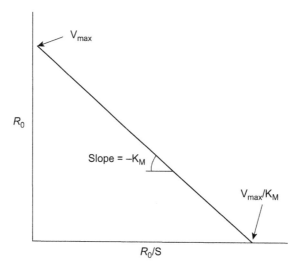

Figure 6.8
The Eadie—Hofstee plot.

These equations compress data at high substrate concentrations, but the relationship between rate and concentration is less obvious than in the Lineweaver—Burk plot. As they only involve one reciprocal term, they tend to give more accurate figures than those obtained in a Lineweaver—Burk plot.

6.1.6 The Haldane Relationship for Reversible Reactions

All reactions are to some degree reversible, and many enzyme-catalyzed reactions can take place in either direction inside a cell. It is therefore interesting to compare the forward and back reactions, especially when the reaction approaches equilibrium, as in an enzyme reactor.

Consider the situation in Scheme 6.2 in which the conversion of substrate to product proceeds via the formation of a single intermediate complex; in the forward direction, it would be regarded as an *ES* complex, whereas in the back reaction it would be regarded as an **EP** complex.

$$S + E \underset{k_{-1}}{\overset{k_1}{\rightleftharpoons}} ES/EP \underset{k_{-2}}{\overset{k_2}{\rightleftharpoons}} P + E$$

Scheme 6.2

The Michaelis—Menten equation in the forward direction at fixed $[E_0]$:

$$R_f = \frac{V_{max}^s [S_0]}{[S_0] + K_m^s} \tag{6.28}$$

where R_f is the initial velocity in the forward direction and

V_{max}^s is the maximum initial velocity in the forward direction.

$$K_m^s = \frac{k_{-1} + k_2}{k_1} \tag{6.29}$$

The Michaelis–Menten equation in the reverse direction at fixed $[E_0]$:

$$R_b = \frac{V_{max}^p [S_0]}{[S_0] + K_m^p} \tag{6.30}$$

where R_b is the initial velocity in the forward direction and

V_{max}^p is the maximum initial velocity in the forward direction.

$$K_m^p = \frac{k_{-1} + k_2}{k_{-2}} \tag{6.31}$$

Haldane derived a relationship between the kinetic and equilibrium constants. At equilibrium, the rates of forward and reverse reactions are equal.

$$k_{-1}[ES] = k_1[E][S] \tag{6.32}$$

$$\frac{[ES]}{[E]} = \frac{k_1}{k_{-1}}[S] \tag{6.33}$$

Under these conditions

$$k_2[ES] = k_{-2}[E][P] \tag{6.34}$$

$$\frac{[ES]}{[E]} = \frac{k_{-2}}{k_2}[P] = \frac{k_1}{k_{-1}}[S] \tag{6.35}$$

$$K_{eq} = \frac{[P]}{[S]} = \frac{k_1}{k_{-1}} \frac{k_2}{k_{-2}} \tag{6.36}$$

But, $V_{max}^s = k_2[E_0]$ and $V_{max}^p = k_{-1}[E_0]$. Thus

$$\frac{V_{max}^s}{V_{max}^p} = \frac{k_2}{k_{-1}} \tag{6.37}$$

Also

$$\frac{K_m^s}{K_m^p} = \frac{(k_{-1} + k_2)}{k_1} \frac{k_{-2}}{(k_{-1} + k_2)} = \frac{k_{-2}}{k_1} \tag{6.38}$$

Thus,

$$K_{eq} = \frac{k_1 k_2}{k_{-1} k_{-2}} = \frac{V_{max}^s K_m^p}{V_{max}^p K_m^s} \quad \text{(The Haldane relationship)} \tag{6.39}$$

If the equilibrium constant is known, this relationship can be used to check the validity of the kinetic constants that have been determined. In general, the equilibrium will favor the metabolically important direction. However, it should be noted that the direction of metabolic flow in cell systems will also be dependent of the concentrations of substrates and products.

This relation is useful in confirming the accuracy of kinetic constants, which are determined by kinetic measurements.

6.1.7 Enzyme Inhibition

Inhibitors are substances that tend to decrease the rate of an enzyme-catalyzed reaction. Although some act on the substrate, the discussion here will be restricted to those inhibitors, which combine directly with the enzyme. Inhibitors have many uses, not only in the determination of the characteristics of enzymes, but also in aiding research into metabolic pathways where an inhibited enzyme will allow metabolites to build up so that they are present in detectable levels. Another important use is in the control of infection where drugs such as sulfanilamides competitively inhibit the synthesis of tetrahydrofolates, which are vitamins essential to the growth of some bacteria. Many antibiotics are inhibitors of bacterial protein synthesis (e.g., tetracycline) and cell-wall synthesis (e.g., penicillin).

There are two types of inhibitors. *Reversible inhibitors* bind to an enzyme in a reversible fashion and can be removed by dialysis (or dilution) to restore full enzyme activity. *Irreversible inhibitors* cannot be removed by dialysis and, in effect, permanently deactivate or *denature* the enzyme.

6.1.7.1 Reversible Inhibition

Reversible inhibition occurs rapidly in a system which is near its equilibrium point, and its extent is dependent on the concentration of enzyme, inhibitor, and substrate. It remains constant over the period when the initial reaction velocity studies are performed. In contrast, irreversible inhibition may increase with time. In simple single-substrate enzyme—catalyzed reactions, there are three main types of inhibition patterns involving reactions following the Michaelis—Menten equation: *competitive*, *uncompetitive*, and *noncompetitive* inhibitions. Competitive inhibition occurs when the inhibitor directly competes with the substrate in forming the enzyme complex. Uncompetitive inhibition involves the interaction of the inhibitor with only the enzyme—substrate complex, whereas noncompetitive inhibition occurs when the inhibitor binds to either the enzyme or the enzyme—substrate complex without affecting the binding of the substrate. The kinetic modifications of the Michaelis—Menten equation associated with the various types of inhibition are shown below.

6.1.8 Competitive Inhibition

Competitive inhibitors often closely resemble some respects of the substrate whose reactions they inhibit and, because of this structural similarity, compete for the same binding site on the enzyme. The enzyme—inhibitor complex either lacks the appropriate reactive groups or is held in an unsuitable position with respect to the catalytic site of the enzyme, which results in a complex that does not react (i.e., gives a dead-end complex). The inhibitor must first dissociate before the true substrate may enter the enzyme and the reaction can take place. An example is malonate, which is a competitive inhibitor of the reaction catalyzed by succinate dehydrogenase. Malonate has two carboxyl groups, like the substrate, and can fill the substrate binding site on the enzyme. The subsequent reaction, however, requires that the molecule be reduced with the formation of a double bond. If malonate is the substrate, this cannot be achieved without the loss of one of the carboxyl groups and therefore no reaction occurs.

The effect of a competitive inhibitor will depend not only upon the inhibitor but also the substrate concentration and the relative affinities of the substrate and inhibitor for the enzyme. Therefore, if one considers the situation where the substrate concentration is low, then the inhibitor will successfully compete but, as the substrate concentration rises, the inhibitor will become less successful in competing with the substrate until, at very high concentrations, one would expect to see very little or no inhibition. Thus, in this situation, V_{max} remains unchanged, but K_M for the substrate is increased, as illustrated below, and the value of the new K_M is designated as K'_M.

The steady-state kinetics of a simple single-substrate, single-binding site, single-intermediate enzyme—catalyzed reaction in the presence of competitive inhibitor are shown in Scheme 6.3.

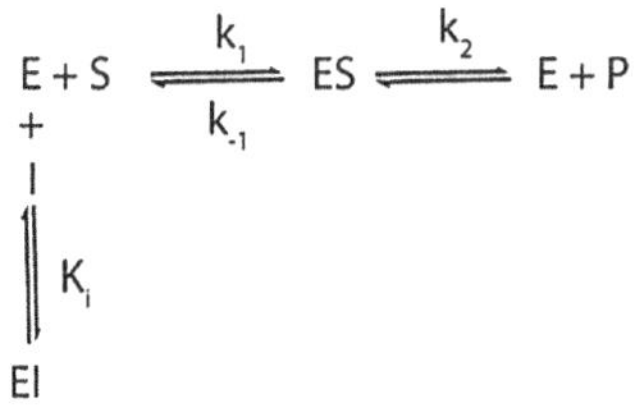

$$E + S \underset{k_{-1}}{\overset{k_1}{\rightleftharpoons}} ES \overset{k_2}{\rightleftharpoons} E + P$$
$$+$$
$$I$$
$$\Big\updownarrow K_i$$
$$EI$$

Scheme 6.3

The equilibrium constant for the reaction between E and I is K_I, where

$$K_i = \frac{[E][I]}{[EI]} \tag{6.40}$$

In this case, K_I is known as the inhibitor constant and the equilibrium between enzyme and inhibitor is almost instantaneous on mixing as $R_0 = k_2\,ES$ (the rate of reaction) is directly dependent on the concentration of the enzyme–substrate complex. The total enzyme present in the inhibitor will be

$$[E_0] = [E] + [ES] + [E][I] \tag{6.41}$$

or, from above, then

$$[E_0] = [E] + [ES] + \frac{[E][I]}{K_i} \tag{6.42}$$

$$= [E]\left(\frac{1 + [I]}{K_i}\right) + [ES] \tag{6.43}$$

$$[E] = \frac{[E_0] - [ES]}{\left(\dfrac{1 + [I]}{K_I}\right)} \tag{6.44}$$

Now substituting for E in the steady-state assumption derivation of the Michaelis–Menten constant (Eq. 6.44) gives

$$K_m = \frac{([E_0] - [ES])[S]}{\left(\dfrac{1 + [I]}{K_i}\right)[ES]} \tag{6.45}$$

Therefore

$$K_m\left(\frac{1 + [I]}{K_i}\right) = \frac{([E_0] - [ES])[S]}{[ES]} \tag{6.46}$$

and

$$[ES] = \frac{[E_0][S]}{[S] + K_m\left(\dfrac{1 + [I]}{K_i}\right)} \tag{6.47}$$

As $R_0 = k_2[ES]$, then

$$R_0 = \frac{k_2[E_0][S_0]}{[S_0] + K_m\left(\dfrac{1 + [I]}{K_i}\right)} \tag{6.48}$$

and as $V_{max} = k_2\,E_0$, then

$$R_0 = \frac{V_{max}[S_0]}{[S_0] + K_m\left(\dfrac{1 + [I]}{K_i}\right)} \tag{6.49}$$

Under conditions, where the concentrations of the inhibitor and the substrate are much greater than that of the enzyme, the above equation holds. Therefore, for simple competitive inhibition, V_{max} is unchanged, while the apparent value of K_m is given by

$$K'_m = K_m \left(1 + \frac{[I]}{K_i}\right) \tag{6.50}$$

The value of K_I is equal to the concentration of a competitive inhibitor, which gives an apparent doubling of the value of K_m. Graphically, a form of the Lineweaver—Burk plot[3] is used (see Section 6.1.6).

$$\frac{1}{R_0} = \frac{K'_m}{V_{max}} \frac{1}{[S_0]} + \frac{1}{V_{max}} \tag{6.51}$$

A plot of $1/R_0$ versus $1/S$ for the reaction in the presence of inhibitor gives the graph illustrated in Fig. 6.9.

6.1.9 Uncompetitive Inhibition

Uncompetitive inhibitors bind only to the enzyme—substrate complex and not to the free enzyme. For example, the substrate binds to the enzyme causing a conformational change, which reveals the inhibitor binding site, or it could bind directly to the enzyme-bound substrate. In neither case does the enzyme compete for the same binding site, so the inhibition cannot be overcome by increasing the substrate concentration. Scheme 6.4 illustrates this uncompetitive behavior.

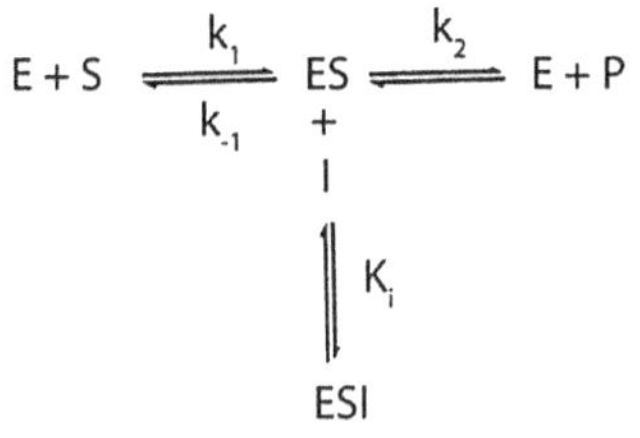

$$E + S \underset{k_{-1}}{\overset{k_1}{\rightleftharpoons}} ES \overset{k_2}{\rightleftharpoons} E + P$$

$$+$$
$$I$$

$$\Big\Updownarrow K_i$$

$$ESI$$

Scheme 6.4

ESI is the dead-end complex; the inhibitor constant $K_I = \frac{[ES][I]}{[ESI]}$

Under steady state conditions

$$K_m = \frac{[E][S]}{[ES]}$$

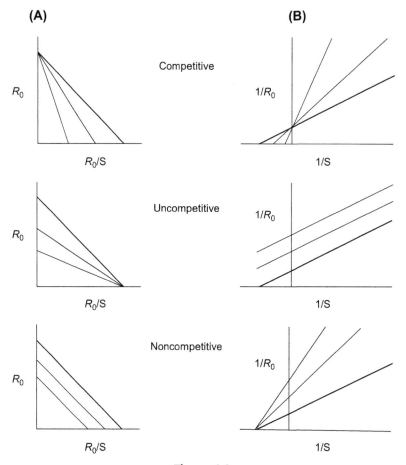

(A)

Competitive

Uncompetitive

Noncompetitive

(B)

Figure 6.9
Graphical representation of enzyme inhibition. (A) Eadie—Hofstee and (B) Lineweaver—Burk plots for different types of inhibition. The *bold line* indicates initial reaction rate in the absence of the inhibitor; the *lighter lines* show initial rates in the presence of inhibitors.

For this system

$$[E_0] = [E] + [ES] + [ESI] \tag{6.52}$$

$$= [E] + [ES] + \frac{[ES][I]}{K_i} \tag{6.53}$$

$$= [E] + [ES]\left(1 + \frac{[I]}{K_i}\right) \tag{6.54}$$

Therefore, on rearrangement and substitution the outcome is

$$R_0 = \frac{V_{max}[S]}{[S]\left(1 + \frac{[I_0]}{K_i}\right) + K_m} \tag{6.55}$$

Dividing throughout by $\left(1 + \frac{[I_0]}{K_i}\right)$ gives

$$R_0 = \frac{\dfrac{V_{max}[S_0]}{\left(1 + \frac{[I_0]}{K_i}\right)}}{S_0 + \dfrac{K_m}{\left(1 + \frac{[I_0]}{K_i}\right)}} \tag{6.56}$$

Eq. (6.56) is now in the same form as the Michaelis–Menten equation, the constants K_m and V_{max} both being divided by a factor $\left(1 + \frac{[I_0]}{K_I}\right)$.

Thus, for uncompetitive inhibition, the apparent values of both K_m and V_{max} are modified.

$$V'_{max} = \frac{V_{max}}{\left(1 + \dfrac{[I_0]}{K_i}\right)} \quad \text{and} \quad K'_m = \frac{K_m}{\left(1 + \dfrac{[I_0]}{K_i}\right)} \tag{6.57}$$

Here V'_{max} is the value of V_{max} in the presence of inhibitor of initial concentration I, of uncompetitive inhibitor, and K'_m is the apparent value of K'_m under the same conditions. An inhibitor concentration equal to K_I will halve the value of both V_{max} and K_m.

The Lineweaver–Burk equation[3] for a reaction in the presence of an uncompetitive inhibitor is

$$\frac{1}{R_0} = \frac{K'_m}{V'_{max}}\frac{1}{[S]} + \frac{1}{V'_{max}} \tag{6.58}$$

and the slope of a Lineweaver–Burk plot is equal to

$$\frac{K'_m}{V'_{max}} = \frac{K_m}{V_{max}}\frac{\left(1 + \dfrac{[I_0]}{K_i}\right)}{\left(1 + \dfrac{[I_0]}{K_m}\right)} = \frac{K_m}{V_{max}} \tag{6.59}$$

The slope of the Lineweaver–Burk plot is not altered by the presence of an uncompetitive inhibitor, but both intercepts change (Fig. 6.9).

6.1.10 Noncompetitive Inhibition

Noncompetitive inhibition is more complicated than either competitive or uncompetitive inhibition. A noncompetitive inhibitor can combine with an enzyme molecule to produce a dead-end complex regardless of whether a substrate molecule is bound or not. Hence the inhibitor must bind at a different site from the substrate. The only case considered here is when the inhibitor destroys the catalytic activity of the enzyme, either binding to the catalytic site or as a result of a conformational change affecting the catalytic site, without affecting the substrate binding site. The situation for a simple single-substrate reaction is shown in Scheme 6.5.

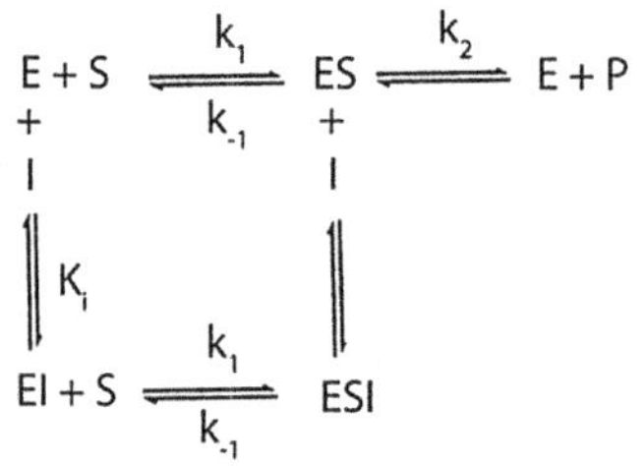

$$
\begin{array}{ccccc}
E + S & \underset{k_{-1}}{\overset{k_1}{\rightleftharpoons}} & ES & \overset{k_2}{\rightleftharpoons} & E + P \\
+ & & + & & \\
I & & I & & \\
\big\updownarrow K_i & & \big\updownarrow & & \\
EI + S & \underset{k_{-1}}{\overset{k_1}{\rightleftharpoons}} & ESI & &
\end{array}
$$

Scheme 6.5

This scheme represents a complex situation for *ES* can be arrived at by alternative routes, making it impossible for an expression of the same form as the Michaelis–Menten equation to be derived using the general steady-state assumption. However, types of noncompetitive inhibition consistent with the Michaelis–Menten type equation and a linear Lineweaver–Burk plot can occur if the rapid equilibrium assumption is valid. In the simplest possible model, involving simple linear noncompetitive inhibition, the substrate does not affect the inhibitor binding. Under these conditions, the reactions $E + I \rightarrow EI$ and $ES + I \rightarrow ESI$ are assumed to have an identical dissociation constant K_I, which is again called the inhibitor constant. The total enzyme concentration is effectively reduced by the inhibitor, decreasing the value of V_{max} but not altering K_m because neither the inhibitor nor the substrate affects the binding of the other.

In the derivation of an initial velocity equation, only the special case where $K_M = K'_S$ is being considered. As before

$$
\frac{[E][S]}{[ES]} = K_m
$$

In the presence of the noncompetitive inhibitor, which will bind equally well to E or to ES

$$
K_i = \frac{[E][I]}{[EI]} = \frac{[ES][I]}{[ESI]} \tag{6.60}
$$

But

$$
[E_0] = [E] + [ES] + [EI] + [ESI] \tag{6.61}
$$

$$
= [E] + [ES] + \frac{[E][I]}{K_i} + \frac{[ES][I]}{K_i} \tag{6.62}
$$

$$
= ([E] + [ES])\left(1 + \frac{[I]}{K_i}\right) \tag{6.63}
$$

and so

$$R_0 = \frac{V_{max}}{\left(1 + \frac{[I_0]}{K_i}\right)} \frac{[S_0]}{([S_0] + K_m)} \tag{6.64}$$

This is in the form of the Michaelis–Menten equation with V_{max} being divided by the factor $(1 + (I/K_I))$. Thus, for a simple linear noncompetitive inhibitor, K_m remains unchanged while V_{max} is altered so that

$$V'_{max} = \frac{V_{max}}{\left(1 + \frac{[I_0]}{K_i}\right)} \tag{6.65}$$

or

$$\frac{1}{V'_{max}} = \frac{1}{V_{max}} \frac{(1 + [I_0])}{K_i} \tag{6.66}$$

where V'_{max} is the value of V_{max} in the presence of concentration I of an inhibitor concentration.

The Lineweaver–Burk equation for simple linear noncompetitive inhibition is

$$\frac{1}{R_0} = \frac{K_m}{V'_{max}} \frac{1}{[S_0]} + \frac{1}{V'_{max}} \tag{6.67}$$

Fig. 6.9 shows a plot of such an inhibition pattern. There are few clear-cut examples of noncompetitive inhibition of a single-substrate reaction, as might be expected from this special case. Normally the inhibitor constants in Scheme 6.5 are different.

6.1.10.1 Mixed Inhibition

In mixed inhibition, a second assumption in the derivation of noncompetitive inhibition must be made—the binding of the substrate and the inhibitor to the enzyme are no longer independent of one another. With Lineweaver–Burk plots of such inhibition phenomena, the slopes and intercepts, and hence K_M and V_{max}, are all affected by the inhibitor. As a result, neither the consistent intercepts nor the constant slopes, which are characteristic of the other forms of inhibition, are obtained.

The effects of different types of inhibition for the Eadie–Hofstee and Lineweaver–Burk plots are shown in Fig. 6.9.

6.1.11 Substrate Inhibition

At very high substrate concentrations, deviations from the classical Michaelis–Menten rate law are observed. In this situation, the initial rate of a reaction increases with increasing substrate concentration until a limit is reached, after which the rate declines with increasing concentration. Substrate inhibition can cause such deviations when two molecules of substrate bind immediately, giving a catalytically inactive form. For example, with succinate dehydrogenase at very high concentrations of the succinate substrate, it is possible for two molecules of substrate to bind to the active site, and this results in nonfunctional complexes. Eq. (6.27) gives one form of modification of the Michaelis–Menten equation.

$$R_0 = \frac{V_{max}[S_0]}{K_m + [S_0] + \frac{[S_0]^2}{K_i}} \tag{6.68}$$

where K_I is the equilibrium constant for substrate in the inhibitory form.

6.1.12 The Kinetics of Two-Substrate Reactions

Most enzymes catalyze reactions involving two interacting substrates, and such reactions show much more complicated kinetics than the simple one-substrate kinetics discussed previously. Enzymes catalyzing bisubstrate reactions include transferase reactions (transferring functional groups from one of the substrates to the other), e.g., aminotransferase and phosphorylases. Another important group consists of the dehydrogenases where NAD(P) or NAD(P)H are also cosubstrates with the carbon substrates, e.g., alcohol dehydrogenase (ADH).[3]

$$CH_3\,CH_2OH + NAD \overset{ADH}{\rightleftharpoons} CH_3CHO + NADH + H^+$$

The kinetic analysis of two-substrate reactions is more complex due to the increased number of possible enzyme–substrate complexes. Consider the following reaction:

$$A + B \overset{Enzyme}{\rightleftharpoons} P + Q$$

A large number of possible enzyme–substrate complexes may form, e.g., the binary complexes *EA, EB, EP*, and *EQ* and ternary complexes *EAB, EPQ, EAQ*, and *EBQ*. Most two-substrate reactions can be grouped into two major classes, based on the reaction sequence in the two-substrate reactions, single-displacement reactions, and double-displacement reactions.

6.1.12.1 Single-Displacement Reactions

In single-displacement reactions, both substrates **A** and **B** simultaneously must be present on the active site of the enzyme to yield a ternary complex *EAB* in order that the reaction may proceed. Single-displacement reactions take place in two forms, random and ordered, and they are distinguished by the way the two substrates bind to the enzyme.

In *random* bisubstrate reactions, either substrate may bind to the enzyme first, indicating that the ternary complex (sometimes called the central complex) *EAB* can be formed equally well in two different ways.

For example,

$$E + A \rightleftharpoons EA$$
$$EA + B \rightleftharpoons EAB$$

or

$$E + B \rightleftharpoons EB$$
$$EB + A \rightleftharpoons EAB$$

In *ordered* single displacements there is a compulsory sequence for the reaction, which dictates that a specific substrate, the leading substrate, must be bound first, before the second, or following substrate, can be bound, as shown below:

$$E + A \rightleftharpoons EA \text{ and then } EA + B \rightleftharpoons EAB$$

In this case, **A** is the leading substrate. Many dehydrogenase reactions, which utilize NAD as a cosubstrate are good examples of ordered single displacements, e.g., malate dehydrogenase (MDH):

$$\text{Malate} + \text{NAD}^+ \overset{\text{MDH}}{\rightleftharpoons} \text{Oxaloacetate} + \text{NADH H}^+$$

In this reaction, NAD must bind first to yield the E−NAD$^+$ complex to which malate then combines to form a ternary complex, E−NAD−malate; the main reaction then takes place to yield NADH and oxaloacetate.

Random bisubstrate reactions can be distinguished from ordered reactions experimentally. The final reaction product can inhibit the overall reaction by competing with only the first (leading) substrate of the reaction. The reaction involving MDH outlined above is ordered and is inhibited by excess NADH, which competes with a normal leading substrate NAD$^+$ for binding to the enzyme. NADH does not, however, compete with the malate.

Such considerations are important when optimizing two-substrate reaction processes; for example, in the increasingly important area of biotransformation reactions.

6.1.12.2 Double-Displacement Reactions

In bisubstrate reactions of the double-displacement type, one substrate must be bound and one product released before the entry of the second substrate. In such reactions, the first substrate reacts with the enzyme to yield a chemically modified form of the enzyme (usually a functional group is changed) and the first product. In the second step, the functional group of the modified enzyme is transferred from the enzyme to the second substrate to form the second product. A good example is the aminotransferase class of enzymes, where an amino group is transferred from an amino acid to the enzyme, from which it is transferred to a keto acid.

6.1.12.3 Kinetic Constant Determination

Determination of the kinetic constant for a bisubstrate reaction is carried out in a similar manner to that for single-substrate reactions. This is achieved by investigating only one substrate at a time, while the other is kept at a set concentration, which is usually its saturation concentration. Thus, to determine the K_m and V_{max} of substrate **A**, **B** is kept constant at a saturating level while the reaction of **A** is investigated at different concentrations. The experimental conditions are then reversed to determine the kinetic constants of **B**. Thus, the kinetic constants for a bisubstrate reaction are determined using two separate kinetic plots, as discussed previously for the conditions where concentrations of **A** or **B** limit the rate of the reaction. Clearly, the conditions under which the rates are determined must be quoted for any determination.

6.1.12.4 The Alberty Equation

As stated above, many two-substrate reactions obey the Michaelis–Menten equation with respect to one substrate while the concentration of the other substrate remains constant. This is true of reactions involving only one site or those involving several sites provided that there is no interaction between sites. Alberty derived the following general equation for the reaction[4]:

$$AX + B \rightleftharpoons BX + A$$

$$R_0 = \frac{V_{max}[A_0][B_0]}{K_m^B[A_0] + K_m^A[B_0] + [A_0][B_0] + K_s^A K_m^B} \tag{6.69}$$

where V_{max} is the maximum possible value of R_0 when **A** and **B** are both saturating, K_m^A is the concentration of **A** which gives $1/2\ V_{max}$ when **B** is saturating, K_m^B is the concentration of **B** which gives $1/2\ V_{max}$ when **A** is saturating, and K_S^B is the dissociation constant for $E + A \rightleftharpoons EA$.

At very large concentrations of B (10–20 times K_m^B) the general equation simplifies to

$$R_0 = \frac{V_{max}[A_0]}{[A_0] + K_m^A} \tag{6.70}$$

or for very large concentrations of **A**

$$R_0 = \frac{V_{max}[B_0]}{[B_0] + K_m^B} \tag{6.71}$$

A good example of this type of simplification occurs when one considers water to be a substrate in hydrolytic reactions where the concentration of water is in vast excess (55 M = 55 kmol/m³), whereas K_m for reactions (although not determined) would be in the region of 10^{-3}–10^{-5} kmol/m³.

6.1.13 The Effects of Temperature and pH on Enzyme Activity

The two major environmental factors that significantly affect the kinetics of enzymes are pH and temperature.

6.1.13.1 The Effect of the Ionization State on Catalytic Activity

The amino acids that make up the primary structure of proteins will change their charge when the pH of the solution is altered due to their acid–base properties (Sections 5.2.1 and 5.3.6). The effects of pH on enzyme-catalyzed reactions can be complex because both K_m and V_{max} may be affected. Here, only the effects on V_{max} are considered, as this usually reflects a single constant rather than several that may be associated within the constant K_m (see Section 6.1.2). It is assumed that pH does not change the limiting step in a multistep process and that the substrate is saturating at all times.

The simplest case, therefore, involves a single ionizing side chain in the enzyme where E^- is the active form while EH^+ and E^{2-} are inactive forms.

$$\underset{\text{inactive}}{-EH^+} \overset{K_{a1}}{\rightleftharpoons} \underset{\text{active}}{-E^-} + H^+ \overset{K_{a2}}{\rightleftharpoons} \underset{\text{inactive}}{E^{2-}} + H^+$$

The acid dissociation constant K_a is given by

$$K_a = \frac{[E][H^+]}{[EH^+]} \tag{6.72a}$$

and therefore

$$\left[EH^+\right] = \frac{[E][H^+]}{K_a} \tag{6.72b}$$

The fraction f of enzyme in the active unprotonated form is given by

$$f = \frac{[E]}{[E] + [EH^+]} = \frac{K_a}{K_a + [H^+]} \tag{6.73}$$

If $(V_{max})_m$ is equal to the maximum rate which can occur for the enzyme in the unprotonated form, then at any pH, the observed V_{max} is given by

$$V_{max} = (V_{max})_m f \tag{6.74}$$

$$V_{max} = (V_{max})_m \frac{K_a}{K_a + [H^+]} \tag{6.75}$$

Similarly, for a second ionizing group which deactivates the active site, the effect on V_{max} at any pH can be shown to be

$$V_{max} = \frac{(V_{max})_m}{1 + \dfrac{H^+}{K_{a1}} + \dfrac{K_{a2}}{H^+}} \tag{6.76}$$

6.1.13.2 The Effects of Temperature on the Kinetics of the Reactions

In general, the effects of changes in temperature on the rate of an enzyme-catalyzed reaction do not provide much useful information about the mechanism of catalysis. However, as with most chemical reactions, the rate of an enzyme-catalyzed reaction increases with increasing temperature and it is therefore useful to be able to predict activity at various temperatures.

Unlike most uncatalyzed reactions as the temperature is raised the rate of an enzyme-catalyzed reaction rises to a maximum and then decreases as the protein is denatured by the heat. Temperature affects not only activity but also the stability of the enzyme because the tertiary structure is particularly susceptible to thermal damage.

For most simple reactions the activation energy is 300 MJ/kmol or less. The activation energy is very low for very fast reactions, and the rate of reaction is then affected less by temperature. However, typically the activation energy is around 50 MJ/kmol, and an increase in temperature from say 25−35°C approximately doubles the rate of reaction, whereas with higher activation energies (around 85 MJ/kmol) the rate can triple over the same temperature range.

The ratio of rate constants of a reaction at two temperatures 10°C apart is called the Q_{10} value of the reaction. This ranges from 1.7 to 2.5 in enzyme-catalyzed reactions.

In the Arrhenius equation, k can be replaced by k_{cat}, and at constant enzyme concentration, E_0 will be directly related to V_{max}.

$$V_{max} = Ae^{-E_a/RT} \tag{6.77}$$

Thus, a standard plot of V_{max} versus $1/T$ gives a straight line with a slope of $\mathbf{E}_a/\mathbf{R}$, allowing the determination of activation energy.

6.1.14 Enzyme Deactivation

Most literature on enzyme kinetics is devoted to initial rate data and the analysis of reversible effects on enzyme activity. In many applications and process settings, however, the rate at which the enzyme activity declines is of critical importance. This is especially true when considering its long-term use in continuous reactors. In such situations the economic feasibility of the process may hinge on the useful lifetime of the enzyme biocatalyst. The focus of this section is on the mechanisms and kinetics of loss of enzyme activity. It should also be recognized that the alteration of protein structure is central to the practical manipulation of proteins (e.g., precipitation, affinity and other forms of protein chromatography, and purification in general).

6.1.14.1 Protein Denaturation

As noted earlier, protein structure is stabilized by a series of weak forces, which often give rise to the properties that are functionally important (models of active sites and substrate binding are discussed above). On the other hand, because active sites involve a set of subtle molecular interactions involving weak forces, they are vulnerable and can be transformed into less active configurations by small perturbations in environmental conditions. It is therefore not surprising that a multitude of physical and chemical parameters may cause perturbations in native protein geometry and structure. Thus, enzyme deactivation rates are usually multifactorial, e.g., enzyme sensitivity to temperature varies with pH and/or ionic strength of the medium.

In most cases the de-activation caused by temperature or other single environmental factors, is a first-order decay process.

Consider the deactivation of active enzyme E_{act} to inactive form E_I, which may be described by a first-order rate constant (k_{de}).

$$E_{act} \xrightarrow{k_{de}} E_I \tag{6.78}$$

Integrating between time $t = 0$ and time t,

$$\int_{E_{act\ t}}^{E_{act\ t}} \frac{dE_{act}}{E_{act}} = \int_0^t -k_{de}dt \tag{6.79}$$

Therefore

$$\ln\frac{E_{act\ t}}{E_{act\ 0}} = -k_{de}t \tag{6.80}$$

or

$$E_{act\ t} = E_{act\ 0}\ e^{-k_{de}t} \tag{6.81}$$

With the reaction equation in this form, the half-life may be determined by setting the enzyme activity at half the initial value.

That is,

$$E_{act\ t} = \frac{E_{act\ 0}}{2} \tag{6.82}$$

Substituting from Eq. (6.81) into Eq. (6.82), then

$$\frac{E_{act\ 0}}{2} = E_{act\ 0}\, e^{-k_{de}t_{1/2}} \tag{6.83}$$

and

$$1/2 = e^{-k_{de}t_{1/2}} \tag{6.84}$$

Therefore

$$t_{1/2} = \frac{\ln 2}{k_{de}} \tag{6.85}$$

Generally, deactivation rates are determined in the absence of substrate, but enzyme deactivation rates can be considerably modified by the presence of substrate and other materials.

The effect of combining the deactivation model with the simple catalytic sequence of the Michaelis–Menten relation is shown below.

$$E_{act} + S \rightleftharpoons E_{act}S \longrightarrow E_{act} + P$$
$$E_{act} \longrightarrow E_I$$

Assuming that the deactivation process is much slower than the reaction represented in Scheme 6.1, and that enzyme E will deactivate faster than the enzyme in the bound state (i.e., ES complex), then Eq. (6.77) may be written as

$$\frac{dE_{act}}{dt} = -Ek_{de} \tag{6.86}$$

but

$$[ES] = [E_0] - [E] \quad \text{(Eq. 6.3)}$$

Replacing E_0 with E_{act} in Eq. (6.3) and eliminating E from Eq. (6.86) give

$$\frac{dE_{act}}{dt} = -k_{de}(E_{act} - ES) \tag{6.87}$$

Substitution for *ES* from Eq. (6.7) and using the definition given in Eq. (6.8) after rearranging, Eq. (6.87) becomes

$$\frac{dE_{act}}{dt} = -\frac{k_{de}E_{act}}{1 + S/K_m} \tag{6.88}$$

This implies that the rates of enzyme deactivation and of substrate conversion can be linked. If *E* and *ES* are deactivated at the same rate, then this rate will be the same as that for the substrate/enzyme preparation. Extending these notions, an enzyme will deactivate at different rates depending on which of the complex forms is present, and the overall deactivation rate will vary according to the proportions of the different forms of the enzyme that are present.

6.1.14.2 Mechanical Forces Acting on Enzymes

Mechanical forces can disturb the elaborate structure of the enzyme molecules to such a degree that deactivation can occur. The forces associated with flowing fluids, liquid films, and interfaces can all cause deactivation. The rate of denaturation is a function of both intensity and exposure time to the flow regime. Some enzymes show an ability to recover from such treatment. It should be noted that other enzymes are sensitive to shear stress and not to shear rate. This characteristic mechanical fragility of enzymes may impose limits on the fluid forces, which can be tolerated in enzyme reactors. This applies when stirring is used to increase mass transfer rates of substrate, or in membrane filtration systems where increasing flux through a membrane can be accompanied by increased fluid shear at the surface of the membrane and within membrane pores. Another mechanical force, surface tension, often causes denaturation of proteins and consequent deactivation of enzymes. Thus, foaming or frothing in protein solutions commonly results in denaturation of protein at the air—water interface.

In the processing context, a combination of mechanical forces and chemical reactions (oxidation, etc.) deactivates enzymes. It should be noted that there is at present no systematic way of improving the stability of an enzyme. Each system has unique properties, and stabilizing agents must be selected in an empirical fashion.

6.2 Immobilization of Biocatalysts

6.2.1 Need for Immobilization

The remarkable catalytic properties of enzymes make them very attractive for use in processes, where mild chemical conditions and high specificity are required. Cheese manufacture has traditionally used rennet, an enzyme prepared from calf stomach, as a specific protease which leads to the precipitation of protein from milk. "Mashing" in the malting of grain for the brewing of beer makes use of β-amylase from germinating grain to hydrolyze starch to produce sugars for the fermentation stage. In both of these examples the enzymes are not recovered from the reaction mixture and a fresh preparation is used for each batch. Similarly, in more modern enzyme reaction applications, such as in biological washing detergents, the enzyme is discarded after single use, but there are, however, situations where it may be desirable to recover the enzyme. This may be because the product is required in a pure state or that the cost of the enzyme preparation is such that single use would be uneconomic. To this end, immobilized biocatalysts have been developed where the original soluble enzyme has been modified to produce an insoluble material which can be easily recovered from the reaction mixture.

Many industrially important microorganisms tend to agglomerate during their growth and form flocs suspended in the culture medium or films that adhere to the internal surfaces of the fermenter. This tendency may or may not be advantageous to the process and is dependent on a variety of parameters such as the pH and ionic strength of the medium and the shear rate experienced in the growth vessel. In some cases, the formation of substantial flocs is essential to the proper operation of the process. In the case of the activated sludge wastewater treatment the settling properties of the flocculated microorganisms are utilized to produce a concentrated stream of biomass for the recycle. The so-called "trickling filter," also in widespread use in wastewater treatment, is reliant on the formation of a film of organisms on the surfaces of its packing material. The operation is not that of a filter, in which material would be removed on the basis of its particle size, but that of a biological reactor in which the waste material forms the substrate for the growth of the microbes in the zoogloeal film. The presence of the film provides a means of retaining a higher microbial concentration in the reactor than would be retained in a comparable stirred-tank fermenter. The formation of flocs and films for the retention of high microbial densities or to facilitate separation of microbes from the growth medium may be desirable in other instances as well. However, in some cases the microbe used may neither be amenable to the natural formation of large flocs nor adhere as surface films, and recourse may be made to the artificial immobilization of microbes.

6.2.2 Types of Immobilization

There are various methods, which have been developed for enzyme and microorganism immobilization and some of these have found commercial application. The two largest-scale industrial processes utilizing immobilized enzymes are the hydrolysis of benzyl penicillin by penicillin acylase and the isomerization of glucose to a glucose–fructose mixture by immobilized glucose isomerase. The immobilization techniques used in general may be broadly categorized as following:

1. *Physical adsorption on to an inert carrier.* The first of these methods has the advantage of requiring only mild chemical conditions so that enzyme deactivation during the

immobilization stage is minimized. The natural formation of microbial flocs and films may be considered to be in this category, although the subsequent adhesion of the microbes to the surface may not be a simple phenomenon. Special materials may be used as supports, which provide the microbes with environments that are particularly amenable to their adhesion; such materials include foam plastics, which provide conditions of low shear in their pores. The process may also be relatively cheap but it does tend to have the drawback that desorption of the enzyme may also occur readily or that the microbial film may slough and be carried into the bulk of the growth medium. The process is dependent on the nature of the specific enzyme or microbe used and its interaction with the carrier and, while it is common in the case of immobilized microbes, it has found only limited application in the case of immobilized enzymes.

2. *Inclusion in the lattices of a polymer gel or in microcapsules.* This method attempts to overcome the problem of leakage by enclosing the relatively large enzyme molecules or microbes in a tangle of polymer gel (one analogy made is that of a football trapped in a heap of brushwood), or to enclose them in a membrane, which is porous to the substrate. It is theoretically possible to immobilize any enzyme or microorganism using these methods but they too have their problems. Some leakage of the entrapped species may still occur, although this tends to be minimal, particularly in the case of microencapsulation of enzymes or respiring but nongrowing cells. The main problem is due to mass transfer limitations to the introduction of the necessarily small substrate molecules into the immobilized structure, and to the slow outward diffusion of the product of the reaction. If the substrate is itself a macromolecule, such as a protein or a polysaccharide, then it will be effectively screened from the enzyme or microbes and little or no reaction will take place.

3. *Covalent binding.* Biological catalysts may be made insoluble and hence immobilized by effectively increasing their size. This can be done either by chemically attaching them to otherwise inert carrier materials or by cross-linking the individuals to form large agglomerations of enzyme molecules or microorganisms. The chemical reagents used for the linking process are usually bifunctional, such as the carbodiimides, and many have been developed from those used in the chemical synthesis of peptides and proteins. The use of a carrier is the most economical in terms of enzyme usage because the local enzyme activity in the cross-linked enzyme will be less of a limitation than the rate of transfer of substrate to the active centers. This has the result that in many cases only about 10% of the original enzyme activity can be realized. The inert carriers used tend to be hydrophilic materials, such as cellulose and its derivatives, but in some cases the debris of the original cells has been used, the cells having been broken and then cross-linked with the enzyme and each other to form large particles. The latter technique has the advantage of missing out some of the purification steps (with their loss of total activity) which are normally associated with enzyme recovery and also avoids the need to satisfy the maintenance requirement of the living cells.

6.2.2.1 Loss of Activity

In general, when enzymes (or microbes) are immobilized for use in engineering systems, a significant decrease in overall activity is observed. The decrease may be ascribed to three effects:

1. *Loss due to deactivation of the catalytic activity by the immobilizing procedure itself.* This includes destruction of the active sites of the enzyme by the reagents used and the obstruction of the active sites by the support material.
2. *Loss of overall activity by diffusional limitation external to the immobilized system.* This refers to the apparent loss in activity when the rate of reaction is controlled by transport of the substrate from the bulk of the solution to the surface of the carrier of the immobilized biocatalyst. This is particularly important when an enzyme is attached to the surface of a carrier or the microbes form a very thin film, with negligible activity within the support.
3. *Loss of overall activity due to diffusional limitation within the immobilized catalyst matrix.* This can clearly arise when gel entrapment is being considered, but it can also occur when enzymes or microbes are covalently attached within pores in the inert carrier.

The consumption or biotransformation of substrate by immobilized microorganisms results in most cases in the growth of the microorganisms. The growth which gives rise to a significant increase of thickness in an established biofilm, occurs at a rate which is essentially slow in comparison with the rates of the diffusion processes. Simultaneously, the attrition of biofilms or flocs arising from the effects of fluid flow tends to maintain their thickness or size, and, overall, the immobilized system can be considered to be in a quasi-steady state when short time intervals are involved. The mathematical similarity of enzyme and microbial kinetics then means that a common set of equations can be used to describe the behavior of both immobilized enzymes and microbial cells. The following discussion is therefore valid for both kinds of immobilized biosystems, which are, in many respects, comparable with the conventional chemical catalysts discussed in Chapter 3.

6.2.3 Effect of Internal Diffusion Limitation

In the case of gel-entrapped biocatalysts, or where the biocatalyst has been immobilized in the pores of the carrier, then the reaction is unlikely to occur solely at the surface. Similarly, the consumption of substrate by a microbial film or floc would be expected to occur at some depth into the microbial mass. The situation is more complex than in the case of surface immobilization because, in this case, transport and reaction occur in parallel. By analogy with the case of heterogeneous catalysis, which is discussed in Chapter 3, the flux of substrate is related to the rate of reaction by the use of an effectiveness factor η. The rate of reaction is itself expressed in terms of the surface substrate concentration, which in many instances will be very close to the bulk substrate concentration. In general, the flux of substrate will be given by

$$R'' = \eta R''_m \frac{S_b}{K_x + S_b} \tag{6.89}$$

where R''_m is V''_{max} (the rate of reaction per unit volume of immobilized catalyst) for an enzyme and $\frac{\mu_m X}{Y_{X/S}}$ for the case of microbes.

The simplest case to consider is that of a uniform microbial film or of an enzyme, which is immobilized uniformly through a slab of supporting material that has infinite area but finite depth. As in the previous discussion the local rate of reaction is assumed to be described by Michaelis—Menten or Monod kinetics, so that at steady state a material balance for any point in the slab gives, using the same nomenclature as before:

$$D_e \frac{d^2 S}{dx^2} = \frac{R''_m S}{K_x + A} \tag{6.90}$$

where D_e is the effective diffusivity of the substrate in the slab and x is the distance measured from the surface of the slab.

This equation may be made dimensionless by putting $\beta = \frac{S}{K_x}$ as before and letting $z = \frac{x}{L}$, where L is the total thickness of the slab. Eq. (6.90) then becomes

$$\frac{d^2 \beta}{dz^2} = \phi^2 \frac{\beta}{1 + \beta} \tag{6.91}$$

where ϕ is the Thiele modulus defined by

$$\phi = L \sqrt{\frac{R''_m}{D_e K_m}} \tag{6.92}$$

for an immobilized enzyme system, and by

$$\phi = L \sqrt{\frac{\mu_m X}{Y_{x/S} D_e K_S}} \tag{6.93}$$

for a microbial film.

Eq. (6.91) can be solved by numerical integration using the boundary values $\beta = \beta_s$ when $z = 0$ and $\frac{d\beta}{dz} = 0$ when $z = 1$ to yield the set of curves shown in Fig. 6.10. The graph shows the overall rate of reaction normalized as $\frac{R''}{R''_m}$, in the same manner as in Fig. 6.12 for the surface-immobilized enzyme, but with ϕ as the parameter.

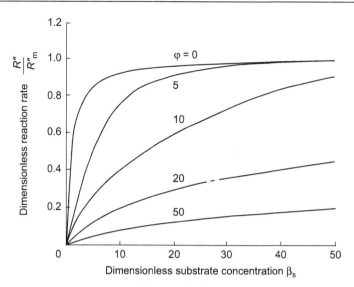

Figure 6.10
The overall rate or reaction for uniformly distributed immobilized biocatalyst in an infinite slab of carrier.

As with the external diffusion limitation, a family of curves is obtained, which shows that the overall rate of reaction decreases with increase in the Thiele modulus (as compared with the Damköhler number for the external diffusion limitation). The rate is essentially under kinetic control at low values of $\phi(\phi < 1)$; that is, there is negligible diffusion limitation. In contrast to the case of external diffusion, it may be seen from the curvature of the lines in Fig. 6.10 that the rate of reaction is always a function of the kinetic parameters, even at the higher values of ϕ.

Horvath and Engasser[5] point out that, when the curves are superficially similar to those produced by the Michaelis–Menten reaction scheme, or Monod kinetics in the case of microorganisms, the functional relationship with the surface concentration is different. This difference may be most marked when the Eadie–Hofstee type plot is considered, as shown in Fig. 6.11. This time the curves obtained for the cases where ϕ is not zero have a sigmoidal shape, even for low values of ϕ.

The effectiveness factor η may be given in terms of the modified Thiele modulus ϕ', which is defined by Atkinson[6] and Atkinson and Mavituna[7] as

$$\phi' = \frac{\beta_s}{(1 + \beta_s)} \frac{\phi}{\sqrt{2(\beta_s - \ln)(1 + \beta_s)}} \tag{6.94}$$

so that

$$\eta = 1 - \frac{\tanh \phi}{\phi} \left(\frac{\phi'}{\tanh \phi'} - 1 \right) \quad \text{for } \phi' \leq 1 \tag{6.95}$$

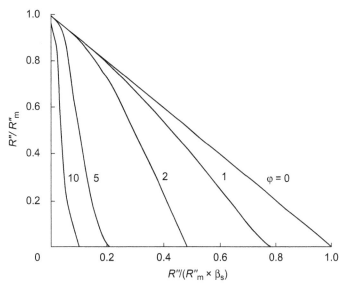

Figure 6.11
Eadie—Hofstee type plot showing departure from Michaelis—Menten kinetics due to internal diffusion limitation.

and

$$\eta = \frac{1}{\phi'} - \frac{\tanh \phi}{\phi}\left(\frac{\phi'}{\tanh \phi'} - 1\right) \quad \text{for } \phi' \geq 1 \tag{6.96}$$

For spherical particle geometry, as in the case of a microbial floc, a pellet of mold or a bead of gel-entrapped enzyme, the expression for the effectiveness factor can again be derived by a procedure similar to that used in Chapter 3 for a spherical pellet of conventional catalyst. A material balance for the substrate across an elementary shell of radius r and thickness dr within the pellet will yield

$$D_e\left(\frac{d^2S}{dr^2} + \frac{2}{r}\frac{dS}{dr}\right) = R_s'' \tag{6.97}$$

where S is the local substrate concentration and R_s'' is the local rate of reaction of the substrate per unit volume of the sphere.

Eq. (6.97) can be converted into a dimensionless form by letting

$$\overline{S} = \frac{S}{S_b} \quad \text{and} \quad \overline{r} = \frac{r}{r_p}$$

where r_p is the radius of the particle. The equation then becomes

$$\frac{d^2\overline{S}}{d\overline{r}^2} + \frac{2}{r}\frac{d\overline{S}}{d\overline{r}} = \frac{R_s'' r_p}{D_e S_b} \tag{6.98}$$

or

$$\frac{d^2\overline{S}}{d\overline{r}^2} + \frac{2}{r}\frac{d\overline{S}}{d\overline{r}} = \phi^2\frac{\overline{S}}{1 + \beta_s\overline{S}}$$

(6.99)

where $\beta_s = \frac{S_s}{K_X}$. The Thiele modulus ϕ is defined as in Eqs. (6.92) and (6.93) but with the thickness L of the film replaced by the radius r_p of the particle.

At low values of S_s the reaction will be pseudo first-order and the solution given in Chapter 3 will apply. The effectiveness factor will in that case be given by

$$\eta = \frac{1}{\phi}\left(\frac{1}{\tanh 3\phi} - \frac{1}{3\phi}\right)$$

(6.100)

Sufficiently large values of S_s will give rise to zero-order kinetics and in that case, it may be shown that

$$\eta = 1 - \left(1 - \frac{6\beta}{\phi^2}\right)$$

(6.101)

In the case where neither of these asymptotic conditions applies, that is the system shows intermediate order, then Eq. (6.99) has to be solved numerically. The boundary conditions for that solution are that $\overline{S} = 1$ when $\overline{r} = 1$ and $\frac{d\overline{S}}{d\overline{r}} = 0$ when $\overline{r} = 0$.

6.2.4 Effect of External Diffusion Limitation

If the activity of the immobilized catalyst is sufficiently high, the reaction which it mediates occurs essentially at the interface between the catalyst and the substrate solution. In the case of the surface immobilized enzyme or a thin microbial film, this will, of course, occur irrespective of the level of activity. Under these conditions the limiting process for transporting substrate from the bulk of the solution to the immobilized enzyme is molecular or convective diffusion through the layer of solution immediate to the carrier. Under steady-state conditions, the rate of reaction at the active sites is equal to the rate at which substrate arrives at the site. This material balance may be written as

$$h_D(S_b - S_s) = R'$$

(6.102)

where h_D is the mass transfer coefficient; S_b is the bulk substrate concentration; S_s is the substrate concentration at the surface; and R' is the rate of reaction per unit surface area of catalyst.

Assuming that the local rate of enzyme reaction follows Michaelis—Menten kinetics or that the microbe film follows Monod kinetics regardless of immobilization, then Eq. (6.102) becomes

$$h_D(S_b - S_s) = \frac{R'_m S_s}{K_x + S_s}$$

(6.103)

where R'_m is the maximal rate of reaction per unit surface area, and K_x is the Michaelis *or* the Monod constant.

The dimensionless substrate concentration may be defined by $\beta = \frac{S}{K_x}$ so that substitution in Eq. (6.103) and rearranging gives

$$(\beta_b - \beta_s) = \frac{\beta_s}{(1 + \beta_s)} \frac{R'_m}{h_D K_x} \qquad (6.104)$$

where the subscripts b and s refer to bulk and surface parameters as before. The dimensionless group $\frac{R'_m}{h_D K_x}$ is the Damköhler number (Da) and Eq. (6.104) may be written as

$$(\beta_b - \beta_s) = \frac{\beta_s}{(1 + \beta_s)} Da \qquad (6.105)$$

The Damköhler number represents the ratio of the maximum rate of reaction to the maximum transport rate of substrate to the surface. A large value for Da therefore indicates that the transport is the limiting step in the consumption of the substrate, whereas a small value would show that the rate of reaction is more important. Fig. 6.12 shows the variation of the rate of reaction with substrate concentration for various values of Da. The rate has been normalized by expressing it as a fraction of R'_m and it may be seen that, where there is no diffusional resistance ($Da = 0$), the form of the plot is that which would be obtained if the enzyme were used in free solution. With increase in Da the observed

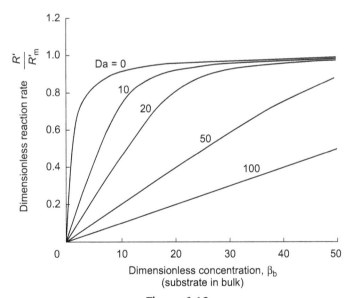

Figure 6.12

Dimensionless plot of overall reaction rate against bulk substrate concentration for a surface immobilized biocatalyst.

rate of reaction decreases and, in particular, the initial slope of the curve (Fig. 6.12) is less. This would be observed as an increase in the apparent value of the affinity constant (K_S or K_m) for the reaction if such a determination were to be carried out disregarding the immobilized nature of the biocatalyst. On the other hand, R'_m would be found to be unchanged, although difficult to determine, because the plot of rate against substrate concentration would be linear for a much greater range of concentrations. The straight line is a reflection of the dominance of diffusional effects in controlling the overall rate, because under these conditions $S_b - S_s \Rightarrow S_b$ and the rate of reaction is then given by $h_D S_b$.

If the rate of reaction obtained with no diffusional restrictions is denoted by the symbol R'_k, then an effectiveness factor may be defined as

$$\eta_e = \frac{R'}{R'_k} \tag{6.106}$$

A plot of η_e against the Damköhler number is shown in Fig. 6.13 with the bulk concentration as parameter. On the graph, it may be seen that η_e is dependent on both β_b and Da, but that three identifiable regions exist. At low values of Da, kinetic control of the reaction is observed and the curves show that η_e approaches unity for most substrate concentrations, whereas at high bulk substrate concentrations the effectiveness factor still approaches unity, even for Damköhler numbers of a 100. The diffusion-controlled domain is indicated on the diagram as the region below the broken line.

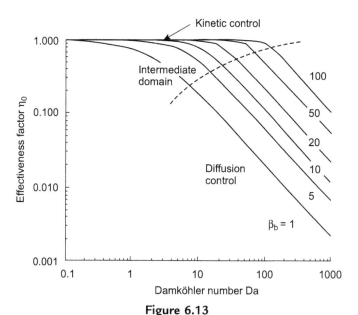

Figure 6.13

Variation of effectiveness factor η_e for a surface immobilized enzyme with Damköhler number.

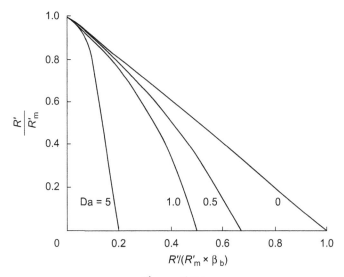

Figure 6.14
Eadie–Hofstee type plot showing departure from Michaelis–Menten kinetics due to external diffusion limitation.

The deviation of the reaction rate R' from the rectangular hyperbola, which would be shown by a true Michaelis–Menten reaction law, is best illustrated by considering the data as represented by an Eadie–Hofstee plot (Fig. 6.14). The original equation for the Michaelis–Menten or Monod kinetics,

$$R' = \frac{R'_m S}{K_x + S} \tag{6.107}$$

is converted to its dimensionless form again by dividing by R'_m and replacing S and K_x by $\beta = \frac{S}{K_x}$ to give

$$\frac{R'}{R'_m} = \frac{\beta}{1 + \beta} \tag{6.108}$$

Then, as previously, inverting the equation and multiplying through by $\frac{R'}{R'_m}$ gives, after rearranging:

$$\frac{R'}{R'_m} = -\frac{R'}{R'_m} \frac{1}{\beta} + 1 \tag{6.109}$$

Now, for Michaelis–Menten or Monod kinetics a plot of $\frac{R'}{R'_m}$ against $\frac{R'}{R'_m} \frac{1}{\beta}$ as shown in Fig. 6.7 would result in a straight line with a slope of -1.

The plot at larger values of Da shows a marked departure from this pattern with the line tending to become vertical and near to the ordinate in the extreme case.

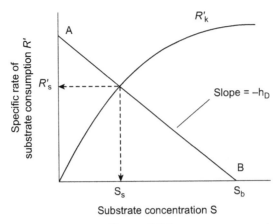

Figure 6.15
Graphical solution of Eq. (6.102).

The rate of reaction of a surface-immobilized biocatalyst may be determined graphically, as shown in Fig. 6.15. The curve marked R'_k is that relating the substrate concentration to the rate of reaction when there is no diffusional limitation (i.e., when the reaction is kinetically controlled). The intersecting line AB is drawn such that B is on the axis where $S_s = S_b$, which represents the condition when the rate of reaction is zero, and the surface substrate concentration is, as a result, the same as that in the bulk. The slope of this line is $-h_D$ and the line cuts the R' axis at the point where $R' = h_D S_b$. The point at which this line intersects the kinetic curve will correspond to the effective rate of reaction and the substrate concentration at the surface.

6.3 Microbial Stoichiometry

6.3.1 Cell Composition

Cell growth involves consumption of nutrients for the synthesis of biomass. The same organic compound can act both as the energy and carbon source for growth and product synthesis in many microorganisms. To examine the growth of cell, it is important to know what the cells are made up of and their chemical composition. Large fraction of existing biological species is made of a few elements—carbon (C), hydrogen (H), oxygen (O), and nitrogen (N). Phosphorous, Sulfur, potassium, calcium, and sodium will be present as trace elements. 70% of total cell constituent is made of water and the remaining is made of dry matter. Hence, cell composition will be typically expressed on dry basis. The content of elements present in microbes can be approximated as follows: carbon, 46%–50%; oxygen, 29%–35%; hydrogen, 6%–7%; and nitrogen, 8%–14%. The variation of element content in different microorganisms depends on the substrate and growth conditions. In general, cell can be considered as a chemical species with the formula $CH_{1.8}O_{0.5}N_{0.2}$ for stoichiometric calculations.

6.3.2 Stoichiometry of Growth and Yields

Growth of microorganisms and formation of products are complex processes that reflect the overall kinetics and stoichiometry of intracellular reactions in a cell. Stoichiometry of growth and product formation can be used for the comparison of theoretical and actual yields, formulation of growth and production medium, and checking the consistency of experimental data. Some stoichiometrically related parameters are defined to describe the growth kinetics of microorganisms. The yield coefficients are defined based on the amount of consumption of substrate or another material.

As the fermentation progresses, the amount of biomass produced is linearly related to the amount of substrate consumed. The biomass yield, $Y_{X/S}$, is expressed as

$$Y_{X/S} = -\frac{\Delta X}{\Delta S} = \frac{g \ of \ cells \ produced}{g \ of \ substrate \ consumed} \tag{6.110}$$

Yield coefficients based on other substrate or product formation may be defined as follows: Yield coefficient based on oxygen consumption:

$$Y_{X/O_2} = -\frac{\Delta X}{\Delta O_2} = \frac{g \ of \ cells \ produced}{g \ of \ oxygen \ consumed} \tag{6.111}$$

Yield coefficient based on product formation (product yield) may be defined as the amount of product formed per gram of substrate consumed:

$$Y_{P/S} = -\frac{\Delta P}{\Delta S} = \frac{g \ of \ product \ produced}{g \ of \ substrate \ consumed} \tag{6.112}$$

Biomass and product yield is influenced by many factors including composition of the medium; nature of carbon and nitrogen sources; fermentation conditions such as pH, temperature, agitation, and aeration rate. Choice of electron acceptor, e.g., oxygen, nitrate, or sulfate, also has a significant consequence on biomass yields. The biomass yield coefficient is reduced substantially in anaerobic cultures.

6.3.3 Elemental Balance

A material balance on cellular reaction can be simply written regardless of the complex and numerous intracellular reactions involved in the growth of cell, when the composition of the substrates and products is known. By assuming, incorporation of atoms of carbon, hydrogen, nitrogen, and other elements into cells or products of microbial reactions are consumed during growth, we can write the following simplified biological conversion with CO_2 and H_2O as the only extracellular products:

$$C_iH_jO_kN_l + aO_2 + bH_xO_yN_z \rightarrow cCH_\alpha O_\beta N_\delta + dCO_2 + eH_2O \tag{6.113}$$

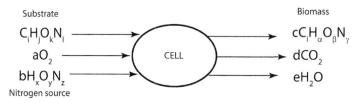

Figure 6.16
Schematic of macroscopic reaction occurring in a microbial cell.

Eq. (6.113) is written on the basis of 1 mol of substrate, where a, b, c, d, and e are stoichiometric coefficients, $C_iH_jO_kN_l$ represents the chemical formula for 1 mol of carbohydrate, $H_xO_yN_z$ is the chemical formula for the nitrogen source, and $CH_\alpha O_\beta N_\delta$ stands for 1 mol of cellular dry biomass. Because the accurate determination of cell composition is difficult, a representative cellular composition of microorganisms can be indicated as $CH_{1.8}O_{0.5}N_{0.2}$. A macroscopic cellular metabolism is depicted in Fig. 6.16. It includes the components that have net exchange with the environment and does not account for ATP and NADH, which are not subject to net exchange with environment but undergo exchange cycles within the cell and the contribution of vitamins and minerals were neglected because these compounds were considered in small quantity.

Simple elemental balances on C, H, O, and N yield the following equations:

$$\text{C balance}: \ i = c + d \tag{6.114}$$

$$\text{H balance}: \ j + bx = c\alpha + 2e \tag{6.115}$$

$$\text{O balance}: \ k + 2a + by = c\beta + 2d + e \tag{6.116}$$

$$\text{N balance}: \ l + bz = c\delta \tag{6.117}$$

The above equations can be solved simultaneously to determine the stoichiometric coefficients. Notice that there are four equations and we have five unknowns (a, b, c, d, and e) to be determined. An additional experimental information is required to determine the stoichiometric coefficients. Respiratory quotient (RQ) is defined as the ratio of moles of CO_2 produced to the moles of O_2 consumed:

$$\text{Respiratory quotient, (RQ)} = \frac{moles\ of\ CO_2\ produced}{moles\ of\ O_2\ consumed} = \frac{d}{a} \tag{6.118}$$

Eqs. constitute five equations to determine five unknowns.

The inconvenience in measuring concentration of water proffers difficulties in H and O balances. As an alternative, the reducing power or available electrons can be used to determine the quantitative relation between substrates and products.

6.3.4 Degree of Reduction

In reactions including formation of extracellular products, more information is required to determine the additional stoichiometric coefficient. Degree of reduction, γ, is defined as the number of equivalents of available electrons per gram of atom carbon in that quantity of material. These available electrons are those that can be transferred to oxygen during oxidation of a compound to CO_2, H_2O, and NH_3. The degrees of reduction for some elements are $C = 4$, $H = 1$, $N = -3$, and $O = -2$. The number of available electrons in the substrate $C_iH_jO_kN_l$ can be calculated as $(4i + j - 2k - 3l)$. The degree of reduction for the substrate, γ_s, is therefore $(4i + j - 2k - 3l)/i$. Similarly, for biomass, the degree of reduction, γ_b, for $CH_\alpha O_\beta N_\delta$ is $(4 + \alpha - 2\beta - 3\delta)$. The degree of reduction for CO_2, H_2O, and NH_3 is zero.

Degree of reduction for other substrates can be calculated as follows:

Ethanol (C_2H_5OH): $2(4) + 6(1) + 1(-2) = 12$, $\gamma = 12/2 = 6$
Glucose ($C_6H_{12}O_6$): $6(4) + 12(1) + 6(-2) = 24$, $\gamma = 24/6 = 4$
Methane (CH_4): $1(4) + 4(1) = 8$, $\gamma = 8/1 = 8$

The higher degree of reduction specifies lower degree of oxidation. For example, $\gamma_{CH_4} > \gamma_{C_2H_5OH} > \gamma_{C_6H_{12}O_6}$.

6.3.5 Electron Balance

In addition to elemental balances, electron–proton balances are required to determine the stoichiometric coefficients in bioreactions. Because water is present in excess quantity and the amount of water formed or used in bioreaction is difficult to measure, the hydrogen and oxygen balances become complicated in practical applications. In such cases, an available electron balance will be used as an additional information to determine the stoichiometric coefficients in cellular reactions. During metabolism of cell, the number of available electrons for transfer to oxygen is conserved by considering the amount of chemical element to be conserved in a balanced growth. By assuming ammonia as the nitrogen source in Eq. (6.113), the available electron balance can be written as

$$i\gamma_S - 4a = c\gamma_B$$

where γ_S and γ_B are the degrees of reduction of substrate and biomass, respectively. In addition to carbon, nitrogen, and available electron balances, parameters such as RQ and yield coefficients would be useful in evaluating the stoichiometric coefficients. Note that the coefficient, c is on a molar basis.

$$Y_{\frac{X}{S}} = \frac{g \; of \; cells \; produced}{g \; of \; substrate \; consumed} = \frac{c \; (MW \; of \; cells)}{(MW \; of \; substrate)} \qquad (6.119)$$

where MW is molecular weight, MW of cells is the molecular weight of biomass formed, and MW of substrate is molecular weight of substrate consumed.

6.3.6 Product Stoichiometry

Consider the aerobic production of a single extracellular product during growth. Eq. (6.113) can be written with the inclusion of product synthesis as given below:

$$C_iH_jO_kN_l + aO_2 + bH_xO_yN_z \rightarrow cCH_\alpha O_\beta N_\delta + dCO_2 + eH_2O + fCH_uO_vN_w \qquad (6.120)$$

where f is the stoichiometric coefficient of the product $CH_uO_vN_w$. An additional relationship between coefficients is required to determine the unknown stoichiometric coefficient of the product. Product yield coefficient $\left(Y_{\frac{p}{s}}\right)$ or product yield from substrate, which is calculated from experimental data serves as an additional parameter to determine the stoichiometric coefficients. Product yield can be calculated as follows:

$$Y_{\frac{p}{s}} = \frac{g\ of\ product\ produced}{g\ of\ substrate\ consumed} = \frac{f\ (MW\ of\ product)}{(MW of\ substrate)} \qquad (6.121)$$

where MW of product represents the molecular weight of the product formed. Eq. (6.120) holds good only for growth-associated product formation and not for production of secondary metabolites or biotransformations. Independent reaction equation must be required to represent growth and product synthesis in case of non−growth-associated product formation reaction.

6.4 Cellular Growth

6.4.1 Growth Curve and Specific Growth Rate

The mathematical description of the rate of growth of a microbial culture frequently makes use of the concept of doubling time and, by implication, an exponential growth pattern. This arises from the premise that the growth rate is directly proportional to the existing population and the proportionality constant is a function of the organism type. In most analyses the cell number, representative of population, is replaced by cell mass. This is both more easily measured and avoids problems in material balances that might arise when cell size varies with growth rate. If then, the mass of cells per unit volume of a batch culture is represented by X, exponential growth of the microbes is given by Malthus' law which may be stated as

$$\frac{dX}{dt} = \mu X \qquad (6.122)$$

where t represents time and μ is the specific growth rate of the culture. The quantity X is frequently referred to as the "microbial density" or the "biomass concentration."

Integration of this equation between the limits X_0 at time $t = 0$ and X at some time t gives

$$\ln\left(\frac{X}{X_0}\right) = \mu t \tag{6.123}$$

or

$$X = X_0\, e^{\mu t} \tag{6.124}$$

that is, the growth is exponential. If the time for the biomass concentration to double is t_d then substitution in Eq. (6.123) gives

$$\ln\left(\frac{2X_0}{X_0}\right) = \mu t_d \tag{6.125}$$

or

$$t_d = \frac{\ln 2}{\mu} \tag{6.126}$$

Exponential growth, however, occurs only for a limited time during the course of the development of a microbial culture with a fixed supply of nutrients. The classic time course of the growth of such a culture is shown in Fig. 6.17, where the logarithm of the microbial density (or alternatively microbial number) is plotted against time.

The culture passes through a series of phases characterized by the rate of growth, starting with a lag phase during which little or no increase in the microbial density occurs. This gives way during the acceleration phase to a period of exponential growth, which

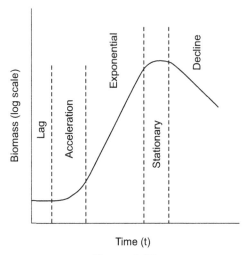

Figure 6.17
Phases of growth of a microbial culture.

continues until the food supply becomes exhausted. It then enters a stationary phase during which the organisms cease to grow and then the final decline phase in which the active mass diminishes as the cell population dies away.

The initial phase, the so-called lag phase, represents a period of time in which subtle but complicated changes occur in the internal organization of the individual cells. The cause of the delay in the development of exponential growth can be traced to a variety of factors, such as change in food type or its concentration, change in pH, or the presence of an inhibitor. Much of the theory relating to microbial growth is restricted in that it deals with cells which are fully adapted to their environment and are growing exponentially. In the case of continuous cultures this requirement is normally fulfilled, but in batch cultures the lag phase assumes more importance because it represents a significant proportion of the time over which the fermentation takes place. The mathematical modeling of the lag phase is a difficult subject, as is the collection of experimental data against which the model can be tested.

It is commonplace to account for the lag by a method attributed to Lodge and Hinshelwood.[8] This involves extrapolating the straight sections of the growth curve when plotted on a logarithmic scale as shown in Fig. 6.18. The vertical dropped from the point of intersection **P** is then used to read off the lag time on the abscissa. Clearly, this is a crude device but it gives a numerical value to the combined lag and so-called "acceleration" phases, which can be used for scheduling calculations.

Exponential growth is the most extensively quantified of the phases of microbial growth and, in the case of growth associated products, is also the most important from an industrial point of view. The vast bulk of the growth of a microbial culture occurs in this

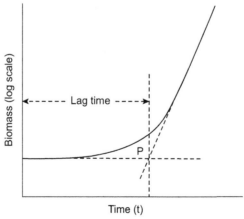

Figure 6.18
Estimation of lag time.

period and, as referred to earlier, is characterized by the periodic doubling of the biomass. This growth, however, cannot be sustained indefinitely and for one reason or another will lead to the stationary phase. Pearl and Reed[9] modified Eq. (6.122) by adding a further term to account for "inhibition" at high biomass concentration.

$$\frac{dX}{dt} = kX - k\gamma X^2 \tag{6.127}$$

The second term assumes that the "inhibition" is proportional to the square of the biomass concentration and this equation, on integration, yields the "logistic" equation:

$$X = \frac{X_0 \, e^{kt}}{1 - \gamma X_0 (1 - e^{kt})} \tag{6.128}$$

As may be seen from Fig. 6.19, a plot of X against t for this equation gives rise to a sigmoidal shaped curve with the section at smaller values of t approximating to an exponential curve.

However, while Eq. (6.128) is not based on any theory, which relates to biological observation other than that the growth curve is sigmoidal, it does serve to present data in a compact form. It can be used to describe the lag, exponential, and stationary phases of microbial growth and the constants involved can be related to some physical features of the fermentation. The constant k is, in fact, the maximal specific growth rate of the culture, and γ is the reciprocal of the final biomass concentration (X_m). The main problem

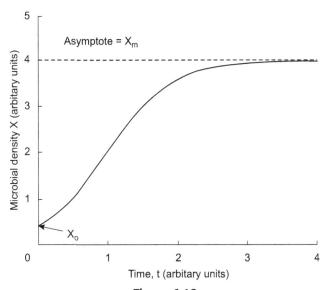

Figure 6.19

Microbial density as predicted by Eq. (6.128).

with its use is its inflexibility and its lack of cognizance of the substrate concentration. Modifications of the equation (by Edwards and Wilke[10]) have been suggested, which involve substituting for the exponential term kt in a polynomial function of the form $a_0 + a_1t + a_2t^2 + \ldots$, which, although successful in terms of accuracy, increases the number of constants required and hides the physical significance of k.

6.4.2 Microbial Growth Kinetics

6.4.2.1 Monod Kinetics

Monod[11] proposed the use of a saturation-isotherm type of equation to relate the growth rate of a microorganism culture to the prevailing feed concentration. This has become known as the "Monod" equation and is usually expressed as

$$\mu = \frac{\mu_m S}{K_S + S} \tag{6.129}$$

where, as before, μ is the specific growth rate, S is the feed or substrate concentration, μ_m is a constant sometimes known as the maximum specific growth rate, and K_S is a constant referred to as the "Monod constant." Fig. 6.20 shows the general form of the relation. Note that the microbe may well require several substrates for its growth to proceed, but it is assumed that all but one are in excess of requirements and the substance to which S relates is the *limiting* substrate component. The value of K_S will be linked to the particular limiting substrate component in any given mixture, and for different substrate mixtures it would also be expected that μ_m would change. It is usual to quote μ_m and K_S for the substance, which is the prime supplier of carbon (and energy) for the organism being grown.

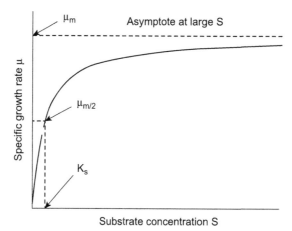

Figure 6.20
The general form of the Monod equation.

The graphical significance of the constants in the Monod equation is identical to the corresponding constants in the Michaelis−Menten relationship for enzyme kinetics (see Section 6.1.2). The specific growth rate initially increases with increasing substrate concentration and reaches a limiting value of μ_m at high substrate levels. In Eq. (6.129) when $S \gg K_S$

$$\mu = \frac{\mu_m S}{K_S + S} \Rightarrow \frac{\mu_m S}{S}$$

or

$$\mu = \mu_m \tag{6.130}$$

Also, when $\mu = \mu_m/2$, the substitution in Eq. (6.129) gives

$$\frac{\mu_m}{2} = \frac{\mu_m S}{K_S + S}$$

or, after division by μ_m and rearranging

$$K_S = S \tag{6.131}$$

i.e., when the specific growth rate is half its maximum value, the limiting substrate concentration is numerically equal to K_S.

Two possible explanations can be readily put forward as to why this form of equation should be suitable for describing the dependence of microbial growth rate on feed concentration. The first of these is that the equation has the same form as the theoretically based Michaelis−Menten equation used to describe enzyme kinetics. The chemical reactions occurring inside a microbial cell are generally mediated by enzymes, and it would be reasonable to suppose that one of these reactions is for some reason slower than the others. As a result the growth kinetics of the microorganism would be expected to reflect the kinetics of this enzyme reaction, probably modified in some way, but in essence having the form of the Michaelis−Menten equation.

The other possible cause of the saturation-isotherm type of response of specific growth rate to feed concentration is that the rate-determining step may be controlled by the transport of the feed through the boundary layers to the microorganism or through the membranes forming the cell wall itself. In these cases, even if the transport is "active," the rate equation will be of the same general form and results in the Monod type equation given above.

6.4.3 Inhibition Kinetics

The important point to note, however, is that the Monod equation is, like that of Pearl and Reed,[9] essentially empirical, and there is no reason to limit equations describing microbial growth rates to this form alone. Indeed, the use of other kinetic equations, which were originally derived for enzyme kinetics, has met with considerable success. Edwards[12] used an equation originally derived for substrate-inhibited enzyme reactions (see Eq. 6.68) to describe the growth rates of microorganisms on feed material which, at high concentrations, was inhibitory to growth. The growth of microbes on, for example, acetate is an example of the use of the rate equation:

$$\mu = \mu_m \frac{S}{K_S + S + \left[\dfrac{S^2}{K_I}\right]} \tag{6.132}$$

where K_I is the "inhibition" constant. In this case it should be observed that while the notation used is similar to that in the Monod equation the significance of the constants has changed. The form of this relation is shown in Fig. 6.21.

The maximum growth rate in this instance does not occur at the highest substrate concentration, but rather at some intermediate value. Differentiation of Eq. (6.132) with respect to S and setting the result equal to zero enables the maximum growth rate to be determined. The maximum rate, μ_{max}, is shown to occur when

$$S = \sqrt{K_S K_I} \tag{6.133}$$

and substitution of this result into Eq. (6.132) leads to

$$\mu_{max} = \frac{\mu_m}{2\sqrt{K_S/K_I} + 1} \tag{6.134}$$

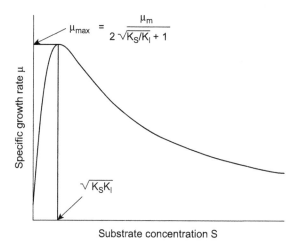

Figure 6.21
Influence of substrate inhibition on kinetics.

The success with which this equation predicts the specific growth rate of a substrate-inhibited fermentation is somewhat varied. Typically, the constants may be adjusted to represent the experimental data with a fair degree of accuracy at low substrate values. But at substrate concentrations above $\sqrt{K_S K_I}$ the predicted growth rate tends to be slightly greater than that observed in practice.

Another important case of inhibition of microbial growth is that of product inhibition. Examples of this are widespread, the effect of ethanol on the fermentation of sugars by yeast being one, and here again the Monod equation does not give an adequate estimate of the specific growth rate. The classic approach to the mathematical description is to consider the product as depressing the value of μ_m and introducing a further pair of parameters into what otherwise would be a Monod-type equation. One expression used (Aiba et al.[13]) is

$$\mu = \frac{\mu_m S}{K_S + S} e^{-k_1 P} \tag{6.135}$$

where P is the concentration of the product and k_1 is a constant. An alternative, also proposed by Aiba et al.,[14] is

$$\mu = \frac{\mu_m S}{K_S + S} \frac{K_P}{K_P + P} \tag{6.136}$$

K_P being the constant of the inhibitory system. In both cases the expressions reduce to the Monod equation when the product concentration is zero.

Table 6.1 shows some of the different expressions that have been used to describe microbial growth kinetics. (Note that the constants used in the table may not fulfill the same function and their units may be different, the table being intended to give some indication of the breadth of forms which have been employed.)

Table 6.1: Expressions used for microbial growth rates.

$\mu = \mu_m \dfrac{S}{K_S + S}$	Monod[11]
$\mu = \mu_m \dfrac{(K_c S)^a}{1 + (K_c S)^a}$	Moser[41]
$\mu = \mu_m \dfrac{1}{1 + \dfrac{K_k X}{S}}$	Contois[42]
$\mu = \mu_m \dfrac{S}{K_S + S + \left(\dfrac{S^2}{K_I}\right)}$	Edwards[41]
$\mu = \mu_m \dfrac{S}{K_S + S} \dfrac{K_P}{K_P + P}$	Aiba et al.[14]

6.4.4 Effect of Maintenance, pH, and Temperature on Growth

6.4.4.1 Maintenance

The most successful and widely used of the equations in Table 6.1 is that due to Monod. Although it may not be universally applicable, it gives a reasonable description of the variation of growth rate with substrate concentration in a surprisingly large number of cases. While it does not allow for the lag phase at the beginning of a batch process, it may be modified by the addition of one extra term to allow for the consumption of cellular material to produce maintenance energy.

$$\mu = \mu_m \frac{S}{K_S + S} - k_d \tag{6.137}$$

The constant k_d is referred to as the endogenous respiration coefficient or the specific maintenance rate. Pirt[15] points out that k_d is proportional to m, the maintenance coefficient.

$$k_d = mY_G \tag{6.138}$$

At high growth rates the endogenous respiration rate has little net effect on the formation of biomass, but in continuous culture with low nutrient levels then endogenous respiration becomes significant and has to be included in the expression for growth rate.

6.4.4.2 pH

While the kinetic equations considered are concerned with substrate and product concentrations in relation to the specific growth rate of microbial cultures, other factors may also have a profound effect on the rate. In general, microorganisms have an optimum pH for growth, usually near to neutrality, although a number of species can tolerate rather hostile environments and even grow in them. The effect of pH on the specific growth rate is conveniently represented in a manner similar to product inhibition, that is, by considering it as a modification of μ_m. Fig. 6.22 shows a typical plot of μ_m against pH (Harwood[16]).

Sinclair and Kristiansen[17] describe the use of another function, previously used to describe enzyme kinetics, to quantify this effect:

$$\mu_m(\text{pH}) = \frac{\mu_m}{1 + \dfrac{[H^+]}{K_{a1}} + \dfrac{K_{a2}}{[H^+]}} \tag{6.139}$$

where K_{a1} and K_{a2} are constants, and the hydrogen ion concentration is represented by $[H^+]$. The specific growth rate μ_m is not necessarily a maximum at the optimum pH and the expression suffers from the fact that it produces a symmetrical curve as shown in

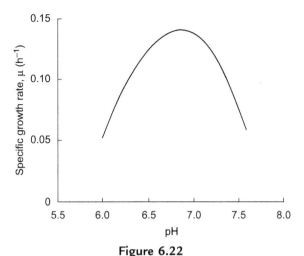

Figure 6.22

Maximal growth rate of *Methylococcus capsulatus* as a function of pH (Harwood[16]).

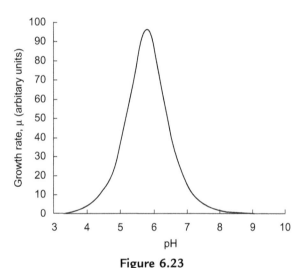

Figure 6.23

Plot of growth rate against pH as predicted by Eq. (6.139).

Fig. 6.23, a pattern not always observed in practice. Wasungu and Simard[18] show an example of this in the variation of the growth rate of *Saccharomyces cerevisiae* at 30°C (Fig. 6.24).

Much of the work carried out to investigate the effect of pH on microbial growth has been done using buffer solutions to stabilize, rather than to control, its value accurately. The internal components of a cell are protected to some extent by active transport mechanisms, as discussed in Section 5.4.6, which are able to generate large pH gradients across the cell

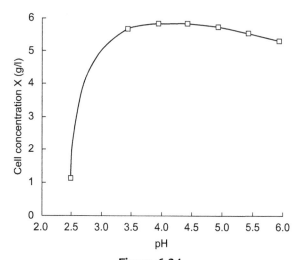

Figure 6.24
Effect of pH on the growth rate of *Saccharomyces cerevisiae* (Wasungu and Simard[18]).

wall boundary. This, clearly, represents an expenditure of energy and results in a change in the overall activity of the cell. It is to be expected then that the pH of the growth medium may have a profound effect on secondary metabolism; e.g., the rate of melanin formation in *Aspergillus nidulans* undergoes a 20-fold change when the external pH is changed from 7 to 7.9 (Rowley and Pirt[19]).

6.4.4.3 Temperature

There is considerable variation in temperature tolerance, and optima between species and three broad categories of microorganisms has been distinguished in this respect. *Psychrophiles* have temperature optima below 20°C, *mesophiles* between 20 and 45°C, and *thermophiles* have their optimum growth temperatures above 45°C (Atkinson and Mavituna[7]). Ryu and Mateles[20] showed that the growth rate of *Eschericha coli* on a nitrogen-limited glycerol medium varied according to an Arrhenius function:

$$\mu(T) = \mu_0 \exp\left\{ -\frac{\mathbf{E}_a}{\mathbf{R}}\left(\frac{1}{T} - \frac{1}{T_0}\right)\right\} \tag{6.140}$$

where $\mu(T)$ is the specific growth rate at temperature T, μ_0 is that at temperature T_0, $\mathbf{R}$ is the gas constant, and $\mathbf{E}_a$ the activation energy, in their particular case having a value of 3.2×10^3 kJ/kmol.

A simplified form of this equation, recommended for use in wastewater treatment (Metcalf and Eddy Inc.[21]) has the form

$$\mu(\theta) = \mu(20)\alpha^{(\theta-20)} \tag{6.141}$$

where $\mu(\theta)$ is the specific growth rate at $\theta°C$, $\mu(20)$ is that at $20°C$, and α is a constant in the range $1.00-1.08$ with a typical value of 1.04. It should be noted that the exponent in this equation is not dimensionless and temperature must be expressed in $°C$.

Temperature is recognized as having an effect on the growth yield, the endogenous respiration rate, and the Monod kinetic parameters K_S and μ_m. Within the temperature range of $25-40°C$, these have been shown to have dependencies which could be accounted for by Arrhenius-type exponential equations (Topiwala and Sinclair[22]). If the temperature-dependent nature of the constants has to be taken into account, Eq. (6.137) must be written as

$$\mu(T) = \mu_m(T)\frac{S}{K_S(T) + S} - k_d(T) \tag{6.142}$$

The relationships proposed were of the form

$$\mu_m(T) = A_1 \exp\left(\frac{-\mathbf{E}_{a1}}{\mathbf{R}T}\right) - A_2 \exp\left(\frac{-\mathbf{E}_{a2}}{\mathbf{R}T}\right) \tag{6.143}$$

$$\frac{1}{K_S(T)} = A_3 \exp\left(\frac{-\mathbf{E}_{a3}}{\mathbf{R}T}\right) \tag{6.144}$$

and

$$k_d(T) = A_4 \exp\left(\frac{-\mathbf{E}_{a4}}{\mathbf{R}T}\right) \tag{6.145}$$

It must be emphasized that these expressions are valid over only a relatively narrow range of temperatures, reflecting the extremely complex nature of the interacting biochemical reactions taking place within the cell. Plots of $\ln(\mu)v(1/T)$ typically give several regions of only local linearity indicating the danger of extrapolating outside the ranges investigated experimentally.

6.4.4.4 Other Factors

Other factors, which may affect the specific growth rate, include the ionic strength of the medium. In the case of the mixed culture used in wastewater treatment, an increase in salinity to that of sea water is accompanied by a sharp drop in the growth rate (Kincannon and Gaudy[23]). This, however, is a transient response and the culture recovers a large proportion of its activity when allowed to acclimatize. The effect is difficult to quantify, as it applies to a mixed culture and is a transient, but it does underline the complexity of environmental influences on growing microorganisms. For the most part therefore the mathematical treatment of biological reactors is restricted to considering the effect of substrate concentration on the growth rate, it being assumed that temperature, pH, ionic strength, and other parameters are constant.

6.4.5 Product Formation

Microbial products may be formed as a result of a variety of processes which occur within the cells of a microbial culture. In some cases, the cell itself may be the desired product, or it may be that the product is formed as the direct consequence of its growth. There is in the latter circumstances a direct link between growth and the accumulation of product. However, there are many important microbial products which are not growth associated, where there may exist mechanisms within the cells, as outlined in Section 5.4.2, which operate to produce a particular material only under certain special circumstances. As a result, the yield coefficient $Y_{\frac{P}{S}}$ is of little value in predicting the course of product accumulation, except when the overall productivity of a complete batch operation is considered.

Gaden[24] proposed a scheme, which grouped fermentations according to the manner in which the microbial product appeared in the broth (Table 6.2). While this is simpler than that put forward later by Deindoerfer,[25] it does form an useful basis from which to develop a quantitative description of the process.

Luedeking and Piret[26] proposed that the formation of microbial products could be described by a two-parameter model. This was based on their studies of lactic acid fermentation using *Lactobacillus delbruekii* in which the accumulation of the acid is not directly linked to the growth rate. The function used was

$$\frac{dP}{dt} = \delta\frac{dX}{dt} + \varepsilon X \qquad (6.146)$$

and the quantities δ and ε were found to be pH dependent. The first term on the right-hand side of the expression relates to that part of the product which *is* growth associated, while the second refers to that portion of the product which results from cellular activity associated with maintenance functions. Dividing Eq. (6.146) by the biomass concentration X gives the relationship in terms of the specific rates, i.e.,

$$\frac{1}{X}\frac{dP}{dt} = \delta\frac{1}{X}\frac{dX}{dt} + \varepsilon \qquad (6.147)$$

Table 6.2: Gaden's classification scheme for fermentations.

Fermentation Type	Characteristic	Example
1	Product formation is directly related to carbohydrate utilization	Ethanol
2	Product formation indirectly related to carbohydrate utilization	Citric acid
3	Product formation apparently not associated with carbohydrate utilization	Penicillin

and substituting for the specific growth rate

$$\frac{1}{X}\frac{dP}{dt} = \delta\mu + \varepsilon \tag{6.148}$$

The values of δ and ε may be estimated by calculating the slope and intercept of the plot of the specific rate of product formation $(1/X)(dP/dt)$ against μ. Clearly, for a purely growth-associated product, ε tend to zero but will dominate the expression for a growth-independent product.

Shu[27] devised a product formation model, which took into account the fact that the individual cells making up the growing culture are not identical and that their metabolic functions vary accordingly. The differentiation of the population in this manner gives rise to a *segregated* model of microbial culture. It is arguably a more realistic approach than that where the biomass is considered as being homogeneous, but suffers from the drawback that it is mathematically complex. In view of the apparent dependence of product formation on the age of the individual cells, Shu's model makes the age distribution the central theme.

If τ is the cell age then the instantaneous rate of product formation under constant environmental conditions is assumed to take the form

$$\frac{dP}{dt} = \sum_i A_i e^{-k_i \tau} \tag{6.149}$$

where A_i and k_i, are positive constants. The concentration of product accumulated in the broth over the lifetime of the cell per unit mass of the cell will be

$$P_t = \int_0^\tau \sum_i A_i e^{-k_i \tau} d\tau \tag{6.150}$$

If the culture contains a range of cells of all possible ages, represented by $c(\tau)$ and the culture time is t then the overall cell concentration will be

$$X_t = \int_0^t c(\tau)d\tau \tag{6.151}$$

and the overall product concentration in the fermenter becomes

$$P = \int_0^t c(\tau)\left\{\int_0^\tau \sum_i A_i e^{-k_i \tau} d\tau\right\}d\tau \tag{6.152}$$

Implementation of the model requires the choice of a suitable form for $c(\tau)$ (for example, $X_t = X_0 e^{k(t-\tau)}$ would be appropriate for exponential growth) as well as the values of A_i and k_i. Shu developed from Eq. (6.152) a double exponential expression to describe penicillin production. The formula is open to criticism for its extreme flexibility and generality but is noteworthy as being comparable with the Ludeking and Piret model in its form and performance.

6.5 Bioreactor Configurations

6.5.1 Enzyme Reactors

Enzymes are frequently used as catalysts to promote specific reactions in free solution. They are typically required in small amounts and are attractive in that they obviate both the need to provide the nutritional support which would be required for microorganisms to perform the same conversion and the possible subsequent removal of those microbes. Furthermore, the enzyme need not necessarily be of microbial origin so that a wider choice of operating conditions and characteristics may be available.

In a batch reactor where substrate is being converted to product by the enzyme, a material balance for the substrate gives

$$-\left\{ \begin{array}{c} \text{Rate of} \\ \text{formation of} \\ \text{substrate} \end{array} \right\} = \left\{ \begin{array}{c} \text{Rate of} \\ \text{enzyme} \\ \text{reaction} \end{array} \right\}$$

If the substrate concentration is S, the volume of the reactor is V, and t is the time, then this becomes

$$-V\frac{dS}{dt} = VR_0 \qquad (6.153)$$

where R_0 is the activity of the enzyme expressed in quantity of substrate used per unit volume in unit time. Thus, if the initial concentration of substrate is S_0 (i.e., at time $t = 0$) and S is the concentration at time t, then these are related by the integral:

$$-\int_{S_0}^{S} dS = \int_{0}^{S} R_0 dt \qquad (6.154)$$

As discussed before, the rate of an enzyme-catalyzed reaction is dependent, not only on the substrate concentration at any instant, but also on the temperature, pH, and the degree of decay of that enzyme. For given values of pH and temperature, the specific rate of reaction v is given, in the simplest case, by the Michaelis–Menten equation:

$$R_0 = \frac{V_{max,t}S}{K_m + S} \qquad \text{(Eq. 6.18)}$$

and the constant $V_{max,t}$ at any time t by

$$V_{max,t} = V_{max,0} \exp(-k_{de}t) \qquad (6.155)$$

where $V_{max,0}$ is the specific activity at the beginning of the batch reaction and k_{de} is the first-order kinetic constant for the decay of the enzyme. Combining these equations gives

$$R_0 = \frac{V_{max,0}S}{K_m + S} \exp(-k_{de}t) \tag{6.156}$$

which may be substituted into the expression derived from the material balance above to give

$$-dS = \frac{V_{max,0}S}{K_m + S} \exp(-k_{de}t)dt \tag{6.157}$$

Separating the variables

$$-\int_{S_0}^{S} \frac{K_m + S}{S} dS = V_{max,0} \int_{0}^{t} \exp(-k_{de}t)dt \tag{6.158}$$

and integrating between the limits shown

$$K_m \ln\left(\frac{S_0}{S}\right) + (S_0 - S) = V_{max,0} \frac{1 - \exp(-k_{de}t)}{k_{de}} \tag{6.159}$$

In the case where the decay of the enzyme is negligible, then $k_{de} \Rightarrow 0$ and $\frac{1-\exp(-k_{de}t)}{k_{de}} \Rightarrow t$, which result in the expression

$$K_m \ln\left(\frac{S_0}{S}\right) + (S_0 - S) = V_{max}t \tag{6.160}$$

These design equations for a free-enzyme batch reaction may be formulated in a similar manner for kinetic schemes other than the case used above. If, for example, the back reaction is significant, then at infinite time an equilibrium mixture of substrate and product would be obtained.

6.5.2 Batch Growth of Microorganisms

The batch growth of microorganisms involves adding a small quantity of the microorganisms or their spores (the seed culture or inoculum) to a quantity of nutrient material in a suitable vessel. In the case of an aerobic fermentation (i.e., a growth process requiring the presence of molecular oxygen) the contents of the vessel (or fermenter) are aerated and the growth of the microorganisms is allowed to proceed. For convenience, the case where the feed material is present in aqueous solution is considered and, furthermore, it is assumed that carbon is contained in the feed and energy source, which is the limiting substrate for the growth of the culture. While for an aerobic culture aeration is of prime importance, the fact that air enters the vessel and leaves enriched in carbon dioxide will be ignored in this discussion and the analysis be focused on the changes occurring in the liquid phase.

After inoculation, assuming no lag phase, the resultant growth can be analyzed by considering the unsteady-state material balances for the substrate and biomass. The general form of this balance for a fermenter is

$$\left\{ \begin{array}{c} \text{Flow of} \\ \text{material} \\ \text{in} \end{array} \right\} + \left\{ \begin{array}{c} \text{Formation by} \\ \text{biochemical} \\ \text{reaction} \end{array} \right\} - \left\{ \begin{array}{c} \text{Flow of} \\ \text{material} \\ \text{out} \end{array} \right\} = \text{Accumulation}$$

Because a batch process is being considered, the flow in and flow out of the fermenter are both zero and the expression reduces to

$$\left\{ \begin{array}{c} \text{Formation by} \\ \text{biochemical} \\ \text{reaction} \end{array} \right\} = \text{Accumulation}$$

So, for the case of the biomass

$$R_X V = \mu V X = \frac{dX}{dt} V \tag{6.161}$$

where μ is the specific growth rate, V is the volume of the vessel, and X is the instantaneous concentration of the biomass. If Y is the overall yield coefficient for the formation of biomass and the limiting substrate concentration is S, then the equivalent expression for substrate is

$$R_S V = \frac{dS}{dt} V \tag{6.162}$$

where R_0 is the rate of conversion of substrate per unit volume of the reactor. Eq. (6.162) makes no assumptions regarding the uniformity of the yield coefficient $Y_{X/S}$, but if that can be taken to be constant, then Eq. (6.110) may be used to relate Eqs. (6.161) and (6.162). This condition is met when μ is large in comparison with m, so that, dispensing with the subscript, the differential form of Eq. (6.110) can be written as

$$Y \frac{dS}{dt} = -\frac{dX}{dt} \tag{6.163}$$

which gives

$$YR_S = -\mu X \tag{6.164}$$

Eq. (6.162) thus becomes

$$-\frac{1}{Y} \mu X = \frac{dS}{dt} \tag{6.165}$$

The yield coefficient may also be expressed in its integral form as

$$Y = \frac{X - X_0}{S_0 - S} \tag{6.166}$$

which can be rearranged as

$$S = S_0 - \frac{X - X_0}{Y} \tag{6.167}$$

If the growth follows the Monod kinetic model, then Eq. (6.129) may be substituted into Eq. (6.161) to give

$$\frac{dX}{dt} = \frac{\mu_m S X}{K_S + S} \tag{6.168}$$

The condition of the fermentation after any time t would then be given by

$$\int_{X_0}^{X} \frac{K_S + S}{\mu_m S} \frac{dX}{X} = \int_{0}^{t} dt \tag{6.169}$$

However, S is a function of X, and substitution using Eq. (6.167) must be made before carrying out the integration. The result is

$$\frac{(K_S Y + S_0 Y + X_0)}{\mu_m (Y S_0 + X_0)} \ln\left(\frac{X}{X_0}\right) + \frac{K_S Y}{\mu_m (Y S_0 + X_0)} \ln\left(\frac{Y S_0}{Y S_0 + X_0 - X}\right) = t \tag{6.170}$$

A similar expression can be obtained for the substrate concentration:

$$\frac{(K_S Y + S_0 Y + X_0)}{\mu_m (Y S_0 + X_0)} \ln\left(1 + \frac{Y(S_0 - S)}{X_0}\right) - \frac{K_S Y}{\mu_m (X_0 + Y S_0)} \ln\left(\frac{S}{S_0}\right) = t \tag{6.171}$$

These rather unwieldy equations can be used to generate a graph showing the changes in biomass and substrate concentrations during the course of a batch fermentation (see Fig. 6.25). Their main disadvantage is that they are not explicit in X and S so that a trial and error technique has to be used to determine their values at a particular value of t.

It is worth noting that the curves obtained in Fig. 6.25 show an inflexion towards the final stages of the fermentation, whereas the broken line showing the values of X generated by Eq. (6.124) shows no such characteristic and predicts that the growth would proceed to give an infinite value of X.

6.5.3 Continuous Culture of Microorganisms

The continuous growth of microorganisms, as with continuous chemical reactions, may be carried out either in tubular fermenters (plug flow) or in well-mixed tank (back-mix)

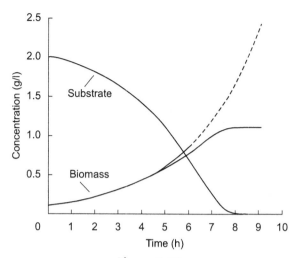

Figure 6.25
The time course of a batch fermentation as predicted by Eqs. (6.170) and (6.171). The parameters used in the calculation were $K_S = 0.015$ g/L, $\mu_m = 0.35$ h^{-1}, $Y = 0.5$, $X_0 = 0.1$ g/L, $S_0 = 2$ g/L.

fermenters. Such fermenters are idealized forms, and a practical fermenter exhibits, to a greater or lesser extent, some of the features of both forms. The continuous stirred-tank fermenter (CSTF) is characterized by containing a homogeneous liquid phase into which nutrients are continuously fed and from which the suspension of microorganisms and depleted feed are continuously removed. Because the fermenter is well mixed, samples taken from any location in the fermenter will be identical and, in particular, the composition of the exit stream will be identical to that of the liquid in the fermenter. While the CSTF can be operated as a *turbidostat*—where the feed is metered in such a way as to maintain a constant biomass concentration (as measured by its turbidity) in the fermenter—it is more commonly used as a *chemostat*. In that case, constant feed conditions are maintained and the system is allowed to attain a steady state.

If a CSTF is considered (Fig. 6.26), which has a volume V, volumetric feed flow rate F, with influent substrate and biomass concentrations S_0 and X_0, respectively, then suppose that the substrate and biomass concentrations in the fermenter are S and X. A material balance can be established over the fermenter in the same manner as for the batch fermenter. This is

$$\left\{ \begin{matrix} \text{Flow of} \\ \text{material} \\ \text{in} \end{matrix} \right\} + \left\{ \begin{matrix} \text{Formation by} \\ \text{biochemical} \\ \text{reaction} \end{matrix} \right\} - \left\{ \begin{matrix} \text{Flow of} \\ \text{material} \\ \text{out} \end{matrix} \right\} = \text{Accumulation}$$

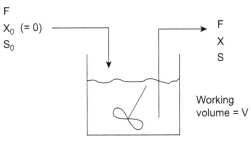

Figure 6.26
The continuous stirred-tank fermenter.

The balance may be carried out for both biomass and substrate, noting that the formation term for the substrate becomes negative because biomass is formed by the consumption of substrate. For the biomass, this becomes

$$FX_0 + R_X V - FX = \frac{dX}{dt} V \tag{6.172}$$

where R_X is the rate of formation of biomass per unit volume of the fermenter. For the substrate, the balance gives

$$FS_0 + R_S V - FS = \frac{dS}{dt} V \tag{6.173}$$

R_S being the corresponding rate of formation of substrate (negative).

Eq. (6.172) may be developed by introducing the dilution rate D, defined by

$$D = \frac{F}{V}$$

so, that at steady state, when $\frac{dX}{dt} = 0$,

$$D(X_0 - X) = -R_X \tag{6.174}$$

Now, the formation rate of biomass R_X is μX so that

$$D(X_0 - X) = -\mu X \tag{6.175}$$

or

$$D = \mu \frac{X}{X - X_0} \tag{6.176}$$

If the feed contains no microorganisms (i.e., it is sterile), then $X_0 = 0$ and the expression, for steady state, becomes

$$D = \mu \tag{6.177a}$$

It should be noted that this result is independent of the type of growth kinetics that the microorganisms follow. If the dependence of specific growth rate on substrate concentration is described by the Monod equation, then substitution from Eq. (6.129) will give for the steady-state condition:

$$D = \frac{\mu_m S}{K_S + S} \tag{6.177b}$$

and rearranging this gives the steady-state substrate concentration as

$$S = \frac{DK_S}{\mu_m - D} \tag{6.178}$$

It is interesting to note that this expression implies that the steady-state substrate concentration is independent of the feed substrate concentration S_0. The biomass concentration under this condition does, however, depend on the value of S_0. The behavior of the biomass may be deduced by similar development of Eq. (6.173) and relating R_S to μ using Eq. (6.164). More conveniently, Eq. (6.166) may be rearranged to give

$$X = X_0 + Y(S_0 - S) \tag{6.179}$$

Then, substituting for S using Eq. (6.178), so that for a CSTF at steady state utilizing sterile feed, the biomass concentration is

$$X = Y\left(S_0 - \left[\frac{DK_S}{\mu_m - D}\right]\right) \tag{6.180}$$

Example 6.1

Calculate the steady-state substrate and biomass concentrations in a continuous fermenter that has an operating volume of 25 L when the sterile feed stream contains limiting substrate at 2000 mg/L and enters the vessel at 8 L/h. The values of K_S and μ_m are 10.5 mg/L and 0.45 h^{-1}, respectively, and the yield coefficient may be taken to be 0.48.

Solution

$$\text{Dilution rate for the fermenter} = \frac{F}{V} = \frac{8.1}{25} = 0.32 \text{ h}^{-1}$$

From Eq. (6.178)

$$S = \frac{0.32 \times 10.5}{(0.45 - 0.32)} = 25.8 \text{ mg/L}$$

and substituting in Eq. (6.179) (remembering that $X_0 = 0$ for sterile feed) gives

$$X = 0 + 0.48 \times (2000 - 25.8) = \underline{948 \text{ mg/L}}$$

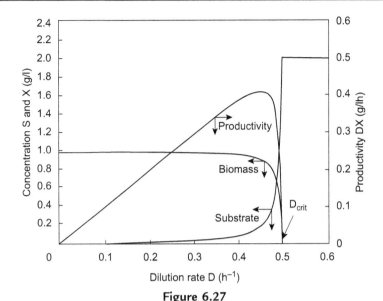

Figure 6.27
Performance curves for continuous stirred-tank fermenter at steady state.

Eqs. (6.178) and (6.180) indicate that the substrate and biomass concentrations will vary in an inverse relationship to each other. At low dilution rates, such that $D << K_S$, then S will be small and X will be large. Increasing the dilution rate will result in larger values of the steady-state substrate concentration and lower values of biomass concentration. The graph shown in Fig. 6.27 has been generated using these two equations and illustrates this behavior. In the graph it may also be seen that, as D approaches the value of μ_m, X becomes infinitesimally small. This condition is referred to as *washout* and the value of D at which it occurs is D_{crit}.

The specific production rate of biomass by the CSTF is the product of the biomass concentration and the volumetric flow rate of feed divided by the volume of the fermenter.

But

$$\frac{F}{V} = D$$

Thus

$$\text{Cell productivity } \frac{FX}{V} = DX \qquad (6.181)$$

The behavior of the cell productivity DX with varying dilution rate and Monod constant K_S is shown in Fig. 6.28. It may be seen that it becomes zero when the dilution rate is D_{crit}, and in addition to that it passes through a maximum when plotted against the

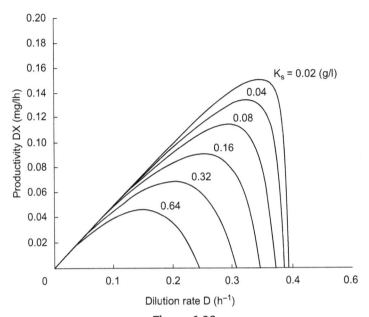

Figure 6.28
Variation of productivity DX with K_S.

dilution rate. The location of this maximum may be obtained by first multiplying Eq. (6.180) by D to give the expression for productivity:

$$DX = DY\left(S_0 - \left[\frac{DK_S}{\mu_m - D}\right]\right) \tag{6.182}$$

Now when DX is a maximum:

$$\frac{d}{dD}\left\{ DY\left(S_0 - \left[\frac{DK_S}{\mu_m - D}\right]\right)\right\} = 0 \tag{6.183}$$

so that on differentiating:

$$S_0 - \frac{(\mu_m - D_{op})2D_{op}K_S + D_{op}^2 K_S}{(\mu_m - D_{op})^2} = 0 \tag{6.184}$$

where D_{op} is the value of the dilution when the productivity is a maximum. The equation may be solved for D_{op} to give

$$D_{op} = \mu_m\left(1 - \sqrt{\frac{K_S}{K_S + S_0}}\right) \tag{6.185}$$

Eq. (6.185) represents an important design criterion for the continuous production of microbial cells, but in arriving at the equation an assumption has been made regarding the value of the yield coefficient Y. The overall yield is a function of the growth rate (and hence the dilution rate). Furthermore, there may be biological factors intervening, possibly triggered by a change in dilution rate, which might alter Y, so some caution needs to be exercised when using Eq. (6.185).

The biomass concentration prevailing in the CSTF at the condition for maximum productivity may be obtained by substitution of the expression for D_{op} into Eq. (6.180). On simplification this gives

$$X_{op} = Y\left(S_0 - \sqrt{K_S(K_S + S_0)} + K_S\right) \tag{6.186}$$

and so the productivity itself is given by

$$D_{op}X_{op} = \mu_m Y\left(1 - \sqrt{\frac{K_S}{K_S + S_0}}\right)\left(S_0 - \sqrt{K_S(K_S + S_0)} + K_S\right) \tag{6.187}$$

It is reasonable to assume that because a high productivity is being considered or sought, then the feed substrate concentration would be high, and in particular $S_0 \gg K_S$ so that the expression for maximum productivity of a CSTF reduces to

$$D_{OP}X_{OP} = \mu_m Y S_0 \tag{6.188}$$

In the case of a batch fermenter the cell productivity is given by the ratio of the biomass produced to the batch cycle time. The latter comprises the actual growth time period t_g and the "unproductive" time t_u, which includes the time required for vessel cleaning, sterilization, lag time, and harvesting time. The growth time may be calculated using Eq. (6.123), assuming again that $S_0 \gg K_S$, so that $\mu \Rightarrow \mu_m$.

Thus

$$t_g = \frac{1}{\mu_m}\ln\left(\frac{X_m}{X_0}\right) \tag{6.189}$$

where X_m is the final biomass concentration which is assumed to occur when all the substrate has been consumed, i.e., $X_m - X_0 = Y_{X/S}S_0$. The productivity may then be obtained by dividing, as outlined above, to give

$$\text{Productivity of batch fermenter} = \frac{Y_{X/S}S_0}{\frac{1}{\mu_m}\ln\left(\frac{X_m}{X_0}\right) + t_u} \tag{6.190}$$

The ratio of the CSTF productivity to that of the batch fermenter is obtained by dividing Eq. (6.188) by Eq. (6.190) to give:

$$\text{Productivity Ratio} = \frac{\mu_m Y_{X/S} S_0 \left(\dfrac{1}{\mu_m} \ln \left(\dfrac{X_m}{X_0} \right) + t_u \right)}{Y_{X/S} S_0} \tag{6.191}$$

$$= \ln \left(\frac{X_m}{X_0} \right) + t_u \mu_m$$

The continuous fermenter will be the more productive whenever this expression is greater than unity and, regardless of the turnround time t_u, this will be the case if $\frac{X_m}{X_0} > 2.27$. Typical values for $\frac{X_m}{X_0}$ lie in the range 8–10 so that, under normal circumstances, a continuous fermenter will give a higher production rate of biomass than a batch unit of the same volume. It should be noted, however, that this may not be the case for product formation where the relationship between growth and formation may not be linear. Problems associated with the maintenance of a monoculture in the fermenter may also cause a batch fermentation to be favored in practice.

Another design criterion for a CSTF is the critical dilution rate above which biomass would be removed from the fermenter at a rate faster than it could regenerate itself and ultimately lead to the absence of biomass in the reactor. This condition is referred to as *washout*. Because there is no biomass present, washout is characterized by the substrate concentration in the reactor becoming equal to that in the feed solution. Rewriting Eqs. (6.177a and b) as

$$\frac{D}{\mu_m} = \frac{S}{K_S + S} \tag{6.192}$$

and then inserting the condition $S \to S_0$ as $D \to D_{crit}$:

$$\frac{D_{crit}}{\mu_m} = \frac{S_0}{K_S + S_0} \tag{6.193}$$

The quantity $\frac{S_0}{K_S + S_0}$ will always be less than unity, and D_{crit} for a CSTF will always be less than μ_m by that factor.

The shape of the performance curve for a CSTF is dependent on the kinetic behavior of the microorganism used. In the case where the specific growth rate is described by the Monod kinetic equation, then the productivity versus dilution rate curve is given by Eq. (6.182) and has the general shape shown by the curve in Fig. 6.28. However, if the specific growth rate follows substrate inhibition kinetics and Eq. (6.132) is applicable, then at steady state, Eq. (6.176) becomes

$$D = \frac{\mu_m S}{K_S + S + S^2/K_I} \tag{6.194}$$

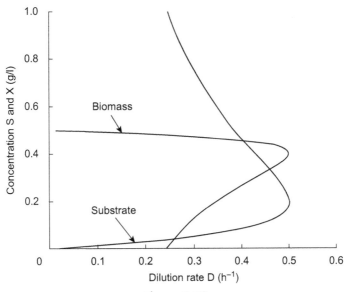

Figure 6.29
Performance curves for a continuous stirred-tank fermenter with substrate inhibition.

This equation is quadratic in S and may be written in the form

$$\frac{1}{K_I}S^2 + \left(1 - \frac{\mu_m}{D}\right)S + K_S = 0 \tag{6.195}$$

As a result, there are two possible values for the substrate concentration at a certain dilution rate and two possible corresponding biomass concentrations. Fig. 6.29 shows the performance curves produced by solving Eq. (6.195) and calculating the biomass concentration using the expression for yield coefficient.

The lower part of the biomass concentration curve in Fig. 6.29 represents a set of solutions for Eq. (6.195), which, while being a physical possibility, nevertheless represents an unstable state because any perturbation will cause the prevailing conditions either to move to a stable steady-state (with a corresponding decrease in S) or to result in washout and failure of the fermenter.

6.5.4 Stirred-Tank Reactor With Recycle of Biomass

It is frequently desirable, particularly in the field of wastewater treatment, to operate a continuous fermenter at high dilution rates. With a simple stirred-tank, this has two effects—one is that the substrate concentration in the effluent will rise, and the other is that such a system in practice tends to be unstable. One solution to this problem is to use a fermenter with a larger working volume, but an alternative strategy is to devise a method

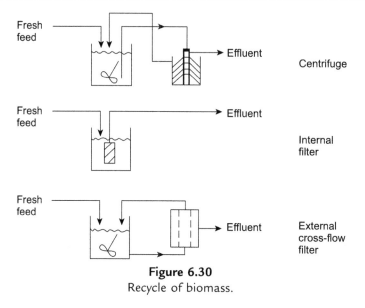

Figure 6.30
Recycle of biomass.

to retain the biomass in the fermenter while allowing the spent feed to pass out. There are several methods by which this may be achieved (see Fig. 6.30), and the net effect is the same in each case, but the analysis might vary according to the configuration. One of the most important methods is that where the effluent from the fermenter is passed to a settler—thickener and the concentrated biomass, or a portion of it, is returned to the growth vessel.

The operation of the system outlined in Fig. 6.31 is analyzed by taking material balances over the fermenter vessel. It is assumed that in this idealized case, there is no biochemical reaction or growth occurring in the separator, so that the substrate concentration S in the

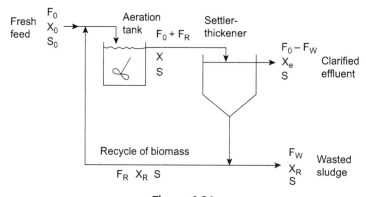

Figure 6.31
Continuous stirred-tank fermenter with settler—thickener and recycle of biomass.

entering stream is the same as that in the clarified liquid effluent stream, recycle stream, and exit biomass rich stream. The material balance then becomes

$$\left\{ \begin{array}{c} \text{Material} \\ \text{in with} \\ \text{fresh feed} \end{array} \right\} + \left\{ \begin{array}{c} \text{Material in} \\ \text{with recycle} \\ \text{stream} \end{array} \right\} + \left\{ \begin{array}{c} \text{Formation by} \\ \text{biochemical} \\ \text{reaction} \end{array} \right\} - \left\{ \begin{array}{c} \text{Output} \\ \text{to} \\ \text{separator} \end{array} \right\} = \text{Accumulation}$$

so that for the substrate

$$F_0 S_0 + F_R S + R_S V - (F_0 + F_R)S = V\frac{dS}{dt} \tag{6.196}$$

and for the biomass

$$F_0 X_0 + F_R X_R + R_X V - (F_0 + F_R)X = V\frac{dX}{dt} \tag{6.197}$$

For the biomass the rate of formation R_X is μX, and under steady-state conditions, $\frac{dX}{dt} = 0$ so that, assuming a sterile feed, Eq. (6.197) may be rewritten as

$$\frac{F_R}{F_0}X_R + \mu X \frac{V}{F_0} - X - \frac{F_R}{F_0}X = 0 \tag{6.198}$$

Now $\frac{F_R}{F_0}$ is the recycle ratio R (based on the fresh feed) and the dilution rate D, also based on the fresh feed, is $\frac{F_0}{V}$ so that substitution into Eq. (6.198) gives

$$\frac{\mu X}{D} = X(1 + R) - X_R \tag{6.199}$$

At this stage the performance of the separator may be taken into account. If the concentrating ability of the separator is represented by ξ and defined by

$$\xi = \frac{X_R}{X} \tag{6.200}$$

then division of Eq. (6.199) by X and substitution gives

$$\frac{\mu}{D} = 1 + R - \xi R \tag{6.201}$$

and after rearranging

$$D = \frac{\mu}{1 - R(\xi - 1)} \tag{6.202}$$

The denominator in Eq. (6.202) is clearly less than unity, provided the separator has any concentrating ability (i.e., works at all). As in the case for the expression linking D and μ (Eq. 6.176) for a simple CSTF, it is independent of the microbial kinetics. If $R = 0$ or $\xi = 1$, then Eq. (6.202) reduces to Eq. (6.176), corresponding to a CSTF at steady state with no recycle. The effect of the recycle is to allow the fermenter to be operated at higher dilution rates than would otherwise be possible.

If the growth rate can be described by Monod kinetics, then using Eq. (6.202) to substitute into Eq. (6.129), the substrate concentration at steady state is given by

$$S = \frac{K_S D (1 - R\xi + R)}{\mu_m - D(1 - R\xi + R)} \quad (6.203)$$

It may also be shown that under steady-state conditions Eq. (6.196) reduces to give

$$D(S_0 - S) = -R_S \quad (6.204)$$

and, if Monod kinetics are introduced with a constant yield coefficient Y, this becomes

$$D(S_0 - S) = \frac{\mu_m S X}{Y(K_S + S)} \quad (6.205)$$

which can be rearranged to give the steady-state biomass concentration as

$$X = \frac{DY(S_0 - S)(K_S + S)}{\mu_m S} \quad (6.206)$$

The conditions for washout with the recycle stream of biomass can be determined by putting $S \to S_0$ as $D \to D_{crit}$ so that with the incorporation of Monod kinetics into Eq. (6.202) gives

$$D = \frac{\mu_m S}{\{1 - R(\xi - 1)\}(K_S + S)} \quad (6.207)$$

and

$$\frac{D_{crit}}{\mu_m} = \frac{S_0}{\{1 - R(\xi - 1)\}(K_S + S_0)} \quad (6.208)$$

This confirms the result anticipated from Eq. (6.202) that the dilution rate for washout will be greater than that for the simple CSTF.

6.5.5 Stirred-Tank Reactors in Series

There are certain circumstances when it may be desirable to operate a series of stirred tanks in cascade, with the effluent of one forming the feed to the next in the chain. In the case of two such fermenters the first will behave in the manner of a simple chemostat and the performance equations of the chemostat will apply. The second tank, however, will not have a sterile feed so that some of the simplifications, which led to those relationships, will not be valid.

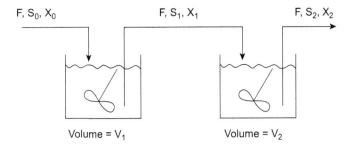

Figure 6.32
Two continuous stirred tanks in series.

As indicated in Fig. 6.32, the feed for the second fermenter will have biomass concentration X_1 and substrate concentration S_1, which, when the first tank has zero biomass in its feed (i.e., $X_0 = 0$), will be given by

$$S_1 = \frac{D_1 K_S}{\mu_m - D_1} \tag{6.209}$$

and from the definition of yield

$$X_1 = Y \left\{ S_0 - \frac{D_1 K_S}{\mu_m - D_1} \right\} \tag{6.210}$$

where D_1 is the dilution rate for the first vessel. Now a material balance for the second vessel will give

$$F X_1 + \mu_2 V_2 X_2 - F X_2 = \frac{dX_2}{dt} V_2 \tag{6.211}$$

At steady state this balance becomes

$$D_2 = \frac{\mu_2 X_2}{X_2 - X_1} \tag{6.212}$$

where D_2 is the dilution rate and μ_2 is the specific growth rate for the second tank. When Monod kinetics is incorporated into this expression it becomes

$$D_2 = \frac{\mu_m S_2}{K_S + S_2} \frac{X_2}{X_2 - X_1} \tag{6.213}$$

A material balance over both fermenters gives the biomass concentration in the second vessel as

$$X_2 = Y(S_0 - S_2)$$

so that substituting for X_2 in Eq. (6.213) gives

$$D_2 = \frac{\mu_m S_2}{K_S + S_2} \frac{(S_0 - S_2)}{\left(\dfrac{D_1 K_S}{\mu_m - D_1} - S_2\right)} \tag{6.214}$$

Solving this for S_2 results in

$$(\mu_m - D_2)S_2^2 + \left\{\frac{D_1 D_2 K_S}{(\mu_m - D_1)} - D_2 K_S - \mu_m S_0\right\}S_2 + \frac{D_1 D_2 K_S^2}{(\mu_m - D_1)} = 0 \tag{6.215}$$

Only one of the roots of this quadratic will give a positive value for the substrate concentration S_2 in the second fermenter in the train.

Example 6.2

Two CSTFs are connected in series, the first having an operational volume of 100 L and that of the second being 50 L. The feed to the first fermenter is sterile and contains 5000 mg/L of substrate, being delivered to the fermenter at 18 L/h. If the microbial growth can be described by the Monod kinetic model with $\mu_m = 0.25$ h^{-1} and $K_S = 120$ mg/L, calculate the steady-state substrate concentration in the second vessel. What would happen if the flow were from the 50-L fermenter to the 100-L fermenter?

Solution

Dilution rate for first fermenter, $D_1 = \frac{18}{100} = 0.18$ h^{-1}

Dilution rate for second fermenter, $D_2 = \frac{18}{50} = 0.36$ h^{-1}

Substitution in Eq. (6.215) gives

$$(0.25 - 0.36)S_2^2 + \left\{\frac{0.18 \times 0.36 \times 120}{(0.25 - 0.18)} - 0.36 \times 120 - 0.25 \times 5000\right\}S_2$$
$$+ \frac{0.18 \times 0.36 \times 120 \times 120}{0.25 - 0.18} = 0$$

which simplifies to

$$-0.11S_2^2 - 1182S_2 + 13,330 = 0$$

from which

Substrate concentration in second vessel $\underline{S_2 = 11-3 \text{ mg/L}}$.

If the order of the fermenters were reversed then the dilution rate for the first vessel would be above D_{crit} and washout would occur in that vessel. The feed would enter the second fermenter unchanged and containing no microorganisms so that the steady-state substrate concentration in it would be given by substitution in Eq. (6.178).

i.e.,

$$S_2 = \frac{0.18 \times 120}{(0.25 - 0.18)} = \underline{309 \text{ mg/L}}$$

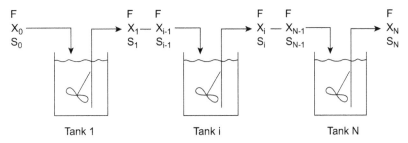

Figure 6.33
Stirred-tank fermenters in series.

The more generalized case of several CSTFs in series may be analyzed by considering N such fermenter vessels, each of volume V and with fresh feed introduced to the first tank at a volumetric flow rate F (see Fig. 6.33). Because no streams enter or leave intermediately, the flow rates between stages and of the final product will also be F.

The subscript flows on the variables indicated on Fig. 6.33 refer to the number of the tank from which that stream flow. The fresh feed has the subscript zero and is distinctive in that it may be sterile, i.e., $X_0 = 0$. In general, a material balance for biomass over the ith tank will give

$$FX_{i-1} - FX_i + V\mu_i X_i = V\frac{dX_i}{dt} \tag{6.216}$$

where μ_i is the specific growth rate in the ith fermenter. At steady state, $\frac{dX_i}{dt} = 0$, so that the above may be rearranged to give

$$X_i = \frac{DX_{i-1}}{D - \mu_i} \tag{6.217}$$

where D is the dilution rate based on the volume of each individual tank. For the last tank in the series, where $i = N$, the biomass concentration is given by

$$X_N = \frac{DX_{N-1}}{D - \mu_N} \tag{6.218}$$

and similarly, for the penultimate tank, where $i = (N - 1)$, Eq. (6.217) becomes

$$X_{N-1} = \frac{DX_{N-2}}{D - \mu_{N-1}} \tag{6.219}$$

so that, by substitution of Eq. (6.219) into Eq. (6.218), X_{N-1} may be eliminated.

Thus

$$X_N = \frac{D^2 X_{N-2}}{(D - \mu_N)(D - \mu_{N-1})} \tag{6.220}$$

This substitution may be repeated for progressively earlier tanks in the series so that the effluent concentration for N growth tanks may be expressed as

$$X_N = \frac{D^{N-1}X_1}{\prod\limits_{k=2}^{N}(D - \mu_k)} \tag{6.221}$$

Note that this expression does not involve the feed to the first fermenter, which may well operate under sterile feed conditions. In that case, and provided that the tanks are of equal size, the washout condition is analogous to the case of the single stirred-tank fermenter, that is

$$\frac{D_{crit}}{\mu_m} = \frac{S_0}{K_S + S_0} \qquad \text{(Eq. 6.193)}$$

because washout of the first tank in the series will inevitably result in washout in each successive vessel.

The performance of a set of fermenters in series may be improved by the inclusion of a recycle stream, as in the case of a single stirred-tank fermenter. The situation for N vessels each of volume V may be analyzed in a similar manner to the case for no recycle.

For a series of fermenters as shown in Fig. 6.34, the material balance for biomass over the ith fermenter gives

$$F_0(1 + R)X_{i-1} - F_0(1 + R)X_i + \mu_i X_i V = \frac{dX_i}{dt}V \tag{6.222}$$

At steady state, Eq. (6.222) reduces to

$$D(1 + R)\frac{(X_i - X_{i-1})}{X_i} = \mu_i \tag{6.223}$$

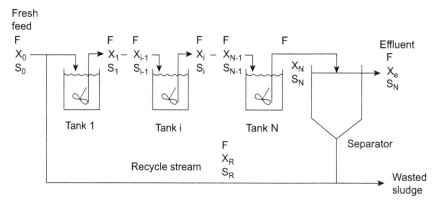

Figure 6.34
Stirred-tank fermenters in series with recycle of biomass.

Rearranging for X_i gives

$$X_i = \frac{DX_{i-1}}{D(1+R) - \mu_i}$$ (6.224)

and, by a similar argument to that used above, it may be deduced that the biomass concentration in the Nth tank will be given by

$$X_N = \frac{D^{N-1}X_1}{\prod_{k=2}^{N}(D(1+R) - \mu_i)}$$ (6.225)

In this case, however, while the fresh feed entering the system may be sterile, the biomass concentration entering the first tank is not zero. If this is represented by X_0', then

$$X_0' = X_R \frac{F_R}{F_R + F_0}$$ (6.226)

$$= X_R \frac{R}{(R+1)}$$ (6.227)

Eq. (6.227) may now be developed a stage further to give

$$X_N = \frac{D^N X_R \frac{R}{(R+1)}}{\prod_{k=1}^{N}(D(+R) - \mu_k)}$$ (6.228)

For incipient washout, in any tank, $\mu_k \Rightarrow \mu_{crit}$ as $D \Rightarrow D_{crit}$. This condition may be applied to Eq. (6.228) so that

$$X_N = \frac{D_{crit}^N X_R \frac{R}{(R+1)}}{[D_{crit}(1+R) - \mu_{crit}]^N}$$ (6.229)

or using $\xi = \frac{X_R}{X_N}$ and rearranging to give D

$$D_{crit} = \frac{\mu_{crit}}{(1+R)\left(1 - \left\{\frac{\xi R}{1+R}\right\}^{1/N}\right)}$$ (6.230)

The washout criterion is typified by $S_k \Rightarrow S_0$, then if Monod kinetics is assumed, as in the previous considerations

$$\mu_{crit} = \frac{\mu_m S_0}{K_S + S_0}$$

Then this can be substituted into Eq. (6.230) to give

$$D_{crit} = \frac{\mu_m S_0/(K_S + S_0)}{(1 + R)\left(1 - \left\{\dfrac{\xi R}{1 + R}\right\}^{1/N}\right)} \qquad (6.231)$$

It may be seen that Eq. (6.231) reduces to Eq. (6.208) when $N = 1$; that is, it agrees with the expression derived for the critical dilution rate for a single stirred-tank fermenter with recycle of biomass.

6.5.6 Plug Flow Fermenters

The plug flow tubular fermenter (PFTF) is in some respects the opposite limiting form of fermenter to the CSTF. In its idealized form, it is characterized by the fact that the liquid phase passes through the fermenter without back-mixing. The fresh feed and inoculum enter at one end of the fermenter and the mixture of feed and growing cells progress in unison toward the exit point. A small portion of the feed behaves as in a batch fermenter from its moment of entry up to the time when it leaves the fermenter, with the difference that the time coordinate of the batch fermentation must be replaced by the time taken to travel the axial distance along the fermenter tube. In practice, the fermenter may not necessarily be a tube and the idealized requirement of no back-mixing is never achieved. Analysis of the operation of the idealized fermenter may be carried out by considering an elementary volume of the fermenter as shown in Fig. 6.35.

As with the batch fermenter, the growth has to be initiated by the addition of an inoculum; in this case it is represented by the stream at a volumetric flow rate of F_I with biomass concentration X_I and substrate concentration S_I. This is mixed with the fresh-feed stream, which has a volumetric flow rate of F_0, biomass concentration X_0, and substrate concentration S_0 to produce the entry stream of flow rate F_A, biomass concentration X_A, and substrate concentration S_A.

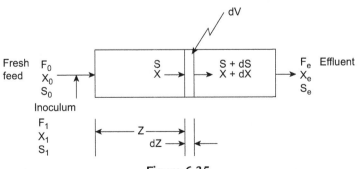

Figure 6.35
The plug-flow fermenter.

The material balance over the elementary volume gives

$$
\left\{\begin{array}{c} \text{Material} \\ \text{entering} \\ \text{by bulk} \\ \text{flow} \end{array}\right\} + \left\{\begin{array}{c} \text{Formation} \\ \text{by} \\ \text{biochemical} \\ \text{reaction} \end{array}\right\} - \left\{\begin{array}{c} \text{Material} \\ \text{leaving} \\ \text{by bulk} \\ \text{flow} \end{array}\right\} = \text{Accumulation} \tag{6.232}
$$

If the biomass concentration at the point of entry to the volume dV is X then this results in

$$
(F_0 + F_1)X + R_X dV - (F_0 + F_1)(X + dX) = \frac{dX}{dt}dV \tag{6.233}
$$

For a fermenter with cross-sectional area a_c the volume term can be replaced by $dV = a_c dZ$ and, at steady state, Eq. (6.233) becomes

$$
R_X a_C dZ - F_A dX = 0 \tag{6.234}
$$

The performance of the fermenter will be given by the integral of

$$
\int_{X_A}^{X_e} \frac{dX}{R_X} = \frac{a_C}{F_A} \int_0^Z dZ \tag{6.235}
$$

Now, the right-hand side of this equation may be readily integrated to give

$$
\int_{X_A}^{X_e} \frac{dX}{R_X} = \frac{a_C Z}{F_A} = \nu \tag{6.236}
$$

where ν is the residence time of the liquid in the fermenter. This expression is similar to that for microbial growth in a batch fermenter (see Eq. 6.169). It would be expected to give a similar integral form, but with ν replacing the fermentation time t, and the boundary conditions being altered. The boundary condition at the entry point is

$$
X_A = \frac{F_0 X_0 + F_I X_I}{F_0 + F_I} \tag{6.237}
$$

At the exit point of the reactor $X = X_e$. The incorporation of Monod kinetics for the growth rate results in

$$
\frac{(K_S Y + S_A Y + X_A)}{\mu_m (Y S_A + X_A)} \ln\left(\frac{X_e}{X_A}\right) + \frac{K_S Y}{\mu_m (Y S_A + X_A)} \ln\left(\frac{Y S_A}{Y S_A + X_A - X_e}\right) = \nu \tag{6.238}
$$

A similar expression can be obtained for the substrate concentration. The material balance for substrate will result in

$$
R_{Sa} c dZ - F_A dS = 0 \tag{6.239}
$$

and integrating

$$\int_{S_A}^{S_e} \frac{dX}{R_S} = v \tag{6.240}$$

where S_A is given by

$$S_A = \frac{F_0 S_0 + F_I S_I}{F_0 + F_I} \tag{6.241}$$

The result for Monod growth kinetics and assuming a constant yield coefficient is

$$\frac{(K_S Y + S_A Y + X_A)}{\mu_m (Y S_A + X_A)} \ln \left(1 + \frac{Y(S_A - S_e)}{X_A} \right) - \frac{K_S Y}{\mu_m (X_A - Y S_A)} \ln \left(\frac{S_e}{S_A} \right) = v \tag{6.242}$$

This integration relies on the fact that S_A and X_A are independent of the exit concentrations, there being no recycle stream in this instance. Plug flow fermenters are, however, operated with recycle of microorganisms (Fig. 6.36), and the condition is not valid for this generalized case.

When recycle of biomass is included, Eq. (6.233) becomes

$$(F_0 + F_R)X + R_X dV - (F_0 + F_R)(X + dX) = \frac{dX}{dt} V \tag{6.243}$$

and again, because $dV = a_c dZ$, at steady state Eq. (6.243) becomes

$$R_X a_c dZ - (F_0 + F_R)dX = 0 \tag{6.244}$$

or

$$R_X a_c dZ - F_0(1 + R)dX = 0 \tag{6.245}$$

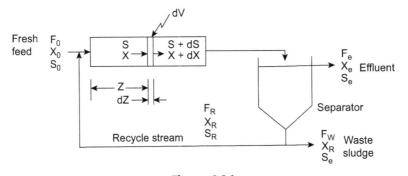

Figure 6.36
The plug-flow fermenter with recycle of biomass.

where the recycle ratio R is defined by

$$R = \frac{F_R}{F_0}$$

The performance of the reactor will then be expressed in terms of the integrated form of this equation

$$\int_{X_A}^{X_e} \frac{dX}{R_X} = \int_0^Z \frac{a_c dZ}{F_0(1+R)} \tag{6.246}$$

Similarly, for the substrate the form is

$$\int_{S_A}^{S_e} \frac{dS}{R_S} = \int_0^Z \frac{a_c dZ}{F_0(1+R)} \tag{6.247}$$

In this case the material balance for the substrate about the point of mixing of the fresh feed and the recycle stream gives

$$S_0 F_0 + S_R F_R = S_A(F_0 + F_R) \tag{6.248}$$

Thus, the exit concentration for the substrate is given by

$$S_A = \frac{S_0 + S_e R}{(1+R)} \tag{6.249}$$

Similarly, for the biomass

$$X_0 F_0 + X_R F_R = X_A(F_0 + F_R) \tag{6.250}$$

or, if $X_R = \xi X_e$, then this may be rearranged to give the other boundary condition, for the biomass as

$$X_A = X_e \frac{\xi R}{1+R} \tag{6.251}$$

The resultant equations are nonlinear, and in this general case numerical solution techniques must be used. However, there exists a special case where an analytical solution may be obtained. If the increase in biomass concentration during flow through the reactor is small then an average value for the biomass concentration, independent of the distance Z along the fermenter, may be used. The material balance for the substrate over the reactor element may then be written as

$$(F_0 + F_R)S + R_S dV - (F_0 + F_R)(S + dS) = \frac{dS}{dt} V \tag{6.252}$$

If the recycle ratio R is defined as $R = \frac{F_R}{F_0}$, then at steady state Eq. (6.252) becomes

$$-F_0(1+R)dS + R_S dV = 0 \tag{6.253}$$

and, because $dV = a_c dZ$, then

$$-F_0(1+R)dS + R_S a_c dZ = 0 \qquad (6.254)$$

For Monod kinetics and constant yield, the reaction rate with respect to the substrate R_S may be replaced by

$$R_S = -\frac{\mu_m S X_A}{Y(K_S + S)} \qquad (6.255)$$

to give

$$F_0(1+R)dS = \frac{-\mu_m S X_A a_C}{Y(K_S + S)}dZ \qquad (6.256)$$

and, because $S = S_I$ when $Z = 0$ and $S = S_e$ at the exit, then for a reactor of length Z

$$\int_{S_A}^{S} \frac{(K_S + S)dS}{S} = \frac{-\mu_m X_A a_C}{YF_0(1+R)} \int_0^Z dZ \qquad (6.257)$$

which becomes

$$(S_I - S_e) + K_S \ln\left[\frac{S_I}{S_e}\right] = \frac{\mu_m X_A a_C Z}{YF_0(1+R)} \qquad (6.258)$$

The definition of the yield coefficient may now be used to derive an expression for the biomass concentration, by substituting

$$Y = \frac{X_e - X_I}{S_I - S_e} \qquad (6.259)$$

it may be shown that

$$(X_e - X_I) = \frac{\mu_m X_A a_C Z}{F_0(1+R)} - YK_S \ln\left[\frac{S_I}{S_e}\right] \qquad (6.260)$$

The validity, or otherwise of the assumption, regarding the change in X over the length of the reactor may now be checked; that is, the difference between X_e and X_I should be small.

6.6 Estimation of Kinetic Parameters

The domination of microbial growth kinetics by the Monod equation has led to the development of techniques to determine the constants K_S and μ_m used in that equation. While μ_m is in fact dependent on other parameters, such as temperature and pH, in the usual case these are specified, and a design procedure requires values of the Monod constants under these conditions. The yield coefficient Y will also be required to link calculations of microbial growth to substrate concentrations.

The measurement of these constants may be carried out using either batch or continuous fermenter experiments, but it should be noted that there is a possibility that different results might be obtained from the different methods. This is due to the fact that the growth conditions in each case are typically very different, with the substrate concentration in the batch fermenter being much higher than that in a continuous fermenter. Complications can arise when adaptation to a particular substrate occurs, with the batch experiment necessarily spanning a transient phase with the continuous flow experiment being performed at steady state. For the present, it will be assumed that the values obtained by each method are identical, and that the constants derived using one configuration will be applicable to calculations on the performance of fermenters operating in the other configuration.

6.6.1 Use of Batch Culture Experiments

One simple, rapid approach to the problem is to perform a batch experiment and ensure that the initial substrate concentration, S_0, is very much greater than the probable value of K_S. This is not an unreasonable condition to satisfy and, for a normal batch fermentation, the substrate concentration would typically be much larger than the value of K_S for most of the growth period. In that case the Monod Eq. (6.129) reduces, as has been shown, to

$$\mu = \mu_m$$

and by combining this with Eq. (6.123)

$$\ln \left[\frac{X}{X_0} \right] = \mu_m t \tag{6.261}$$

or

$$\ln X = \mu_m t + \ln X_0 \tag{6.262}$$

This shows that a plot of $\ln X$ against t will have a slope equal to μ_m. The experiment, therefore, simply involves making measurements of the biomass concentration X at a series of times during the growth phase.

Such a plot is shown in Fig. 6.37 for which the data were generated using Eq. (6.170). The true values of the constants are therefore known and the performance of the technique may be judged by comparing its results with these. In Fig. 6.37, the generated data are represented by crosses and the dotted line shows the result of a linear regression fit. Clearly, the straight line is not a very good representation of the data, there being a marked deviation toward the end of the growth period. This is due to the fact that, by this time, the substrate concentration has declined sufficiently to become comparable with K_S and the assumption that $\mu = \mu_m$, as explained above, is no longer valid. The linear regression line for the data points tends to give an underestimate of the value of μ_m and a

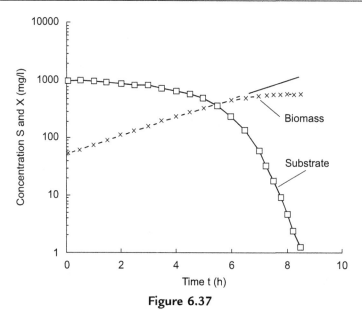

Figure 6.37
Logarithmic plot of biomass concentration against time for a batch fermentation.

more realistic value would be given by the slope of a line based on the earlier points only. This is not therefore a satisfactory method of determination; not only does a subjective assessment have to be made of the extent of the data to be taken into account, but it also ignores that part of the curve which is influenced by the value of K_S, and consequently does not offer any estimate of its value.

Gates and Marlar[28] devised a graphical procedure to determine the kinetic constants, which took into account the data in the final stages of the growth phase of a batch fermentation. As a result, their procedure can be used to estimate values of μ_m, K_S, and the yield coefficient Y. The method involves rewriting Eq. (6.171), which is the complementary expression to the integrated form of the Monod Eq. (6.170), but refers to the substrate:

$$\frac{1}{t}\ln\left[\frac{S}{S_0}\right] = \frac{1}{t}\left[1 + \frac{X_0 + YS_0}{K_S Y}\right]\ln\left[Y\frac{(S_0 - S)}{X_0} + 1\right] - \frac{(X_0 + S_0 Y)}{K_S Y}\mu_m \qquad (6.263)$$

At this stage, the following substitutions may be made:

$$a = \frac{Y}{X_0}$$

and

$$C = 1 + \frac{X_0 + YS_0}{K_S Y}$$

$$G = S_0 - S$$

and

$$R = \frac{X_0 + YS_0}{K_S Y}\mu_m$$

The expression becomes

$$\frac{1}{t}\ln\left[\frac{S}{S_0}\right] = \frac{C}{t}\ln(ag+1) - r \tag{6.264}$$

From this equation it may be seen that a plot of $\frac{1}{t}\ln\left[\frac{S}{S_0}\right]$ against $\frac{1}{t}\ln(ag+1)$, (Fig. 6.38), should produce a straight-line graph with slope f and intercept r. It is worthwhile noting at this stage that Eq. (6.263), and hence Eq. (6.264), does not explicitly use the value of X,

	□	×	◇	
a	0.0095	0.01	0.00105	1/mg
K_s	19.7	15.0	11.7	mg/l
μ_m	0.412	0.400	0.389	h⁻¹
Y	0.475	0.500	0.525	–

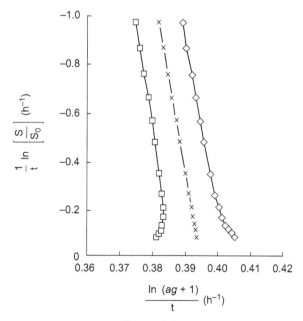

Figure 6.38

Graphical determination of the Monod kinetic constants using the method of Gates and Marlar.[28]

the biomass concentration, at various times during the fermentation. The parameter measured during the course of the experiment is the substrate concentration. Only the initial value of the biomass X_0 and one other value at a later stage in the fermentation (together with the corresponding value of S) are required for an estimate of the yield coefficient to be made. This requirement has the practical advantage that measurement of substrate concentration is normally easier and more accurate than that of biomass. A value of the yield coefficient Y is estimated from the data gathered, and a series of values computed for q and $\frac{1}{t}\ln\left[\frac{S}{S_0}\right]$. At this stage, only an approximate value of a is available, but this is sufficient for a first attempt at a solution by calculating the corresponding series of $\frac{1}{t}\ln(ag + 1)$ values and plotting the graph described above. The graph is then inspected for linearity, and at this stage, considerable curvature would be expected at values derived from the later measurements in the experiment. A trial-and-error technique is then embarked upon, with the value of a being adjusted to improve progressively the quality of the "best" straight line through the points generated.

When the optimum value of a is obtained, the kinetic parameters can be calculated from the following relationships:

$$\mu_m = \frac{r}{f - 1}$$

$$K_S = \frac{1/a + S_0}{p - 1}$$

and

$$Y = aX_0$$

Fig. 6.38 shows the result of three calculations of this nature using the same data as that used in the estimation of μ_m in Fig. 6.37. The optimum value of a in this case is 0.01 L/mg corresponding to μ_m of 0.4 h^{-1}, K_S of 15.0 mg/L, and a yield coefficient of 0.5. Changing a to 0.0095 or to 0.01,05 l/mg may be seen to give distinctly curved lines which, as such, may be rejected. The values of K_S derived from this method are, as expected, much more sensitive to the quality of the line fit than those of μ_m.

Ong[29] proposed that the best straight line, and the quality of that line could most readily be determined by its correlation coefficient. In this way, the problem would be translated into one of one-dimensional optimization, that is, maximizing the value of the correlation coefficient by adjusting the value of a in Eq. (6.264). One problem with this approach is that the method itself requires very high-quality data for reliable results to be obtained. Fig. 6.39 shows the effect of performing the same calculations as above on the same data but with a random error of $\pm 1\%$ imposed on the "measured" value of S. It now becomes

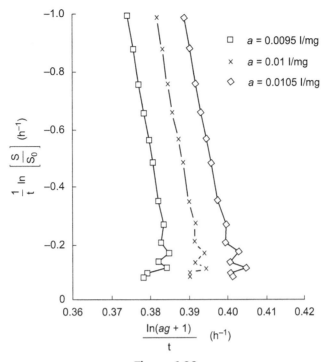

Figure 6.39
Graphical determination of the Monod kinetic constants using the method of Gates and Marlar but with error imposed on the experimental values.

quite difficult to distinguish which is the "best" straight line, and relying on the regression coefficient does in fact leads to the wrong choice in this case.

A limitation of the methods described so far is that they have assumed a constant overall yield coefficient and do not allow the endogenous respiration coefficient k_d (or alternatively the maintenance coefficient, m) to be evaluated. Eq. (6.123) shows that the overall yield, as measured when monitoring a batch reactor, is affected by the growth rate and has the greatest impact when the growth rate is low. Consequently, it is desirable to be able to estimate the values of k_d or m, so that the yield coefficient reflects the true growth yield. An equivalent method would be one where the specific rates of formation of biomass and consumption of substrate were determined independently, again without the assumption of a constant overall yield coefficient.

The approach made by Gates and Marlar is unsuitable for use with Eq. (6.137) because the result of the material balance for the biomass in the batch fermentation is

$$\frac{1}{X}\frac{dX}{dt} = \mu_m \frac{S}{K_S + S} - k_d \quad \text{(Eq. 6.137)}$$

Eq. (6.137) cannot be integrated analytically. Esener et al.,[30] however, propose a method of estimating k_d which makes use of a fed-batch fermentation. In such an experiment the growth is arranged initially to proceed in the same manner as in a batch fermentation but, at a later stage, when exponential growth has been well established, more feed is introduced into the growth vessel. This differs from a continuous fermentation in that no material is simultaneously withdrawn from the vessel. As with the simple batch fermentation, the initial substrate concentration is arranged to be much larger than the value of K_S so that the specific growth rate will be equal to μ_m. Thus, measurements of biomass concentration and substrate concentration at this stage will allow the estimation of μ_m and Y.

The experiment is then continued by adding feed to the fermentation broth, at a known flow rate, which declines in a preset manner (usually linearly). This will cause the system to come to a net zero growth rate at some stage. Fig. 6.40 shows the variation of the total biomass and substrate in the fermenter with time, noting that in this case the feed has been started at the beginning of the fermentation.

The material balance for the biomass in the fed-batch fermentation gives

$$\frac{d(XV)}{dt} = \frac{\mu_m SXV}{K_S + S} - k_d XV \tag{6.265}$$

and for the substrate

$$\frac{d(SV)}{dt} = FS_0 - \frac{\mu_m SXV}{Y(K_S + S)} \tag{6.266}$$

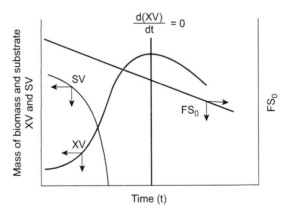

Figure 6.40
The variation of total substrate and biomass in a fed-batch fermentation.

Now, at the time when $\frac{d(XV)}{dt} = 0$ and $\frac{d(SV)}{dt} = 0$, Eq. (6.266) becomes

$$YFS_0 = \frac{\mu_m SXV}{K_S + S} \tag{6.267}$$

and combining this with Eq. (6.139) gives

$$k_d = \frac{YFS_0}{XV} \tag{6.268}$$

Hence k_d can be determined knowing the volume V at the time when $d/(XV)/dt$ is zero.

Knights[31] devised a procedure for determining the kinetic constants of a batch fermentation system that involves monitoring both the biomass and substrate concentrations, but without assuming a constant yield coefficient. Exponential growth in a batch fermenter represented by Eq. (6.122) can be written as

$$\mu = \frac{1}{X}\frac{dX}{dt} = \frac{\mu_m S}{K_S + S} \tag{6.269}$$

for the biomass, and the material balance for the substrate gives the specific rate of reaction ψ as

$$\psi = \frac{-1}{X}\frac{dS}{dt} = \frac{\psi_m S}{K_S + S} \tag{6.270}$$

The quantity ψ_m is the constant corresponding to μ_m but relating to the consumption of substrate. The analytic simultaneous solution of these equations is not possible unless the yield is constant, but the integration of Eq. (6.270) leads to

$$F + \frac{K_S}{S_0}\ln\left[\frac{1}{1-F}\right] = \frac{\psi m}{S_0}\int_0^t X dt \tag{6.271}$$

where

$$F = \frac{S_0 - S}{S_0}$$

The biomass concentration X is a function of time so that the right-hand side of Eq. (6.271) cannot be integrated directly, but the form of the integral will be that of the logistic equation (see Eq. 6.128). If, therefore, that integral is replaced by the logistic expression then, after rearranging, Eq. (6.271) becomes

$$F + \frac{K_S}{S_0}\ln\left[\frac{1}{1-F}\right] = \frac{\psi_m}{\mu_m S_0}\frac{X_0}{\chi_m}\ln\left[\chi_m e^{(\mu_m t)} - \chi_m + 1\right] \tag{6.272}$$

Here χ_m is the ratio X/X_m, where X_m is a specified maximum biomass concentration. At this stage the substitutions

$$\mathbf{F}_f = \mathbf{F} + \frac{K_S}{S_0} \ln\left[\frac{1}{1-F}\right] \tag{6.273}$$

and

$$\mathbf{T} = \ln\left[\chi_m e^{(\mu_m t)} - \chi_m + 1\right] \tag{6.274}$$

will produce

$$\mathbf{F}_f = \frac{\psi_m}{S_0} \frac{X_0}{\chi_m} \mathbf{T} \tag{6.275}$$

Thus it may be seen that a plot of $\mathbf{F}_f$ against $\mathbf{T}$ will lie on a straight line passing through the origin.

The experimental data gathered during the exponential growth phase of a batch reaction are used initially to provide an approximate value for μ_m by analyzing the $\ln(X)v$ time graph as described before. This will probably give a low estimate of μ_m, which will, however, be adequate to serve as a starting value for further calculations. As with the method of Gates and Marlar, Knights' procedure now devolves into a trial-and-error technique. A series of values is chosen for K_S/S_0 and for χ_m, and then the experimental data are used to plot a set of curves computed using Eq. (6.275). The true values of μ_m, ψ_m, and K_S are calculated using the combination of parameters, which produces the best agreement with a straight-line plot.

6.6.2 Use of Continuous Culture Experiments

In contrast to the batch fermentation—based methods of determining kinetic constants, the use of a continuous fermenter (Fig. 6.41) requires more experiments to be performed, but the analysis tends to be more straightforward. In essence, the experimental method

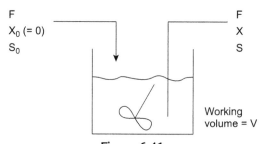

Figure 6.41
The chemostat used for kinetic parameter estimation.

involves setting up a CSTF to grow the microorganisms on a sterile feed of the required substrate. The feed flow rate is adjusted to the desired value which, of course, must produce a dilution rate below the critical value for washout, and the system is allowed to reach steady state. Careful measurements of the microbial density X, the substrate concentration S, and the flow rate F are made when a steady state has been achieved, and the operation is then repeated at a series of suitable dilution rates.

It has been shown (Eq. 6.173) that the material balance for substrate across a CSTF gives

$$FS_0 + R_S V - FS = \frac{dS}{dt} V \quad \text{(Eq. 6.173)}$$

and at steady state it reduces to

$$D(S_0 - S) = -R_S \tag{6.276}$$

If the yield coefficient Y for the conversion of substrate into microbial cells is assumed to be constant, then when Monod kinetics is applicable the material balance becomes

$$D(S_0 - S) = \frac{\mu_m S X}{Y(K_S + S)} \tag{6.277}$$

Inversion of this equation gives the linearized form

$$\frac{X}{D(S_0 - S)} = \frac{K_S Y}{\mu_m} \frac{1}{S} + \frac{Y}{\mu_m} \tag{6.278}$$

It may be seen that a plot of $\frac{X}{D(S_0-S)}$ against $\frac{1}{S}$ (Fig. 6.42) will give a straight line and evaluating its slope and intercept on the abscissa will allow the value of K_S to be determined.

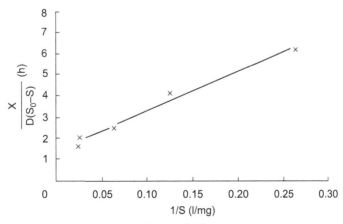

Figure 6.42
Plot of $\frac{X}{D(S_0-S)}$ against $\frac{1}{S}$ for chemostat culture.

The evaluation of μ_m first requires the value of Y to be determined. Where the maintenance requirements of the culture may be ignored, then the yield coefficient for growth is equal to the overall observed yield coefficient and will be given by

$$Y = \frac{X}{S_0 - S} \tag{6.279}$$

because the use of sterile feed means that X_0 will be zero. The mean value of Y calculated at each dilution rate used in the experiments will thereafter allow μ_m to be determined from the graph.

However, inspection of Fig. 6.27 suggests that the operation of a chemostat at dilution rates other than those approaching washout conditions results in growth at low values of substrate concentration S. For this condition the endogenous respiration rate becomes significant and $Y_G \neq Y_{X/S}$. It therefore becomes more realistic to express the specific growth rate μ as

$$\mu = \frac{\mu_m S}{K_S + S} - k_d \quad \text{(Eq. 6.137)}$$

and from Eqs. (6.177a and b)

$$D = \frac{\mu_m S}{K_S + S} - k_d \tag{6.280}$$

Eq. (6.277) may be rearranged to give

$$\frac{\mu_m S}{K_S + S} = YD \frac{(S_0 - S)}{X} \tag{6.281}$$

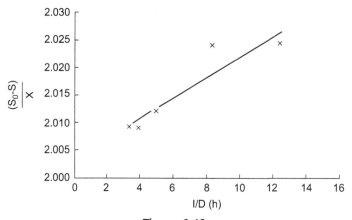

Figure 6.43
Kinetic parameter determination with chemostat.

so that substitution of this result in Eq. (6.280) produces

$$\frac{S_0 - S}{X} = \frac{k_d}{YD} + \frac{1}{Y} \qquad (6.282)$$

Thus, a plot of $\frac{S_0-S}{X}$ against $1/D$ will produce a straight-line graph as shown in Fig. 6.43 with slope k_d/Y and intercept $1/Y$. This allows the determination of both k_d and Y and, in conjunction with the calculations associated with Fig. 6.42, the complete set of kinetic parameters for the microbial growth.

Example 6.3

The steady-state substrate and biomass concentrations for a CSTF operated at various dilution rates are given below. Given that the fresh feed concentration is 700 mg/L, calculate the values of the Monod constants μ_m and K_S, the yield coefficient Y, and the endogenous respiration coefficient k_d.

Dilution Rate (h^{-1})	Substrate Concentration (mg/L)	Biomass Concentration (mg/L)
0.30	45	326
0.25	41	328
0.20	16	340
0.12	8.0	342
0.08	3.8	344

Solution

From the data given above the values of $(S_0 - S)/X$, $X/D(S_0 - S)$ may be calculated:

$\frac{1}{D}$	$\frac{S_0-S}{X}$	$\frac{1}{S}$	$\frac{X}{D(S_0-S)}$
3.33	2.0092	0.0222	1.659
4.00	2.0091	0.0244	1.991
5.00	2.0118	0.0625	2.485
8.33	2.0234	0.1250	4.118
12.50	2.0238	0.2632	6.176

Graphs can then be plotted of $(S_0 - S)/X$ against $1/D$ (Fig. 6.43) and $X/D(S_0 - S)$ (Fig. 6.42) against $1/S$ and their slopes and intercepts determined. Now for Fig. 6.43

$$\text{intercept} = \frac{1}{Y} = 2.0033$$

Therefore

$$\text{yield coe } Y = 1/2.0033 = 0.5$$

$$\text{slope} = \frac{k_d}{Y} = 0.0018(h)$$

Therefore

$$k_d = 0.0018/2.0033 = 0.0009 \text{ h}^{-1}$$

From Fig. 6.42

$$\text{intercept} = \frac{Y}{\mu} = 1.448 \text{ h}$$

Therefore

$$\mu_m = 0.50/1.448 = 0.34 \text{ h}^{-1}$$

$$\text{slope} = \frac{K_s Y}{\mu_m} = 18.48 \text{ mg/L h}$$

Therefore

$$K_S = 18.48 \times 0.34/0.5 = \underline{12.8 \text{ mg/L}}$$

6.7 Non—Steady-State Microbial Systems
6.7.1 Predator—Prey Relationships

Natural food chains frequently start at the level of the microbe and gradually larger and more sophisticated animals enter into the sequence. The starting point is usually a photosynthetic microorganism, which grows in the presence of a supply of minerals and carbon but uses light as its energy source. In a defined situation, the starting point need not be at the photosynthetic level, but in any case the next species in the chain will regard the prey as its food and energy source. In the case of the activated sludge waste-treatment process, such interactions can be very important in limiting the growth of unflocculated microorganisms. These prey microbes consume the suspended waste and are themselves consumed by a variety of predators, which also exist in the sludge.

The modeling of real food webs can be an exceedingly complicated task, but to illustrate the basic technique, a situation may be defined where a continuous stirred-tank biological reactor contains two species, one the predator and the other the prey. The food for the prey is assumed to enter as the sterile feed streams to the reactor, so that the predator may only consume the prey which grows in the reactor. Material balances can be drawn up for the process in much the same way as has been done for the earlier cases of microbial growth. Three relationships can be obtained from

$$\left\{ \begin{array}{c} \text{Material} \\ \text{in with} \\ \text{feed} \end{array} \right\} + \left\{ \begin{array}{c} \text{Rate of} \\ \text{Formation} \\ \text{by reaction} \end{array} \right\} - \left\{ \begin{array}{c} \text{Rate of} \\ \text{removal} \end{array} \right\} = \text{Accumulation}$$

which, for the substrate becomes

$$D(S_0 - S) - \frac{\mu_{mj}SX_j}{Y_j(K_{sj} + S)} = \frac{dS}{dt} \tag{6.283}$$

for the prey

$$-DX_j + \frac{\mu_{mj}SX_j}{K_{sj} + S} - \frac{\mu_{mk}X_jX_k}{Y_k(K_{sk} + X_k)} = \frac{dX_j}{dt} \tag{6.284}$$

and for the predator

$$-DX_k + \frac{\mu_{mk}X_jX_k}{K_{sk} + X_j} = \frac{dX_k}{dt} \tag{6.285}$$

where the subscript j refers to the prey and i to the predator.

For steady-state operation, the right-hand sides of these equations become zero; however, the more interesting solutions are obtained under unsteady-state conditions. The integration of the differential equations may be carried out numerically after suitable values for the constants have been chosen. Several workers (Lotka,[32] Tsuchiya et al.[33] inter alia) have studied this condition, typically using bacteria as prey and amoebae as predators. Figs. 6.44–6.46 show the results of solving the equations. Changing the conditions, particularly the feed concentration and the dilution rate, can cause the system to react differently, stable steady states being possible, as well as washout of the predator or of both species.

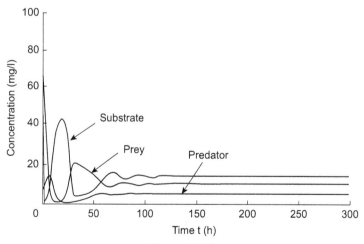

Figure 6.44
Damped oscillations in population of both species leading to steady state.

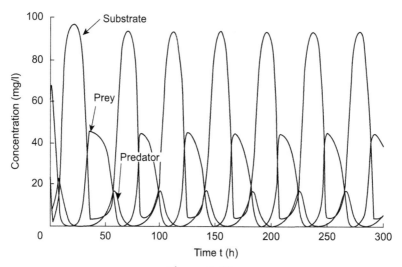

Figure 6.45
Sustained oscillations in populations.

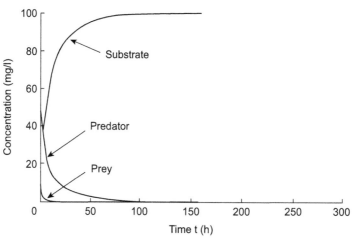

Figure 6.46
Washout of both species.

6.7.2 Structured Models

The preceding consideration of the growth of microbial cultures implies the assumption that the biomass may be taken to be a homogeneous material. Typically, two material balances have been carried out, one for the substrate and the other for the biomass, and these have resulted in a pair of simultaneous differential equations which may be solved to describe the behavior of the culture. This has the implication that during growth each cell experiences the same conditions as every other cell in the culture, that all components of the microbial cell increase at the same rate and that the function of the cell is invariant. Such growth is referred to as *balanced* growth and the models involved as being *unstructured* and *unsegregated*. While these models are still the mainstay of the design and analysis of biological reactors, they do not make much use of the wealth of biological knowledge referred to in earlier sections of this chapter. The models themselves are relatively simple mathematically and have the advantage of being amenable to algebraic manipulation, justifying their use whenever possible.

When a more detailed analysis of microbial systems is undertaken, the limitations of unstructured models become increasingly apparent. The most common area of failure is that where the growth is not exponential as, for example, during the so-called lag phase of a batch culture. Mathematically, the analysis is similar to that of the interaction of predator and prey, involving a material balance for each component being considered.

One of the earliest structured models is that put forward by Williams,[34] who proposed that the material of a cell could be divided into two categories. One of these is referred to as the active component, the other being the structural component. The model considered that all the cells in the fermentation broth were identical and substrate was incorporated initially into the active component and thence was used to form the structural component. The second, structural, component controlled the observed growth of the culture in that doubling of that component would be a necessary and sufficient condition for the cells to divide.

Ramkrishna et al.[35] proposed a similar model at about the same time—this too was an unsegregated model, which also divided the biomass into two *compartments*. They referred to the material in the two compartments as **G**-*mass* and **D**-*mass*, respectively, and suggested that these materials were formed in parallel. They also proposed that the microorganism produced a toxic substance, which inhibited its growth. They produced a set of differential equations obtained from material balance considerations to describe the behavior of such a system in both batch and continuous culture. For batch culture

$$
\left.
\begin{aligned}
\frac{dC_G}{dt} &= R_G - \kappa_{IG}C_G I \\
\frac{dC_D}{dt} &= R_D - \kappa_{ID}C_D I \\
\frac{dG}{dt} &= -b_1 R_G - b_2 R_D \\
\frac{dI}{dt} &= b_3 R_D + b_4 \kappa_{IG}C_G I
\end{aligned}
\right\}
\qquad (6.286)
$$

where C_G is the concentration of **G**-mass, C_D is the concentration of **D**-mass, I is the concentration of inhibitor, k_{IG} and k_{ID} are rate constants, and b_1, b_2, b_3, and b_4 are stoichiometric constants.

The rates R_G and R_D are given by

$$R_G = \frac{\mu_G S C_D}{(K_{SG} + S)} \frac{C_G}{(K_{GG} + C_G)}$$

$$R_D = \frac{\mu_D S C_D}{(K_{SD} + S)} \frac{C_G}{(K_{GD} + C_G)}$$

(6.287)

where μ_G and μ_D are rate constants and K_{SD}, K_{SG}, K_{GG}, and K_{GD} are affinity constants. Solving these equations numerically using the values for the constants as given in Table 6.3 produced the graph of the batch fermentation shown in Fig. 6.47.

The success of the model may be seen from the shape of the curve shown in Fig. 6.47. In the initial stages of growth, a lag phase is predicted, which duly gives way to the exponential growth period. After this, the calculations indicate that a stationary phase is attained which is followed by a decline in the active biomass concentration.

This model may be criticized in two respects, respectively, that the materials referred to are not quantifiable in a precise way (Esener et al.[36]) and that there is no account taken of the dilution effect on the intracellular material by the expansion of the biomass (Fredrickson[37]). In respect of the latter item, a distinction must be drawn between the concentrations of the components in the culture and those in the biomass itself. For a batch culture, a material balance for an active material, which forms a fraction f of the

Table 6.3: Constants used in the model of Ramkrishna et al.[35]

| Constant | Customary Biological Units | | SI Units | |
	Value	Units	Value	Units
μ_G	0.5	H^{-1}	0.14×10^{-3}	s^{-1}
μ_D	2.5	H^{-1}	0.69×10^{-3}	s^{-1}
K_{SG}	0.2	g/L	0.2	kg/m^3
K_{SD}	0.1	g/L	0.1	kg/m^3
b_1	8		8	
b_2	2		2	
k_{IG}	150	g/h	42×10^{-6}	kg/s
k_{ID}	70	g/h	19×10^{-6}	kg/s
K_{GG}	30×10^{-6}	g/L	30×10^{-6}	kg/m^3
K_{GD}	5×10^{-6}	g/L	5×10^{-6}	kg/m^3
b_3	20×10^{-6}		20×10^{-6}	
b_4	26.7×10^{-3}		26.7×10^{-3}	

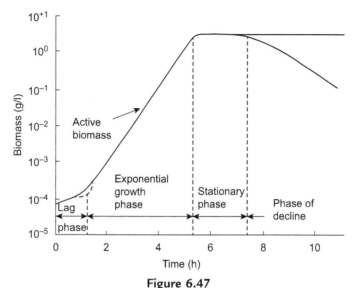

Figure 6.47
Batch growth curve produced by the model of Ramkrishna et al.[35]

growing biomass, is undertaken and its concentration increases from X to $X + dX$ in a time period dt, then

$$\left\{ \begin{array}{c} \text{Final} \\ \text{activity} \end{array} \right\} - \left\{ \begin{array}{c} \text{Initial} \\ \text{activity} \end{array} \right\} = \left\{ \begin{array}{c} \text{Rate of generation} \\ \text{of active material} \end{array} \right\} \times \left\{ \begin{array}{c} \text{Time} \\ \text{period} \end{array} \right\}$$

which may be stated mathematically as

$$\left\{ fX + \frac{d}{dt}(fX)dt \right\} - fX = Xwdt \tag{6.288}$$

where w is some function, which describes the specific rate of generation of f (which may or may not be dependent on f itself). Simplifying Eq. (6.288)

$$X\frac{df}{dt} + f\frac{dX}{dt} = Xw \tag{6.289}$$

Rearranging gives

$$\frac{df}{dt} = w - f\frac{1}{X}\frac{dX}{dt} \tag{6.290}$$

and, substituting for the specific growth rate

$$\frac{df}{dt} = w - f\mu \tag{6.291}$$

The last term in Eq. (6.291) represents the dilution of active component f, by the expansion of the biomass. Esener et al.[36] also present a two-compartment model, which takes this effect into account, and they emphasize the need to devise the theory so that it can be tested by experiment. In their model, they identify a "**K**" compartment of the biomass that comprised the RNA and other small cellular molecules. The other compartment contained the larger genetic material, enzymes, and structural material. The model assumes that the substrate is absorbed by the cell to produce, in the first instance, **K** material, and thence it is transformed into **G** material. Additionally, the **G** material can be reconverted to **K** material, a feature intended to account for the maintenance requirement of the microorganism. A series of material balances for the cellular components during growth in a CSTF produced the following differential equations.

For the substrate

$$\frac{dS}{dt} = -\frac{\psi_m SX}{K_S + S} + D(S_0 - S) \tag{6.292}$$

$$\frac{dX}{dt} = Y_{S/K}\frac{\psi_m SX}{K_S + S} + \left(Y_{G/K} - 1\right)k_k M_k M_G X - DX \tag{6.293}$$

$$\frac{dM_K}{dt} = (1 - M_K)Y_{S/K}\frac{\psi_S S}{K_S + S} - k_K M_K M_G\left[1 + \left(Y_{G/K} - 1\right)C_k\right] + m_G M_G \tag{6.294}$$

and

$$\frac{dM_G}{dt} = -\frac{dM_K}{dt} \tag{6.295}$$

where M_G is the mass fraction of **G** compartment in the biomass, M_K is the mass fraction of **K** compartment in the biomass, k_K is the rate constant for **K** consumption, m_G is the maintenance rate for **G** compartment, $Y_{K/S}$ is the yield coefficient for conversion of substrate to **K** compartment, and $Y_{G/K}$ is the yield coefficient for conversion of **K** compartment to **G** compartment.

The model was used by Esener et al.[30] to calculate the outcome of both fed-batch and continuous culture experiments, and the results were compared with the experimental values.

The curves shown in Fig. 6.48 indicate the ability of the model to give a reasonable description of the performance of the fermentations. The authors of the paper, however, draw attention to the fact that the prediction of the RNA component was nowhere as accurate and concluded that the model had failed. This, they pointed out, was a necessary test to prevent the model from becoming merely a curve-fitting exercise as opposed to a mechanistic model.

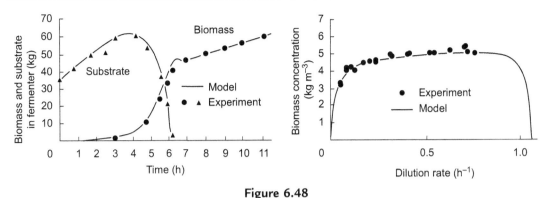

Figure 6.48
Comparison of theoretical and experimental results using the model of Esener et al.[36]

Other workers (Iamanaka et al.,[38] Papageorgakopoulu and maier,[39] inter alia) have developed more intricate structured models for the description of microbial growth. These models in particular have been based on a mechanistic approach, taking into account a series of cellular functions and products. A set of six simultaneous differential equations arise in the treatment of Iamanaka et al.[38] and nine in that of Papageorgakopoulu and Maier,[39] each having a total of over 20 constants associated with them. While both models produced good agreement between calculated and experimental values, their complicated nature makes them rather unwieldy and of limited use for design purposes.

6.8 Further Design Considerations

The primary function of a fermenter is to provide conditions, which are conducive to microbial growth. Assuming that the fermenter contains the appropriate nutrients, the most important conditions, which must be provided, or controlled, are temperature, pH, and dissolved oxygen. The last item is included because the vast majority of industrially important fermentations (with the exception of brewing beer and wine) involve the aerobic growth of microorganisms. Fermenters, as with other items of process engineering equipment, may vary considerably in their design according to their particular application. The stirred-tank and PFTFs which have been considered are idealized forms, which in some respects are extreme cases, which nevertheless serve as suitable design models for real installations. However, the ideal forms considered do not give much insight into how the operational requirement of a fermenter sets it apart from any other chemical reactor.

The constructional details of a fermenter vary considerably according to its intended application, particularly with regard to the scale of its operation and the necessity or otherwise sterilization before the initiation of microbial growth. In broad terms, small fermenters for laboratory use would be constructed mainly of borosilicate glass with some

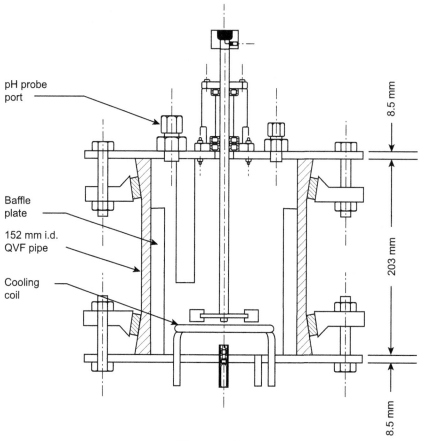

Figure 6.49
A typical laboratory fermenter.

parts, such as sensor ports and heating surfaces, in stainless steel. Such a fermenter is illustrated in Fig. 6.49, although in other designs the base may also be of glass. A vessel of this size may be easily cleaned when dismantled and is conveniently sterilized by placing the assembled unit (often containing the substrate necessary for the subsequent experiment) in a steam autoclave.

Fermenters with a capacity of over about 10 L are too heavy to sterilize in autoclaves. While they may still be laboratory sized, they have to be constructed so that they may be sterilized in situ. They become, as a consequence, pressure vessels, and the extensive use of glass becomes impractical and the preferred material of construction is a stainless steel. Seals are typically of silicone or other synthetic rubber or fluorinated plastics, with borosilicate glass being retained for sighting windows. This format is retained for vessels which are far larger than the laboratory scale, and Fig. 6.50 outlines the construction of a typical industrial *deep tank* fermenter.

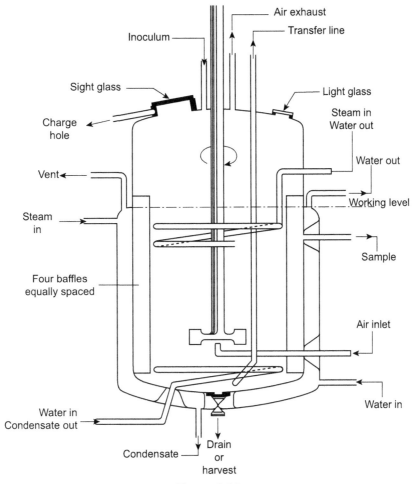

Figure 6.50
The deep tank fermenter.

This design is extensively used for the growth of single cultures of microbes, but it is by no means the only form of practical fermenter. Larger installations, particularly those intended for the growth of shear-sensitive organisms, may rely on the aeration system for mixing the contents. In the *air lift* design, the difference in bulk density between aerated and unaerated fermenter broth is used to induce circulation within the vessel. Microbial growth vessels intended for wastewater treatment differ even more dramatically, largely due to the fact that the constraint of maintaining a single species while excluding all others (*mono culture*) has been removed. They are usually open tanks, frequently constructed of concrete and often employing surface-aeration equipment, which also serve to mix the contents.

6.8.1 Aseptic Operation

A fermenter used for the growth of a specified microorganism must usually offer a containment facility for the microbial culture, normally such that contamination from outside is avoided, although the retention of the cultured organisms within the fermenter is also a frequent requirement. Monoculture of microbes is the rule rather than the exception and the fermenter vessel, and the feed, must be sterilized before the introduction of the seed microorganisms. The entry of wild organisms from the outside world generally has an adverse effect on the progress of a fermentation. The contaminating species may compete with the desired organism for the nutrients available, or may consume it or its products as substrate, or may spoil the fermentation by releasing toxic products into the growth medium. For these reasons it is usually necessary not only to destroy all residual microbes in the fermenter (and the medium) before the growth is initiated, but also to ensure that the air entering during the fermentation is free of microbial contamination by passing it through a suitable filtration system beforehand. Similarly, the adventitious introduction of microbes or their spores during sampling or during the transfer of additives to the broth must be avoided, and this alone considerably complicates the design of a fermenter.

Fig. 6.51 shows a piping arrangement suitable for the aseptic transfer of sterilized material from a holding vessel into a fermenter. The arrangement of valves allows the transfer lines to be flushed with steam and sterilized before the vessels are connected. The condensate formed is removed via the steam traps, which are then isolated before the sterile liquid is transferred.

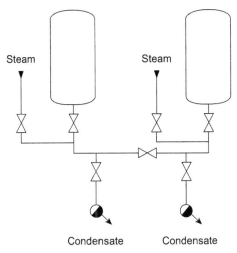

Figure 6.51
Aseptic transfer of material to a fermenter.

6.8.2 Aeration

Many microorganisms are capable of respiring in the absence of molecular oxygen (anaerobic conditions) and, in some cases, this is a necessary condition for either their well-being, or necessary for them to produce the desired product. On the other hand, some microorganisms are *obligate aerobes* and cease to function when deprived of oxygen. Fulfilling the oxygen requirement of a microbial fermentation presents another series of problems in the design of a fermenter. The solubility of oxygen in pure water at a typical fermentation temperature of 30°C is 7.7 mg/L, and this figure may be depressed by the presence of solutes such as carbohydrates and mineral salts. The respiration rate of the culture is usually such that the residual oxygen in the broth represents less than a minute's supply for the cells, and consequently, the oxygen has to be continuously replenished during the fermentation.

The respiration rate of an aerobic microbe is dependent on the dissolved oxygen concentration in the growth medium. The specific volumetric oxygen uptake rate (OUR), (Q_{O_2}, the mass of oxygen consumed by unit mass of cells in unit time), when plotted against dissolved oxygen concentration, exhibits an initial rapid rise to a plateau level, as shown in Fig. 6.52. The point at which the rather sharp inflexion in the curve occurs defines the "critical" dissolved oxygen concentration, above which the dissolved oxygen concentration does not limit the rate of reaction. The specific respiration rate is related to the specific growth rate by the yield coefficient for oxygen utilization Y_{O_2} as

$$Q_{O_2} = \frac{\mu}{Y_{O_2}} \tag{6.296}$$

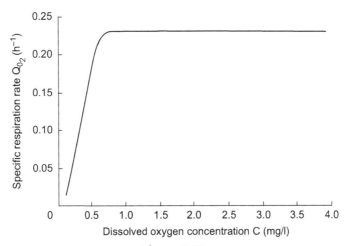

Figure 6.52

Oxygen consumption rate of yeast at various oxygen concentrations.

The value for the "critical" oxygen concentration tends to be rather small and this makes its direct measurement difficult. Operational fermenters are frequently equipped with oxygen-sensing probes so that control systems can be used to ensure that the dissolved oxygen concentration is well above the critical value, typically above 20% of the saturation value. The practical problem related to aerobic fermenters is one of mass transfer and, in some cases, an answer has been found by increasing the driving force for the dissolution process. This has been achieved by raising the partial pressure of oxygen in the gas being sparged into the fermenter either by using oxygen-enriched air or increasing the overall gas pressure. Generally, the overall mass transfer coefficient for the dissolution process is maximized, by a combination of efficient stirring of the broth and sparging of the gas used.

The dissolution of a sparingly soluble gas in a liquid is discussed in Volumes 1 and 2, and the reader is referred to the relevant sections (Volume 1, Chapter 10[39a] and Volume 2, Chapter 12[39b]) for detailed discussion of the mechanisms involved. In the case of the aeration of a fermenter, the controlling resistance to mass transfer is usually in the liquid phase so that it is appropriate to describe the process mathematically as

$$\frac{dC}{dt} = K_L a (C^* - C) \tag{6.297}$$

where C is the *mass* concentration of the gas in the liquid phase, C^* is the equilibrium (saturation) gas *mass* concentration, K_L is the overall mass transfer coefficient, and a is the interfacial area per unit volume of liquid.

The interfacial area per unit volume a is, again, a difficult quantity to measure in a fermenter because it depends on the number and size of the air bubbles entrained in the fermentation broth. These, in turn, are dependent on such factors as the aeration rate, the rheology of the broth at that instant, and the presence or otherwise of surfactants. As a result, the quantity $K_L a$ tends to be treated as a single entity in experimental measurements of oxygen transfer rates in fermenters.

The simplest method of determining $K_L a$ is to monitor the oxygen concentration during the aeration of a fermenter containing water or growth medium, which has initially been rendered oxygen free. The removal of the oxygen can be effected by *gassing out* using a nitrogen supply or by the addition of an oxygen-consuming agent, such as sodium sulfite, to the liquid. This is the so-called *static* method.

Eq. (6.297) may be integrated to give

$$\ln\left(\frac{C^*}{C^* - C}\right) = K_L a t \tag{6.298}$$

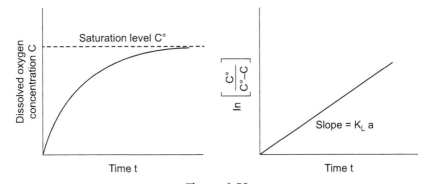

Figure 6.53
Measurement of K_La using the static method.

where C is the dissolved oxygen concentration at any time t. K_La may be determined from the slope of a plot of this equation, as shown in Fig. 6.53.

The static method of estimating K_La for a fermenter is not always satisfactory. The use of pure water instead of the fermentation medium can lead to inaccuracies, and the deoxygenation of a complex broth with sodium sulfite can result in undefined changes to the broth and be expensive to carry out on a large scale. An alternative dynamic method of determining K_La involves monitoring the course of a perturbation to the dissolved oxygen concentration during a fermentation. This is brought about by temporarily ceasing and then resuming the aeration (Fig. 6.54).

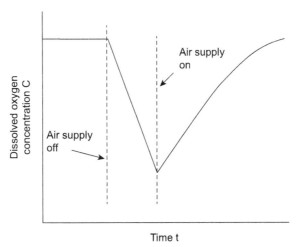

Figure 6.54
The dynamic method of determining K_La.

If during the beginning of the experiment the fermentation is at a quasi-steady state, that is the growth rate is such that the biomass concentration X may be considered to be constant during the experiment, then a material balance for the oxygen gives

$$\left\{ \begin{array}{c} \text{Oxygen} \\ \text{uptake} \\ \text{rate by} \\ \text{microbes} \end{array} \right\} + \left\{ \begin{array}{c} \text{Oxygen} \\ \text{transfer} \\ \text{rate} \end{array} \right\} = \left\{ \begin{array}{c} \text{Accumulation rate} \\ \text{of oxygen in} \\ \text{fermenter} \end{array} \right\}$$

This becomes

$$-Q_{O_2}X + K_La(C^* + C) = \frac{dC}{dt} \tag{6.299}$$

During the period when the aeration is suspended, the slope of the curve will be determined by the OUR so that the first term in Eq. (6.299) may be evaluated. When aeration is resumed, the curve will be that described by the whole of the equation rearranged as

$$C = C^* - \frac{1}{K_La}\left(\frac{dC}{dt} + Q_{O_2}X\right) \tag{6.300}$$

A plot of the concentration C against $\left(\frac{dC}{dt} + Q_{O_2}X\right)$ will give a straight line with slope $-1/K_La$.

While this method has its advantages over the static method, it also presents some practical difficulties. One of these is that the dissolved oxygen concentration must not be allowed to fall below the critical value C_{crit} as defined above, because this may be accompanied by a change in the metabolism of the microbe, which is reflected in a change in Q_{O_2}.

Many correlations have been proposed for the estimation of K_La, and there is an extensive literature on the subject. Among the most useful are those relating it to the power input to the stirrer and the aeration rate. These generally take the form

$$K_La = \text{const} \times (\text{power})^n \times (\text{aeration rate})^w$$

where n and w are usually fractions. A typical example is that given by Middleton[40] for noncoalescing solutions:

$$K_La = 1.2\left(\frac{P_g}{V_L}\right)^{0.7}(u_S)^{0.6} \tag{6.301}$$

where P_g is the power input under gassed conditions (W), V_L is the liquid volume (m^3), and u_s is the superficial velocity of the air (m/s^{-1}).

This expression is quoted as having a typical error of $\pm 10\%$ (Middleton[40]). It is not dimensionally consistent and the value of the coefficient therefore depends on the units being used.

6.8.3 Special Aspects of Biological Reactors

As mentioned earlier, the use of biological reactors for wastewater treatment is widespread and some very large scale units exist. Because such installations are intended primarily to purify the wastewater, and do not, as such, result in a saleable product, they are not expected to show an operational profit in the same way as a conventional process. Some recovery of operational costs may be realized by the use of anaerobic digesters whose off-gas can be used to reduce the requirement for natural gas or to generate electricity on-site. Some waste streams, such as whey from cheese making, have been utilized to produce potable alcohol and edible protein, again taking advantage of the potential of biological reactors to make valuable end products. In contrast to wastewater treatment, biological reactors can be used to manufacture fine chemicals, particularly pharmaceuticals, and other very valuable materials. These benefits have traditionally been sufficient to offset some of the less desirable aspects of biochemical reactors which, might otherwise render the process uneconomic.

Many of the materials produced by biological reactors are chemically labile and may, in conventional terms, be present in aqueous solutions of low concentrations. Mild chemical and physical conditions in the reactor are essential for their formation and survival, and such conditions must usually be maintained in the subsequent stages of the process. As a result, biochemical processes tend to have multiple separation stages, each one operating at near ambient temperature, or below if chillers are used. Centrifugation, micro- and ultrafiltration, solvent extraction, precipitation, and species-specific adsorption are unit operations frequently used in the separation stages. Although the high value of the product may make it economically feasible to use multiple stages, there is still an incentive to reduce their number in the overall process. Biological reactors, which combine reaction and separation in one stage, are currently under development. The use of immobilized enzymes and microorganisms is an example of such an approach. Another is the inclusion of a microfilter in a fermenter, which allows the removal of the desired product while retaining the biomass in the bioreactor.

The near-ambient temperatures used in most enzyme reactors and fermenters are themselves not normally determining factors in the design specification of the vessels concerned. The preferred method of sterilizing fermenters is the use of steam at temperatures above $100°C$ (typically $120-140°C$) and hence at pressures above atmospheric. This presents little problem for small laboratory units because they may be conveniently sterilized in a steam autoclave so that there is no net pressure difference between the inside and the outside of the vessel. On the other hand, pilot and production scale fermenters are pressure vessels, which must be designed to the same standards as pressure vessels used in conventional chemical manufacturing processes. Heat transfer during microbial growth conditions can present some problems because the temperature differences involved can then be small. While the high surface area to volume ratio of bench-scale units means that they may actually require heating during operation, the metabolic heat evolved in large fermenters frequently necessitates the use of cooling jackets or coils and the circulation of large quantities of cooling water.

Dealing with clean or sterile materials is not a problem peculiar to processes involving biochemical reactors, nor is the use of material of non-Newtonian rheology, which may give rise to persistent, unwanted foams. Established techniques of chemical engineering are usually applicable, albeit allowing for some of the constraints mentioned above. However, the advantages of the biochemical reactors often outweigh these and other, similar, problems, and their increased use in the process industries in future may be expected.

Nomenclature

		Units in SI System	Dimensions in M, N, L, T, θ
A	Concentration of A	kg/m^3	ML^{-3}
A, A_1, A_2, A_4	Arrhenius constants	s^{-1}	T^{-1}
A_3	Arrhenius constant	m^3/kg	M^{-1}L^3
A_i	Constant in Eq. (6.148)	kg/m^{3}s^1	ML^{-3}T^{-1}
a	Surface area per unit volume	m^{-1}	L^{-1}
a	Parameter in Eq. (6.264)	m^3/kg	M^{-1}L^3
a_c	Cross-sectional area	m^2	L^2
B	Concentration of B	kg/m^3	ML^{-3}
C	Concentration of gas in liquid phase	kg/m^3	ML^{-3}
C^*	Saturation concentration of gas in liquid	kg/m^3	ML^{-3}
C_G	Concentration of G-mass	kg/m^3	ML^{-3}
C_D	Concentration of D-mass	kg/m^3	ML^{-3}
D	Dilution rate	s^{-1}	T^{-1}
Da	Damköhler number	—	—
D_{crit}	Critical dilution rate for washout	s^{-1}	T^{-1}
D_e	Effective diffusion coefficient	m^2/s	L^{2}T^{-1}
D_{op}	Dilution rate for maximum biomass production	s^{-1}	T^{-1}
D_1	Dilution rate for CSTF 1	s^{-1}	T^{-1}
D_2	Dilution rate for CSTF 2	s^{-1}	T^{-1}
E_a, E_{a1},...,E_{a4}	Activation energy	J/kmol	MN^{-1}L^{2}T^{-2}
E	Enzyme concentration	kg/m^3	ML^{-3}
E_{act}	Concentration of active enzyme	kg/m^3	ML^{-3}

		Units in SI System	Dimensions in M, N, L, T, θ
E_{actt}	Active enzyme concentration at time t	kg/m^3	ML^{-3}
E_{act0}	Initial active enzyme concentration	kg/m^3	ML^{-3}
E_i	Concentration of inactive enzyme	kg/m^3	ML^{-3}
E_{tot}	Total enzyme concentration	kg/m^3	ML^{-3}
$[EA]$	Concentration of enzyme–substrate complex with substance A	kg/m^3	ML^{-3}
$[EAB]$	Concentration of ternary complex of enzyme with substances **A** and **B**	kg/m^3	ML^{-3}
$[EB]$	Concentration of enzyme–substrate complex with substance **B**	kg/m^3	ML^{-3}
$[EH^+]$	Concentration of protonated enzyme	kg/m^3	ML^{-3}
$[EP]$	Concentration of enzyme product complex with substance **P**	kg/m^3	ML^{-3}
$[ES]$	Enzyme–substrate concentration	kg/m^3	ML^{-3}
$[ESI]$	Enzyme–substrate–inhibitor concentration	kg/m^3	ML^{-3}
F	Volumetric flow rate	m^3/s	L^3T^{-1}
F_0	Volumetric flow rate of feed	m^3/s	L^3T^{-1}
F_R	Volumetric flow rate of recycle stream	m^3/s	L^3T^{-1}
F	Dimensionless substrate concentration	—	—
F$_f$	Dimensionless function defined in Eq. (6.257)	—	—
f	Fraction of biomass or enzyme	—	—
$\Delta G^{0\prime}$	Gibbs free energy at pH 7	J/kmol	$MN^{-1}L^2T^{-2}$
H^+	Proton concentration	kg/m^3	ML^{-3}
h_D	Mass transfer coefficient	m/s	LT^{-3}
I	Concentration of inhibitor	kg/m^3	ML^{-3}
i	ith item in series	—	—
K_{GG}	Affinity constant for **G** compartment	kg/m^3	ML^{-3}
K_{GD}	Affinity constant for **D** compartment	kg/m^3	ML^{-3}
K_I, K_i	Inhibition constants	kg/m^3	ML^{-3}
K_L	Mass transfer coefficient	m/s	LT^{-1}
K_m	Michaelis constant	kg/m^3	ML^{-3}
K_p	Product inhibition constant	kg/m^3	ML^{-3}
K_S	Monod constant	kg/m^3	ML^{-3}
K_{sj}, K_{si}	Monod constants for prey and predator	kg/m^3	ML^{-3}
K_S	Enzyme–substrate equilibrium constant	m^3/kg	$M^{-1}L^{-3}$
K_{SG}	Affinity constant for **G** compartment	kg/m^3	ML^{-3}
K_{SD}	Affinity constant for **D** compartment	kg/m^3	ML^{-3}
$K_S(T)$	Monod constant at temperature T	kg/m^3	ML^{-3}
K_x	Michaelis *or* Monod constant	kg/m^3	ML^{-3}
K_a	Equilibrium constants for protonated and unprotonated enzyme	kg/m^3	ML^{-3}

Continued

		Units in SI System	Dimensions in M, N, L, T, θ
K_m'	Modified Michaelis constant	kg/m^3	ML^{-3}
K_m^A	Michaelis constant for substance **A**	kg/m^3	ML^{-3}
K_m^B	Michaelis constant for substance **B**	kg/m^3	ML^{-3}
K_S^A	The dissociation constant for the reaction E + A $\rightleftharpoons$ EA	kg/m^3	ML^{-3}
K_{eq}	Equilibrium constant for enzyme reaction	kg/m^3	ML^{-3}
K_m^P	Michaelis constant for substance **P**	kg/m^3	ML^{-3}
K_m^S	Michaelis constant for substance **S**	kg/m^3	ML^{-3}
k	Rate constant	s^{-1}	T^{-1}
k_1	Rate constant for forward reaction for the formation of ES from E and S	m^3/kg s	$M^{-1}L^3T^{-1}$
k_2	Rate constant for forward reaction decomposition from ES	s^{-1}	T^{-1}
k_{-1}	Rate constant for back reaction first step	s^{-1}	T^{-1}
k_{-2}	Rate constant for back reaction second step	m^3/kg s	$M^{-1}L^3T^{-1}$
k_{cat}	Rate constant for decomposition of ES complex	s^{-1}	T^{-1}
k_{cat}^A	Rate constant for decomposition of enzyme substrate complex involving substance **A**	s^{-1}	T^{-1}
k_{cat}^B	Rate constant for decomposition of enzyme substrate complex involving substance **B**	s^{-1}	T^{-1}
k_1	Rate constant in logistic equation	s^{-1}	T^{-1}
k_d	Endogenous respiration rate	s^{-1}	T^{-1}
$k_d(T)$	Endogenous respiration rate at temperature T	s^{-1}	T^{-1}
k_{de}	Enzyme decay coefficient	s^{-1}	T^{-1}
k_K	Rate constant for **K** compartment	s^{-1}	T^{-1}
k_{IG}, k_{ID}	Rate constants for deactivation of **G** and **D**	kg/s	MT^{-1}
L	Biological film thickness	m	L
M_G	Mass fraction of **G** in biomass	—	—
M_K	Mass fraction of **K** in biomass	—	—
m	Specific maintenance requirement	s^{-1}	T^{-1}
m_G	Maintenance rate constant for **G** compartment	s^{-1}	T^{-1}
N	Total number	—	—
n_1, n_2	Indices in expression for $K_L a$	—	—
P	Product concentration	kg/m^3	ML^{-3}
$\mathbf{P}_G$	Power input under gassed condition	W	ML^2T^{-3}
Q	Concentration of substance **Q**	kg/m^3	ML^{-3}
Q_{O_2}	Specific respiration rate	s^{-1}	T^{-1}

		Units in SI System	Dimensions in M, N, L, T, θ
Q_{10}	Ratio of rates at 10K temperature difference	—	—
g	Parameter in Eq. (6.264)	kg/m^3	ML^{-3}
R_0	Initial rate of reaction	kg/m^3 s	ML^{-3}T^{-1}
R_0^A	Initial rate of reaction of substance **A**	kg/m^3 s	ML^{-3}T^{-1}
R_0^B	Initial rate of reaction of substance **B**	kg/m^3 s	ML^{-3}T^{-1}
R	Universal gas constant	8314 J/kmol K	MN^{-1}L^{2}T^{-2}θ^{-1}
R	Recycle ratio	—	—
R	Rate of reaction per unit volume of reactor	kg/m^3 s	ML^{-3}T^{-1}
R_D, R_G	Rates of reaction of **D** and **G** materials	kg/m^3 s	ML^{-3}T^{-1}
R_S	Rate of substrate reaction	kg/m^3 s	ML^{-3}T^{-1}
R_S''	Local rate of reaction of substrate per unit volume of sphere	kg/m^3 s	ML^{-3}T^{-1}
R_x	Rate of biomass reaction	kg/m^3 s	ML^{-3}T^{-1}
R'	Rate of reaction per unit area of biocatalyst	kg/m^3 s	ML^{-2}T^{-1}
R_m'	Maximal rate of reaction based on unit area of biocatalyst	kg/m^2 s	ML^{-2}T^{-1}
R''	Rate of reaction per unit volume of biocatalyst	kg/m^3 s	ML^{-3}T^{-1}
R_m'	Saturation constant based on unit volume of biocatalyst	kg/m^3 s	ML^{-3}T^{-1}
R_k'	Rate of reaction per unit area of biocatalyst with no diffusional limitation	kg/m^2 s	ML^{-2}T^{-1}
r	Radius	m	L
r_p	Radius of particle	m	L
$\bar{r}$	Dimensionless radius	—	—
r	Parameter in Eq. (6.264)	s^{-1}	T^{-1}
$[S]$	Substrate concentration	kg/m^3	ML^{-3}
$\bar{S}$	Dimensionless substrate concentration	—	—
$[S_A]$	Average substrate concentration	kg/m^3	ML^{-3}
$[S_e]$	Effluent substrate concentration	kg/m^3	ML^{-3}
$[S_G]$	Concentration of substrate used for growth	kg/m^3	ML^{-3}
$[S_0]$	Initial or feed substrate concentration	kg/m^3	ML^{-3}
$[S_1]$, $[S_2]$	Substrate concentration in the first and second vessels	kg/m^3	ML^{-3}
$[S_b]$	Substrate concentration in bulk of liquid	kg/m^3	ML^{-3}
$[S_s]$	Substrate concentration at surface	kg/m^3	ML^{-3}
T	Absolute temperature	K	θ
T	Function defined in Eq. (6.274)	—	—

Continued

		Units in SI System	Dimensions in M, N, L, T, θ
t	Time	s	T
t_d	Doubling time	s	T
t_g	Productive fermentation time	s	T
t_u	Unproductive fermentation time	s	T
u_s	Superficial gas velocity	m/s	LT^{-1}
V	Volume	m^3	L^{-1}
V_1, V_2	Volume of first and second vessels	m^3	L^3
V_L	Liquid volume	m^3	L^3
V_{max}	Enzyme velocity constant	kg/m^3 s	$ML^{-3}T^{-1}$
$(V_{max})_m$	Maximum enzyme velocity constant in unprotonated form	kg/m^3 s	$ML^{-3}T^{-1}$
$V_{max,0}$	Initial enzyme velocity constant	kg/m^3 s	$ML^{-3}T^{-1}$
V''_{max}	Enzyme velocity constant based on unit volume of immobilised biocatalyst	kg/m^3 s	$ML^{-3}T^{-1}$
V^S_{max}	Maximum rate of reaction involving substance **S**	kg/m^3 s	$ML^{-3}T^{-1}$
V^P_{max}	Maximum rate of reaction involving substance **P**	kg/m^3 s	$ML^{-3}T^{-1}$
w	Specific rate of generation of biomass fraction	s^{-1}	T^{-1}
X	Biomass concentration	kg/m^3	ML^{-3}
X_0	Initial or feed biomass concentration	kg/m^3	ML^{-3}
X_A	Average biomass concentration	kg/m^3	ML^{-3}
X_J	Concentration of prey	kg/m^3	ML^{-3}
X_L	Concentration of predator	kg/m^3	ML^{-3}
X_{op}	Biomass concentration at optimum dilution rate	kg/m^3	ML^{-3}
X_e	Effluent biomass concentration	kg/m^3	ML^{-3}
X_R	Biomass concentration in recycle stream	kg/m^3	ML^{-3}
X_1, X_2	Biomass concentration in vessels 1 and 2	kg/m^3	ML^{-3}
X_i	Biomass concentration in the *i*th vessel	kg/m^3	ML^{-3}
X_m	Final biomass concentration	kg/m^3	ML^{-3}
x	Distance	m	L
Y	Yield coefficient	—	—
Y_G	Yield coefficient for growth	—	—
Y_J	Yield coefficient for growth of prey	—	—
Y_L	Yield coefficient for growth of predator	—	—
$Y_{P/S}$	Yield coefficient for product formation	—	—
Y_1, Y_2	Yield coefficient for the first and second vessels	—	—
$Y_{G/S}$	Yield coefficient for formation of **G** from **S**	—	—

		Units in SI System	Dimensions in M, N, L, T, θ
$Y_{K/S}$	Yield coefficient for formation of **K** from **S**	—	—
$Y_{G/K}$	Yield coefficient for formation of **G** from **K**	—	—
$Y_{X/S}$	Overall yield coefficient for growth	—	—
$Y_{i/S}$	Yield coefficient for the *i*th product from **S**	—	—
Y_{O_2}	Yield coefficient for oxygen utilisation	—	—
Z	Length of reactor	m	**L**
z	Dimensionless distance	—	—
α	Empirical constant in Eq. (6.138)	—	—
β	Dimensionless substrate concentration	—	—
β_b	Bulk dimensionless substrate concentration	—	—
β_s	Surface dimensionless substrate concentration	—	—
χ	Dimensionless biomass concentration	—	—
δ	Coefficient in Eq. (6.146)	—	—
ε	Coefficient in Eq. (6.146)	s^{-1}	T^{-1}
ξ	Biomass concentration ratio	—	—
ϕ	Thiele modulus	—	—
ϕ'	Modified Thiele modulus	—	—
γ	Inverse of maximum biomass concentration	m^3/kg	$M^{-1}L^3$
η	Effectiveness factor	—	—
η_e	Effectiveness factor for surface reaction	—	—
μ	Specific growth rate	s^{-1}	T^{-1}
μ''	Specific growth rate based on unit area	s^{-1}	T^{-1}
μ''	Specific growth rate based on unit volume	s^{-1}	T^{-1}
μ_0	Specific growth rate at reference temperature	s^{-1}	T^{-1}
μ_m	Maximal specific growth rate	s^{-1}	T^{-1}
μ_J	Maximal specific growth rate of prey	s^{-1}	T^{-1}
μ_L	Maximal specific growth rate of predator	s^{-1}	T^{-1}
$\mu_m(T)$	Maximal specific growth rate at temperature T	s^{-1}	T^{-1}
μ'_m	Maximal specific growth rate based on unit area	s^{-1}	T^{-1}
μ''_m	Maximal specific growth rate based on unit volume	s^{-1}	T^{-1}
μ_{max}	Maximum observed specific growth rate	s^{-1}	T^{-1}

Continued

		Units in SI System	Dimensions in M, N, L, T, θ
Ψ	Specific rate of substrate consumption	s^{-1}	T^{-1}
Ψ_m	Maximum specific rate of substrate consumption	s^{-1}	T^{-1}
τ	Cell age	s	T
θ	Celsius temperature	$^\circ C$	θ
ν	Residence time	s	T
Subscripts			
0	Initial value		
b	Bulk		
i	ith component		
n	nth component		
op	Optimum		
s	Surface		
t	Value at time t		

References

1. Michaelis L, Menten ML. Die Kinetik der Invertinwirkung [The kinetics of invertin action]. *Biochem Z* 1913;**49**:333.
2. Fersht A. *Enzyme structure and mechanism*. 2nd ed. New York: W. H. Freeman and Co.; 1985.
3. Leninger AL. *Biochemistry*. 2nd ed. New York: Worth Publishers Inc.; 1975.
4. Palmer T. *Understanding enzymes*. 2nd ed. Chichester: Ellis Horwood Publishers; 1985.
5. Horvath C, Engasser J-M. External and internal diffusion in heterogeneous enzyme systems. *Biotechnol Bioeng* 1974;**16**:909.
6. Atkinson B. *Biochemical reactors*. Pion; 1974.
7. Atkinson B, Mavituna F. *Biochemical engineering and biotechnology handbook*. 2nd ed. Basingstoke: Macmillan Publishers; 1991.
8. Lodge RM, Hinshelwood CN. Physicochemical aspects of bacterial growth. *J Chem Soc* 1943;**213**:148.
9. Pearl R, Reed LJ. Growth equation with inhibition factor leading to logistic equation. *Proc Natl Acad Sci USA* 1920;**6**:275.
10. Edwards VH, Wilke CR. Mathematical representation of batch culture data. *Biotechnol Bioeng* 1968;**10**:205.
11. Monod J. *Recherches sur des Croissances des Cultures Bacteriennes*. Paris: Hermann et Cie; 1942.
12. Edwards VH. The influence of high substrate concentration on microbial kinetics. *Biotechnol Bioeng* 1970;**12**:679.
13. Aiba S, Shoda M, Nagatani M. Kinetics of product inhibition in alcohol fermentation. *Biotechnol Bioeng* 1968;**10**:845.
14. Aiba S, Humphrey AE, Millis NF. *Biochemical engineering*. 2nd ed. New York: Academic Press; 1973.
15. Pirt SJ. *Principles of microbe and cell cultivation*. Oxford: Blackwell Scientific Publications; 1975.
16. Harwood JH. *Studies on the physiology of methylococcus capsulatis growing on methane* [Ph.D. thesis]. University of London; 1970.
17. Sinclair CG, Kristiansen B. Fermentation kinetics and modelling. In: Bu'Lock JD, editor. Milton Keynes: Open University Press; 1987.

18. Wasungu KM, Simard RE. Growth characteristics of baker's yeast in ethanol. *Biotechnol Bioeng* 1982;**24**:1125.

19. Rowley BI, Pirt SJ. Melanin production by *Aspergillus nidulans* in batch and chemostat cultures. *J Gen Microbiol* 1972;**72**:553.

20. Ryu DD, Mateles RI. Transient response of continuous cultures to changes in temperature. *Biotechnol Bioeng* 1968;**10**:385.

21. Metcalf and Eddy Inc. Tchobanoglous G, Burton FL. *Wastewater engineering: treatment, disposal and reuse.* 3rd ed. New York: McGraw-Hill; 1991.

22. Topiwala HH, Sinclair CG. Temperature relationship in continuous culture. *Biotechnol Bioeng* 1971;**13**:795.

23. Kincannon DF, Gaudy AF. Response of biological waste treatment systems to changes in salt concentrations. *Biotechnol Bioeng* 1968;**10**:483.

24. Gaden EL. Fermentation kinetics and productivity. *J Biochem Microbiol Tech* 1959;**1**:413.

25. Deindoerfer FH. Fermentation kinetics and model processes. *Adv Appl Microbiol* 1955;**2**:321.

26. Luedeking R, Piret EL. A kinetic study of the lactic acid fermentation. Batch process at controlled pH. *J Biochem Microbiol Technol Eng* 1959;**1**:393.

27. Shu P. Mathematical models for the product accumulation in microbial processes. *J Biochem Microbiol Technol Eng* 1961;**3**:95.

28. Gates WE, Marlar JT. Graphical analysis of batch culture data using the Monod expressions. *J Water Pollut Cont Fed* 1968;**40**:R469.

29. Ong SL. Least-squares estimation of batch culture kinetic parameters. *Biotechnol Bioeng* 1983;**25**:2347.

30. Esener AA, Roels JA, Kossen NWF. Theory and applications of unstructured growth models: kinetic and energetic aspects. *Biotechnol Bioeng* 1983;**25**:2803.

31. Knights AJ. *Determination of the biological kinetic parameters of fermenter design from batch culture data* [Ph.D. thesis]. University of Wales; 1981.

32. Lotka AJ. Undamped oscillations derived from the law of mass action. *J Am Chem Soc* 1920;**42**:1595.

33. Tsuchiya HM, Drake JF, Jost JL, Fredrickson AG. Predator-prey interactions of *Dictyoselium discoidem* and *Eschericha coli* in continuous culture. *J Bacteriol* 1972;**110**:1147.

34. Williams FM. A model of growth dynamics. *J Theor Biol* 1967;**15**:190.

35. Ramkrishna D, Fredrickson AG, Tsuchiya HM. Dynamics of microbial propagation. *Biotechnol Bioeng* 1967;**9**:129.

36. Esener AA, Veerman T, Roels JA, Kossen NWF. Modeling of bacterial growth; formulation and evaluation of a structured model. *Biotechnol Bioeng* 1982;**24**:1749.

37. Fredrickson AG. Structured models. *Biotechnol Bioeng* 1976;**18**:1481.

38. Iamanaka T, Kaieda T, Sato K, Tacughi H. Optimisation of α-galactosidase production by mold. *J Ferment Technol* 1972;**50**(9):633.

39. Papageorgakopoulu H, Maier WJ. A new modeling technique and computer simulation of computer growth. *Biotechnol Bioeng* 1974;**26**:275;
39a. Coulson JM, Richardson JF. *Coulson and Richardson's Chemical Engineering: Fluid Flow, Heat Transfer and Mass Transfer,* vol.1. 1999 [Chapter 10: Mass Transfer];
39b. Coulson JM, Richardson JF. *Coulson and Richardson's Chemical Engineering: Particle Technology and Seperation Processes,* vol. 2. 1996 [Chapter 12: Absorption of Gases].

40. Middleton JC. Gas-liquid dispersion and mixing. In: Harney N, Edwards MF, Nienow AW, editors. *Mixing in the process industries.* 2nd ed. Oxford: Butterworth Heinemann; 1992. p. 322.

41. Moser H. *The dynamics of bacterial populations in the chemostat.* Carnegie Institute Publ. No. 614; 1958.

42. Contois DE. Kinetics of microbial growth. Relationship between population density and specific growth rate of continuous culture. *J Gen Microbiol* 1959;**21**:40.

Further Reading

Bailey JE, Ollis DF. *Biochemical engineering fundamentals*. 2nd ed. New York: McGraw-Hill; 1986.

Cano RJ, Colome JS. *Microbiology*. St. Paul, Minnesota: West Publishing Co.; 1986.

Palmer T. *Understanding enzymes*. 4th ed. Chichester: Ellis Horwood Limited; 1995.

Doran MP. *Bioprocess engineering principles*. 1st ed. San Diego: Academic Press Limited; 1995.

Segal IH. *Biochemical calculations*. 2nd ed. New York: John Wiley; 1976.

Stanbury PF, Hall S, Whitaker A. *Principles of fermentation technology*. 3rd ed. Oxford: Pergamon Press; 2016.

Berg JM, Tymoczko JL, Stryer L. *Biochemistry*. 5th ed. New York: Freeman; 2002.

Index

'*Note*: Page numbers followed by "f" indicate figures, "t" indicate tables.'

Printed in the United States
By Bookmasters